Automotive Engineering Fundamentals

Other SAE titles of interest:

Automotive Technology, Third Edition
By M.J. Nunney
(Order No. R-242)

Bosch Automotive Handbook
(Order No. BOSCH5)

**The Future of the Automotive Industry:
Challenges and Concepts for the 21st Century**
Edited by Ralf Landmann, Heiko Wolters, Wolfgang Bernhart,
and Holger Karsten
(Order No. R-291)

The Motor Vehicle, 13th Edition
By T.K. Garrett
(Order No. R-298)

For more information or to order a book, contact SAE at
400 Commonwealth Drive, Warrendale, PA 15096-0001;
phone (724) 776-4970; fax (724) 776-0790;
e-mail CustomerService@sae.org;
website http://store.sae.org.

Automotive Engineering Fundamentals

Richard Stone

and

Jeffrey K. Ball

Warrendale, Pa.

All rights reserved. No part of this publication may be reproduced, stored in a retrieval system, or transmitted, in any form or by any means, electronic, mechanical, photocopying, recording, or otherwise, without the prior written permission of SAE.

For permission and licensing requests, contact:

SAE Permissions
400 Commonwealth Drive
Warrendale, PA 15096-0001 USA
E-mail: permissions@sae.org
Tel: 724-772-4028
Fax: 724-772-4891

Library of Congress Cataloging-in-Publication Data

Stone, Richard.
　　Automotive engineering fundamentals / Richard Stone and Jeffrey K. Ball.
　　　p. cm.
　　Includes bibliographical references and index.
　　ISBN 0-7680-0987-1
　　1. Automobiles—Design and construction. I. Ball, Jeffrey K. II. Title.

TL240.S853 2004
629.2'3—dc22

2004041782

SAE
400 Commonwealth Drive
Warrendale, PA 15096-0001 USA
E-mail: CustomerService@sae.org
Tel: 877-606-7323 (inside USA and Canada)
            724-776-4970 (outside USA)
Fax: 724-776-1615

Copyright © 2004   Richard Stone and Jeffrey K. Ball

ISBN 0-7680-0987-1

SAE Order No. R-199

Printed in the United States of America.

Table of Contents

Preface xiii

Acknowledgments xv

Chapter 1—Introduction and Overview 1
    1.1   Beginnings 1
    1.2   Growth and Refinement 6
    1.3   Modern Development 9
    1.4   Overview 16

Chapter 2—Thermodynamics of Prime Movers 17
    2.1   Introduction 17
    2.2   Two- and Four-Stroke Engines 17
    2.3   Indicator Diagrams and Internal Combustion Engine
          Performance Parameters 20
    2.4   Otto and Diesel Cycle Analyses 23
          2.4.1   The Ideal Air Standard Otto Cycle 24
          2.4.2   The Ideal Air Standard Diesel Cycle 25
          2.4.3   Efficiencies of Real Engines 30
    2.5   Ignition and Combustion in Spark Ignition and Diesel Engines 32
    2.6   Sources of Emissions 37
          2.6.1   Simple Combustion Equilibrium 37
          2.6.2   Unburned Hydrocarbons (HC) and Nitrogen Oxides (NOx)
                 in Spark Ignition Engines 41
          2.6.3   Unburned Hydrocarbons (HC), Nitrogen Oxides (NOx),
                 and Particulates in Compression Ignition Engines 45
    2.7   Fuel and Additive Requirements 45
          2.7.1   Abnormal Combustion in Spark Ignition Engines 48
          2.7.2   Gasoline and Diesel Additives 48
    2.8   Gas Exchange Processes 50
          2.8.1   Valve Flow and Volumetric Efficiency 50
          2.8.2   Valve Timing 55
          2.8.3   Valve Operating Systems 58
          2.8.4   Dynamic Behavior of Valve Gear 60
    2.9   Engine Configuration 64
          2.9.1   Choosing the Number of Cylinders 64
          2.9.2   Balancing of the Primary and Secondary Forces
               and Moments 68
    2.10 Fuel Cells 79
          2.10.1 Solid Polymer Fuel Cells (SPFC) 79
          2.10.2 Solid Polymer Fuel Cell (SPFC) Efficiency 81
               2.10.2.1 Activation Losses 83

                2.10.2.2 Fuel Crossover and Internal Currents 85
                2.10.2.3 Ohmic Losses 87
                2.10.2.4 Mass Transfer Losses 87
                2.10.2.5 Overall Response 88
        2.10.3 Sources of Hydrogen for Solid Polymer Fuel Cells
                (SPFC) 88
                2.10.3.1 Steam Reforming (SR) 89
                2.10.3.2 Partial Oxidation (POX) Reforming 90
                2.10.3.3 Autothermal Reforming (AR) 90
                2.10.3.4 Carbon Monoxide Clean-Up and Solid Polymer
                          Fuel Cell (SPFC) Operation on Reformed Fuel 91
                2.10.3.5 Hydrogen Storage 92
        2.10.4 Hydrogen Fuel Cell Systems 93
  2.11 Concluding Remarks 97
  2.12 Problems 97

Chapter 3—Spark Ignition Engines 101
  3.1 Introduction 101
  3.2 Spark Ignition and Ignition Timing 101
        3.2.1 Ignition System Overview 101
        3.2.2 The Ignition Process 105
        3.2.3 Ignition Timing Selection and Control 107
  3.3 Mixture Preparation 109
  3.4 Combustion System Design 113
        3.4.1 Port Injection Combustion Systems 113
        3.4.2 Direct Injection Spark Ignition (DISI) Combustion Systems 116
  3.5 Emissions Control 120
        3.5.1 Development of the Three-Way Catalyst 121
        3.5.2 Durability 124
        3.5.3 Catalyst Light-Off 125
        3.5.4 Lean-Burn NOx-Reducing Catalysts, "DENOx" 126
  3.6 Power Boosting 127
        3.6.1 Variable Valve Timing and Induction Tuning 127
        3.6.2 Supercharging 128
  3.7 Engine Management Systems 132
        3.7.1 Introduction 132
        3.7.2 Sensor Types 134
                3.7.2.1 Crankshaft Speed/Position and Camshaft Position 134
                3.7.2.2 Throttle Position 136
                3.7.2.3 Air Flow Rate 136
                3.7.2.4 Inlet Manifold Absolute Pressure 137
                3.7.2.5 Air Temperature and Coolant Temperature 137
                3.7.2.6 Air-Fuel Ratio 137
                3.7.2.7 Knock Detector 140

Table of Contents

vii

- 3.8 Engine Management System Functions 142
  - 3.8.1 Ignition Timing 142
  - 3.8.2 Air-Fuel Ratio Control 143
  - 3.8.3 Exhaust Gas Recirculation (EGR) Control 144
  - 3.8.4 Additional Functions 144
  - 3.8.5 Concluding Remarks on Engine Management Systems 146
- 3.9 Conclusions 147
- 3.10 Questions 147

Chapter 4—Diesel Engines 149
- 4.1 Introduction 149
- 4.2 Direct and Indirect Injection Combustion Chambers 150
- 4.3 Fuel Injection Equipment 152
  - 4.3.1 Pump–Line–Injector (PLI) Systems 153
  - 4.3.2 Electronic Unit Injectors (EUI) 155
  - 4.3.3 Common Rail (CR) Fuel Injection Systems 156
- 4.4 Diesel Engine Emissions and Their Control 157
  - 4.4.1 Diesel Engine Emissions 157
  - 4.4.2 Diesel Engine Emissions Control 158
    - 4.4.2.1 Exhaust Gas Recirculation (EGR) 158
    - 4.4.2.2 Particulate Traps 159
- 4.5 Turbocharging 161
  - 4.5.1 Introduction 161
  - 4.5.2 Turbocharger Performance 164
  - 4.5.3 Turbocharged Engine Performance 169
- 4.6 Diesel Engine Management Systems 172
- 4.7 Concluding Remarks 175
- 4.8 Examples 177
- 4.9 Problems 185

Chapter 5—Ancillaries 189
- 5.1 Introduction 189
- 5.2 Lubrication System 189
  - 5.2.1 Bearings 189
    - 5.2.1.1 Anti-Friction Bearings 190
    - 5.2.1.2 Guide Bearings 190
    - 5.2.1.3 Thrust Bearings 191
    - 5.2.1.4 Journal Bearings 192
  - 5.2.2 Engine Lubricants 195
  - 5.2.3 Lubrication of Journal Bearings 197
- 5.3 Vehicle Cooling Systems 202
  - 5.3.1 Coolant 206
- 5.4 Drive Belts 208
  - 5.4.1 Flat Belt Drives 208
  - 5.4.2 V-Belts 212

5.5 Air Conditioning Systems 213
5.5.1 Overview 213
5.5.2 Thermodynamic Performance and Operation 215
5.5.3 Coefficient of Performance (CoP): 216
5.5.4 Air Conditioning System Performance 222
5.6 Generators, Motors, and Alternators 223
5.6.1 Fundamentals 223
5.6.2 Practical Alternators 227
5.6.3 Practical Starter Motors 231
5.7 Conclusions 233

Chapter 6—Transmissions and Driveline 235
6.1 Introduction 235
6.2 Friction Clutches 236
6.2.1 Torque Capability of an Axial Clutch 239
6.2.1.1 Uniform Pressure: $p = p_a$ 240
6.2.1.2 Uniform Wear 242
6.3 Gear Theory 243
6.3.1 Straight-Tooth Spur Gears 244
6.3.2 Helical Spur Gears 244
6.3.3 Straight-Tooth Bevel Gears 245
6.3.4 Spiral Bevel Gears 246
6.3.5 Hypoid Gears 246
6.4 Manual Transmissions 249
6.4.1 Transmission Power Flows 251
6.4.1.1 First Gear 251
6.4.1.2 Second Gear 251
6.4.1.3 Third Gear 252
6.4.1.4 Fourth Gear 252
6.4.1.5 Reverse 253
6.4.2 Synchronizer Operation 254
6.5 Automatic Transmissions 255
6.5.1 Fluid Couplings and Torque Converters 256
6.5.2 Planetary Gears 261
6.5.3 Planetary Gear-Set Torque Converter 265
6.5.4 Simpson Drive 267
6.5.4.1 Power Flow in First Gear 268
6.5.4.2 Power Flow in Second Gear 270
6.5.4.3 Power Flow in Third Gear 270
6.5.4.4 Power Flow in Reverse 271
6.5.5 Hydraulic Control System 272
6.6 Continuously Variable Transmissions (CVT) 275
6.6.1 Introduction 275
6.6.2 Van Doorne Continuously Variable Transmission (CVT) 275
6.6.3 Torotrak Continuously Variable Transmission (CVT) 277

6.7	Driveshafts		281
	6.7.1	Hooke's Joints	281
	6.7.2	Shaft Whirl	286
6.8	Differentials		290
6.9	Four-Wheel Drive (FWD) and All-Wheel Drive (AWD)		293
	6.9.1	Part-Time Four-Wheel Drive (4WD)	294
	6.9.2	On-Demand Four-Wheel Drive (4WD)	295
	6.9.3	Full-Time Four-Wheel Drive (4WD)	295
	6.9.4	All-Wheel Drive (AWD)	295
6.10	Case Study: The Chrysler 42LE Automatic Transaxle		296
	6.10.1	Configuration	296
	6.10.2	Planetary Gear Set	296
	6.10.3	Chain Transfer Drive	299
	6.10.4	Control System	299
6.11	Problems		299

Chapter 7—Steering Systems and Steering Dynamics 303

7.1	Introduction		303
7.2	Steering Mechanisms		303
	7.2.1	Worm Systems	305
	7.2.2	Worm and Sector	305
	7.2.3	Worm and Roller	305
	7.2.4	Recirculating Ball	307
	7.2.5	Rack and Pinion Steering	308
	7.2.6	Power Steering	308
7.3	Steering Dynamics		311
	7.3.1	Low-Speed Turning	311
	7.3.2	High-Speed Turning	312
	7.3.3	Effects of Tractive Forces	318
7.4	Wheel Alignment		320
	7.4.1	Camber	320
	7.4.2	Steering Axis Inclination (SAI)	320
	7.4.3	Toe	321
	7.4.4	Caster	323
	7.4.5	Wheel Alignment	324
7.5	Steering Geometry Errors		324
7.6	Front-Wheel-Drive Influences		327
	7.6.1	Driveline Torque	327
	7.6.2	Loss of Cornering Stiffness Due to Tractive Forces	329
	7.6.3	Increase in Aligning Torque Due to Tractive Forces	329
7.7	Four-Wheel Steering		330
	7.7.1	Low-Speed Turns	331
	7.7.2	High-Speed Turns	332
	7.7.3	Implementation of Four-Wheel Steering	333

7.8	Vehicle Rollover		337
	7.8.1	Quasi-Static Model	337
	7.8.2	Quasi-Static Rollover with Suspension	337
	7.8.3	Roll Model	339
7.9	Problems		343

Chapter 8—Suspensions 345
- 8.1 Introduction 345
- 8.2 Perception of Ride 345
- 8.3 Basic Vibrational Analysis 347
  - 8.3.1 Single-Degree-of-Freedom Model (Quarter Car Model) 347
  - 8.3.2 Two-Degrees-of-Freedom Model (Quarter Car Model) 351
  - 8.3.3 Two-Degrees-of-Freedom Model (Half Car Model) 354
- 8.4 Suspension System Components 363
  - 8.4.1 Springs 363
    - 8.4.1.1 Leaf Springs 363
    - 8.4.1.2 Torsion Bars 364
    - 8.4.1.3 Coil Springs 365
    - 8.4.1.4 Pneumatic (Air) Springs 368
  - 8.4.2 Dampers (Shock Absorbers) 371
- 8.5 Suspension Types 372
  - 8.5.1 Solid Axle Suspensions 373
    - 8.5.1.1 Hotchkiss Suspensions 373
    - 8.5.1.2 Four-Link Suspensions 374
    - 8.5.1.3 de Dion Suspensions 374
  - 8.5.2 Independent Suspensions 375
    - 8.5.2.1 Short-Long Arm Suspensions (SLA) 375
    - 8.5.2.2 MacPherson Struts 375
    - 8.5.2.3 Trailing Arm Suspensions 376
    - 8.5.2.4 Multi-Link Suspensions 378
    - 8.5.2.5 Swing Arm Suspensions 379
- 8.6 Roll Center Analysis 379
  - 8.6.1 Wishbone Suspension Roll Center Calculation 381
  - 8.6.2 MacPherson Strut Suspension Roll Center Calculation 382
  - 8.6.3 Hotchkiss Suspension Roll Center Calculation 382
  - 8.6.4 Vehicle Motion About the Roll Axis 382
- 8.7 Active Suspensions 391
- 8.8 Conclusions 396

Chapter 9—Brakes and Tires 397
- 9.1 Introduction 397
- 9.2 Braking Dynamics 399
- 9.3 Hydraulic Principles 402
- 9.4 Brake System Components 403
  - 9.4.1 Master Cylinder 403
  - 9.4.2 Power Assistance 404

        9.4.3   Combination Valve 405
                 9.4.3.1   Proportioning Valve 406
                 9.4.3.2   Pressure Differential Switch 406
                 9.4.3.3   Metering Valve 406
   9.5 Drum Brakes 406
        9.5.1   Analysis of Drum Brakes 409
        9.5.2   Example 412
   9.6 Disc Brakes 414
        9.6.1   Disc Brake Components 414
                 9.6.1.1   Brake Disc 414
                 9.6.1.2   Brake Pads 416
                 9.6.1.3   Caliper 416
        9.6.2   Disc Brake Analysis 417
        9.6.3   Heat Dissipation from Disc Brakes 419
   9.7 Antilock Brake Systems (ABS) 421
   9.8 Tires 424
        9.8.1   Tire Construction 425
        9.8.2   Tire Designations 426
        9.8.3   Tire Force Generation 429
   9.9 Summary 433
   9.10 Problems 433

Chapter 10—Vehicle Aerodynamics 435
   10.1 Introduction 435
   10.2 Essential Aerodynamics 436
        10.2.1   Introduction, Definitions, and Sources of Drag 436
        10.2.2   Experimental Techniques 445
   10.3 Automobile Aerodynamics 450
        10.3.1   The Significance of Aerodynamic Drag 450
        10.3.2   Factors Influencing Aerodynamic Drag 452
   10.4 Truck and Bus Aerodynamics 456
        10.4.1   The Significance of Aerodynamic Drag 456
        10.4.2   Factors Influencing Aerodynamic Drag 456
   10.5 Aerodynamics of Open Vehicles 461
   10.6 Numerical Prediction of Aerodynamic Performance 463
   10.7 Conclusions 464
   10.8 Examples 465
   10.9 Discussion Points 469

Chapter 11—Transmission Matching and Vehicle Performance 473
   11.1 Introduction 473
   11.2 Transmission Matching 473
        11.2.1   Selecting the Engine Size and Final Drive Ratio for
                 Maximum Speed 474
        11.2.2   Use of Overdrive Ratios to Improve Fuel Economy 477

        11.2.3 Use of Continuously Variable Transmissions (CVT)
               to Improve Performance 479
        11.2.4 Gearbox Span 482
    11.3 Computer Modeling 486
        11.3.1 Introduction 486
        11.3.2 ADVISOR (ADvanced VehIcle SimulatOR) 488
    11.4 Conclusions 491

Chapter 12—Alternative Vehicles and Case Studies 495
    12.1 Electric Vehicles 495
        12.1.1 Introduction 495
        12.1.2 Battery Types 496
            12.1.2.1 Lead-Acid Batteries 498
            12.1.2.2 Nickel-Cadmium (NiCd) Batteries 498
            12.1.2.3 Nickel-Metal Hydride (NiMH) Batteries 499
            12.1.2.4 Lithium Ion (Li-Ion)/Lithium Polymer Batteries 499
        12.1.3 Types of Electric Vehicles 500
        12.1.4 Conclusions About Electric Vehicles 502
    12.2 Hybrid Electric Vehicles 502
        12.2.1 Introduction 502
        12.2.2 Dual Hybrid Systems 505
    12.3 Case Studies 507
        12.3.1 Introduction 507
        12.3.2 The Vauxhall 14-40 507
            12.3.2.1 Introduction 507
            12.3.2.2 Specifications 508
            12.3.2.3 Engine Design and Performance 508
            12.3.2.4 Engine Performance 513
            12.3.2.5 Vehicle Design and Performance 517
            12.3.2.6 Conclusions 521
        12.3.3 The Toyota Prius 521
        12.3.4 Modeling the Dual Cconfiguration 522
    12.4 Conclusions 524

Chapter 13—References 525

Index 541

About the Authors 595

Preface

This book arose from a need for an automotive engineering textbook that included analysis, as well as descriptions of the hardware. Specifically, several courses in systems engineering use the automobile as a basis. Additionally, many universities are now involved in collegiate design competitions such as the SAE Mini Baja and Formula SAE competitions. This book should be helpful to such teams as an introductory text and as a source for further references. Given the broad scope of this topic, not every aspect of automotive engineering could be covered while keeping the text to a reasonable and affordable size.

The book is aimed at third- to fourth-year engineering students and presupposes a certain level of engineering background. However, the courses for which this book was written are composed of engineering students from varied backgrounds to include mechanical, aeronautical, electrical, and astronautical engineering. Thus, certain topics that would be a review for mechanical engineering students may be an introduction to electrical engineers, and vice versa. Furthermore, because the book is aimed at students, it sometimes has been necessary to give only outline or simplified explanations. In such cases, numerous references have been made to sources of other information.

Practicing engineers also should find this book useful when they need an overview of the subject, or when they are working on particular aspects of automotive engineering that are new to them.

Automotive engineering draws on almost all areas of engineering: thermodynamics and combustion, fluid mechanics and heat transfer, mechanics, stress analysis, materials science, electronics and controls, dynamics, vibrations, machine design, linkages, and so forth. However, automobiles also are subject to commercial considerations, such as economics, marketing, and sales, and these aspects are discussed as they arise.

Again, to limit the scope of this project, several important automotive engineering concepts are notable for their absence. Two examples notable for their absence are manufacturing and structural design and crashworthiness. Neither of these topics was omitted because the topics were deemed unimportant. Rather, they did not fit the particular curriculum this book targeted. In short, topics that have been omitted are not intended to slight the importance of the topics, but choices had to be made in the scope of the text.

The book has been organized to flow from the source of power (i.e., engine) through the drivetrain to the road. Chapter 1 is a brief and selective historical overview. Again, topics for Chapter 1 had to be limited to keep the scope reasonable, and the intent was to show the progression of automotive engineering over the last 100 years. Undoubtedly, readers will find several topics absent from the historical overview. Again, the absences are not intended to minimize the importance of any development, but to limit the size of Chapter 1.

Chapter 2 contains an overview of the thermodynamic principles common to internal combustion engines and concludes with an extensive discussion of fuel cell principles and their systems. The differing operations of spark ignition engines and compression ignition engines are discussed in Chapters 3 and 4, respectively. Because many diesel engines now employ forced induction, the topic of turbo- and supercharging is discussed in Chapter 4 as well. Chapter 5 covers the ancillary systems associated with the engine and includes belt drives, air conditioning, and the starting and charging systems.

Transmissions and drivelines are the topic of Chapter 6. This chapter includes discussion and analysis of both manual and automatic transmissions, driveshaft design, and four- and all-wheel-drive systems. The steering system is discussed in Chapter 7 and includes basic techniques for analyzing vehicle dynamics and rollover. The suspension system is discussed in Chapter 8, and basic models are provided as first-order analysis tools. The suspension system is another topic that is worthy of a textbook in itself, but Chapter 8 provides students and practicing engineers with several references to more detailed models and analysis techniques. Brakes and tires are the topic of Chapter 9, and Chapter 10 discusses vehicle aerodynamics.

Because computer modeling is becoming increasingly important for the automotive engineer, Chapter 11 discusses matching transmissions to engines and provides a link to a computer model that is useful for predicting overall vehicle performance. Chapter 12 concludes the book with two case studies chosen to highlight the advances made in automotive engineering over the last century. The first case study is the Vauxhall 14-40, a vehicle that was studied extensively by Sir Harry Ricardo in the 1920s. As a point of comparison, the second case study is the Toyota Prius, which represents cutting-edge technology in a hybrid vehicle.

The material in the book has been used by the authors in teaching an automotive systems analysis course and as part of a broad-based engineering degree course. These experiences have been invaluable in preparing this manuscript, as has been the feedback from the students. The material in the book comes from numerous sources. The published sources have been acknowledged, but of greater importance have been the conversations and discussions with colleagues and researchers involved in all areas of automotive engineering, especially when they have provided us with copies of relevant publications.

We welcome criticisms or comments about the book, either concerning the details or the overall concept.

Richard Stone
Jeff Ball
Autumn 2002

Acknowledgments

The following figures in this book first appeared in *Introduction to Internal Combustion Engines, Third Edition*, by Richard Stone, published by Palgrave Macmillan in 1999:

- Figures 2.2, 2.4 through 2.7, 2.9, 2.10, 2.12 through 2.15, 2.17 through 2.20, 2.23 through 2.28, and 2.30 through 2.32.

- Figures 3.1 through 3.6, 3.9, 3.10, 3.12 through 3.14, 3.17, 3.18, and 3.26.

- Figures 4.2, 4.3, 4.9, and 4.11 through 4.19.

- Figures 5.8 and 5.9.

Chapter 1

Introduction and Overview

1.1 Beginnings

In June 1895, the Honorable Evelyn Henry Ellis arrived at Southampton from Paris and proceeded to drive his freshly crafted Panhard et Lavassor motor vehicle to his country home—a distance of 56 miles. He thus made history as the first person to drive an automobile in England. He also covered the distance in 5 hours and 32 minutes, excluding stops, which gave him an average speed of 9.84 mph (Womack et al., 1991). In doing so, he entered the history books as the first automotive lawbreaker, because the legal speed limit in England at the time was 4 mph. This speed was mandated by what was known as the "Flag Law." In addition to limiting the speed of self-propelled vehicles, the Flag Law required the operator to have a runner precede the vehicle, waving a red flag to warn pedestrians of the approach of the vehicle. At night, the red flag was replaced by a red lantern.

However, Mr. Ellis was not by nature a lawbreaker, and his extreme speed had a purpose. Mr. Ellis was, in fact, a member of Parliament, and by 1896, he had successfully encouraged Parliament to repeal the Flag Law. The new law increased the national speed limit to 12 mph and dispensed with the flagman. To celebrate their victory, Mr. Ellis and several enthusiasts organized an "Emancipation Run" from London to Brighton on November 14, 1896 (*Autocar*, 1996), and many of the vehicles engaged promptly violated the new speed limit.

Although the Flag Law in England gives some insight into the general public's hesitation over this new technology, this hesitation faded rapidly. The first automotive magazine, *Autocar*, began publication in 1895—the same year as the first British auto show (*Autocar*, 1996). The British automotive industry rose quickly to prominence, led by Daimler in 1896, and Ford and Vauxhall in 1903. Over a few decades, this industry would spawn some of the most coveted makes of cars in the world, such as Rolls-Royce, Bentley, MG, Triumph, and Jaguar.

Meanwhile, across the Atlantic, the arrival of the automobile in the United States was greeted with a strange mixture of loathing and curiosity. The clanking, hissing monsters of the late 1800s often were met by cries of "Get a horse!" Many states also passed legislation that required automobile operators to take their cars apart and hide them in the woods when a horse approached (Clymer, 1950). Several states considered laws requiring drivers to stop every ten minutes and fire a Roman candle as a warning, but no record exists that such laws were actually passed. U.S. President Woodrow Wilson proclaimed the automobile to be "such an ostentatious display of wealth that it would stimulate socialism by inciting envy of the rich" (Rae, 1965). The general public's reaction also ranged to great curiosity. In 1896, the Barnum and Bailey circus displayed a Duryea vehicle in its sideshow, and the vehicle received more attention than the usual sideshow fare of bearded ladies and so forth (May, 1975).

It also is an odd fact of history that the United States had to reinvent the automobile for itself. The Europeans had solved the problem of powering a vehicle with an internal combustion engine in the 1880s, and France took the early lead in automobile production in the 1890s (May, 1975). It is generally accepted that automobile development in the United States until the turn of the century was 10 years behind the Europeans (Rae, 1965). Why this occurred is a mystery, because the Unitd States certainly had access to European developments and the requisite mechanical and engineering talent. One possible explanation is the daunting prospect of automobile travel in a land of vast distances with poor roads.

Despite the less than enthusiastic response to the automobile, the idea slowly caught on. Exactly who was the first to drive an automobile in the United States is a point of contention. Frank and Charles Duryea successfully drove a single-cylinder car through the streets of Springfield, Massachusetts, in 1893, and this is generally regarded as the first operation of an automobile in the United States (May, 1975). This claim ignores several early experiments that have been regarded by historians as unproductive.

One example of the misfortunes of early automotive engineers is provided by the experiences of Albert and Louis Baushke of Benton Harbor, Michigan, and is outlined by May (1975). Together with William O. Worth, they received a patent for a gasoline engine on June 17, 1895. Their idea was to use the engine to power a horseless carriage, an idea on which they claimed to have worked since 1884. The local newspaper, the Benton Harbor *Palladium*, caught wind of their efforts and, by November 1895, wrote that their vehicle was "ready for tests of speed, safety, convenience, and practicability." The Baushkes announced the formation of the Benton Harbor Motor Carriage Company, and the *Palladium* enthusiastically predicted fame and fortune "when these motor carriages are turned out in quantities for the market." A January 1896 story reported a successful test run of the vehicle at speeds of "from 1 to 23-1/2 miles per hour."

What happened next is somewhat murky, but on February 8, 1896, the *Palladium* reported that Mr. Worth claimed that the Baushkes had failed to produce a practical engine for his carriage. The story went on to say that the earlier reports by the *Palladium* regarding the performance of the vehicle were false, and that the vehicle actually had remained in the factory, "a subject of ridicule and a spectacle of folly." Nothing more was heard from the Baushkes, although Mr. Worth continued his efforts in the automobile industry. He attempted another vehicle with Henry W. Kellogg of Battle Creek, Michigan, and together they formed the Chicago Motor Vehicle Company, with Worth as president and Kellogg as treasurer and superintendent. A picture of a delivery vehicle appeared on company letterhead, but no record exists that the company actually produced any vehicles. Henry Kellogg's 1918 obituary makes no mention of his career as an automotive executive, further attesting to the company's lack of success. These unfortunate men are only a few of the early pioneers who failed in their attempts to produce practical automobiles.

Even the year in which automobile production began in the United States is debated. Some historians declare 1896 as the first year of U.S. auto production because the Duryea brothers produced 13 identical cars for sale to customers that year. Other historians claim that 1897 is the rightful "first year," as it marked the first year of major production by several producers, including Pope electrics, Stanley steamers, and Olds and Winton gasoline-powered vehicles (Rae, 1965).

Leaving for now the debate over whom was first to the historians, it can be safely stated that by the turn of the century, the fledgling U.S. automotive industry was firmly established, and public acceptance of the car was on the rise. As the new century dawned, the prospective automobile buyer was presented with a dizzying array of choices: electric, steam, or gasoline power. If the choice was gasoline, should it be air-cooled or water-cooled? Four-stroke or two-stroke? Electric, friction, or chain transmission? Part of the reason for the numerous choices is that from the turn of the century through World War I, automobile companies sprouted like weeds in a flower bed. Unfortunately, many of them disappeared just as quickly (Rae, 1965).

By the end of World War I, the supremacy of the gasoline-powered engine was assured, but at the turn of the century, this was not a given. Colonel Albert A. Pope, founder of the Pope Manufacturing Company, predicted the imminent demise of the gasoline engine because, "You can't get people to sit over an explosion" (Rae, 1965). The fact that the Pope Manufacturing Company produced an electric vehicle called the Columbia undoubtedly biased his assessment.

Steam-powered cars had strong support at this time. Thanks to the railroad industry, there was a wealth of experience with steam engines. The steam engine of that period also produced more power and did not require a complicated transmission, and numerous "experts" were quite confident that ordinary people would never learn how to shift gears. The success of the Stanley steamer also added credence to the arguments in support of steam power. However, steam power had some significant disadvantages. First, there was an ever present fear of boiler explosions, despite the weight of evidence against such failures. A lightweight steam engine that operated with pressures of 600 psi also required skilled maintenance, thus making it unsuitable for mass consumption (Rae, 1965). Finally, although sources of soft water were abundant in the Northeast, steam travel through the desert Southwest of the United States would have required construction of a water supply infrastructure similar to the railroad stations in existence at that time (Rae, 1965).

This period from 1900 to World War I saw great strides in automotive production and design. Ransom Olds began production of the Curved Dash Olds in 1901, and it became the first truly successful vehicle in the United States. Henry Leland, founder of Cadillac, became renowned for precision parts. In 1908, the Royal Automobile Club of England selected three Cadillacs at random from a shipment of eight. The three cars were disassembled, the parts were thoroughly mixed, and three cars were reassembled. For this, Henry Leland and Cadillac received the Dewar Trophy, the highest award for automotive achievement (*Motor Trend*, 1996).

This period also saw the application of electrics to vehicles. Several methods of ignition were used in early gasoline engines, including hot tubes and sparks. Until 1912, spark ignition was provided by a trembler coil, as shown in Fig. 1.1. The system used a set of contacts that responded to the magnetic field in the primary coil, and these contacts made and broke the primary circuit (Johnston, 1996). The resulting action of the contacts was a sort of vibratory motion, hence the name trembler coil. The demise of the trembler coil began in 1908 when Charles Kettering developed the breaker point, or Kettering, ignition system shown in Fig. 1.2. This system used cam-driven contacts to interrupt the primary circuit, which resulted in a single spark being produced to ignite the mixture rather than the steady stream of sparks produced by the trembler coil.

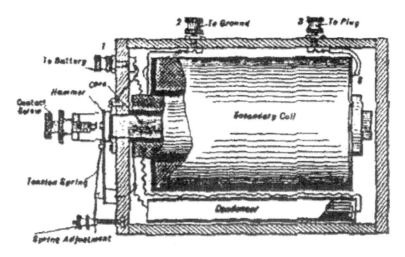

Figure 1.1. *Trembler coil (Johnston, 1996).*

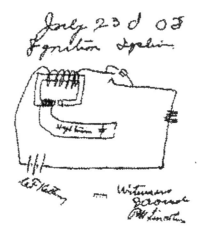

Figure 1.2. *Kettering's sketch of the breaker-point ignition system (Johnston, 1996).*

A second major electrical innovation of this period was the electric starter. Until this time, engines were started with a hand crank at the front of the vehicle. The process required the operator to manually retard the ignition timing, usually with a lever on the steering column. If the operator failed to do this, the crank handle could kick back and cause serious injury to the operator. Byron Carter, builder of the Cartercar and a friend of Henry Leland, stopped to assist a lady who was having difficulty starting her car. The handle kicked back, breaking Carter's jaw. Gangrene set in, and he died several days later (Rae, 1965). Henry Leland was determined that such accidents would not happen again, and he directed Kettering, an engineer with Cadillac, to develop a solution. Kettering's solution was the electric starter, a system that remains in use to this day. Obviously, the starting and ignition systems produced by Kettering required a power source, and during this time, he also was busy developing a generator-battery system for electrical power.

One of the biggest developments during this period was the mass production system. Henry Ford did not invent the moving assembly line—he claimed his inspiration was a meat packing plant where he watched hog carcasses being disassembled as they moved past workers on a chain (*Motor Trend*, 1996). Nor did he invent interchangeable parts. His success was spawned by his application of both to the manufacture of automobiles. Ford was a shrewd individual and realized he could not implement an entire assembly line for a car all at once. Instead, in 1913, he set up a moving assembly line to make magnetos. Rather than having a single worker spend 20 minutes assembling a magneto, he had a conveyor move the assemblies past a series of workers, each of whom performed one or two steps in the process. Once perfected, his assembly line could produce a magneto in 5 minutes. Ford continued to improve his assembly line until, by October 1913, an entire Model T could be assembled in slightly less than 3 hours. By April 1914, assembly time on the Model T had dropped to only 93 minutes.

Ford constantly looked for ways to save time. He found that he could eliminate a bracket by extending the frame slightly. Because the bracket took a worker a minute to install, this saved 3,300 hours of assembly labor over a run of 200,000 cars. This also was the motivation behind Ford's statement that the customer could have any color he or she wanted, as long as it was black. By 1917, Ford's line was moving at such a rapid pace that production was slowed by the time it took for the paint to dry on the body. Ford found that black Japan enamel was the only paint that would dry quickly enough for his line to maintain its pace (*Motor Trend*, 1996).

Ford's success with the Model T was due to three factors. First, the car was designed for the mass production assembly line. As already noted, he continually tinkered with his design to shave time off the assembly process. As a result, by 1914, he was able to produce 200,000 cars while reducing his payroll from 14,336 to 12,880 employees (*Motor Trend*, 1996).

Ford's second stroke of genius was to design the Model T for the roads of the day. Having grown up on a farm, Ford appreciated the fact that a vehicle needed to be able to traverse rough, unimproved terrain, which basically described most of the roads of the day. His vehicle had a high ground clearance and a fairly flexible frame that enabled the wheels to maintain contact with the ground in rough terrain.

Finally, on January 5, 1914, Ford announced that the standard wage for a Ford worker was $5 per day, and the standard shift was reduced from 10 hours to 8 hours. Ford was not being altruistic; he was being shrewd. The 8-hour shift meant that the factory could run 3 shifts 24 hours per day instead of 2 shifts for 20 hours per day. Until that time, cars really were a conveyance for the wealthy. With the huge wage Ford paid, he created a middle class of consumers who could afford to buy the cars they built. Thus, he created his own market for his product, and he became both rich and famous as a result.

While Ford was busy making a car for the common man, William Crapo Durant was busy trying to harness several automakers into one corporation. Durant knew very little about manufacturing in general or the car business in particular, but he was a dynamic businessman who was not averse to taking risks. He began his career by taking control of the Buick Motor Company in 1904 and promptly returned it to profitability (Rae, 1965). In 1908, he began negotiations to buy four companies, including REO, the Olds Motor Works, and the Ford Motor Company. Talks fell through when Henry Ford demanded payment in cash, but Durant continued his quest and eventually Olds joined the Durant stable. In 1909, Durant added the crown jewel to his mix—Cadillac (May, 1975). Durant also gained control of several lesser companies, but his claim to fame was his success in organizing this disparate bunch of companies into General Motors.

Durant made another attempt to buy Ford in 1910, and this time Henry Ford compromised on his demand for cash. Durant needed $2 million in a hurry, and all seemed to be going according to plan. However, at the last minute, the National City Bank of New York, which had promised the money, withdrew the offer under the direction of its loan committee, and the opportunity was lost. The year 1910 also saw a dip in demand for autos in general, but especially for Durant's collection of high-priced, low-volume Buicks, Oldsmobiles, Oaklands, and Cadillacs. The board of directors was concerned that Durant's policies had left GM overextended due to its rapid expansion, and Durant was unceremoniously dumped.

Dumped, but not finished. In 1911, Durant teamed with Louis Chevrolet to form the Chevrolet Motor Car Company (Rae, 1965). They produced a car for the masses and, by 1915, were challenging the dominance of the Ford Model T with their Chevrolet 490—so named because it was supposed to sell for $490. The success of the Chevrolet company led Durant to offer the company in exchange for GM stock, which at that time was not paying dividends. With support from the DuPont family, the deal went through in 1916, and Durant again found himself in control of GM, where he remained until the ascension of Alfred Sloan in 1923.

1.2 Growth and Refinement

By 1920, the car was a common fixture on both sides of the Atlantic, and automakers began to focus on improved performance for their vehicles. Cadillac had introduced the V-8 engine in 1915 (Fig. 1.3), and by 1916, eighteen companies were producing V-8s (Rinschler and Asmus, 1995). Packard introduced the straight eight in 1923, and by 1930, Cadillac introduced its 7.4L V-16. Engine performance was greatly improved with the development of the turbulent head by Harry Ricardo shortly after World War I (Fig. 1.4). The turbulent head aided combustion

Figure 1.3. *The Cadillac V-8 of 1915 (Rinschler and Asmus, 1995).*

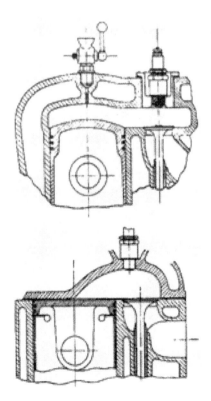

Figure 1.4. *The Ricardo turbulent head (bottom), compared to a standard L-head of the period (top) (Rinschler and Asmus, 1995).*

and allowed engines to operate at a higher compression ratio, a definite advantage given the low octane rating of fuels at the time. More details on the Ricardo head are given in the case study of the Vauxhall 14-40 presented in Chapter 12.

This period also saw significant improvements in braking, lighting, tires, and windshields, and there was a dramatic shift in buyer preference toward closed cars. Until 1920, most cars were open-topped vehicles, which had obvious negative implications for driving in bad weather. The enclosed car isolated the occupants from rain, snow, and dust, but it also provided advantages in safety. Early closed vehicles had roofs made of fabric-covered wooden frames. In an accident, occupants sometimes were ejected through the roof (Yanik, 1996). Thus, work began on developing a steel roof. This was no small feat, as initial attempts with flat steel roofs produced a drumming sound when traveling. Harley Earl solved the problem by curving the roof, and GM put his invention into production as the "Turret Top" in 1935 (Fig. 1.5). The enclosed, all-steel vehicles also prompted the first use of safety to market automobiles, and rollover tests such as those shown in Fig. 1.6 were used as advertisements to demonstrate the safety and sturdiness of such vehicles.

The 1930s also saw the advent of crash testing. General Motors used a test driver standing on the running board, who would direct the car down a hill toward a wall, jumping off at the last moment (Yanik, 1996). The only analysis that could be made at that time was to observe the resulting damage. Of course, the main event in the 1930s was the Great Depression, which brought a huge drop in demand for cars. Automakers were forced into a survival mode, and many automakers did not survive this period, notably Marmon, Peerless, Duesenberg, Cord, Auburn, Graham, Hupp, and Stutz.

Figure 1.5. *Fisher Body plant manufacturing "Turret Tops" in 1935 (Yanik, 1996).*

Figure 1.6. *A rollover test in the 1930s, which demonstrated the advantages of an all-steel enclosure (Yanik, 1996).*

1.3 Modern Development

World War II brought a halt to auto production in the United States as automakers switched to wartime materiel production. However, the halt was only temporary. At the conclusion of the war, not a single U.S. factory had been bombed. The same could not be said for Europe. Thus, American engineers in 1946 could immediately update their products, and the U.S. factories began churning out vehicles quickly. The four-year hiatus in auto production also created a pent-up demand for new vehicles, which spurred enormous growth in the U.S. economy.

The 1950s found the U.S. auto industry leading the world, and the cars reflected this general attitude. Cars were bedecked with ever more chrome trim, and tailfins rose in height until the 1959 Cadillac presented a practical limit to fin height. The Corvette was introduced in 1953 and has continued as "America's Sports Car" to this day. In 1955, Chevrolet introduced the now-famous "small-block" V-8. Initially, this was a 265-cubic-inch carbureted engine advertised at 180 hp (*Autocar*, 1996). The impact of this engine cannot be overstated. Until that point, the fastest cars also were the most expensive. Thus, a Cadillac could outrun a Buick, and so on down the cost ladder. The small-block V-8, under the workings of a skilled engine tuner, suddenly was enabling bargain-priced Chevys to outperform Cadillacs and Lincolns. Even today, 1950s-era cars with small-block Chevy engines are solid performers at the track (Fig. 1.7).

The good times for U.S. automakers continued into the 1960s, which saw the "horsepower race" begin in earnest. As early as 1963, every manufacturer had a 426- or 427-cubic-inch engine on its option lists and advertised horsepower ratings that climbed above 400 hp. However, these numbers were "gross" ratings, meaning that when the engine was run on the dynamometer, all accessories were removed, including alternators, air conditioning compressors, oil and water pumps, and so forth. The Society of Automotive Engineers (SAE) finally stepped in with engine test standards and mandated all horsepower ratings to be given as SAE net. In

Figure 1.7. *A 1955 small-block Chevrolet staged at a drag strip. Courtesy of Mr. Martin Bowe.*

other words, the engine on the test stand was required to be identical to the installed engine—all accessories and pumps were to be driven by the engine.

The 1960s also saw increased attention placed on automotive safety. General Motors pioneered the collapsible steering wheel column, which absorbed energy in an impact rather than spearing the driver, and the innovation soon appeared on other makes (Yanik, 1996). Other safety-related innovations included the clutch/starter interlock, auto-locking doors, and seat belts. In 1962, GM developed a high-speed impact sled at its Milford proving grounds (Yanik, 1996). The sled allowed controlled simulation of accidents, and engineers at GM went on to develop the head injury criteria (HIC) as a method of predicting when head injury was likely to occur (Yanik, 1996).

On September 9, 1967, U.S. President Lyndon B. Johnson signed into law the National Traffic and Motor Vehicle Safety Act, ushering in the era of government regulation of the automobile industry. The act went into effect on January 1, 1968, and contained 19 standards covering accident avoidance, crash protection, and post-crash survivability (Crandall et al., 1986). By 1974, the number of standards had grown to 46, but their effect was beginning to be felt, as shown in Fig. 1.8. This figure illustrates the number of highway deaths in the United States per 100 million vehicle miles.

Also during the postwar era, import cars began to make a showing in the United States. Initially, the imports tended to be sports cars brought home by U.S. servicemen, with two-seat British roadsters being a particular favorite. However, the 1970s brought new challenges to the automotive industry in the form of oil shortages. As the price of gasoline soared, consumers desperately wanted more fuel-efficient cars than Detroit was producing. Sales of imports rose. Consumers bought them for their economy but then stayed with them for their quality, particularly the Japanese vehicles. The Japanese auto industry made an attempt to broach the U.S. market in 1958, when Toyota introduced the Toyota Crown. The car was woefully underpowered, rattled at highway speed, and tended to boil over in the heat of Southern

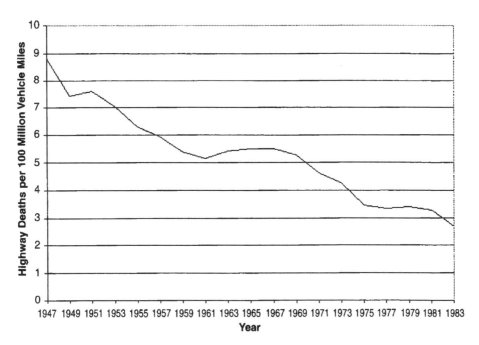

Figure 1.8. *Highway deaths per 100 million vehicle miles in the United States (Crandall et al., 1986).*

California (Ingrassia and White, 1995). Toyota retreated from the U.S. market in 1960, but it was far from defeated.

Toyota then looked to W. Edwards Deming for guidance. Deming was a management consultant who mixed rigorous statistical and measuring methods with a management philosophy that gave more power and responsibility to the workers (Ingrassia and White, 1994). The result of Toyota's implementation of Deming's methods was the lean production system. The details of this manufacturing system are beyond the scope of this text but are examined in depth in the book by Womack et al. (1991). Nevertheless, the impact of the Japanese on worldwide auto manufacturing cannot be overemphasized, and a brief explanation of the lean production system is in order.

Until this point, all major manufacturers employed the mass production system. The system depended on economies of scale and a constantly moving assembly line to produce cheap but profitable cars. The implications of this are numerous, but for the purpose of example, a few implications will be examined in the areas of the factory and designing the car. First, because the system depended on a constantly moving line, parts were stockpiled in the factory to ensure a ready supply at all times. This resulted in huge factories, with the extra space being used to store the excess capacity of parts. Furthermore, if a particular batch of parts was defective, the line workers were expected to attach the parts as best they could. Workers were never able to stop the assembly line; such a prerogative rested solely with management. This technique required a team of reworkers at the end of the line who would tear into the car to fix any defects.

To design and produce a car in the mass production system, several different departments must work together, such as marketing, powertrain, chassis, and manufacturing. Within the system, engineers would be assigned to work on specific projects, but those would not be their only projects. Furthermore, they were still responsible primarily to their functional chief as opposed to the vehicle project manager. Thus, the project manager on a particular model found that he had the responsibility for developing the model but did not have the authority required to move the process. These managers were in a position of coordinating the efforts of a disparate group rather than managing a cohesive team.

Adding to the turmoil was the fairly sequential nature of the process. For example, the engineers designing the car often would do so in isolation from the manufacturing engineers. Thus, when the design was passed to manufacturing, it often was returned as a "no build," meaning that the design could not be built with current manufacturing tools. The design engineers then would have to redesign the vehicle before passing the updated version to manufacturing. This cycle could be repeated several times, with an accompanying slippage in the timetable. Thus, it often would require five or more years to bring a new vehicle into production, with an associated large increase in the cost of doing so.

To the Japanese, such practices were *muda*, or waste. They recognized that storing weeks' worth of parts in the factory greatly increased overhead costs. Thus, they worked with their suppliers so that parts were delivered to the factory "just in time." Lean production factories thus had only a few hours' worth of parts available on hand. Furthermore, if a worker discovered a defective part, that worker was able to immediately stop the line. Workers, managers, and engineers would then try to discover the reason behind the defect, using a process known as the "five whys" (Womack et al., 1991). The logic behind this was that simply passing defects down the line was wasteful because it required a team of reworkers. A better solution was to get to the source of the defect and fix it, thus removing the problem permanently. Suppliers also were involved in the process because they were the ones who produced the parts. Because the system still required a constantly moving assembly line, increased pressure was placed on suppliers to provide parts with no defects, precisely when those parts were needed. The result of this process was to produce economical cars of extremely high quality.

As for designing the car, the Japanese took the sensible step of forming teams from all functional departments under the authority of the product manager. The engineers from all departments, with manufacturing, marketing, styling, and so forth, worked side by side throughout the product development process. As a result, "no build" situations could be resolved on the spot, significantly reducing the time and expense required to design a new vehicle. In fact, by the 1980s, Toyota's development cycle was down to 36 months (Womack et al., 1991). The lean production system has since been adopted by all U.S. producers and can rightly be called a revolution in the auto industry. Again, this short synopsis should not be construed as minimizing Japanese contributions, and the interested reader is referred to Womack's book for a complete discussion of the lean production system.

Returning to the 1970s, U.S. automakers faced a serious challenge from the imports, as well as increasing government regulation of fuel economy and emissions. The pace of legislation and the solutions found by automakers to keep pace are discussed in Section 3.5.

Another interesting facet of postwar automobile production is the divergent paths taken by the U.S. and European auto industries. In Europe, the mainstream vehicle became smaller and lighter and emphasized handling. In the United States, the mainstream vehicle became large and powerful and emphasized straight-line speed and stability. One reason for this disparity in design is found in the road systems developed on the two continents. In Europe, the road system predated the automobile by several centuries. The roads that existed thus were designed for pedestrian traffic or, at best, horse-drawn traffic. When the car arrived, common sense dictated that the existing roads should be covered with asphalt. This resulted in "narrow, winding roads, blind turns, and hidden entrances" (Olley, 1946). The nature of the roads thus required "small, bantam-weight cars with the agility of a dancer and what is know as 'flashing performance' " (Olley, 1946).

Conversely, in the United States, the car preceded the road system. Road designers thus were at liberty to select both the preferred path between points as well as the width of the road itself. The interstate highway system that was developed in the 1950s is a prime example. The highways between cities were built as straight as possible and were constructed with multiple, divided lanes, each lane being approximately 12 feet wide. Furthermore, the distances between cities are significantly longer than those in Europe. The distance covered in driving across the state of Texas on Interstate 10 is only a few miles less than the distance from Lands End to John O'Groat in the United Kingdom—a favored trip for cyclists because it is the longest trip one can take within the United Kingdom. This implies that the design of cars in the United States "departed from the qualities of nimbleness or handiness…the emphasis is now all on directional stability" (Olley, 1946).

Regarding the size and power of American engines, this has everything to do with what the motorist pays for fuel. Contrary to popular belief, the U.S. driver pays approximately the same amount for a liter of fuel as a motorist in Europe. The large price discrepancy is due solely to the level of taxation placed on fuel by the respective governments. Figure 1.9 shows the levels of taxation.

As Fig. 1.9 shows, the cost of a liter of gasoline is roughly $0.30 in the United States and Europe, with the exception being Japan, where the cost of gasoline is $0.44 per liter. Another way of looking at this data is to calculate the percentage of the fuel cost devoted to taxes, as shown in Fig. 1.10.

As shown in Fig. 1.10, the United Kingdom has the highest level of fuel taxation, at 75%, whereas the United States has the lowest, at 26%. Whether a high taxation level is good or bad is a political debate and is beyond the scope of this text. From a motorist's perspective, the low taxation level generally is applauded. However, the drawback to the U.S. taxation policy is that market fluctuations in the price of crude oil are drastically reflected at the gas pump. For example, during the summer of 2001, gasoline prices in Colorado averaged nearly $2.00 per gallon for unleaded fuel. By the fall of 2001, the price had fallen to near $1.10 per gallon. This has a profound effect on product planners in the U.S. auto industry. When gas prices neared their peak, the demand for large vehicles with V-8 engines dropped, with prices for used vehicles of such size. Conversely, as the price of gasoline falls, the demand for such large vehicles again rises. This makes forecasting difficult for any auto manufacturer, and

14 | *Automotive Engineering Fundamentals*

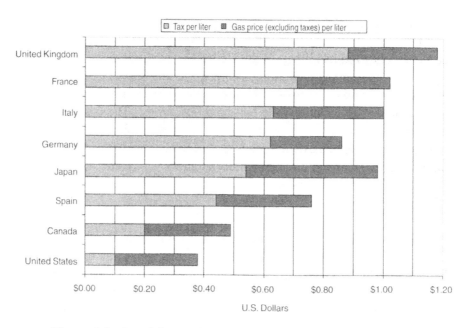

Figure 1.9. *Breakdown of gasoline prices, as of September 2000. (International Energy Agency, "Energy Prices and Taxes, Quarterly Statistics")*

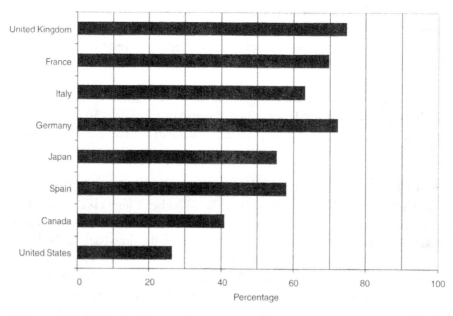

Figure 1.10. *Percentage of gasoline cost due to taxes, as of September 2000. (International Energy Agency, "Energy Prices and Taxes, Quarterly Statistics")*

one needs to look only at the U.S. auto industry in the late 1970s to understand the impact of failing to predict market trends.

The effect of emissions legislation on American cars was a reduction in compression ratio, which led to a decrease in performance. The market conditions of the late 1970s nearly caused the U.S. "Big Three" automakers to go under, and Chrysler resorted to a $1.5 billion loan guaranteed by the U.S. government to stay alive. Times improved, in large part due to advances in technology. By the end of the 1980s, the carburetor was replaced by electronically controlled fuel injection systems. This represents the latest revolution in automotive design—the increasing use of digital electronics to control all aspects of the functions of cars. Today's cars perform better than their predecessors of the 1960s, while getting better fuel economy and producing far fewer emissions. Digital computer control has allowed the implementation of safety devices such as antilock brake systems (ABS), stability and traction control systems, and air bags. Computer aided design (CAD) and finite element analysis (FEA) have allowed engineers to create stronger, lighter bodies that are designed to absorb energy in an impact while protecting the occupants. One example of the advances in automotive engineering is brought out by comparing the performance of new vehicles in tests such as the standing quarter mile, as shown in Table 1.1. The performance of the average minivan today is comparable to the performance cars of the late 1950s. Such performance also depends on the great advances in transmission technology and, above all, tire technology.

**TABLE 1.1
PERFORMANCE COMPARISON
(MOTOR TREND, 1999)**

Year/Model	Engine	Transmission	0–60 mph (sec)	1/4 Mile (sec/mph)
1955 Chrysler 300	331 in.^{3} 300 hp	Three-Speed Automatic	10.0	17.6/82.0
1956 Chevrolet Corvette	283 in.^{3} 245 hp	Two-Speed Automatic	11.6	17.9/77.5
1957 Chevrolet Bel Aire	283 in.^{3} 270 hp	Two-Speed Automatic	9.9	17.5/77.5
1957 Maserati 2000-GT	122 in.^{3} 150 hp	Four-Speed Manual	12.5	18.9/77.1
1957 Jaguar 3.4	210 in.^{3} 210 hp	Four-Speed Manual	10.7	17.7/78.2
1958 Porsche Speedster	96 in.^{3} 88 hp	Four-Speed Manual	10.1	18.3/73.7
1958 Ford T-Bird	352 in.^{3} 300 hp	Three-Speed Automatic	10.5	16.8/81.3
1999 Honda Odyssey	212 in.^{3} 210 hp	Four-Speed Automatic	9.3	16.9/82.6

1.4 Overview

The purpose of this book is to give automotive engineering students a basic understanding of the principles involved with designing a vehicle. Naturally, any attempt to provide a manual for the complete, up-to-date design of a car would result in a huge book that would be unaffordable to the average college student. Thus, this work focuses on "first principles," be they the principles of thermodynamics, machine design, dynamics, or vibrations, with a bit of heat transfer and material properties added to the mix.

The book attempts to take a logical approach to the car and starts with the front end—namely, the engine. The engine chapters (Chapters 2 through 5) begin with thermodynamic principles and proceed through spark ignition and compression ignition engines. Chapter 5 is concerned with the accessories driven by the engine, such as the lubrication system, cooling system, belt drives, and air conditioning. Chapter 6 picks up at the flywheel and continues through the transmission and driveline. Chapters 7 and 8 delve into steering systems, steering dynamics, and suspension systems and their analysis. The complexity of these particular topics requires the use of complex models for analysis. However, the reader is reminded again of the introductory nature of this work. Thus, all analyses in these chapters use highly simplified models to illustrate basic principles. Direction is given in these chapters toward books of a more specialized nature. Chapter 9 covers brakes and tires, including drum brakes, disc brakes, and antilock brake systems (ABS). Chapter 10 introduces vehicle aerodynamics, and Chapter 11 is devoted to computer modeling of vehicle performance. Finally, the book concludes with a chapter on alternative vehicles and provides two case studies. The first case study is of the 1922 Vauxhall 14-40, a cutting-edge vehicle in its day. This is compared to a modern vehicle that represents current cutting-edge technology, the 1998 Toyota Prius.

In addition to providing an overview of some of the techniques used in automotive engineering, it is hoped that the student will come away from this book with an appreciation for the automobile as a system. The modern automobile is more than the sum of its parts. Each subsystem must work in harmony with the others, and the modern automotive market is quick to discern vehicles that are merely a collection of independently produced parts. The engine designer can ill afford to neglect the design of the transmission, for history is replete with amateur engine tuners who do a marvelous job with the engine, only to promptly destroy their driveline with their additional torque.

Chapter 2

Thermodynamics of Prime Movers

2.1 Introduction

This chapter concentrates on reciprocating internal combustion (IC) engines because gas turbines normally are not considered for automotive use. Although the adjective "reciprocating" precludes Wankel engines, the thermodynamic operation of Wankel engines is no different from reciprocating engines. This chapter concludes with an introduction to fuel cells. Enormous effort is being devoted to applying fuel cells to vehicles. Therefore, it is important for engineers not only to understand how they work, but to see how their efficiency compares with conventional engines. The treatment of reciprocating engines covers their mechanical operation, their representation by air standard cycles, and their ignition and combustion characteristics (and thus the necessary fuel requirements), and it ends with a discussion of the gas exchange processes. The treatment of gas exchange includes superchargers and turbochargers because these are applicable to both gasoline and diesel engines. Comprehensive treatments of internal combustion engines can be found in Ferguson (2001), Heywood (1988), and Stone (1999). Larminie and Dicks (2000) provides a good introduction to fuel cells.

2.2 Two- and Four-Stroke Engines

Internal combustion engines usually operate on either the four-stroke (one power stroke every two revolutions) or two-stroke (one power stroke every revolution) mechanical cycle. The four-stroke operating cycle can be explained by reference to Fig. 2.1.

1. **The induction stroke**. The inlet valve is open, and the piston travels down the cylinder, drawing in a charge of air. In the case of a spark ignition engine, the fuel usually is premixed with the air.

2. **The compression stroke**. Both valves are closed, and the piston travels up the cylinder. In the case of compression ignition engines, the fuel is injected toward the end of the compression stroke. As the piston approaches top dead center (tdc), ignition occurs either by means of a spark or by auto-ignition.

3. **The expansion, power, or working stroke**. Combustion propagates throughout the charge, raising the pressure and temperature, and forcing the piston downward. At the end of the power stroke, as the piston approaches bottom dead center (bdc), the exhaust valve opens, and the irreversible expansion of the exhaust gases is termed "blow-down."

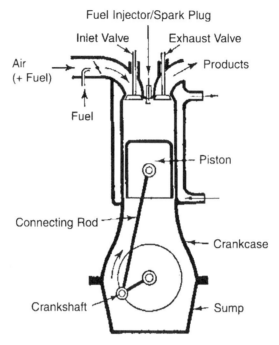

Figure 2.1. *A four-stroke cycle engine. Adapted from Rogers and Mayhew (1967).*

4. **The exhaust stroke.** The exhaust valve remains open, and the piston travels up the cylinder and expels most of the remaining gases. At the end of the exhaust stroke, when the exhaust valve closes, some exhaust gas residuals will remain. These will dilute the next charge.

The four-stroke cycle sometimes is summarized as "suck, squeeze, bang, and blow." Because the cycle is completed only once every two revolutions, the valve gear (and any in-cylinder fuel injection equipment) must be driven by mechanisms operating at half engine speed. Some of the power from the expansion stroke is stored in a flywheel, to provide the energy for the other three strokes.

The two-stroke cycle eliminates the separate induction and exhaust strokes, so that between the expansion and compression processes, a scavenging process occurs. The simplest scavenging arrangement is under-piston scavenging, and this system can best be explained with reference to Fig. 2.2. In the case of compression ignition engines, the fuel is injected toward the end of the compression stroke.

1. **The compression stroke (Fig. 2.2a).** The piston travels up the cylinder, compressing the trapped charge. If the fuel is not pre-mixed, the fuel is injected toward the end of the compression stroke; ignition should again occur before top dead center. Simultaneously, the underside of the piston is drawing in a charge through a reed valve.

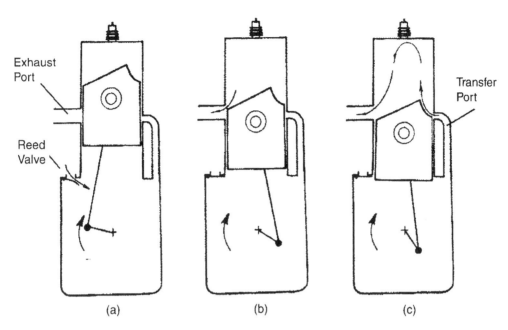

Figure 2.2. *A two-stroke engine with under-piston scavenging; (a), (b), and (c) are defined in the text (Stone, 1999).*

2. **The power stroke.** The burning mixture raises the temperature and pressure in the cylinder and forces the piston downward. The downward motion of the piston also compresses the charge in the crankcase. As the piston approaches the end of its stroke, the exhaust port is uncovered (Fig. 2.2b), and blow-down occurs. When the piston is even closer to bottom dead center (Fig. 2.2c), the transfer port also is uncovered, and the compressed charge in the crankcase expands into the cylinder. Some of the remaining exhaust gases are displaced by the fresh charge. Because of the flow mechanism, this is called loop scavenging. As the piston travels up the cylinder, first the transfer port is closed by the piston, and then the exhaust port is closed.

For a given size of engine operating at a particular speed, a two-stroke engine will be more powerful than a four-stroke engine because the two-stroke engine has twice as many power strokes per unit time. Unfortunately, the efficiency of a two-stroke engine is likely to be lower than that of a four-stroke engine, and there is the difficulty of controlling the gas exchange processes when they are not undertaken with separate strokes of the piston. The problem with two-stroke engines is ensuring that the induction and exhaust processes occur efficiently, without suffering charge dilution by the exhaust gas residuals. The spark ignition engine is particularly troublesome because at part throttle operation, the crankcase pressure can be less than atmospheric pressure. This leads to poor scavenging of the exhaust gases, and a rich air-fuel mixture becomes necessary for all conditions, with an ensuing low efficiency (Section 2.5).

These problems can be overcome in two-stroke direct injection by supercharging engines (either with spark ignition or compression ignition), so that the air pressure at the inlet to the crankcase is greater than the exhaust back-pressure. This ensures that when the transfer port is opened, efficient scavenging occurs. If some air passes straight through the engine, it does not lower the efficiency because no fuel has so far been injected. Two-stroke engines are not widely used in automotive applications, and even with two-wheeled vehicles, emissions legislation is reducing their prevalence. Thus, they will not be discussed further here, but additional information can be found in Stone (1999).

2.3 Indicator Diagrams and Internal Combustion Engine Performance Parameters

Much can be learned from a record of the cylinder pressure and volume. The results can be analyzed to reveal the rate at which work is being done by the gas on the piston, and the rate at which combustion is occurring. In its simplest form, the cylinder pressure is plotted against volume to give an indicator diagram.

Figure 2.3 is an indicator diagram from a spark ignition engine operating at part throttle, with an inset to clarify the pressure difference between the exhaust stroke and the induction stroke—the pumping loop. The shaded area in Fig. 2.3 represents the work done on the piston by the gases during the expansion stroke. For the change in volume shown, this is greater than the work done on the gases during the compression process. The difference in areas at a given volume increment will represent the net work done on the piston by the gases. Thus, the area enclosed by the compression and expansion processes (the power loop) is proportional to the work done on the piston by the gas. The pumping loop is enclosed by processes in an anti-clockwise direction, and it can be seen that this represents the net work done by the piston on the gases.

The term *indicated work* is used to define the net work done on the piston per cycle, but it can either include or exclude the pumping loop. In North America, it tends to exclude the pumping work. These ambiguities can be avoided by using gross and net as qualifiers:

$$\text{Net indicated work, } W_i = \text{power loop} - \text{pumping loop} = \int p dV \quad (2.1)$$

and

$$\text{Net indicated work, } W_i = \text{gross indicated work} - \text{pumping work} \quad (2.2)$$

This in turn leads to the definition of a fictional pressure, the indicated mean effective pressure (imep), $\bar{p}_i$, which is defined by

$$W_i = \bar{p}_i \times V_s \quad (2.3)$$

where V_s = swept volume.

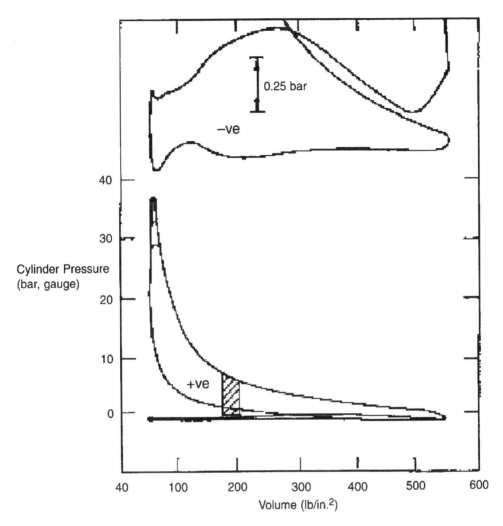

Figure 2.3. *The pressure–volume or indicator from a Rover M16 engine operating at 2000 rpm, with an enlargement of the pumping loop; bmep = 3.8 bar, and imep = 4.6 bar (including the pumping work of 0.45 bar pmep). Adapted from Stone (1999).*

The imep is a hypothetical pressure that would produce the same indicated work if it were to act on the piston throughout the expansion stroke. The concept of imep is useful because it describes the thermodynamic performance of an engine, in a way that is independent of engine size and speed and frictional losses.

Unfortunately, not all the work done by the gas on the piston is available as shaft work because there are frictional losses in the engine. These losses can be quantified by the brake mean effective pressure (bmep, $\bar{p}_b$), a hypothetical pressure that acts on the piston during the expansion stroke and would lead to the same brake work output in a frictionless engine. In other words,

$$\dot{W}_b = \bar{p}_b \times V_s \times n \quad \text{or} \quad \bar{p}_b = \dot{W}_b / (V_s \times n) \qquad (2.4)$$

where n (cycle/s) will be

$$N \text{ (rpm)}/120 \text{ for four-stroke engines}$$

and

$$N \text{ (rpm)}/60 \text{ for two-stroke engines}$$

Here, V_s is the swept volume of the entire engine.

The difference between the net imep and bmep is due to friction, and this leads to the definition of the frictional mean effective pressure, fmep:

$$\text{fmep} = \text{net imep} - \text{bmep}$$

or

$$\text{fmep} = \text{gross imep} - \text{bmep} - \text{pmep} \tag{2.5}$$

where pumping mean effective pressure,

$$\text{pmep} = \text{gross imep} - \text{net imep}$$

For a modern spark ignition engine, the frictional loss can be assumed to depend on only the mean piston speed, v_p (m/s):

$$\text{fmep} = 700 - 0.5\, v_p + 0.33\, v_p^2 \text{ (kPa)} \tag{2.6}$$

The mechanical efficiency η_{mech} can be defined in a number of ways:

$$\eta_{\text{mech}} = \frac{\dot{W}_b}{\dot{W}_i} = \frac{\eta_b}{\eta_i} = \frac{\bar{p}_b}{\bar{p}_i} \tag{2.7}$$

where

$$\dot{W}_i = \bar{p}_i \times V_s \times n \tag{2.8}$$

and

$$\eta_b = \dot{W}_b / (\dot{m}_f \times CV) \quad \text{or} \quad \eta_i = \dot{W}_i / (\dot{m}_f \times CV) \tag{2.9}$$

with CV being the calorific value of the fuel, and $\dot{m}_f$ the mass flow rate of the fuel.

Another way of characterizing engine efficiency is in terms of the specific fuel consumption (sfc), because this enables the direct computation of the rate at which fuel will be used. The sfc is a reciprocal of efficiency.

Brake specific fuel consumption, bsfc = m_f / W_b

Indicated specific fuel consumption, isfc = m_f / W_i

(2.10)

The units might be MJ (fuel)/kWh (work) or kg (fuel)/kWh (work) – 1 kWh ≡ 3.6 MJ.

The final parameter to be defined here is the volumetric efficiency of the engine (η_v); the ratio of actual air flow to that of a perfect engine is

$$\eta_v = \frac{\dot{V}_a}{V_s \times n} \qquad (2.11)$$

where V_s = swept volume and n (cycle/s) will be

N (rpm)/120 for four-stroke engines

and

N (rpm)/60 for two-stroke engines

In general, it is quite easy to provide an engine with extra fuel; therefore, the power output of an engine will be limited by the amount of air that is admitted to an engine. The relationship between the output of an engine and its volumetric efficiency is developed in Section 2.8.1.

The volumetric efficiency is reduced by fluid friction, convective heating during induction, mixing with the hot residual gases remaining in the cylinder, and throttling in the induction or exhaust system.

The volumetric efficiency is enhanced by induction tuning and evaporative cooling when air-fuel mixtures are prepared in the induction system.

2.4 Otto and Diesel Cycle Analyses

Regardless of whether an internal combustion engine operates on a two-stroke or four-stroke cycle and whether it uses spark ignition or compression ignition, it follows a mechanical cycle rather than a thermodynamic cycle. However, the thermal efficiency of such an engine is assessed by comparison with the thermal efficiency of air standard cycles because of the similarity between the engine indicator diagram and the state diagram of the corresponding hypothetical cycle. These cycles are useful because they explain why the efficiency of both engine types increases with load and why the diesel engine efficiency falls less rapidly than that of a spark ignition engine as the load is reduced.

2.4.1 The Ideal Air Standard Otto Cycle

The Otto cycle typically is used as a basis of comparison for spark ignition and high-speed compression ignition engines. The cycle consists of four non-flow processes, as shown in Fig. 2.4.

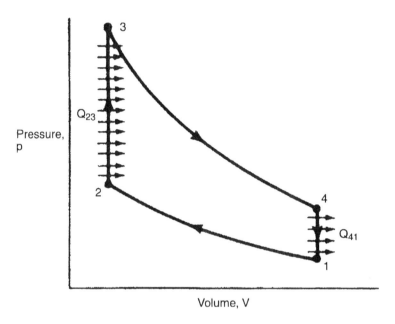

Figure 2.4. *The air standard Otto cycle (Stone, 1999).*

The compression and expansion processes are assumed to be adiabatic (i.e., no heat transfer) and reversible, and thus isentropic. The processes are as follows:

1→2 Isentropic compression of air through a volumetric compression ratio $r_v = V_1/V_2$

2→3 Addition of heat Q_{23} at constant volume

3→4 Isentropic expansion of air to the original volume

4→1 Rejection of heat Q_{41} at constant volume to complete the cycle

The efficiency of the Otto cycle is

$$\eta_{Otto} = \frac{W}{Q_{23}} = \frac{Q_{23} - Q_{41}}{Q_{23}} = 1 - \frac{Q_{41}}{Q_{23}} \tag{2.12}$$

By considering air as a perfect gas, we have constant specific heat capacities. For mass m of air, the heat transfers are

$$Q_{23} = mc_v (T_3 - T_2) \qquad (2.13)$$

$$Q_{41} = mc_v (T_4 - T_1)$$

Thus,

$$\eta_{Otto} = 1 - \frac{T_4 - T_1}{T_3 - T_2} \qquad (2.14)$$

For the two isentropic processes 1→2 and 3→4, $TV^{\gamma-1}$ is a constant. Thus,

$$\frac{T_2}{T_1} = \frac{T_3}{T_4} = r_v^{\gamma-1} \qquad (2.15)$$

where γ is the ratio of gas specific heat capacities, c_p/c_v. Thus,

$$T_3 = T_4 r_v^{\gamma-1} \quad \text{and} \quad T_2 = T_1 r_v^{\gamma-1} \qquad (2.16)$$

Substituting into Eq. 2.14 gives

$$\eta_{Otto} = 1 - \frac{T_4 - T_1}{r_v^{\gamma-1}(T_4 - T_1)} = 1 - \frac{1}{r_v^{\gamma-1}} \qquad (2.17)$$

The value of η_{Otto} depends on the compression ratio, r_v, and not the temperatures in the cycle. To make a comparison with a real engine, only the compression ratio must be specified. The variation in η_{Otto} with compression ratio is shown in Fig. 2.5 with that of η_{diesel}.

2.4.2 The Ideal Air Standard Diesel Cycle

The diesel cycle has heat addition at constant pressure, instead of heat addition at constant volume, as in the Otto cycle. With the combination of a high compression ratio (to cause self-ignition of the fuel) and constant-volume combustion, the peak pressures would be very high. In large compression ignition engines such as marine engines, fuel injection sometimes is arranged so that combustion occurs at approximately constant pressure to limit the peak pressures.

The four non-flow processes constituting the cycle are shown in the state diagram (Fig. 2.6). Again, the best way to calculate the cycle efficiency is to calculate the temperatures around the cycle. To do this, it is necessary to specify the cutoff ratio or load ratio, α:

26 — *Automotive Engineering Fundamentals*

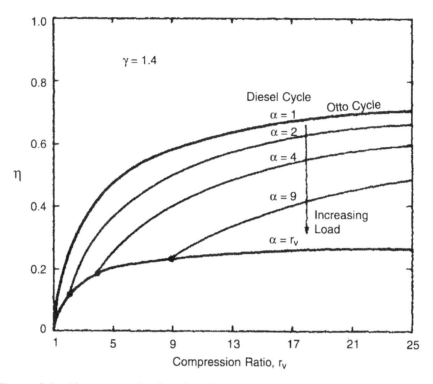

Figure 2.5. *The air standard cycle efficiency for the Otto cycle and diesel cycle at different compression ratios (Stone, 1999).*

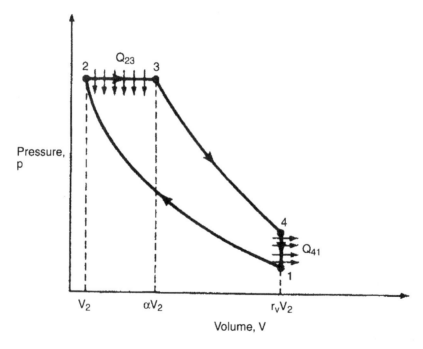

Figure 2.6. *The air standard diesel cycle (Stone, 1999).*

$$\alpha \equiv V_3/V_2 \tag{2.18}$$

The processes are all reversible, and as with the Otto cycle, the compression and expansion processes are assumed to be adiabatic (i.e., no heat transfer) and thus isentropic. The processes in the diesel cycle are as follows:

1→2   Isentropic compression of air through a volume ratio V_1/V_2, the volumetric compression ratio r_v

2→3   Addition of heat Q_{23} at constant pressure while the volume expands through a ratio V_3/V_2, the load or cutoff ratio α

3→4   Isentropic expansion of air to the original volume

4→1   Rejection of heat Q_{41} at constant volume to complete the cycle

The efficiency of the diesel cycle, η_{diesel}, is

$$\eta_{diesel} = \frac{W}{Q_{23}} = \frac{Q_{23} - Q_{41}}{Q_{23}} = 1 - \frac{Q_{41}}{Q_{23}} \tag{2.19}$$

By treating air as a perfect gas, we have constant specific heat capacities. For mass m of air, the heat transfers are

$$Q_{23} = mc_p(T_3 - T_2)$$
$$Q_{41} = mc_v(T_4 - T_1) \tag{2.20}$$

Note that the process 2→3 is at constant pressure. Substitution of Eq. 2.20 into Eq. 2.19, and recalling that γ is the ratio of gas specific heat capacities (c_p/c_v), gives

$$\eta_{diesel} = 1 - \frac{1}{\gamma}\frac{T_4 - T_1}{T_3 - T_2} \tag{2.21}$$

For the isentropic process 1→2, $TV^{\gamma-1}$ is a constant; therefore,

$$T_2 = T_1 r_v^{\gamma-1} \tag{2.22}$$

For the constant pressure process 2→3,

$$\frac{T_3}{T_2} = \frac{V_3}{V_2} = \alpha \quad \text{thus} \quad T_3 = \alpha r_v^{\gamma-1} T_1 \tag{2.23}$$

For the isentropic process 3→4, $TV^{\gamma-1}$ is a constant

$$\frac{T_4}{T_3} = \left(\frac{V_3}{V_4}\right)^{\gamma-1} = \left(\frac{\alpha}{r_v}\right)^{\gamma-1}$$

Thus,

$$T_4 = \left(\frac{\alpha}{r_v}\right)^{\gamma-1} \quad T_3 = \alpha r_v^{\gamma-1}\left(\frac{\alpha}{r_v}\right)^{\gamma-1} \quad T_1 = \alpha^\gamma T_1 \tag{2.24}$$

Substituting for all the temperatures in Eq. 2.21 in terms of T_1 using Eqs. 2.22 to 2.24 gives

$$\eta_{\text{diesel}} = 1 - \frac{1}{\gamma} \times \frac{\alpha^\gamma - 1}{\alpha r_v^{\gamma-1} - r_v^{\gamma-1}} = 1 - \frac{1}{r_v^{\gamma-1}}\left[\frac{\alpha^\gamma - 1}{\gamma(\alpha - 1)}\right] \tag{2.25}$$

At this stage, it is worth making a comparison between the air standard Otto cycle efficiency (Eq. 2.17) and the air standard diesel cycle efficiency (Eq. 2.25).

The diesel cycle efficiency is less convenient. It is not solely dependent on compression ratio, r_v, but also is dependent on the load ratio α. The two expressions are the same, except for the term in square brackets

$$\left[\frac{\alpha^\gamma - 1}{\gamma(\alpha - 1)}\right]$$

The load ratio lies in the range $1 < \alpha < r_v$ and thus is always greater than unity. Consequently, the expression in square brackets is always greater than unity, and the diesel cycle efficiency is lower than the Otto cycle efficiency for the same compression ratio. This is shown in Fig. 2.5, where efficiencies have been calculated for a variety of compression ratios and load ratios. There are two limiting cases. The first is, as $\alpha \to 1$, then $\eta_{\text{diesel}} \to \eta_{\text{Otto}}$.

The second limiting case is when $\alpha \to r_v$ and point 3→4 in the cycle, and the expansion is wholly at constant pressure; this corresponds to maximum work output in the cycle. Figure 2.5 shows that as the load *increases*, with a fixed compression ratio the efficiency reduces.

However, because the compression ratio of a compression ignition engine usually is greater than for a spark ignition engine, the diesel engine usually is more efficient. Because the diesel cycle efficiency depends on the load ratio, it is worthwhile to estimate the values that this variable can take.

The load ratio is found by considering the work done by the cycle and using this to find the imep of the air standard diesel cycle.

The net work in the cycle,

$$W_{net} = Q_{23} - Q_{41} = imep \times V_s \qquad (2.26)$$

where the swept volume (V_s) can be expressed as

$$V_s = (r_v - 1)V_2$$

Rearranging Eq. 2.26 and substituting Eq. 2.20, and using Eqs. 2.22 to 2.24 for the temperatures, gives

$$\begin{aligned} imep \times V_s &= m\left[c_p(T_3 - T_2) - c_v(T_4 - T_1)\right] \\ &= mc_v T_1 \left[\gamma\left(\alpha r_v^{\gamma-1} - r_v^{\gamma-1}\right) - \left(\alpha^\gamma - 1\right)\right] \end{aligned} \qquad (2.27)$$

Note also the equation of state

$$p_1 V_1 = mRT_1$$

and that

$$V_1 = r_v V_2$$

gives

$$imep \times V_2(r_v - 1) = (p_1 r_v V_2/R) c_v \left[\gamma\left(\alpha r_v^{\gamma-1} - r_v^{\gamma-1}\right) - \left(\alpha^\gamma - 1\right)\right]$$

As

$$c_v/R = c_v/(c_p - c_v) = 1/\left[(c_p/c_v) - 1\right] = 1/(\gamma - 1)$$

then

$$\text{imep} = \frac{p_1 r_v}{(\gamma-1)(r_v-1)}\left[\left(\alpha r_v^{\gamma-1} - r_v^{\gamma-1}\right) - \alpha^\gamma - 1\right] \qquad (2.28)$$

Consider an engine with a compression ratio of 15, and evaluate the imep for a range of load ratios (assuming $\gamma = 1.4$):

α	1.0	1.5	2.0	2.5
imep/$\bar{p}_i$	0.0	3.49	6.67	9.63

Thus, for a typical diesel engine, the load ratio is in the range 2.0 to 2.5 at full load. An inspection of Fig. 2.5 shows that the air standard Otto cycle efficiency at a compression ratio of 10 to 1 is comparable with the diesel cycle efficiency at a compression ratio of 15 to 1 with a load ratio of 2. Of course, a spark ignition engine does not have instantaneous combustion, nor does a diesel engine have constant pressure combustion. Therefore, for both types of engines, the efficiency will fall between that predicted by the Otto cycle and the diesel cycle. For this reason, a dual cycle can be analyzed, in which some of the heat is added at constant volume, and some of it is added at constant pressure. The efficiency of the dual cycle falls between that of the diesel and Otto cycle efficiencies, and the complexity of the analysis is difficult to justify when practical engine efficiencies are, in any case, approximately half the ideal cycle values.

2.4.3 *Efficiencies of Real Engines*

The efficiencies of real engines are below those predicted by the ideal air standard cycles for several reasons. Most significantly, the gases in internal combustion engines do not behave perfectly with a ratio of heat capacities remaining at 1.4.

The fuel-air cycle efficiency allows for the non-perfect thermodynamic behavior of the gases (the heat capacities are allowed to vary with composition and temperature, but not pressure) and the effect of dissociation at the high temperatures encountered in engines. In Fig. 2.7, the equivalence ratio is defined as

$$\phi = \text{stoichiometric air-fuel ratio/actual air-fuel ratio}$$

where stoichiometric means the quantity of air just sufficient for complete combustion.

Consider a spark ignition engine with a compression ratio of 10, for which the Otto cycle efficiency predicts an efficiency of 60% and the fuel-air cycle predicts an efficiency of 47% for stoichiometric operation. In reality, such an engine might have a full throttle brake efficiency of 30%, and this means 17 percentage points must be accounted for, perhaps as follows:

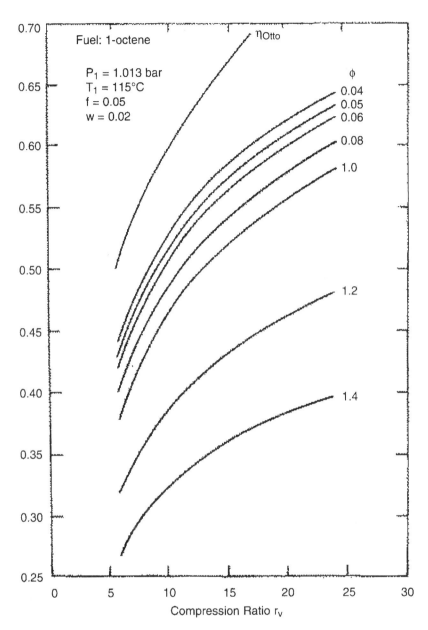

Figure 2.7. *Variation of efficiency with compression ratio for a constant-volume fuel-air cycle with 1-octene fuel for different equivalence ratios. Adapted from Taylor (1985a).*

Percentage points mechanical losses (friction)	3
Finite speed of combustion:	
20° 10–90% burn	1
40° 10–90% burn	3
Blow-by and unburned fuel in the exhaust	1
Cycle-by-cycle variations in combustion	2
Exhaust blow-down and gas exchange	2
Heat transfer	7

The finite speed of combustion accounts for the rounding of the indicator diagram in Fig. 2.3, which is in contrast to the constant volume heat addition of Fig. 2.4. Figure 2.3 also shows that for three successive cycles, there is quite a significant variation in the maximum pressure (28–37 bar), but that the variation in the imep (which is proportional to the enclosed area) is much smaller. If all the cycles had the largest value of imep, and there were no cycle-by-cycle variations in combustion, then the engine efficiency would be higher.

The exhaust blow-down process also leads to a rounding of the indicator diagram, because the exhaust valve starts to open before the piston reaches the end of its stroke. This can be seen more clearly in the inset of Fig. 2.3. Also, even at wide open throttle (WOT), work still will be done by the piston to overcome fluid friction in expelling the exhaust gases and inducting the fresh charge.

2.5 Ignition and Combustion in Spark Ignition and Diesel Engines

Spark ignition (SI) engines usually have pre-mixed combustion, in which a flame front initiated by a spark propagates across the combustion chamber through the unburned mixture. Compression ignition (CI) engines normally inject their fuel toward the end of the compression stroke, and the combustion is controlled primarily by diffusion. More specifically,

a. The fuel is injected and starts to vaporize and mix with air.

b. At the end of the ignition delay period, the flammable mixture formed during the delay period ignites and burns very rapidly (giving the characteristic diesel knock).

c. After the period of rapid combustion, the combustion process is governed by the rate at which the air and fuel mix—*diffusion (controlled) combustion*.

For most hydrocarbon fuels at ambient conditions, the maximum laminar burning velocity is less than 0.5 m/s. The density ratio between the burned and unburned mixture is approximately 5, but the resulting laminar flame speed of approximately 2.5 m/s is not sufficient for combustion to complete in the available time. In all internal combustion engines, turbulence increases the effective flame front area, so that the turbulent burning velocity is at least an order of magnitude higher than the laminar burning velocity. Fortunately, the turbulence intensity increases approximately linearly with engine speed. This means that an engine can operate over a wide speed range, with the turbulent combustion period occupying an almost fixed crank angle period (in the region of 30° ca) at full throttle.

The ignition delay period (compression ignition engines) and the early burn period (the transition from a laminar flame front generated by the spark to fully turbulent combustion) occupy an almost constant time period. This means that as the engine speed is increased, the fuel injection or ignition must be advanced. In spark ignition engines, the early burn period is dominated by laminar combustion because the flame front is small compared to the turbulence scale, and the turbulence will cause displacement of the flame kernel, as opposed to wrinkling of a larger flame front.

Whether combustion is pre-mixed (as in SI engines) or diffusion controlled (as in CI engines) has a major influence on the range of air-fuel ratios (AFRs) that will burn. In pre-mixed combustion, the AFR must be close to stoichiometric—the AFR value that is chemically correct for complete combustion. In practice, dissociation and the limited time available for combustion will mean that even with the stoichiometric AFR, complete combustion will not occur.

Most hydrocarbon fuels have close to a 2:1 hydrogen/carbon (H/C) atomic ratio. Because the mass of the carbon atom is 12 times that of the hydrogen atom, slight variations in the H/C ratio have only a small effect on the gravimetric composition of the fuel. Thus, for most hydrocarbon fuels, the gravimetric stoichiometric AFR is close to 14.5:1. (Note that when liquid or solid fuels are being used, the AFR is invariably gravimetric, and frequently this must be inferred. With gaseous fuels, it is common to use a volumetric or molar AFR.)

With pre-mixed combustion, the AFR must be close to stoichiometric because the air-fuel mixture is essentially homogeneous. In contrast, with diffusion combustion, much weaker AFRs can be used (i.e., an excess of air) because around each fuel droplet will be a range of flammable AFRs. With AFRs approaching stoichiometric or rich of stoichiometric, diffusion combustion will become very incomplete—diffusion processes are comparatively slow, and it is impossible for full utilization of the oxygen in the air. The consequential incomplete combustion is characterized by the formation of soot and a smoky exhaust. Typical ranges for the (gravimetric) air-fuel ratio are as follows:

| CI (diesel) | $18 < AFR < 80$ |
| SI (gasoline) | $10 < AFR < 20$ |

The permissible range of AFRs has a major effect on the way the output of SI and diesel engines can be regulated. Diesel or compression ignition (CI) engines can make use of quality governing, in which the air flow is unthrottled, and the quantity of fuel injected is varied. In contrast, SI engines can achieve only a small variation in output by means of AFR control; thus, it is necessary to reduce the flow of air (as well as the fuel) by throttling. The pressure drop across the throttle plate dissipates work that has been provided by the piston. This is illustrated by Fig. 2.3, which shows the indicator diagram from an SI engine operating at part load. The pumping loop in Fig. 2.3 has an area that equates to a pumping mean effective pressure (pmep) of 0.45 bar. The engine is operating at

$$\text{a net imep of 4.6 bar,}$$

for which

$$\text{bmep} = 3.8 \text{ bar (full-load bmep would be approximately 10 bar)}$$

Therefore,

$$\text{fmep} = 0.8 \text{ bar}$$

As the load is reduced at constant speed,

a. The pmep will increase, and
b. The fmep (principally a function of speed) will remain almost constant

Thus, these two losses will become a bigger fraction of the engine output as the output is reduced, and the mechanical efficiency will fall. Ultimately, when the bmep is zero, both the brake efficiency and the mechanical efficiency will become zero.

In the diesel engine with no throttling, there will be a negligible change to the pmep, and the brake and mechanical efficiencies will fall less rapidly as the load is reduced. Indeed, as the load is reduced in the diesel engine, the combustion becomes more rapid, and the air-fuel ratio becomes weaker (Fig. 2.7). Both contribute to an increase in the indicated efficiency. The limit on the AFR with a diesel means the bmep will be limited to approximately 7.5 bar for a naturally aspirated diesel engine (compared with 10 bar for an SI engine). In a diesel engine, there is a limited amount of time for mixing of the fuel and air. If there is any unburned fuel, this will be a source of particulate emissions. Thus, the fueling is limited to approximately 80% of the stoichiometric value in a diesel engine, whereas with a homogeneous charge engine, a mixture rich of stoichiometric can be used to give the maximum output.

To summarize, diesel engines have a higher maximum efficiency than spark ignition engines for three reasons:

1. The compression ratio is higher.
2. During the initial part of compression, only air is present.
3. The air-fuel mixture is always weak of stoichiometric.

Also, the fall in part-load efficiency of a diesel engine is moderated by the following:

1. The absence of throttling
2. The weakening air-fuel mixture
3. The shorter duration combustion

The relative efficiencies of a spark ignition engine and a diesel engine are shown in Fig. 2.8, which uses data from a Rover K series spark ignition engine and a Volkswagen turbocharged Lupo diesel engine. A midrange speed has been chosen for each engine, which encompasses the maximum efficiency for each engine. Figure 2.8 also shows that turbocharging can more than double the output of a diesel engine because the bmep can be seen to be approximately 16 bar.

The influence of the AFR on spark ignition engine efficiency or its reciprocal, the brake specific fuel consumption (bsfc) and output (bmep), at full throttle, is shown in Fig. 2.9. For

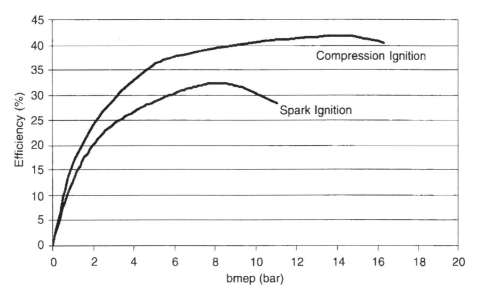

Figure 2.8. *The effect of load on the efficiency of a turbocharged diesel (compression ignition) engine and a spark ignition engine.*

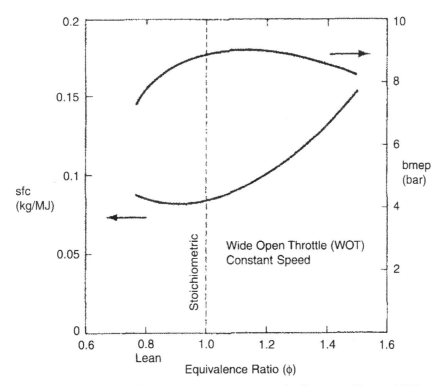

Figure 2.9. *The effect of AFR on output and efficiency (Stone, 1999).*

maximum bmep, the mixture should be approximately 10% rich of stoichiometric; this gives the "best oxygen utilization." If any more fuel is added, it acts as a diluent. With a stoichiometric mixture, dissociation means that there would be approximately 1% oxygen in the exhaust. The minimum bsfc occurs approximately 10% weak of stoichiometric. Weakening the mixture increases the indicated efficiency until the combustion stability limit is reached. However, because the bmep is falling, the fmep becomes more significant (i.e., the mechanical efficiency falls), and the result of these two competing effects is a maximum brake efficiency (or minimum bsfc) at approximately 10% weak of stoichiometric.

The data from a fixed throttle setting (such as Fig. 2.9) can be cross-plotted to give the fishhook curves of Fig. 2.10. A comparison of points A and B indicates that the fuel consumption at part load is minimized by a weak mixture with a slightly more open throttle. When full throttle is reached, the output can be increased only by making the mixture richer. This accounts for the fall in efficiency of the spark ignition engine in Fig. 2.8 as its maximum bmep is reached. The slight fall in diesel engine efficiency as maximum load is reached in Fig. 2.8 is due to the combustion becoming less efficient.

As the throttle is progressively closed, the inlet manifold pressure falls, and the exhaust residuals level at the end of the exhaust stroke increases. This leads to a reduction in the range of flammable air-fuel ratios. Emissions legislation does not give a free choice on the air-fuel ratio, but discussion of this is in the next section.

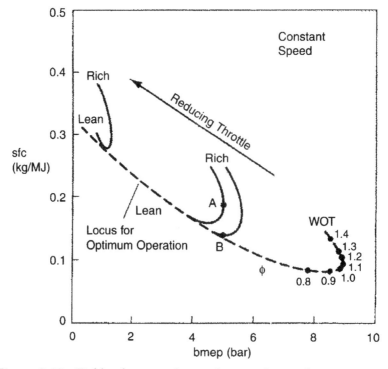

Figure 2.10. *Fishhook curves from a four-stroke spark ignition engine with varying throttle but constant speed (Stone, 1999).*

2.6 Sources of Emissions

2.6.1 *Simple Combustion Equilibrium*

The exhaust from internal combustion engines leaves at a comparatively low temperature (well below 1500 K). At these temperatures, the reaction rates are comparatively low, and the composition of the products will not change much with further cooling. In other words, the products of combustion are not at equilibrium, and for the exhaust from internal combustion engines operating rich of stoichiometric, both carbon monoxide and hydrogen will be present. The relative proportions correspond to the equilibrium of the water–gas reaction at a temperature of approximately 1800 K.

$$CO_2 + H_2 \Leftrightarrow CO + H_2O$$

It will now be shown that this can be a useful way of estimating the exhaust gas composition, perhaps as the first guess in a more comprehensive equilibrium analysis.

Consider now the combustion of an arbitrary fuel ($C_\alpha H_\beta O_\gamma N_\delta$) with air (assumed to be 79.05% nitrogen). If the products of combustion are limited to carbon dioxide (CO_2), carbon monoxide (CO), water vapor (H_2O), hydrogen (H_2), oxygen (O_2), and nitrogen (N_2), then the generalized combustion equation with air can be written as

$$C_\alpha H_\beta O_\gamma N_\delta + \lambda(\alpha + \beta/4 - \gamma/2)\left\{O_2 + \frac{79.05}{20.95}N_2\right\} \rightarrow$$
$$a_1 CO_2 + a_2 CO + a_3 H_2O + a_4 H_2 + a_5 O_2 + a_6 N_2 \quad (2.29)$$

where the excess air ratio (λ, which is equal to $1/\phi$) is unity for stoichiometric reactions, and greater than unity for weak mixtures.

Four atomic balances can be written:

$$\text{C balance:} \quad \alpha = a_1 + a_2 \quad (2.30)$$

$$\text{N balance:} \quad \delta + \lambda(\alpha + \beta/4 - \gamma/2)2 \times \frac{79.05}{20.05} = 2a_6 \quad (2.31)$$

$$\text{H balance:} \quad \beta = 2a_3 + 2a_4 \quad (2.32)$$

$$\text{O balance:} \quad \gamma + \lambda(\alpha + \beta/4 - \gamma/2)2 = 2a_1 + a_2 + a_3 + 2a_5 \quad (2.33)$$

With six unknowns and only four simultaneous equations, then two further equations are needed. A convenient simplification is to assume no oxygen in the products of rich combustion, and no hydrogen or carbon monoxide in the products of weak combustion. In other words,

Rich mixtures ($\lambda < 1$): $\qquad a_5 = 0 \qquad$ (2.34)

Weak mixtures ($\lambda > 1$): $\qquad a_2 = a_4 = 0 \qquad$ (2.35)

Stoichiometric mixtures ($\lambda = 1$): $\qquad a_2 = a_4 = a_5 = 0 \qquad$ (2.36)

For rich mixtures, another equation is required, and this is provided by the water–gas equilibrium:

$$CO_2 + H_2 \Leftrightarrow CO + H_2O \qquad (2.37)$$

for which the equilibrium constant is K

$$K = \frac{a_2 a_3}{a_1 a_4} \approx 3.5 \qquad (2.38)$$

Simultaneous solution of Eqs. 2.30 to 2.36 and Eq. 2.38 yields the results summarized in Table 2.1.

For the rich products of combustion in Table 2.1, the compositions have been written with variable a_2 included. This now can be eliminated by the use of the equilibrium defined in Eq. 2.38 to give a quadratic equation in which the solution required is

TABLE 2.1
SIMPLIFIED PRODUCTS OF COMBUSTION

Species	i	Weak ($\lambda > 1$)	Rich ($\lambda < 1$)
CO_2	1	α	$\alpha - a_2$
CO	2	0	a_2
H_2O	3	$\beta/2$	$\gamma + \lambda(\alpha + \beta/4 - \gamma/2)2 - 2\alpha + a_2$
H_2	4	0	$\beta/2 - \gamma - \lambda(\alpha + \beta/4 - \lambda/2)2 + 2\alpha - a_2$
O_2	5	$(\lambda - 1)(\alpha + \beta/4 - \gamma/2)$	0
N_2	6	$\lambda(\alpha + \beta/4 - \gamma/2)79.05/20.95 + \delta/2$	$\lambda(\alpha + \beta/4 - \gamma/2)79.05/20.95 + \delta/2$

$$0 < a_2 < a$$

$$a_2 = \frac{-b + \sqrt{b^2 - 4ac}}{2a} \quad (2.39)$$

where

$a = K - 1$

$b = (-2K + 2\lambda K - 2\lambda + 2)\alpha + (-K + \lambda K - \lambda)\beta/2 + (1 - \lambda)(K - 1)\gamma$

$c = \alpha K(1 - \lambda)(2\alpha + \beta/2 - \gamma)$

Solution of Eq. 2.39 requires a knowledge of the equilibrium constant (K), and this can be found in many tabulations, such as Howatson et al. (1991). If the constraints of no carbon monoxide being present with weak mixtures and no oxygen being present with rich mixtures are removed, then it is necessary to consider an additional equilibrium, namely,

$$CO + \frac{1}{2}O_2 \Leftrightarrow CO_2 \quad (2.40)$$

This leads to simultaneous polynomial equations that invariably must be solved iteratively to satisfy both energy and equilibrium equations.

The variations in the composition of the major exhaust species from a spark ignition engine are shown in Fig. 2.11, for which further explanation is needed concerning the NOx (nitrogen oxides) and HC (unburned hydrocarbon) emissions.

The main elementary reaction oxidizing CO to CO_2 is

$$OH + CO \rightarrow CO_2 + H$$

However, Gardiner (2000) notes that this reaction is slow compared to other radical reactions involved in combustion processes. In consequence, as the cylinder contents cool and reaction rates fall, the oxidation of carbon monoxide will lag behind its equilibrium value, and the carbon monoxide will appear to have frozen at the equilibrium value for a higher temperature. The carbon monoxide emissions lie between those predicted for equilibrium at peak pressure and the equilibrium values at exhaust valve opening. They tend to be closer to the maximum pressure values for rich mixtures, and closer to the exhaust valve opening values for weak mixtures. The CO emissions can be modeled by specifying a "freezing" temperature at which the equilibrium concentration is evaluated. More accurate models should include the kinetics, and these are reviewed in detail by Heywood (1988). However, the main determinant of

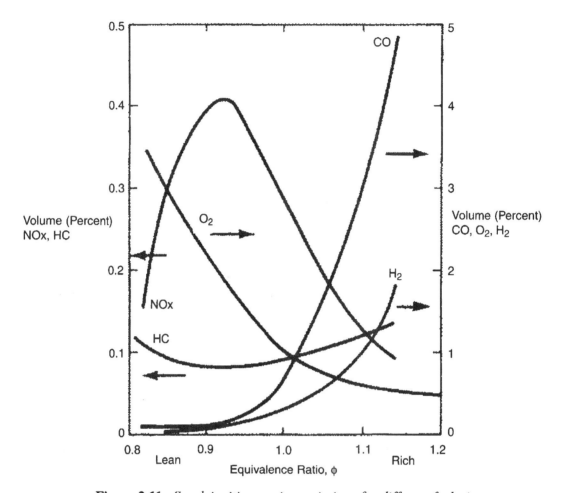

Figure 2.11. *Spark ignition engine emissions for different fuel-air equivalence ratios. Courtesy of Johnson Matthey.*

carbon monoxide emissions is the air-fuel ratio. In multi-cylinder engines operating at stoichiometric, the inter-cylinder variation in air-fuel ratio can have the biggest effect on the carbon monoxide emissions. This can be illustrated by a simple example, a two-cylinder engine operating with an overall stoichiometric mixture:

Case A. Both cylinders are stoichiometric; then the exhaust will contain approximately 0.84% CO.

Case B. Each cylinder is operating 5% away from stoichiometric (one rich, the other lean); then the exhaust will contain approximately ([1.90 + 0.22]/2=) 1.06% CO, which is comparable to the equilibrium CO level at maximum pressure.

Thus, for overall stoichiometric operation, inter-cylinder mixture maldistribution is likely to be as significant as correctly modeling the CO kinetics.

2.6.2 Unburned Hydrocarbons (HC) and Nitrogen Oxides (NOx) in Spark Ignition Engines

The remaining emissions to be discussed are the HC and NOx. Figure 2.12 illustrates the sources of these.

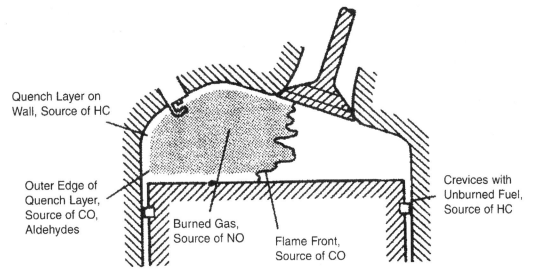

Figure 2.12. *Source of emissions in a spark ignition engine. Adapted from Mattavi and Amann (1980).*

The unburned hydrocarbon (HC) emissions arise from several sources:

a. Short-circuiting, in which unburned mixture flows into the exhaust during the valve overlap period.

b. Oil film and crevice absorption/desorption. Absorption (and flow into the crevices) occurs as the cylinder pressure rises, and desorption (and flow out of the crevices) occurs as the pressure is falling, by which time the burned gas temperature can be too low for complete oxidation to occur in the available time.

c. Misfire (failure to ignite), or partial burn—either because combustion is too slow or started too late, or because the flame front was extinguished by high rates of strain.

d. Poor mixture preparation during transients. With a cold start, an over-rich mixture must be used to ensure a flammable air-fuel mixture, but fuel droplets inducted into the cylinder will not all be fully oxidized. As an engine warms up, this tendency will be reduced rapidly. However, during load-changing transients, there is a tendency (even with fuel-injected engines) for rich and weak air-fuel ratio excursions.

Hochgreb (1998) provides an excellent overview of emissions from spark ignition engines, with a critical review of the mechanisms by which all emissions are formed, and with full details of submodels for use in engine simulations, including the various modes of hydrocarbon absorption/desorption. Approximately 92% of the fuel is oxidized during propagation of the flame front. Of the remaining 8% (which has been absorbed into crevices and so forth), approximately half will be oxidized after being desorbed into the burned gases in the cylinder. This leaves approximately 4% of the fuel entering the exhaust port, where approximately half will be oxidized. Thus, approximately 2% of the original fuel enters the catalyst. With an oxidation efficiency of 95%, only approximately 0.1% of the original fuel leaves the engine as unburned hydrocarbons.

The mixture of nitric oxide (NO) and nitrogen dioxide (NO_2) is referred to as NOx. Nitric oxide usually is by far the most dominant nitrogen oxide formed during combustion. However, subsequent further oxidation in the environment leads to nitrogen dioxide, and the nitrogen dioxide reacts with the non-methane hydrocarbons in the presence of ultraviolet (UV) light to form photochemical smog. Thus, although the major part of NOx will be NO, when emissions are calculated on a specific basis (e.g., g/MJ, g/kWh), it is assumed that all the nitric oxide is oxidized to nitrogen dioxide.

Nitric oxide is formed in flames by three mechanisms: thermal, prompt, and nitrous oxide. The thermal mechanism is based on the extended Zeldovich mechanism.

$$O + N_2 \Leftrightarrow NO + N \tag{2.41}$$

$$N + O_2 \Leftrightarrow NO + O \tag{2.42}$$

$$N + OH \Leftrightarrow NO + H \tag{2.43}$$

Equations 2.41 and 2.42 were identified by Zeldovich (1946). Equation 2.43 was added by Lavoie et al. (1970) because it contributes significantly—this was identified by spectroscopic means. The rate constants for the thermal mechanism are slow, and NO formation (and decomposition) is significant only when there is a high enough temperature (say, above 1800 K) and sufficient time. Thus, the thermal NO mechanism is assumed to occur in the hot combustion gases, in which it can be taken that all other species are in equilibrium. In other words, the combustion reactions are assumed to occur very quickly compared with the thermal NO mechanism, so that the NO kinetics can be decoupled from combustion. The short time during engine combustion means that there is not enough time for equilibrium levels of NO to be attained; therefore, predictions of NO emissions must be made from kinetically based models.

Measurements of NO in the burned gases do not extrapolate to zero at the flame. This implies that NO is formed in the flame, by the so-called prompt mechanism. The prompt mechanism is significant when there is fuel-bound nitrogen, or when the combustion temperatures are so low as to make the thermal mechanism negligible. The prompt mechanism is governed by

$$CH+N_2 \rightarrow HCN+N \tag{2.44}$$

The nitrous oxide mechanism is important at low temperatures and depends on a termolecular reaction, where M denotes any other molecule

$$N_2 + O + M \rightarrow N_2O + M \tag{2.45}$$

with the subsequent decomposition to nitric oxide. Correa (1992) notes that the nitrous oxide submechanism is significant with lean ($\lambda > 1.6$) pre-mixed laminar flames.

It can be shown that the rate of NO production is given by

$$\frac{d[NO]}{dt} = \frac{2R_1\left\{1-\left([NO]/[NO]_e\right)^2\right\}}{1+\left([NO/NO]_e\right)R_1/(R_2+R_3)} \tag{2.46}$$

where the equilibrium concentrations are denoted by $[\ ]_e$, and the three reaction rate variables (R_1, R_2, and R_3) are defined by

$$R_1 = k_1^+[O]_e[N_2]_e = k_1^-[NO]_e[N]_e \tag{2.47}$$

$$R_2 = k_2^+[N]_e[O_2]_e = k_2^-[NO]_e[O]_e \tag{2.48}$$

$$R_3 = k_3^+[N]_e[OH]_e = k_3^-[NO]_e[H]_e \tag{2.49}$$

The thermal NO differential equation (Eq. 2.46) can then be solved simultaneously with any other equations that are defining the combustion process.

In spark ignition engines, there is negligible fuel-bound nitrogen, and it is usual to consider only the thermal NO mechanism. The combustion process is modeled by dividing the chamber contents into a minimum of two regions: the unburned gas and the burned gas, separated by a thin region. It is assumed that the combustion process occurs instantaneously and completely. Subsequently, the burned gas is assumed to remain in thermodynamic equilibrium for the following species: CO, CO_2, H_2O, H_2, OH, O, N_2, and O_2. Also computed is the equilibrium value of the nitric oxide (denoted here as NO_e). The kinetically controlled value of the nitric oxide level can then be computed by solving Eq. 2.46 (denoted here by NOx). The NO kinetics have been decoupled from the combustion reaction, and this is acceptable because the hydrocarbon oxidation kinetics are orders of magnitude faster than the NO formation kinetics.

Figure 2.13 shows the output from a spark ignition engine simulation, in which a single burned gas zone has been assumed. The pressure rise (initially due solely to the piston motion) leads to isentropic compression of the unburned gas because no heat transfer has been allowed. When combustion starts, the burned gas temperature is plotted, and the changes in burned and unburned gas temperatures both reflect the change in pressure. The equilibrium NO (NO_e) "appears" immediately after the start of combustion, and its strong temperature dependence means that it rises rapidly to a maximum (corresponding to p_{max} and $T_{b,max}$), and falls rapidly as the pressure and temperature fall. The kinetically controlled value of the nitric oxide (NOx) starts at zero and always lags behind the equilibrium value (NO_e). When the burned gas temperature has fallen to a temperature of approximately 2000 K, the NO kinetics are so slow that the kinetically calculated value of the nitric oxide (NOx) remains almost constant. (The composition is said to "freeze.") As the burned gas temperature falls, the equilibrium value of the nitric oxide decreases rapidly to a level that is much below that of the kinetically predicted value.

The formation of nitric oxide (NO) requires both a high temperature and oxygen. As the mixture is weakened from stoichiometric, two competing effects occur: first, the oxygen concentration in the burned gas rises, but second, the temperature falls. The overall result

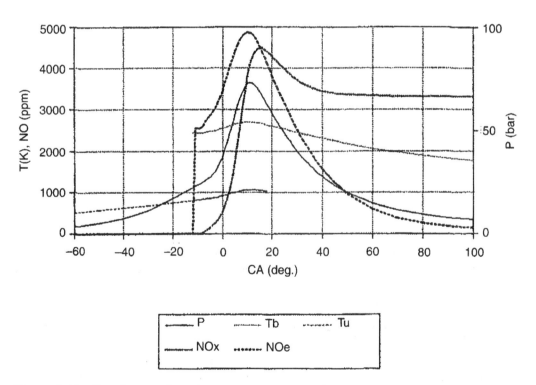

Figure 2.13. *Simulation of spark ignition engine combustion with a single burned zone, showing pressure (P), burned gas temperature (Tb), unburned gas temperature (Tu), and the equilibrium (NOe) and kinetic (NOx) predictions of the nitric oxide concentrations as a function of the crank angle (Stone, 1999).*

(Fig. 2.11) is that the maximum NO emissions occur approximately 10% weak of stoichiometric—the air-fuel ratio that also coincides with the maximum brake efficiency of a naturally aspirated spark ignition engine.

2.6.3 Unburned Hydrocarbons (HC), Nitrogen Oxides (NOx), and Particulates in Compression Ignition Engines

The three sources of hydrocarbon (HC) emissions in diesel engines are as follows:

a. Fuel that is introduced too late into the reaction zone, such as from the tip of the injector nozzle (the sac volume), or fuel that impinged on the combustion chamber walls

b. Over-diluted mixture that occurs at the extremities of the fuel spray (made worse by a long ignition delay period)

c. Fuel that does not burn fully in the rich mixture zones

With diesel engine combustion, it is essential to remember that a wide range of air-fuel ratios are present, and these extend beyond the weak and rich mixture flammability limits. The HC emissions will be present in the gaseous phase and as part of the soot that is a major component of the particulates.

Particulates are any substance apart from water that can be collected by filtering diluted exhaust at a temperature of 325 K. Particulates include sulfates and fuel that has been partially pyrolyzed, as well as high molar mass hydrocarbons that have been condensed. The black smoke associated with a poorly regulated diesel engine consists of carbon particles produced by the thermal decomposition (pyrolysis) of hydrocarbons within the rich part of the air-fuel mixture during the diffusion-controlled combustion stage. The carbon agglomerates into particles that are visible as smoke in the exhaust.

Diesel exhaust particulates will comprise carbon (20–50%), sulfates (5–15%), unburned fuel (10–30%), unburned lubricant (10–20%), and unknown (~10%). The composition will depend on the engine, its operating point, and the fuel being used (sulfur and other inorganic content).

The mechanism of NOx formation is essentially the same as for spark ignition engines, except that some diesel fuels will contain nitrogenous compounds, so that the prompt NO mechanism also should be considered.

2.7 Fuel and Additive Requirements

Gasoline is a mixture of hydrocarbons (with 4 to approximately 12 carbon atoms) and a boiling point range of approximately 30–200°C. Diesel fuel is a mixture of higher molar mass hydrocarbons (typically 12 to 22 carbon atoms), with a boiling point range of approximately 180–380°C. Fuels for spark ignition engines should vaporize readily and be resistant to self-ignition, as indicated by a high octane rating. In contrast, fuels for compression ignition engines should self-ignite readily, as indicated by a high cetane number. Straight chain molecules (especially long alkanes) are prone to self-ignition, whereas branched molecules and those based on the benzene ring (aromatics) are resistant to self-ignition. Figure 2.14 indicates that fuels with a high octane rating have a low cetane rating, and vice versa.

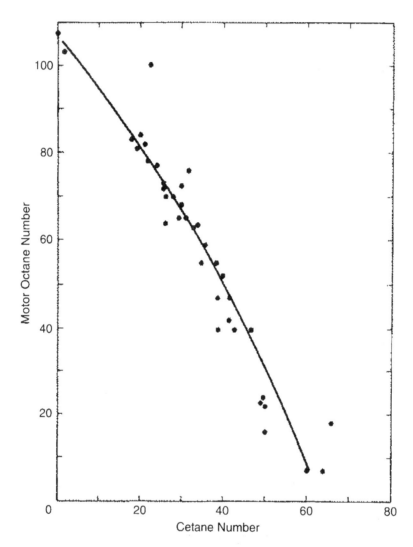

Figure 2.14. *Relationship between the cetane number and the octane number for petroleum-derived fuels. Adapted from Taylor (1985b).*

The octane or cetane rating of a fuel is established by comparing its ignition quality with respect to reference fuels in CFR (Co-operative Fuel Research) engines, according to internationally agreed standards (*ASTM Standards Volume 05.04—Test Methods for Rating Motor, Diesel and Aviation Fuels*).

Octane Rating	0	100
Fuel	n-heptane (C_7H_{16})	iso-octane (C_8H_{18})
		2,2,4 trimethyl pentane

Cetane Rating	15	100
Fuel	heptamethylnonane ($C_{16}H_{34}$)	n-cetane ($C_{16}H_{34}$)
	(iso-cetane)	

Figure 2.15 shows that mean molecular mass of a fuel affects its volatility, density, and gravimetric calorific value. In particular, note that on a volumetric basis, the calorific value of diesel fuel is approximately 15% higher than gasoline.

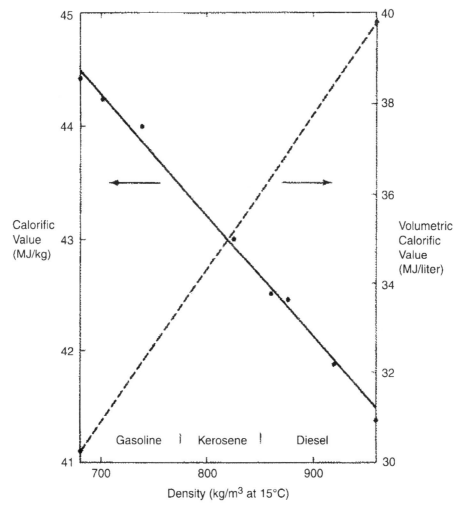

Figure 2.15. *Variation in calorific values for different fuel mixtures. With acknowledgment to Blackmore and Thomas (1977).*

The volatility of gasoline is subject to legislation with seasonal and geographical specifications. This is to avoid unnecessary evaporative loss in hot climates and to ensure that a sufficiently volatile fuel is used in cold climates to promote cold-engine starting and to minimize the operation on enriched mixtures during warm-up.

2.7.1 Abnormal Combustion in Spark Ignition Engines

Pre-ignition and auto-ignition (also known as self-ignition) must not be confused, because they are separate but sometimes coupled phenomena.

Pre-ignition is when the mixture is ignited before the spark occurs, usually from a hot spot such as the exhaust valve, the spark plug electrodes, or combustion chamber deposits. The earlier ignition leads to higher cylinder pressures and thus higher temperatures in the unburned gas. If the unburned gas temperature is high enough for sufficient time, then the unburned mixture can auto-ignite.

Auto-ignition is the spontaneous ignition of unburned gas ahead of the flame front. Because the adiabatic core of the unburned mixture will be at much the same temperature, then when part of the mixture auto-ignites, the remainder will follow rapidly—not least because of the rise in pressure. The resulting combustion is very rapid, and this can lead to pressure oscillations within the combustion chamber and structural excitation. The structural vibrations sometimes can be heard as a tinkling noise known as "knock." Auto-ignition is not necessarily severe enough to be heard as "knock." This phenomenon is incorrectly referred to as detonation, which is the supersonic propagation of a combustion wave. Knock can be eliminated by retarding the ignition timing, unless it is caused by pre-ignition, as described next.

When auto-ignition occurs, the resulting pressure oscillations disrupt the thermal boundary layer, and this can lead to overheating of key components. If the rise in temperature causes pre-ignition, then the earlier ignition will exacerbate the knock and lead to even earlier pre-ignition. This phenomenon is known as runaway knock, which invariably results in engine failure.

Finally, there is a phenomenon known as "running-on," which is the engine continuing to operate after the ignition has been turned off. This is caused by some form of surface ignition (usually from combustion deposits) in a similar manner to pre-ignition. It can occur only with carbureted engines and is usually associated with deposits from leaded fuel.

2.7.2 Gasoline and Diesel Additives

Legislation is now restricting the use of organo-metallic compounds for improving the octane rating of gasoline. Consequently, they are not covered here, but a discussion of their use, the other additives that must be used in association with them, and the consequences of their withdrawal are discussed in Stone (1999). The most significant additives are detergents and antioxidants, but corrosion inhibitors, metal deactivators, biocides, anti-static additives, demulsifiers, dyes and markers, and anti-icing additives also are used. These are discussed in detail by Owen and Coley (1995).

Antioxidants are needed in gasoline to inhibit the formation of gum, which usually is associated with the unsaturated hydrocarbons in fuel. Formation of gum can interfere with the operation of fuel injectors. Detergents are added to reduce the deposits in fuel injectors, the inlet manifold, and the combustion chamber. Surfactants inhibit the formation of deposits in the injectors and the inlet manifold, but a different mechanism is needed to combat valve and port deposits because these deposits are associated with higher temperatures. High-boiling-point, thermally stable, oily materials such as polybutene are used, and these appear to dissolve the deposits.

Diesel additives to improve the cetane number will be discussed first, followed by additives to lower the cold filter plugging point temperature, then additives that are used with low-sulfur fuels, and finally other additives.

The most widely used ignition-improving additive currently is 2-ethyl hexyl nitrate (2EHN), because of its good response in a wide range of fuels and comparatively low cost (Thompson et al., 1997). Adding 1000 ppm of 2EHN will increase the cetane rating by approximately 5 units. In some parts of the world, legislation limits the nitrogen content of diesel fuels, because although the mass of nitrogen is negligible to that available from the air, fuel-bound nitrogen contributes disproportionately to nitric oxide formation. Under these circumstances, peroxides can be used, such as ditertiary butyl peroxide (Nandi and Jacobs, 1995).

Diesel fuel contains molecules with approximately 12 to 22 carbon atoms, and many of the higher molar mass components (e.g., cetane, $C_{16}H_{34}$) would be solid at room temperature if they were not mixed with other hydrocarbons. Thus, when diesel fuel is cooled, a point will be reached at which the higher molar mass components will start to solidify and form a waxy precipitate. As little as 2% wax out of the solution can be enough to gel the remaining 98%. This will affect the pouring properties and (more seriously at a slightly higher temperature) block the filter in the fuel-injection system. These and other related low-temperature issues are discussed comprehensively by Owen and Coley (1995), who point out that as much as 20% of the diesel fuel can consist of higher molar mass alkanes. It would be undesirable to remove these alkanes because they have higher cetane ratings than many of the other components. Instead, use is made of anti-waxing additives that modify the shape of the wax crystals.

Wax crystals tend to form as thin "plates" that can overlap and interlock. Anti-waxing additives do not prevent wax formation. They work by modifying the wax crystal shape to a dendritic (needle-like) form, and this reduces the tendency for the wax crystals to interlock. The crystals are still collected on the outside of the filter, but they do not block the passage of the liquid fuel. The anti-waxing additives in commercial use are copolymers of ethylene and vinyl acetate, or other alkene-ester copolymers. The performance of these additives varies with different fuels, and the improvement decreases as the dosage rate is increased. It is possible for 200 ppm of additive to reduce the cold filter plugging point (CFPP) temperature by approximately 10 K.

Additives can be used with low-sulfur diesel fuels to compensate for their lower lubricity, lower electrical conductivity, and reduced stability. To restore the lubricity of a low-sulfur fuel to that of a fuel with 0.2% sulfur by mass, then a dosage on the order of 100 mg/L is needed. Care is required in the selection of the additive, if it is not to interact unfavorably with other additives (Batt et al., 1996).

Electrical conductivity usually is not subject to legislation, but if fuels have a very low conductivity, then there is the risk of a static electrical charge being built up. If a road tanker, previously filled with gasoline, is being filled with diesel, then there is the possibility of a flammable mixture being formed. The conductivity of untreated low-sulfur diesel fuels can be less than 5 pS/m (Merchant et al., 1997). Conductivities greater than 100 pS/m can be obtained by adding a few parts per million of a chromium-based static dispersant additive. Low-sulfur fuels and fuels that have been hydro-treated to reduce the aromatic content also are prone to the formation of hydroperoxides. These are known to degrade neoprene and nitrile rubbers, but this can be prevented by using antioxidants such as phenylenediamines (suitable only in low-sulfur fuels) or hindered phenols (Owen and Coley, 1995).

Other additives used in diesel fuels are detergents, anti-ices, biocides, and anti-foamants.

Detergents (e.g., amines and amides) are used to inhibit the formation of combustion deposits. Most significant are deposits around the injector nozzles, which interfere with the spray formation. Deposits then can lead to poor air-fuel mixing and particulate emissions. A typical dosage level is 100–200 ppm.

Anti-ices (e.g., alcohols or glycols) have a high affinity for water and are soluble in diesel fuel. Water is present through contamination and as a consequence of humid air above the fuel in vented tanks being cooled below its dewpoint temperature. If ice formed, it could block both fuel pipes and filters.

Biocides act against anaerobic bacteria that can form growths at the water/diesel interface in storage tanks. These are capable of blocking fuel filters.

Anti-foamants (10–20 ppm silicone-based compounds) facilitate the rapid and complete filling of vehicle fuel tanks.

2.8 Gas Exchange Processes

2.8.1 Valve Flow and Volumetric Efficiency

The volumetric efficiency has already been defined by Eq. 2.11 as

$$\eta_V = \frac{\dot{V}_a}{V_s \times n} \qquad (2.11)$$

where V_s = swept volume and n (cycle/s) will be

$$N \text{ (rpm)}/120 \text{ for four-stroke engines}$$

or

$$N \text{ (rpm)}/60 \text{ for two-stroke engines}$$

It can be shown that the output of an engine is related directly to its volumetric efficiency, by considering the mass of fuel that is burned (m_f), its calorific value (CV), and the brake efficiency of the engine

$$m_f = \frac{\eta_v \times V_s \times \rho_a}{\text{AFR}} \tag{2.50}$$

where ρ_a is the density of the air.

The amount of brake work produced in each cycle (W_b) is given by

$$W_b = \text{bmep} \times V_s \quad \text{or} \quad W_b = m_f \times \eta_b \times CV \tag{2.51}$$

Combining Eqs. 2.50 and 2.51, and rearranging, gives

$$\text{bmep} = \eta_b \times \eta_v \times \rho_a \times \frac{CV}{\text{AFR}} \tag{2.52}$$

A similar equation can be derived for the imep. However, in both cases, remember that using a very low AFR (a rich mixture) will lead to a fall in the brake efficiency, as shown by the increase in the brake specific fuel consumption as shown in Fig. 2.9.

In four-stroke engines, the inlet and outlet processes are usually controlled by poppet valves (Fig. 2.1), the operation of which is invariably controlled by camshafts. Because the pressure drop across the inlet valve must be as small as possible (to maximize the volumetric efficiency), it is usual to have inlet valves that are approximately 10% larger in diameter than the exhaust valves. The most notable exception is when an odd number of valves are in each cylinder, such as two inlet valves and one exhaust valve, or on rare occasions, three inlet valves and two exhaust valves. The flow performance of the valves (the discharge coefficient, C_D) is usually measured in steady flow rigs, because the flow performance is not strongly dependent on the flow rate (as measured by the Reynolds number). The discharge coefficient (C_D) is the ratio of the effective flow (A_e) to the geometric flow area (A) that gives the same flow rate for frictionless flow.

$$C_D = A_e/A \tag{2.53}$$

Great care is needed in using discharge coefficients because there are many different ways of defining the geometric flow area. These include the port area, the minimum flow area (whose position varies with valve lift), or the curtain area (the product of the valve circumference and lift).

Figure 2.16 shows the flow characteristics of a sharp-edged inlet valve, in terms of the discharge coefficient based on both the valve curtain area (C_D) and the port area (C'_D). These two discharge coefficients can be related by means of the effective flow area:

$$A_e = C_D L_v \pi D_v = C'_D \pi D_v^2 / 4 \tag{2.54}$$

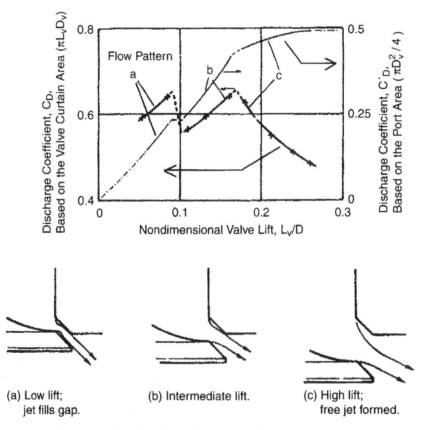

Figure 2.16. *The flow characteristics of a poppet valve. Adapted from Annand and Roe (1974).*

The effective flow area reaches a maximum when the nondimensional valve lift is approximately 0.25, because the smallest flow area is now the annular area between the valve seat and its stem. At low lift (Fig. 2.16a), the jet fills the gap and adheres to both the valve and the seat. At an intermediate lift (Fig. 2.16b), the flow will break away from one of the surfaces. At high lift (Fig. 2.16c), the jet breaks away from both surfaces to form a free jet. The transition points will depend on whether the valve is opening or closing.

When the discharge coefficient results from Fig. 2.16 are combined with some typical valve lift data, then it is possible to plot the effective flow area as a function of camshaft angle. The finite mass of the valve gear means that the valves cannot be opened or closed instantaneously. Figure 2.17 combines valve lift data with flow data. Note that there is a broad maximum for the effective flow area. This is a consequence of the effective flow area being limited by the annular area between the valve stem and the seat, and being almost independent of the valve lift. Also shown in Fig. 2.17 is a "high-performance" or "sport" cam profile. This has an increased lift and valve open duration. However, the valve lift curve has been scaled to give the same maximum valvetrain acceleration. Increasing the valve lift has not significantly increased the maximum effective flow area, but a consequence of the longer duration valve event is an increase in the width of the maximum. In other words, the extended duration, rather than the increased lift, will lead to an improvement in the flow performance of the valve.

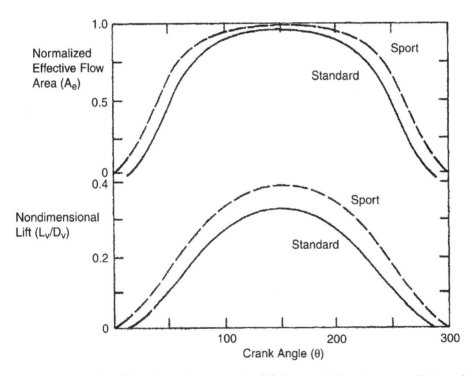

Figure 2.17. *The effect of combining valve lift data with discharge coefficient data to give the effective flow area as a function of cam angle (Stone, 1999).*

If the pressure ratio across the inlet valve becomes too high, a rapid fall in the volumetric efficiency will occur. Taylor (1985a) characterizes the flow by an inlet Mach index, Z, which is the Mach number of a notional air velocity. When the effective flow area is averaged, it can be divided by a reference area (based on the valve diameter D_v), to give a mean flow coefficient

$$\bar{C}_D' = \frac{\bar{A}_e}{\pi D_v^2/4} \tag{2.55}$$

The mean rate of change of the volume depends on the mean piston speed, v_p

$$v_p \times \pi B^2 / 4 \tag{2.56}$$

This leads to a notional mean velocity, which can be divided by the speed of sound, c, to give the Mach index, Z,

$$Z = \frac{v_p \times \pi B^2/4}{c \times \bar{C}'_D \times \pi D_v^2/4} = \frac{B^2}{D_v} \frac{v_p}{c \times \bar{C}'_D} \tag{2.57}$$

For a fixed valve timing, the volumetric efficiency is principally a function of the Mach index. To maintain an acceptable volumetric efficiency, the Mach index should be less than approximately 0.6, as illustrated in Fig. 2.18. Because the bore, stroke, valve diameter, and valve lift all are geometrically linked, then as the mean piston speed increases, so does the air velocity past the inlet valve. Thus, for geometrically similar engines, the Mach index depends on only the mean piston speed. In practice, the volumetric efficiency will be modified by the effects of inlet system pressure pulsations.

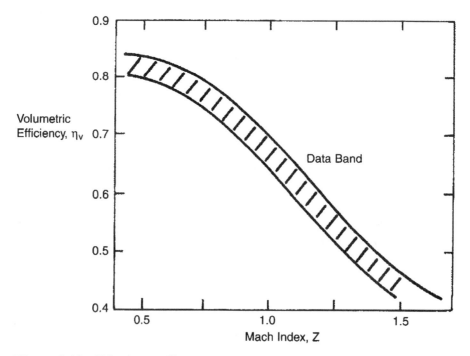

Figure 2.18. *Volumetric efficiency as a function of Mach index (Stone, 1999).*

For a given bmep, it is possible to show that the maximum torque depends on the swept volume, whereas the maximum power depends on the total piston area (A)

$$\text{Brake power} = \text{bmep LAn} \tag{2.4}$$

where n is the number of firing strokes per second.

For a four-stroke engine,

$$n = v_p/(4L)$$

where v_p is the mean piston velocity. Thus,

$$\text{Brake power} = \text{bmep LA } v_p/(4L) = \text{bmep } v_p A/4 \tag{2.58}$$

For a four-stroke engine, the number of firing strokes per second also is given by

$$n = \frac{1}{2}\omega/(2\pi) \tag{2.59}$$

Note that the total swept volume, V_s, is given by

$$V_s = LA \tag{2.60}$$

Substituting Eqs. 2.59 and 2.60 into Eq. 2.58 gives

$$\text{Brake power} = \text{bmep LAn }\omega/(4\pi) = \text{bmep } V_s \,\omega/(4\pi) \tag{2.61}$$

Because torque = power/angular velocity, then division of Eq. 2.61 by the angular velocity (ω) gives

$$\text{Torque} = \text{bmep } V_s/(4\pi) \tag{2.62}$$

2.8.2 Valve Timing

Figure 2.19 shows two timing diagrams. The first diagram (Fig. 2.19a) is typical of an automotive compression ignition engine or conventional spark ignition engine, whereas Fig. 2.19b is typical of a high-performance spark ignition engine. The greater valve overlap in the second case provides a better overall flow performance, for the reasons shown in Fig. 2.17. Highly turbocharged diesel engines also can use large valve overlap periods, but the valve overlap at top dead center (tdc) often is limited by the piston to cylinder-head clearance.

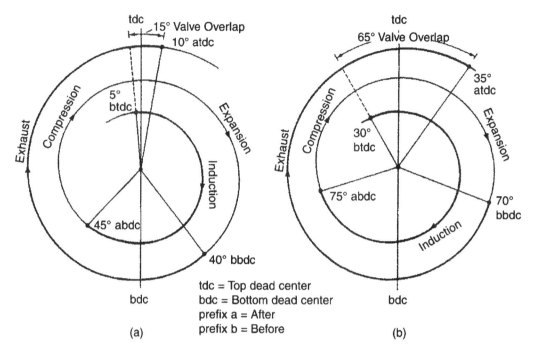

Figure 2.19. *Valve timing diagrams for (a) an automotive diesel engine or conventional spark ignition engine, and (b) a tuned spark ignition engine.*

Also, the inlet valve must close soon after bottom dead center (bdc); otherwise, the reduction in compression ratio may make cold starting too difficult. The exhaust valve opens approximately 40° before bottom dead center (bbdc) to ensure that all the combustion products have sufficient time to escape. This entails a slight penalty in the power stroke, but 40° bbdc (before bottom dead center) represents only approximately 12% of the engine stroke. Remember also that 5° after starting to open, the valve may be only 1% of fully open; after 10°, 5% of fully open; and not fully open until 120° after starting to open.

There are many tradeoffs in selecting the valve timing for internal combustion engines. Early opening of the exhaust valve leads to a reduction in the effective expansion ratio and expansion work, but this is compensated for by reduced exhaust stroke pumping work. When an engine is being optimized for high-speed operation, this leads to the use of earlier exhaust valve opening; the usual timing range is 40–60° bbdc. In the case of turbocharged engines, some of the expansion work that is lost by earlier opening of the exhaust valve is recovered by the turbine.

Exhaust valve closure is invariably after top dead center (atdc), and the higher the boost pressure in turbocharged engines, or the higher the speed for which the engine performance is optimized, then the later the exhaust valve closure. The exhaust valve usually is closed in the range 5–30° atdc. The aim is to avoid any compression of the cylinder contents toward the end of the exhaust stroke. The exhaust valve closure time does not seem to affect the level of residuals trapped in the cylinder or the reverse flow into the inlet manifold. However, for engines with in-manifold mixture preparation, a late exhaust valve closure can lead to fuel entering the exhaust manifold directly.

The inlet valve is opened before top dead center (btdc), so that by the start of the induction stroke, there is a large effective flow area. Engine performance is fairly insensitive to inlet valve opening in the range 10–25° btdc. For turbocharged engines at their rated operating point, the inlet manifold pressure is greater than the cylinder pressure, which in turn is above the exhaust manifold pressure. Under these circumstances, even earlier inlet valve opening (earlier than 30° btdc) leads to good scavenging. However, at part load, for a turbocharged engine or a throttled engine, early inlet valve opening leads to high levels of exhaust residuals and back-flow of exhaust into the inlet manifold. The results of this are most obvious with spark ignition engines, because the increased levels of exhaust residuals lead to increased cycle-by-cycle variations in combustion.

Inlet valve closure is invariably after bottom dead center (abdc) and typically around 40° abdc, because at bottom dead center, the cylinder pressure is still usually below the inlet manifold pressure. This is in part a consequence of the slider crank mechanism causing the maximum piston velocity to occur after 90° bbdc. Figure 2.20 illustrates the influence of inlet valve closure angle on the volumetric efficiency. A simple model has been used here, which ignores compressibility and dynamic effects. The mean piston speed has been used as a variable because it defines engine speed in a way that does not depend on the engine size. Figure 2.20

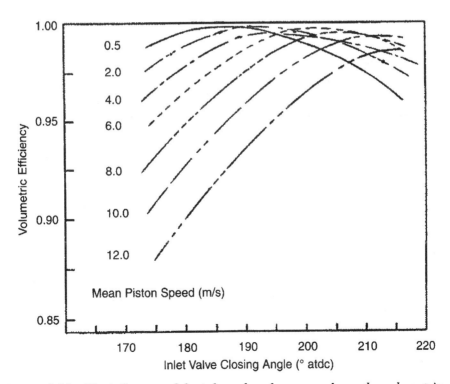

Figure 2.20. *The influence of the inlet valve closure angle on the volumetric efficiency for different piston speeds (Stone, 1999).*

shows that at low speeds, a late inlet valve closure reduces the volumetric efficiency. In contrast, at high speeds, an early inlet valve closure leads to a greater reduction in volumetric efficiency, which limits the maximum power output.

In addition to the individual valve events, the valve overlap period (during which both the inlet and exhaust valves are open) also affects engine performance, especially spark ignition engines operating at part load. In spark ignition engines with large valve overlap, the part throttle and idling operation suffers because the reduced induction manifold pressure causes back-flow of the exhaust. Furthermore, full load economy is poor because some unburned mixture can pass straight through the engine when both valves are open at top dead center. These problems are avoided in a turbocharged engine with in-cylinder fuel injection.

The level of exhaust residuals trapped in the cylinder has a significant effect on the cycle-by-cycle variations in combustion and the emissions of NOx. As with exhaust gas recirculation, high levels of exhaust residuals lead to lower emissions of NOx, and greater cycle-by-cycle variations in combustion. The level of residuals increases with the following:

a. Decreasing absolute inlet manifold pressure
b. Reducing compression ratio
c. Increasing valve overlap
d. Decreasing speed
e. Increasing exhaust back-pressure

2.8.3 Valve Operating Systems

In engines with overhead poppet valves (OHV—overhead valves), the camshaft is either mounted in the cylinder block, or in the cylinder head (OHC—overhead camshaft). Figure 2.21a shows an overhead valve engine in which the valves are operated from the camshaft, via cam followers, pushrods, and rocker arms. This is a cheap solution because the drive to the camshaft is simple (either gear or chain), and the machining is in the cylinder block. In a "V" engine, this arrangement is particularly suitable because a single camshaft can be mounted in the valley between the two cylinder banks.

In overhead camshaft (OHC) engines (Fig. 2.21b), the camshaft can be mounted either directly over the valve stems, or it can be offset. When the camshaft is offset, the valves are operated by rockers, and the valve clearances can be adjusted by altering the pivot height or, as in the case of the exhaust valves in Fig. 2.21b, different thickness shims can be used. For the inlet valves in Fig. 2.21b, the cam operates on a follower or "bucket." The clearance between the follower and the valve end is adjusted by a shim. Although this adjustment is more difficult than in systems using rockers, it is much less prone to change. The spring retainer is connected to the valve spindle by a tapered split collet. The valve guide is a press-fit into the cylinder head, so that it can be replaced when worn. Valve seat inserts are used, especially in engines with aluminum alloy cylinder heads, to ensure minimal wear. Normally, poppet valves rotate to even out any wear and to maintain good seating. This rotation can be promoted if the center of the cam is offset from the valve axis. Invariably, oil seals are placed at the top of the

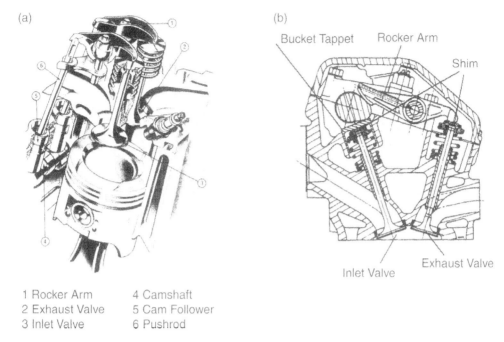

Figure 2.21. *Overhead valve arrangements: (a) Ford V-6 Essex engine, and (b) Triumph Dolomite Sprint overhead camshaft arrangement from the 1970s.*

valve guide to restrict the flow of oil into the cylinder. This is most significant with overhead cast-iron camshafts, which require a copious supply of lubricant. When the valves are not in line (as in Fig. 2.21b), it is more usual to use two camshafts because this gives more flexibility on valve timing and greater control if a variable valve timing system is to be used.

The use of four valves per combustion chamber is quite common in high-performance spark ignition engines and is used increasingly in compression ignition engines. The advantages of four valves per combustion chamber are larger valve throat areas for gas flow, smaller valve forces, and a larger valve seat area. Smaller valve forces occur because a lighter valve with a less stiff spring can be used. This also will reduce the hammering effect on the valve seat when the valve closes. The larger valve seat area is important because this is how heat is transferred (intermittently) from the valve head to the cylinder head. In the case of diesel engines, four valves per cylinder allow the injector to be placed in the center of the combustion chamber, which facilitates the development of low-emission combustion systems.

To reduce maintenance requirements, it is now common to use some form of hydraulic lash adjuster (also known as a hydraulic lifter or tappet), an example of which is shown in Fig. 2.22. This consists of a piston/cylinder arrangement that is pressurized by engine lubricant. However, when the cam starts to displace its follower, a sudden rise in pressure occurs in the lower oil chamber. This causes a check valve (a ball loaded by a weak spring) to close, so that the cam motion then is transmitted to the valve. There is always a small leakage flow from the lash adjuster so that the valve will always seat properly, even when there is a reduction

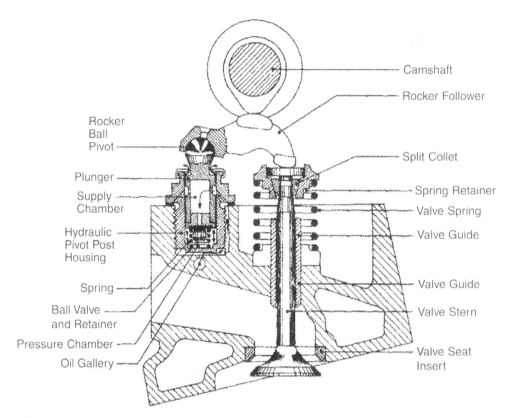

Figure 2.22. *A cam-over-rocker system with a hydraulic lash adjuster (or tappet). Adapted from Heisler (1995).*

in the clearances within the valvetrain. The lash adjusters can be incorporated into the follower of the overhead valve arrangement (Fig. 2.21a) or the bucket tappet of Fig. 2.21b, or the pivot post of a cam-over-rocker system (Fig. 2.22). A disadvantage of this simple substitution is an increase in frictional losses because the cam follower will always be loaded when sliding on the cam base circle. Friction can be reduced by using a roller follower on the rocker of the system in Fig. 2.22, and this cam-over-rocker system also minimizes the mass of the moving valvetrain components. A hydraulic lash adjuster reduces the stiffness of the valvetrain, which will reduce the maximum speed limit for the valve gear.

The drive to the camshaft usually is by chain or toothed belt. Gear drives also are possible but tend to be expensive, noisy, and cumbersome with overhead camshafts. The advantage of a toothed belt drive is that it can be mounted externally to the engine, and the rubber damps out torsional vibrations that otherwise might be troublesome.

2.8.4 Dynamic Behavior of Valve Gear

The theoretical valve motion is defined by the geometry of the cam and its follower. The actual valve motion is modified because of the finite mass and stiffness of the elements in the valvetrain. Figure 2.23 shows the theoretical valve lift, velocity, and acceleration. The lift is the integral of the velocity, and the velocity is the integral of the acceleration. Before the valve starts to move, the clearance must be taken up. The clearance in the valve drive mechanism ensures that the valve can fully seat under all operating conditions, with sufficient margin to allow for the bedding-in of the valve and any change of clearance caused by differential thermal expansion. To control the impact stresses as the clearance is taken up, the cam is designed to give an initially constant valve velocity. This portion of the cam should be large enough to allow for the different clearances during engine operation. The impact velocity typically is limited to 0.5 m/s at the rated engine speed.

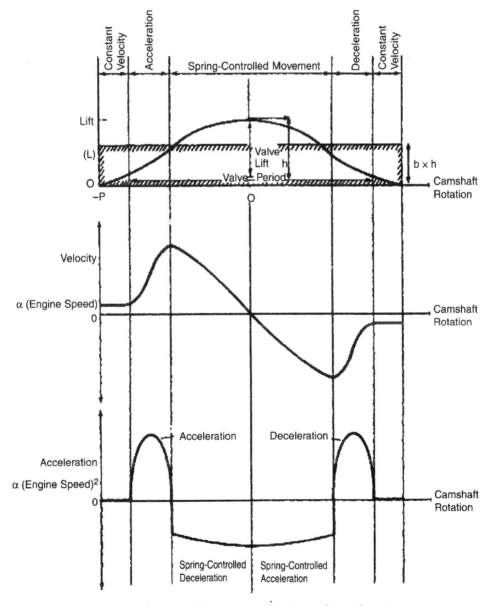

Figure 2.23. *Displacement, velocity, and acceleration of a cam-operated valve (Stone, 1999).*

The next stage is when the cam accelerates the valve. The cam could be designed to give a constant acceleration, but this would give rise to shock loadings, owing to the theoretically instantaneous change of acceleration. A better practice is to use a function that causes the acceleration to rise from zero to a maximum, and then to fall back to zero. Both sinusoidal and polynomial functions are appropriate examples. As the valve approaches maximum lift, the deceleration is controlled by the valve spring. As the valve starts to close, its acceleration is provided by the valve spring. The final deceleration is controlled by the cam, and the same considerations apply. Finally, the profile of the cam should be such as to give a constant closing velocity to limit the impact stresses.

Camshaft design is a complex area, but one that is critical to the satisfactory high-speed performance of internal combustion engines. A widely used type of cam is the polydyne cam, which uses a polynomial function to define the valve lift as a function of cam angle and selects coefficients that avoid harmonics that might excite valve spring oscillations

$$L_v = f(\theta) = a + a_1\theta + a_2\theta^2 + a_3\theta^3 + ... + a_i\theta^i + ... \tag{2.63}$$

in which some values of a_i can be zero.

For a constant angular velocity of ω, differentiation gives

Velocity:
$$L'_v = f'(\theta) = \omega\left(a_1 + 2a_2\theta + 3a_3\theta^2 + ... + ia_i\theta^{i-1} + ...\right) \tag{2.64}$$

Acceleration:
$$L''_v = f''(\theta) = \omega^2\left(2a_2 + 6a_3\theta + ... + i(i-1)a_i\theta^{i-2} + ...\right) \tag{2.65}$$

Jerk:
$$L'''_v = f'''(\theta) = \omega^3\left(6a_3 + ... + i(i-1)(i-2)a_i\theta^{i-3} + ...\right)$$

The dependence of the velocity on ω, the acceleration on ω^2, and the jerk on ω_3 explain why problems at high speeds can occur with valve gear. It is common practice to have the valve lift arranged symmetrically about the maximum lift, as shown in Fig. 2.23, and this is automatically satisfied if only even powers of θ are used in Eq. 2.63. This also ensures that the jerk term will be zero at the maximum valve lift (h). Asymmetric lift profiles are used with hydraulic lash adjusters to ensure that the check valve closes rapidly, whereas a gentle closing ramp is still needed to avoid high-impact loads between the valve and its seat.

The valve lift "area," A_θ, is a widely used concept to give an indication of the ability of the camshafts to admit flow. With reference to Fig. 2.23,

$$A_\theta = \int L_v d\theta = 2bph \tag{2.66}$$

where b represents the effective mean height of the valve lift as a fraction of h, and A_θ has units of radians times meters.

The valve lift characteristics also will be influenced by the stiffness of the valve spring, as this must control the deceleration prior to the maximum lift, as well as the acceleration that occurs after the maximum lift. Ideally, the spring force should be uniformly greater than the required acceleration force at the maximum design speed. The valve acceleration is given by Eq. 2.65, and the mass should be referred to the valve axis. For a pushrod-operated valve system,

$$\text{Equivalent mass, } m_e = \frac{\text{tappet + pushrod mass}}{(r_r/r_v)^2} + \frac{\text{polar inertia of rocker}}{r_v^2} + \text{valve mass} + \frac{\text{spring mass}}{3} \tag{2.67}$$

where

r_x = radius from the rocker axis to the cam line of action
r_v = radius from the rocker axis to the valve axis

In practice, when either a rocker arm or a finger-follower system is used, the values of the radii r_r and r_v will change. Due account can be taken of this to convert the valve lift to cam lift, but remember that the equivalent mass (in Eq. 2.67) also will become a function of the valve lift.

Additional allowances in the spring load must be made for possible overspeeding of the engine and friction in the valve mechanism. The force (F) at the cam/tappet interface is given by Eq. 2.68:

$$F = \left(m_e \times L_v'' + F_o + kL_v + F_g\right) \times r_r/r_v \tag{2.68}$$

where

F_o = valve spring pre-load
F_g = gas force on the valve head (normally significant only for the exhaust valve)
k = valve spring stiffness

Figure 2.24 shows the force at the cam/tappet interface for a range of speeds, with the static force from the valve spring (corresponding to 0 rpm). At low speeds, the maximum force occurs at maximum valve lift, because the valve spring force dominates. As the engine speed is increased, the acceleration terms dominate, and the largest force occurs just after the occurrence of the maximum acceleration. As speed is increased further, the maximum force will increase due to the cam-controlled acceleration, while the minimum force will decrease, and a speed will be reached at which the contact force at the tappet becomes zero. In other words,

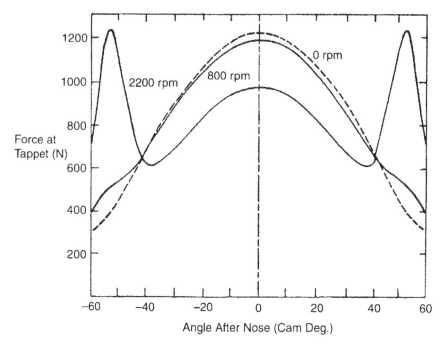

Figure 2.24. *The force at the cam/tappet interface for a range of speeds (Stone, 1999).*

at this speed, there is insufficient force from the spring to decelerate the valve at the desired rate, and separation will occur between the cam and its tappet.

This simple model ignores dynamic effects within the spring, as well as the finite stiffness of other components in the valvetrain. Thus, it will overestimate the maximum operating speed of the valve gear. In the case of the valve spring, prior to the critical speed being reached, there will be inter-coil vibrations known as surge. The natural frequency of the valve spring should be at least an order of magnitude higher than the camshaft frequency. However, because the motion of the cam is complex, high harmonics are present, and these can excite resonance of the valve spring. When this occurs, the spring no longer obeys the simple force/displacement law, and the spring force will fall, so that the practical maximum speed is below the expected critical speed.

2.9 Engine Configuration

2.9.1 Choosing the Number of Cylinders

After the type and size of engine have been determined, the number and disposition of the cylinders must be decided. The main constraints influencing the number and disposition of the cylinders are as follows:

1. The number of cylinders needed to produce a steady output

2. The minimum swept volume for efficient combustion (say, 400 cm³)

3. The number and disposition of cylinders for satisfactory balancing

4. The number of cylinders needed for an acceptable variation in the torque output

For a four-stroke engine with five or more cylinders, there can always be a cylinder generating torque. Figure 2.25 shows the variation in the instantaneous torque associated with different numbers of cylinders. A six-cylinder four-stroke engine is, of course, equivalent to a three-cylinder two-stroke engine.

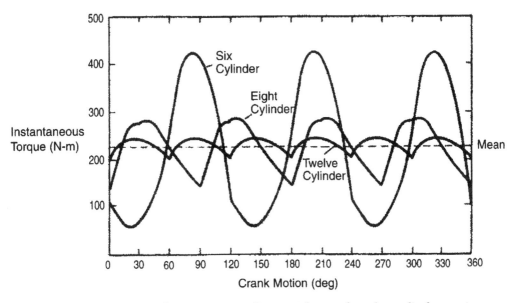

Figure 2.25. *Torque characteristics of six-, eight-, and twelve-cylinder engines. Adapted from Campbell (1978).*

The most common engine types are the straight or in-line, the "V" (with various included angles), and the horizontally opposed (Fig. 2.26).

The "V" engines form a very compact power unit. A more compact arrangement is the "H" configuration (in effect, two horizontally opposed engines with the crankshafts geared together), but this is an expensive and complicated arrangement that has had limited use. Whatever the arrangement, it is unusual to have more than six or eight cylinders in a row because torsional vibrations in the crankshaft then become much more troublesome. In multi-cylinder engine configurations other than the in-line format, it is advantageous if a single crankpin can be used for a connecting rod to each bank of cylinders. This makes the crankshaft simpler,

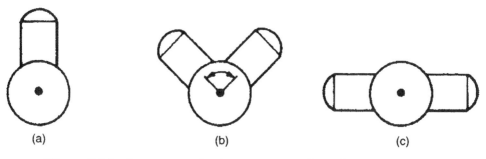

Figure 2.26. *Common engine layouts: (a) in-line, (b) Vee or "V," and (c) horizontally opposed (Stone, 1999).*

reduces the number of main bearings, and facilitates a short crankshaft that will be less prone to torsional vibrations. Nonetheless, the final decision on the engine configuration also will be influenced by marketing, packaging, and manufacturing constraints.

When deciding on an engine layout, two interrelated aspects must be considered: the engine balance, and the firing interval between cylinders. The following discussions will relate to four-stroke engines because these have only a single firing stroke in each cylinder once every two revolutions. An increase in the number of cylinders leads to smaller firing intervals and smoother running; however, with more than six cylinders, the improvements are less noticeable. Normally, the crankshaft is arranged to give equal firing intervals, but this is not always the case. Sometimes a compromise is made for better balance or simplicity of construction. For example, consider a twin-cylinder horizontally opposed four-stroke engine with a single throw crankshaft. The engine is reasonably balanced, but the firing intervals are 180°, 540°, 180°, 540°, and so forth.

When calculating the engine balance, the connecting rod is treated as two masses concentrated at the center of the big end and the center of the little end (Fig. 2.27). For equivalence of a connecting rod of mass m,

$$m = m_1 + m_2 \quad \text{and} \quad m_1 r_1 = m_2 r_2 \qquad (2.69)$$

The mass m_2 can be considered as part of the mass of the piston assembly (piston, rings, piston pin, and so forth) and be denoted by m_r, the reciprocating mass. The mass referred to the big end should be added to the big-end journal (the bearing surface) on the crankshaft, and the crankshaft should be in static and dynamic balance.

As a simple example to illustrate the difference between static and dynamic balance, consider a planar crankshaft for an in-line four-cylinder engine, as shown diagrammatically in Fig. 2.28. If the rotating masses are not balanced at each big-end journal, then the out-of-balance mass can be denoted by m. By taking moments and resolving at any point on the shaft, it can be seen that there is no resultant moment or force from the individual centripetal forces $mr\omega^2$; the absence of any moments is the condition for dynamic balance. Inspection of Fig. 2.28

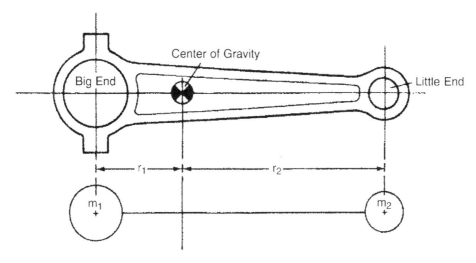

Figure 2.27. *A connecting rod and its kinematic equivalent model (Stone, 1999).*

shows that dynamic balance is independent of static balance. If the first two crank throws were up and the second two were down, then there would be static balance but not dynamic balance. Although the arrangement in Fig. 2.28 is in dynamic balance, considerable bending stresses can be introduced by the centripetal forces at each big-end journal. Figure 2.29 shows counterbalance masses added to the webs on each side of the big-end journals, so that these bending stresses can be eliminated.

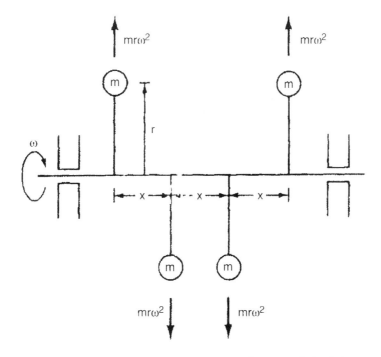

Figure 2.28. *Balancing of a four-cylinder in-line crankshaft (Stone, 1999).*

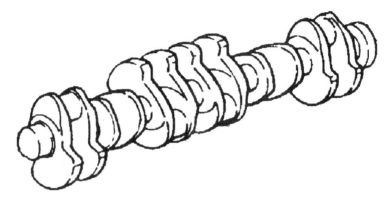

Figure 2.29. *A multi-cylinder crankshaft in which the balance masses on the crank webs provide dynamic balancing for each throw. Adapted from Lilly (1984).*

2.9.2 Balancing of the Primary and Secondary Forces and Moments

The treatment of the reciprocating mass is more involved. If the connecting rod were infinitely long, the reciprocating mass would follow simple harmonic motion, producing a primary out-of-balance force that acts along the cylinder axis. However, the finite length of the connecting rod introduces higher harmonic forces.

Figure 2.30 shows the geometry of the crank-slider mechanism, when there is no offset between the little-end (also called the gudgeon-pin or piston-pin) axis and the cylinder axis. The little-end position is given by

$$x = R\cos\theta + L\cos\phi \qquad (2.70)$$

Inspection of Fig. 2.30 indicates that

$$R\sin\theta = L\sin\phi \qquad (2.71)$$

and recalling that $\cos\phi = \sqrt{(L - \sin^2\phi)}$, then substitution of Eq. 2.71 into Eq. 2.70 gives

$$x = R\left(\cos\theta + L/R\sqrt{\left\{1 - (R/L)^2 \sin^2\theta\right\}}\right) \qquad (2.72)$$

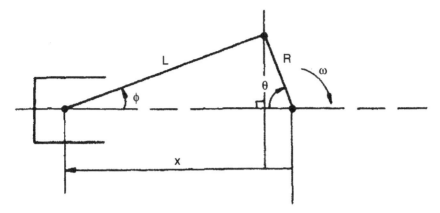

Figure 2.30. *Crank-slider mechanism geometry (Stone, 1999).*

The binomial theorem can be used to expand the square root term in Eq. 2.72,

$$x \approx R\left(\cos\theta + L/R\left[1 - \frac{1}{2}(R/L)2\sin 2\theta - \frac{1}{8}(R/L)4\sin 4\theta + ...\right]\right) \quad (2.73)$$

The powers of sinθ can be expressed as equivalent multiple angles

$$\sin^2\theta = \frac{1}{2} - \frac{1}{2}\cos 2\theta \quad \text{and} \quad \sin^4\theta = \frac{3}{8} - \frac{1}{2}\cos 2\theta + \frac{1}{8}\cos 4\theta \quad (2.74)$$

Substituting the results from Eq. 2.74 into Eq. 2.73 gives

$$x = R\left\{\cos\theta + L/R\left[1 - \frac{1}{2}(R/L)^2\left(\frac{1}{2} - \frac{1}{2}\cos 2\theta\right) - \frac{1}{8}(R/L)^4\left(\frac{3}{8} - \frac{1}{2}\cos 2\theta + \frac{1}{8}\cos 4\theta\right) + ...\right]\right\} \quad (2.75)$$

The geometry of engines is such that $(R/L)^2$ is invariably less than 0.1, in which case it is acceptable to neglect the $(R/L)^4$ terms, because inspection of Eq. 2.75 shows that these terms will be at least an order of magnitude smaller than the $(R/L)^2$ terms. The approximate position of the little end is thus

$$x \approx R\left\{\cos\theta + L/R\left[1 - \frac{1}{2}(R/L)^2\left(\frac{1}{2} - \frac{1}{2}\cos 2\theta\right)\right]\right\} \quad (2.76)$$

Equation 2.76 can be differentiated once to give the piston velocity, and a second time to give the acceleration. (In both cases, the line of action is the cylinder axis.)

$$dx/dt \approx -R\omega\left[\sin\theta + \frac{1}{2}(R/L)\sin 2\theta\right]$$

and

$$d^2x/dt^2 \approx -R\omega^2\left[\cos\theta + (R/L)\cos 2\theta\right] \tag{2.77}$$

This leads to an axial force

$$F_r \approx m_r R\omega^2\left[\cos\theta + (R/L)\cos 2\theta\right] \tag{2.78}$$

where

F_r	=	axial force due to the reciprocating mass
m_r	=	equivalent reciprocating mass
ω	=	angular velocity, $d\theta/dt$
R	=	crankshaft throw
L	=	connecting-rod length
$\cos\theta$	=	primary term
$\cos 2\theta$	=	secondary term

In other words, there is a primary force varying in amplitude with crankshaft rotation, and a secondary force varying at twice the crankshaft speed. Both of these forces act along the cylinder axis. Referring to Fig. 2.28 for a four-cylinder in-line engine, it can be seen that the primary forces will have no resultant force or moment.

By referring to Fig. 2.31, it can be seen that the primary forces for this four-cylinder engine are 180° out of phase and thus cancel. However, the secondary forces will be in phase, and this causes a resultant secondary force on the bearings, which acts in parallel to the cylinder axes. Because the resultant secondary forces have the same magnitude and direction, there is no secondary moment but a resultant force of

$$4m_r r\omega^2 (R/L)\cos 2\theta \tag{2.79}$$

For multi-cylinder engines in general, the phase relationship between the cylinders will be more complex than in the four-cylinder in-line engine. For cylinder n in a multi-cylinder engine,

$$F_{r,n} \approx m_{r,n} R\omega^2\left[\cos(\theta + \alpha_n) + (R/L)\cos 2(\theta + \alpha_n)\right] \tag{2.80}$$

where α_n is the phase separation between cylinder n and the reference cylinder.

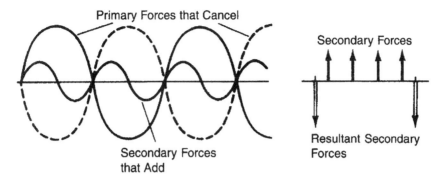

Figure 2.31. *Secondary forces for the crankshaft shown in Fig. 2.28.*

It is then necessary to evaluate all the primary and secondary forces and moments, for all cylinders relative to the reference cylinder, to find the resultant forces and moments. A comprehensive discussion on the balancing and firing orders of multi-cylinder in-line and "V" engines can be found in Taylor (1985b).

In multi-cylinder engines, the cylinders and their disposition are arranged to eliminate as many of the primary and secondary forces and moments as possible. Complete elimination is possible for in-line five-, six-, or eight-cylinder engines, horizontally opposed eight- or twelve-cylinder engines, and twelve- or sixteen-cylinder "V" engines. Primary forces and moments can be balanced by masses running on the crankshaft and a contra-rotating countershaft at engine speed. Secondary out-of-balance forces and moments can be balanced by two contra-rotating countershafts running at twice the engine speed (Fig. 2.32). The counterbalance masses must be phased so that the piston and both masses are closest to the cylinder head at the same time. If the cylinder axis is vertical (as assumed in Fig. 2.32), then it can be seen how the horizontal components of the centripetal forces from the balance masses will cancel. Such systems are rarely used because of the extra cost and mechanical losses involved, but examples can be found on engines with inherently poor balance, such as in-line three-cylinder engines or four-cylinder "V" engines. In vehicular applications, the transmission of vibrations from the engine to the vehicle structure is minimized by the careful choice and placement of flexible mounts.

Engine balancing is best illustrated by examples, and the two examples here are for a single-cylinder engine and a three-cylinder engine.

Example 2.1 A single-cylinder engine has a reciprocating mass of 1.53 kg, the stroke is 89 mm, and the connecting rod length/crank throw ratio (L/R) is 3.6. What product of mass (m) and eccentricity (e) should be used for the primary and secondary balance masses, and how should they be deployed?

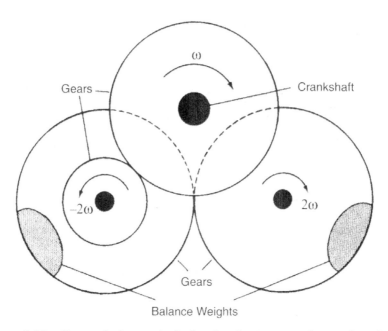

Figure 2.32. *Counterbalance shafts for the elimination of secondary forces.*

Equation 2.78 gives the standard approximation for the axial primary and secondary forces

$$F_r \approx m_r \omega^2 R (\cos\theta + 1\cos 2\theta)$$

where

θ = crankshaft rotation measured from top dead center (tdc)
m_r = reciprocating mass, 1.53 kg
R = crank throw = stroke/2 = 0.089/2 = 0.0445 m
R/L = 1/3.6

The primary out-of-balance force $\left(F_p = m_r\omega^2 r\cos\theta\right)$ can be balanced by a pair of counter-rotating balance masses at the same angular velocity (ω) as the crankshaft. As shown in Fig. 2.33, which assumes a vertical cylinder axis, the centripetal force $m_p e_p \omega^2$ can be resolved into horizontal and vertical components, and with two counter-rotating balance masses, the horizontal components of the forces cancel.

Resolving vertically, and equating with the primary out-of-balance force (Eq. 2.78),

$$F_p = m_r\omega^2 R \cos\theta = 2\left(m_p e_p\right)\omega^2 \cos\theta$$

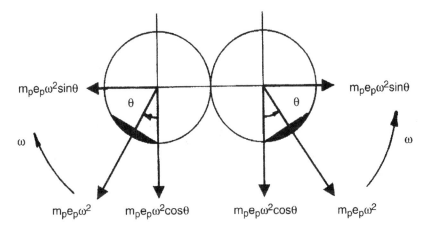

Figure 2.33. *The use of two counter-rotating balance masses to eliminate the horizontal components of the centripetal forces.*

or

$$m_p e_p = m_r R/2 = 1.53 \times 0.0445/2 = 34 \text{ kg mm}$$

The secondary out-of-balance force $\left(F_s = m_r \omega^2 (R/L) R \cos\theta\right)$ can be balanced by a pair of counter-rotating balance masses at twice the angular velocity (ω) of the crankshaft. As shown in Fig. 2.34, two counter-rotating balance masses are needed to eliminate the horizontal components of their forces.

$$F_s = m_r \omega^2 (R/L) R \cos 2\theta = 2(m_s e_s)(2\omega)^2 \cos 2\theta$$

or

$$m_s e_s = m_r R(R/L)/8 = 1.53 \times 0.0445 \times (1/3.6)/8 = 2.4 \text{ kg mm}$$

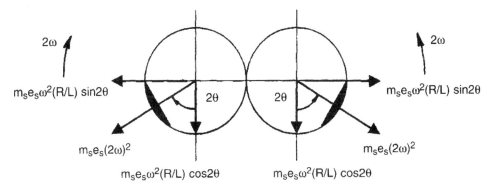

Figure 2.34. *The use of two counter-rotating balance masses at twice crankshaft speed to eliminate the secondary out-of-balance forces.*

The balance mass position has been measured from bottom dead center (bdc), while the piston position has been measured from top dead center (tdc). Thus, when the piston is at tdc, both the primary and secondary balance masses must be in their datum positions (that is, as far from the piston as possible).

Example 2.2 A three-cylinder two-stroke engine uses uniflow scavenging through overhead valves. The reciprocating mass for each cylinder is 0.3 kg, the stroke is 70 mm, the connecting-rod length/crank-throw ratio (L/R) is 3.5, and the inter-cylinder bore spacing is C. Show that, with a 120° firing interval, the primary and secondary forces are balanced. If the balance masses for eliminating the primary moments are a distance D from the bore of the center cylinder, what product of mass (m) and eccentricity (e) should be used, and what will the unbalanced secondary moments be? How should the balance masses be configured?

Equation 2.78 (the standard approximation for the axial primary and secondary forces) must be modified for a multi-cylinder engine, by introducing ϕ_n, the phase separation between cylinder n and the reference cylinder

$$F_{r,n} \approx m_r \omega^2 R \left(\cos[\theta + \phi_n] + R/L \cos 2[\theta + \phi_n] \right)$$

where

θ = crankshaft rotation measured from top dead center (tdc), r/L = 1/3.5
m_r = reciprocating mass, 0.3 kg
R = crank throw = stroke/2 = 0.070/2 = 0.035 m

With the crank throws 120° apart and only three cylinders, the firing order of the cylinders does not matter. As shown in Fig. 2.35, cylinder number 1 will be the reference cylinder, and as drawn, it is at top dead center ($\theta = 0$). At this stage, the counterbalance masses in planes 4 and 5 should be ignored.

The phasor approach will be used here, in which if the polygon of forces is closed, then the forces will be balanced for all positions of the reference crank (that is, all values of θ). The primary and secondary forces can act only along a cylinder axis. Therefore, for a particular value of θ, only the force components along the cylinder axis must be balanced. However, a new value of θ is equivalent to rotating the polygon of forces. Again, only the force components along the cylinder axis must be considered. Therefore, if the original polygon of forces is closed, the forces will be balanced for any crankshaft position.

Considering first the primary out-of-balance forces

$$F_{p,n} = m_r \omega^2 R \cos[\theta + \phi_n]$$

it can be seen that they are balanced (Fig. 2.36).

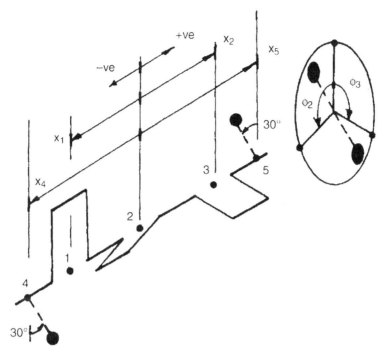

Figure 2.35. *Isometric sketch of the crankshaft.*

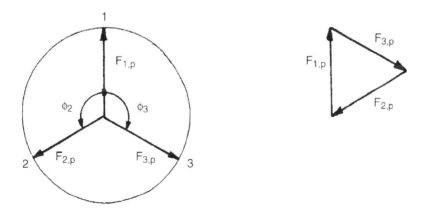

Figure 2.36. *Primary forces in an in-line three-cylinder engine.*

Considering next the secondary out-of-balance forces

$$F_{s,n} = m_r \omega^2 (R/L) R \cos 2[\theta + \phi_n]$$

it can be seen that they too are balanced (Fig. 2.37).

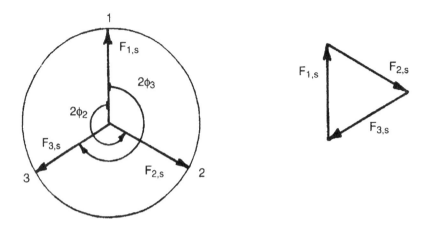

Figure 2.37. *Secondary forces in an in-line three-cylinder engine.*

Greater care is needed with the evaluation of the out-of-balance moments because it is easy to make mistakes in determining the direction of a moment. Also, note that the shape of the polygon of moments depends on the point about which the moments are taken, although the resultant moment (if any) will be the same. This will be illustrated by taking moments about both points 1 and 2.

The double-headed arrow convention will be used here to denote moments, and looking along the arrow toward its head would be a clockwise moment. The positive direction along the crankshaft axis must be defined, as shown in the isometric sketch of the crankshaft (Fig. 2.35). In the phasor diagrams, the moments and forces shown with broken lines should initially be ignored because they arise from the balance masses.

For the primary moments, taking moments about point 2 gives the phasor diagram shown in Fig. 2.38.

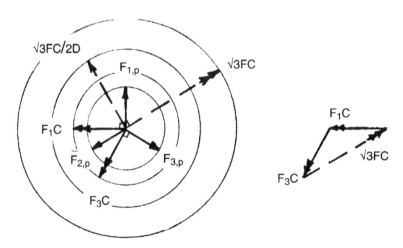

Figure 2.38. *Primary moments about point 2 in an in-line three-cylinder engine.*

Remember that the direction of the moment is at right angles to the force by which it was produced, and that its direction depends on the position of the forces plane relative to the reference plane. In this particular case, the distance $x_2 = -C$.

For the primary moments, taking moments about point 1 gives the phasor diagram shown in Fig. 2.39 (noting that $x_3 = 2C$).

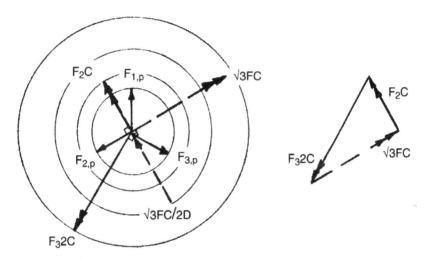

Figure 2.39. *Primary moments about point 1 in an in-line three-cylinder engine.*

In both polygons of moments, it can be seen that the moment required to close the polygon (and thus balance the moments) is the same in magnitude and direction. The simplest way that a moment can be balanced, without introducing any out-of-balance forces, is by a pair of balance masses that are separated along the crankshaft axis and phased 180° apart.

The out-of-balance moment is

$$\left(\sqrt{3}\right)FC$$

Thus, the force from each balance mass must be

$$\left(\sqrt{3}\right)FC/(2D)$$

because the masses shown in Fig. 2.35 are 2D apart.

The plane containing these masses is, of course, perpendicular to the direction of the moment. This leads to the orientation of the balance masses that is shown in the original sketch of the crankshaft (Fig. 2.35). As might be expected from symmetry, these balance masses are in a plane that is perpendicular to the crank throw for the center cylinder.

However, only the component of force along the cylinder axes is required, whereas the rotating balance masses will produce a moment that has components perpendicular to the cylinder axis. Thus, it is necessary to introduce a counter-rotating shaft that contains another pair of balance masses, which will be a mirror image of balance masses on the crankshaft. (The use of counter-rotating shafts was illustrated in Example 2.1.)

Finally, the force from each balance mass must be

$$\left(\sqrt{3/4}\right) m_r \omega^2 R \cos\theta\, C/D = \left(m_p e_p\right) \omega^2 \cos\theta$$

or

$$m_p e_p = \left(\sqrt{3/4}\right) C/D\; m_r R = \left(\sqrt{3/4}\right) C/D \times 0.3 \times 0.035 = 4.55\; C/D\; \text{kg mm}$$

Both masses on the crankshaft must lie in a plane perpendicular to that containing the number 2 cylinder crank throw. The mass at the number 1 cylinder end must be phased 150°ca from the number 1 crank, and the mass at the number 3 cylinder end must be phased 150°ca from the number 3 crank. The balance shaft must counter-rotate at engine speed, and the two balance masses must be the mirror images of those on the crankshaft.

For the secondary out-of-balance moment, it can be seen that the secondary force from cylinder number 2 lies in the direction of the primary force from cylinder number 3 and vice versa. Taking moments about cylinder number 1 thus gives the polygon of moments shown in Fig. 2.40.

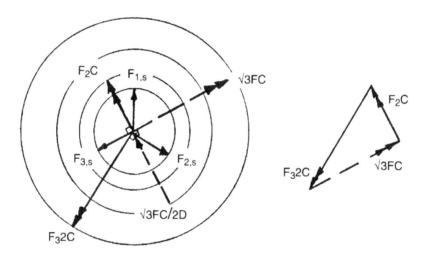

Figure 2.40. *Secondary moments about cylinder number 1 in an in-line three-cylinder engine.*

Noting that the secondary force is

$$F_s = m_r \omega^2 (R/L) R \cos 2\theta$$

gives

$$M_s = \left(\sqrt{3}\right) m_r \omega^2\, CR(R/L) \cos 2\theta$$

With "V" engine configurations, it is necessary to use one bank of cylinders as a datum, and to define the phasing of the pistons within each bank relative to a single piston in the datum cylinder bank. The moments and forces from each bank then can be resolved into their horizontal and vertical components and added. Heisler (1995) provides numerous worked examples for in-line, horizontally opposed, and "V" engines, including "V" engines with offset crank pins. In general, it is desirable to have equal firing intervals; if a V-8 is designed with a 60° (as opposed to 90°) included angle to give a narrow engine, then offsetting the adjacent crank pins by 30° will give equal firing angles. Taylor (1985b) also gives a comprehensive treatment of engine balance, including radial engines. Table 2.2 summarizes the engine balance for several of the more common configurations.

It is interesting to note that for a five-cylinder in-line engine, it is possible to eliminate both the primary and secondary moments and forces by adopting an unequal cylinder spacing. However, the overall length becomes greater than for a six-cylinder engine, thereby eliminating a key advantage of a five-cylinder engine.

2.10 Fuel Cells

Every major vehicle manufacturer has prototype fuel cell powered vehicles, but the major issue currently is the cost of those vehicles. However, Ashley (2001) reports predictions that the cost will fall to $50–$60 per kilowatt when production of 300,000 units per year is achieved, which is about twice the cost of a conventional powerplant. Solid polymer fuel cells (SPFC) are favored for automotive applications, although solid oxide fuel cells (SOFC) are being developed for auxiliary power units because they can operate more directly on conventional hydrocarbon fuels. In contrast, the SPFC requires hydrogen, which either must be stored onboard or produced by reforming a conventional fuel. Only an overview of fuel cells will be provided here, but a comprehensive treatment is provided by Larminie and Dicks (2000). The following material covers the operation and construction of SPFC, how the hydrogen can be provided, and what factors determine the efficiency.

2.10.1 Solid Polymer Fuel Cells (SPFC)

Figure 2.41 shows the sandwich-like construction of a solid polymer fuel cell (SPFC). The electrodes of the SPFC are porous to allow the reactant gases, which are distributed on the electrode by the flow field plates, to diffuse to the catalyst and membrane. The flow field plates also collect the current from the electrodes. Because the open circuit voltage of a fuel

TABLE 2.2
SUMMARY OF ENGINE BALANCE FOR DIFFERENT CONFIGURATION ENGINES, WITH CYLINDER SPACING a, CRANK THROW r, AND CONNECTING ROD LENGTH l

Number of Cylinders	Arrangement of Crank Pins[1]	Firing Interval (°ca)	Primary Forces $m_r\omega^2 R$	Primary Moments $m_r\omega^2 Ra$	Secondary Forces $m_r\omega^2 R^2/L$	Secondary Moments $m_r\omega^2 R^2/L$
In-Line Cylinder Engines						
1	1	720	$V = \cos\theta$	$M = 0$	$V = \cos 2\theta$	$M = 0$
2	1–2	180, 540	$V = 0$	$M = \cos\theta$	$V = 2\cos 2\theta$	$M = 0$
3	1–2–3	240	$V = 0$	$M = \sqrt{3}\sin\theta$	$V = 0$	$M = \sqrt{3}\sin 2\theta$
4	1,4–2,3	180	$V = 0$	$M = 0$	$V = 4\cos 2\theta$	$M = 0$
5	1–5–2–3–4	144	$V = 0$	$M = 0.449\cos\theta$	$V = 0$	$M = 4.98\cos 2\theta$
6	1,6–2,5–3,4	120	$V = 0$	$M = 0$	$V = 0$	$M = 0$
8	1,8 – 4,5 – 2,7 – 3,6	90	$V = 0$	$M = 0$	$V = 0$	$M = 0$
Opposed Cylinder Engines						
2	1,2	180, 540	$V = 2\cos\theta$	$M = 0$	$V = 0$	$M = 0$
2	1–2	360	$V = 0$	$M = \cos\theta$	$V = 0$	$M = \cos 2\theta$
4	1,4–2,3	180	$V = 0$	$M = 0$	$V = 0$	$M = 2\cos 2\theta$
4	1,2–3,4	180	$V = 0$	$M = 2\cos\theta$	$V = 0$	$M = 0$
6	1–4–5–2–3–6	120	$V = 0$	$M = 0$	$V = 0$	$M = 0$
8	1,2,7,8–3,4,5,6	180	$V = 0$	$M = 0$	$V = 0$	$M = 0$
8	1,2–3,4–7,8–5,6	90	$V = 0$	$M = 6\cos\theta - 2\sin\theta$	$V = 0$	$M = 0$
V Cylinder Engines[2]						
2	1,2 2α block angle	2α; 720 – 2α	$V = 2\cos\theta \cos^2\alpha$; $H = 2\sin\theta \sin^2\alpha$	$M = 0$	$V = 2\cos 2\theta \cos^2\alpha \cos 2\alpha$; $H = 2\sin 2\theta \sin^2\alpha \sin 2\alpha$	$M = 0$
4	1–3–2–4 V60°, 4 throws	180	$V = 0$	$M = \cos\theta$	$V = 2\sqrt{3}\cos 2\theta$	$M = 3\cos 2\theta$
6	1–4–5–2–3–6 V60°, 6 throws	120	$V = 0$	$M = 3/2\cos\theta$	$V = 0$	$M = 3/2\cos 2\theta$
8	1,2,7,8–3,4,5,6 V90°	90	$V = 0$; $H = 0$	$M = 0$	$V = 0$; $H = 4\sqrt{2}\cos 2\theta$	$M = 0$
8	1,2–5,6–7,8–3,4 V90°	90	$V = 0$; $H = 0$	$M = 0$	$V = 0$; $H = 0$	$M = 0$
12	1,2,11,12–5,6,7,8–3,4,9,10 V60°	60	$V = 0$; $H = 0$	$M = 0$	$V = 0$; $H = 0$	$M = 0$

Notes:
1. The crank pins are distributed equally around a circle; 1,4–2,3 means that crank pins 1 and 4 are at 12 o'clock when pins 2 and 3 are at 6 o'clock. Consecutively numbered crank pins can, of course, share a journal.
2. Odd-numbered pins will correspond to 1 bank of cylinders, and even numbers to the other bank.

cell is only approximately 1 volt, many cells (~250) are assembled back to back, to form a stack. This is achieved by using bipolar plates, in which the flow field plates have gas passages on each side, so they connect to a cathode on one side and an anode on the other side.

The electrochemical reactions are described by the following:

$$\begin{aligned} \text{Anode} &\quad - \quad 2H_2 \rightarrow 4H^+ + 4e \\ \text{Cathode} &\quad - \quad O_2 + 4H^+ + 4e^- \rightarrow 2H_2O \end{aligned} \qquad (2.81)$$

To provide electricity, the electrons must flow through the external circuit; therefore, the electrolyte separating the anode and cathode must prevent the flow of electrons but permit the flow of hydrogen ions (H^+). Because hydrogen ions are also known as protons, the electrolyte often is referred to as a proton exchange membrane (PEM). The PEM is bonded to the

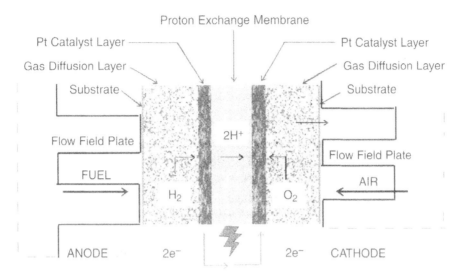

Figure 2.41. *Solid polymer fuel cells (SPFC), showing the proton exchange membrane (PEM) and the membrane electrode assembly (MEA).*

electrodes with a catalyst at the interface to form a membrane electrode assembly (MEA). The PEM is based on PTFE (polytetrafluoroethylene—Teflon®), with side chains terminating in sulfonic acid (HSO_3). The HSO_3 group is ionically bonded, with the end of the side chain being SO_3^-, and the H^+ (proton) being held by strong electrostatic attraction. Such polymers can be referred to as ionomers. The sulfonic acid is strongly hydrophilic. Thus, it tends to absorb water, and within these moist regions, the protons are relatively weakly bonded and are able to drift through the ionomer when there is a voltage gradient. This is analogous to having delocalized electrons in a metal, which can permit electrons to flow through the material when connected to an external circuit with a potential difference

Solid polymer fuel cells operate at approximately 80°C (176°F). Therefore, to increase the rates of the electrochemical reactions, it is necessary to use very active catalysts. Platinum or a platinum-based catalyst is used at the anode, and platinum is the only catalyst that can withstand the corrosive environment at the cathode. The catalyst particles are supported on particles of carbon black to increase the dispersion and therefore the surface area for the electrochemical. The SPFC is able to reach its operating temperature (~80°C) relatively quickly from ambient conditions, and that characteristic makes it suitable for transportation applications. It also produces power at high densities, which allows it to satisfy the volume and mass restrictions in a vehicle. Ballard Power Systems' SPFC have achieved power densities of 1000 W/ℓ and 700 W/kg for fuel cell stacks operating on hydrogen and air at practical conditions for vehicles.

2.10.2 Solid Polymer Fuel Cell (SPFC) Efficiency

As with any thermodynamic system, the maximum work that can be extracted from a given change of state is the change in the Gibbs energy. In the case of an electrical system, the work done is the product of voltage and charge. (Power is, of course, the product of voltage and current, and current is the rate of flow of charge.) Thus,

Maximum work $\quad\quad\quad \Delta G = -zFE \quad\quad\quad$ (2.82)

where

ΔG = difference in molar Gibbs energy per kilomole of reactants
z = number of electrons per kilomole of reactants
F = Faraday constant, 96.49×10^{-6} C/kmol; the charge on a kilomole of electrons
E = reversible open circuit voltage
  −ve as the charge on an electron is negative

Equation 2.81 shows that two electrons are released when a kilomole of hydrogen is reduced to two protons, and at 80°C, when the reactants enter and leave in the gaseous phase, $\Delta G° = -226.1$ MJ/kmol H_2 (the ° denotes a datum pressure of 1 bar). Rearranging Eq. 2.82 gives

$$E° = -226.1 \times 10^{-6} / \left(-2 \times 96.49 \times 10^{-6}\right) = 1.17 \text{ V}$$

When a kilomole of reactants is converted to products, then the energy released will correspond to the change in enthalpy, ΔH, and according to the First Law of Thermodynamics, the energy not converted to work will leave as heat. If the cell operating voltage is V, then the quantity of energy leaving as heat, Q, also will include that due to the voltage loss within the cell

$$Q = (\Delta G - \Delta H) + \Delta G (E - V)/E \quad\quad\quad (2.83)$$

Equation 2.83 can be simplified to

$$Q = \Delta H (E_H - V), \quad \text{where} \quad E_H = \Delta H / (-z F) \quad\quad\quad (2.84)$$

and the fuel cell efficiency is

$$\eta = V/E_H \quad\quad\quad (2.85)$$

When the reactants enter and leave in the gaseous phase at 80°C, $\Delta H = -237.5$ MJ/kmol H_2, and $E_H = 1.23$ V.

It will be seen later that when a fuel cell is operating at maximum power, then approximately equal quantities of heat and work will leave the system. This places much greater demands on the vehicle cooling system because, although the ratio of heat to work is comparable with that of an internal combustion engine, there will be a much smaller temperature difference between the heat exchanger and the environment if the cell is operating at 80°C.

The pressure dependence of Gibbs energy (G = H − TS) is due to the pressure dependence of entropy. For the generalized reaction,

$$aA + bB \leftrightarrow cC + dD$$

this leads to a pressure dependence on the reversible fuel cell voltage, which is described by the Nernst equation

$$E = E° - \frac{RT}{zF} \ln \frac{P_A^a P_B^b}{P_C^c P_D^d} \tag{2.86}$$

where

$$E° = \frac{-\Delta G}{zF}$$

z = number of electrons
P = partial pressure of the species

The reversible open circuit voltage (E) has only a slight dependence on pressure. However, more significantly, raising the pressure increases the mass transfer rates so that the specific output of a cell can be increased.

Of course, the reversible open circuit voltage is never obtained, and the reasons for the voltage losses are discussed next. These losses are called variously overvoltage or overpotential, polarization, irreversibility, losses, or voltage drop. There are four principal causes of these voltage drops: activation losses, fuel crossover and internal currents, ohmic losses, and mass transport or concentration losses.

2.10.2.1 Activation Losses

In 1905, Tafel found (by experiments) that the voltage loss at an electrode followed a similar mathematical model for many electrochemical reactions. He developed plots known as Tafel plots (Fig. 2.42), which are plots of voltage loss against log current. Tafel noted these curves could be modeled by the following equation:

$$V = A \ln \left(\frac{i}{i_0} \right) \tag{2.87}$$

The constant A can be found from the gradient of the measured voltage and i_0 the intersect with the X axis.

This equation is valid when the current flowing (i) is greater than i_0. The gradient of a Tafel plot A typically is approximately 0.03 V for an SPFC and is dependent on the speed of the

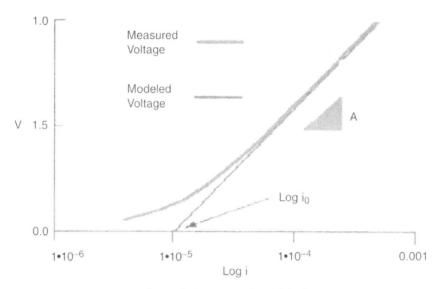

Figure 2.42. *Sample Tafel plot.*

reaction. The more significant Tafel variable is known as the exchange current i_0. This factor is linked to the reversibility of the electrode reactions. If no current is flowing, then the hydrogen electrode reaction may be assumed not to be taking place. However, in reality, the reaction is occurring reversibly in both directions at an equilibrium rate. If this equilibrium current flow is occurring at a high rate, it will be easier to shift the reaction to one side to allow a current to flow. If this equilibrium reaction is occurring at a low rate, it is more difficult to shift the equilibrium rate, and this means it is desirable for i_0 to be as large as possible. This increase in i_0 is achieved by changing the material of the electrode (Table 2.3). The surface area of an electrode also plays a large part.

TABLE 2.3
i_0 FOR THE HYDROGEN ELECTRODE
(BLOOM, 1981)

Metal	I_0 (A·cm^{-2})
Silver (Ag)	4.0×10^{-7}
Nickel (Ni)	6.0×10^{-6}
Platinum (Pt)	5.0×10^{-4}

Figures for the oxygen electrode are much smaller (10^{-8} A/cm^2; Appelby and Foulkes, 1989) and therefore can be ignored. If this voltage loss is taken into account, the theoretical voltage current graph is given by the theoretical voltage (E) minus the activation loss (Tafel equation)

$$V = E - A \ln\left(\frac{i}{i_0}\right) \tag{2.88}$$

This voltage model (Eq. 2.88) is plotted in Fig. 2.43.

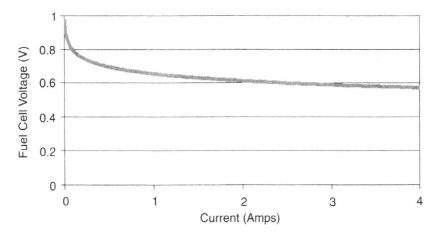

Figure 2.43. *Theoretical model of fuel cell voltage using the Tafel loss equation. (E = 1.2 V, A = 0.06 V, and i_0 = 0.04 mA·cm^{-2})*

2.10.2.2 Fuel Crossover and Internal Currents

The electrolyte in a fuel cell is designed to allow only the mobile ion species to conduct across it. In the case of an SPFC, the PEM allows the H$^+$ ions to conduct. The PEM is never ideal and will allow some of the hydrogen fuel and also the electrons to conduct across the membrane. These two leaks have a noticeable effect on the current voltage curve.

Figure 2.44 shows how an electron conducting through the electrolyte bypasses the external circuit and does no electrical work. Similarly, if a hydrogen molecule passes across the electrolyte, it will either react to form water or disperse in the oxidant flow, wasting two electrons.

With low-temperature fuel cells such as the SPFC, the open circuit voltage is smaller than the theoretical value of 1.2 V. This open circuit voltage drop is caused by the fuel and electron leakage. This leakage acts as a current, flowing even when the cell is open circuit; thus, the current flow is not zero but approximately 2 mA/cm^2 (Larminie and Dicks, 2000). This current flow leads to an activation loss of 0.23 V

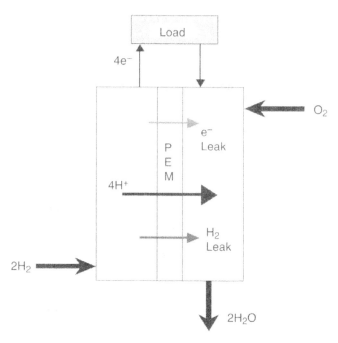

Figure 2.44. *Diagram showing fuel crossover and internal current losses in a solid polymer fuel cell (SPFC).*

$$A = 0.06 \text{ V}$$

$$i_0 = 0.04 \text{ mA}$$

$$V_{\text{activation loss}} = 0.06 \ln\left(\frac{2}{0.04}\right) = 0.23 \text{ V}$$

This loss leads to a lower open circuit voltage of approximately 0.97 V. This loss can be modeled mathematically by adding an extra current term (i_n) into the activation loss equation (and any other loss equation containing i), giving a theoretical voltage equation of

$$V = E - A \ln\left(\frac{i + i_n}{i_0}\right) \tag{2.89}$$

Using different membrane materials and designing them to resist fuel leakage and electron conduction can reduce this loss. When there is a large activation loss (as with SPFC), then a small change in leakage has a large effect on the open circuit voltage.

2.10.2.3 Ohmic Losses

These voltage losses occur due to resistance (r) in the fuel cell, the electrodes, electrolyte, and cell connections, and simply follow Ohms law:

$$V_{loss} = i.r \qquad (2.90)$$

2.10.2.4 Mass Transfer Losses

These losses occur mainly when mixed fuels or oxidants are used, such as air to provide oxygen as the oxidant, or a hydrogen/carbon dioxide fuel mixture. Here, there are gas species that do not take an active part in the fuel cell reaction. At the surface of the PEM, the hydrogen and oxygen are used up in the production of water. This reduces the concentration of the fuel and oxidant gases at the active membrane surface. These hydrogen and oxygen depleted regions are replenished by diffusion of the reactant gases through the inactive species. This depleted region causes a reduction in the partial pressures of the hydrogen and oxygen. It can be shown that pressure has an effect on the fuel cell voltage by considering the Nernst equation; therefore, mass transfer or concentration losses can be modeled by the following equation (which takes negative values):

$$\text{Voltage drop due to mass transfer} = B \ln\left(1 - \frac{i}{i_1}\right) \qquad (2.91)$$

The effect of mass transfer losses means that at high current drains, the voltage suddenly reaches a cutoff point, where the current drawn cannot increase and the voltage collapses, as shown in Fig. 2.45.

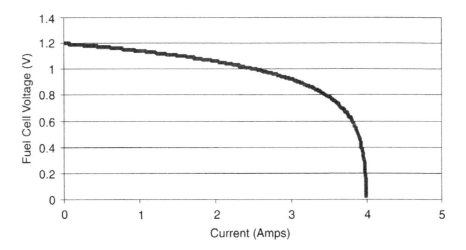

Figure 2.45. *Theoretical current/voltage plot, showing mass transfer losses ($E = 1.2$ V, $B = 0.2$ V, and $i_1 = 4$ A).*

This effect also occurs with pure gases due to pressure drops in the fuel cell system because there is a maximum flow rate to the membrane surface. If the current drawn demands a greater flow rate than this, then the voltage will drop off because the required gas flow rate cannot be supplied. Also, water vapor is being produced at the cathode; therefore, oxygen must diffuse through the water.

2.10.2.5 Overall Response

The overall response can be modeled by combining the loss terms in Eqs. 2.88 to 2.91:

$$V = E - (i+i_n)r - A \ln\left(\frac{i+i_n}{i_0}\right) + B \ln\left(1 - \frac{i+i_n}{i_1}\right) \tag{2.92}$$

The high capital cost and bulk of fuel cells means that they frequently are operated at the maximum power density, giving a cell voltage of approximately 0.6 V for solid polymer fuel cells.

Using the data in Table 2.4 results in the voltage current characteristic (often called a polarization curve) as shown in Fig. 2.46. Also plotted in Fig. 2.46 are the power density and efficiency (based on the lower calorific value of hydrogen), which shows how the maximum power density occurs at 3.74 A/cm² with an operating point of 0.5 V and an efficiency of 0.41.

**TABLE 2.4
COEFFICIENTS TO DEFINE
LOSSES IN FUEL CELLS**

Constant	SPFC
E/volts	1.2
i_n/mA·cm^{-2}	2
r/kΩ·cm^{-2}	60 ×10^{-6}
i_0/mA·cm^{-2}	0.02
A/volts	0.03
B/volts	0.08
I_1/mA·cm^{-2}	5000

2.10.3 Sources of Hydrogen for Solid Polymer Fuel Cells (SPFC)

Hydrogen can be produced by electrolysis. Although this should be reversible, in practice it has an efficiency of only approximately 80%. The losses are for the same reasons as there are losses in fuel cells. Using electricity, of course, raises the question of emissions when the electricity has been generated from fossil fuels.

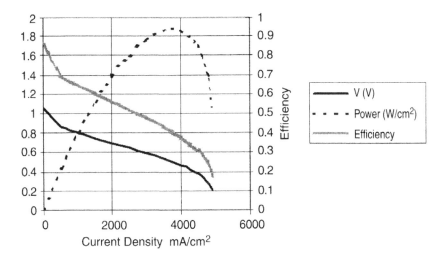

Figure 2.46. *Polarization curve, power density, and efficiency of a solid polymer fuel cell (SPFC).*

A more practical source, and one that has better energy utilization, is fuel reforming, of which there are three systems: steam reforming, partial oxidation reforming, and autothermal reforming. Each of these will be discussed in turn.

2.10.3.1 Steam Reforming (SR)

Widely used throughout the petrochemical industry, steam reforming (SR) endothermically combines a vaporized hydrocarbon with steam over a catalyst (often nickel supported on alumina) at high temperatures (700–1000°C).

$$C_xH_y + 2xH_2O \leftrightarrow xCO + (y/2 + 2x)H_2 \Rightarrow CO, CO_2, H_2, H_2O \qquad (2.93)$$

During steam reforming, the water–gas shift reaction also takes place, thereby converting the hydrogen in water directly to hydrogen gas

$$CO + H_2O_{(g)} \leftrightarrow CO_2 + H_2 \qquad (2.94)$$

The most common design of SR for industrial use is the tubular reformer, a furnace containing several tubes filled with catalysts through which the reactants pass. The reaction inside the tubes obtains heat from a flame external to the tubes. Tubular reformers present one main drawback for small PEM systems. They are not easily miniaturized.

2.10.3.2 Partial Oxidation (POX) Reforming

In a partial oxidation (POX) reaction, the hydrocarbon is reacted with an amount of oxygen below the stoichiometric amount

$$C_xH_y + x/2\,O_2 \leftrightarrow xCO + y/2\,H_2 \qquad (2.95)$$

During partial oxidation, a small fraction of steam may be added, both to prevent carbon deposition on reformer surfaces and to reduce carbon monoxide production

$$CO + H_2O_{(g)} \leftrightarrow CO_2 + H_2 \Rightarrow CO,\ CO_2,\ H_2,\ H_2O \qquad (2.96)$$

The primary disadvantage of the POX reaction is that it may waste a significant portion of the energy in the fuel as heat, if this heat is not recovered. Its primary advantages are as follows:

1. It is exothermic and therefore self-sustaining.

2. It can be used on fuels with a variety of chemical components because catalysts are not required.

A POX reaction may be either non-catalytic or catalytic.

2.10.3.3 Autothermal Reforming (AR)

Autothermal reforming (AR) combines the processes of steam reforming (SR) and partial oxidation (POX). By combining the hydrocarbon with a mixture of oxygen and steam, the reformer uses the exothermic reaction of the hydrocarbon and oxygen to provide heat for the endothermic reaction of the hydrocarbon and steam.

$$C_xH_y + zH_2O + (x - z/2)O_2 \leftrightarrow xCO_2 + (z + y/2)H_2 \Rightarrow CO,\ CO_2,\ H_2,\ H_2O \qquad (2.97)$$

For methane, the change in enthalpy for the preceding reaction is zero when z is equal to 1.115. The primary benefits of autothermal reforming are as follows:

1. Its high thermal efficiency (as a result of combining reactions)

2. Its compactness (as a result of avoiding a separate combustion stream to provide heat as with a steam reformer)

3. Its rapid startup and quick load following (as a result of combining reactions)

The primary disadvantage is its lower hydrogen yield—only 42 to 48% hydrogen content in the output gas stream (dry basis)—compared with steam reforming, which produces 75 to 80%

hydrogen (dry basis). The low hydrogen yield results from the intake of nitrogen, which dilutes the output gas.

Onboard reformation is invariably by autothermal reformation because of its rapid startup, whereas stationary production of hydrogen usually is by steam reformation because of its higher thermodynamic efficiency. In the short-term, methanol (CH_3OH) is the fuel favored for onboard reformation because this occurs at approximately 300°C (572°F) compared with approximately 900°C (1652°F) required for gasoline.

2.10.3.4 Carbon Monoxide Clean-Up and Solid Polymer Fuel Cell (SPFC) Operation on Reformed Fuel

The hydrogen produced by fuel reformation is also mixed with other gases, and a typical composition resulting from methanol reformation is 43% H_2, 15% CO_2, 21% H_2O, 3.3% CO, and 17.4% N_2. The presence of other gases lowers the partial pressure of hydrogen at the anode, and this results in approximately a 10% reduction in the maximum power output of an SPFC. However, a fuel cell that is not operating on pure hydrogen needs to be supplied with significantly more hydrogen than is utilized. The gas mixture will flow through the channels in the bipolar plate, and as the hydrogen is utilized from the gas mixture, the partial pressure of the hydrogen will fall along the length of the channel. This would result in a reduced voltage being generated, and any higher potential generated elsewhere on the plate would be "pulled down" to this potential with resistive losses within the plate. By providing a surplus of hydrogen, this "dilution" loss is reduced, albeit at the expense of unreacted hydrogen leaving the fuel cell. In the absence of fuel crossover and internal currents, then the current is linearly proportional to the molar consumption rate of hydrogen

$$I = z F M \qquad (2.98)$$

where M is the molar flow rate.

If a fuel cell is operating with 25% extra hydrogen (a typical value), this is referred to as a lambda (λ) of 1.25. The equation for the fuel cell efficiency (Eq. 2.85) now must be modified to

$$\eta = V/(E_H \lambda) \qquad (2.99)$$

Similarly with the cathode, when it is operating on air (as opposed to oxygen), then it likewise must be supplied with excess oxygen, and the same definition of lambda (λ) (Eq. 2.98) is used. With the cathode, the only penalty of providing excess air is the need to compress the air when a fuel cell is operating above ambient pressure. However, some of the compression work can be recovered if the exit gases from the anode and cathode are reacted together (to raise their temperature) and then expanded in an expander (e.g., a turbine).

A more serious problem with a fuel cell operating on hydrogen produced from fuel reformation is that any carbon monoxide (CO) is preferentially adsorbed onto the surface of the catalyst, thereby preventing the hydrogen from reacting. Any more than a few parts per million of carbon monoxide will "poison" the catalyst; therefore, it is essential to remove the carbon monoxide. The equilibrium level of carbon monoxide is determined by the water–gas shift reaction

$$CO + H_2O_{(g)} \leftrightarrow CO_2 + H_2 \qquad (2.94)$$

Fortunately, the equilibrium of this system moves to the right as the temperature is reduced, which reduces the quantity of CO present and increases the level of hydrogen. However, as the temperature is reduced, the rates of reaction reduce. Even if a catalyst were able to assist in achieving equilibrium, the carbon monoxide level would still be too high. Instead, the carbon monoxide level is reduced by selective oxidation, which relies on operating a platinum-based catalyst in a narrow temperature range within which the oxidation rates for carbon monoxide are much higher than those for hydrogen. Selective oxidation relies on a small quantity of air being added to the reformate, and the mixture being passed through a selective oxidation catalyst and then cooled, before the process is repeated in subsequent stages in which further oxidation and cooling occur. One advantage of using hydrogen produced by onboard reformation is that the moisture levels are high enough not to need additional humidification.

2.10.3.5 Hydrogen Storage

Many people believe that onboard fuel reformation is only a short-term solution before hydrogen is supplied at filling stations. The options for hydrogen storage are as a liquid (at −253°C [−423°F]), as a high-pressure gas (250 bar or higher), or adsorbed into a material. Norbeck et al. (1996) provide a comprehensive overview of using hydrogen for surface transportation, and Larminie and Dicks (2000) provide a useful overview. The options are summarized next.

Liquid storage must minimize the boil-off loss (typically approximately 1% per day) and have a means of handling this safely. The liquid density is only 71 kg/m^3, and with a net calorific value of 120 MJ/kg, the energy density is only 8.5 MJ/L, ignoring the mass and volume of the storage system. (This is approximately a factor of four lower than any hydrocarbon.) The volume of the tank typically is double the volume of the liquid, and its mass might be four to five times the mass of the hydrogen. Therefore, the energy storage density typically is approximately 24 MJ/kg or 4.25 MJ/L. Also, a significant amount of energy is needed to liquefy hydrogen (approximately 16% of its calorific value in a thermodynamically reversible process). Thus, the overall "well-to-wheel" efficiency of a hydrogen-fueled system is inherently poor.

At 250 bar, hydrogen has a density of approximately 22.5 kg/m^3. By the time the volume and mass of the tank are considered, the energy storage density typically is approximately 4 MJ/kg or 1.7 MJ/L. This is lower than that for liquid hydrogen, but the work for reversible isothermal compression to 250 bar is only approximately 6% of its calorific value, which is much lower than the energy required for liquefaction. Furthermore, the irreversibilities associated with liquefaction are likely to be much greater than those for gas compression.

The final option is to use a material that absorbs hydrogen. The simplest approach is to use active charcoal to adsorb hydrogen. For a given temperature and pressure, the mass of hydrogen per liter is greater than that for the compressed gas, despite the volume occupied by the carbon. Young (1992) reports that a liter of activated carbon at 150 K and a pressure of 56 bar can store 14 g (0.5 oz) of hydrogen. Related to this is the use of carbon nanofibers that are claimed to store up to three times their own mass of hydrogen. However, tests have been limited only to samples of a few grams, so clearly more work is needed (NEL, 1999). A well-established technique for storing hydrogen is the use of metal hydrides such as titanium iron hydride ($TiFeH_2$). There is a reversible reaction, controlled by pressure, in which

$$TiFeH_2 \Leftrightarrow TiFe + H_2 \qquad (2.95)$$

Hydrogen is absorbed when the system pressure is raised, and it is released when the pressure is reduced. When hydrogen is absorbed, the reaction is exothermic; thus, it is necessary to cool the system. Conversely, when hydrogen is released, heat is absorbed; therefore, it is necessary to heat the hydride store. This provides a useful safety feature because if a sudden loss of pressure occurred, the rate of hydrogen release is governed by the heat supply. This is a very efficient way to store hydrogen, with 1 kg (2.2 lb) stored in 9.8 liters (2.6 gal) (a greater "density" than with liquid hydrogen). However, the system is very heavy, with only about 1.9% hydrogen by mass.

Larminie and Dicks (2000) suggest the data in Table 2.5 for different hydrogen storage systems.

TABLE 2.5
ENERGY DENSITIES OF DIFFERENT HYDROGEN STORAGE SYSTEMS (ADAPTED FROM LARMINIE AND DICKS, 2000)

Method	Gravimetric Storage Efficiency, % Mass Hydrogen	Volumetric Storage Efficiency, Mass of Hydrogen per Unit Volume (kg/L)
Pressurized Gas	0.7–3.0	0.015
Reversible Metal Hydride	0.65	0.06
Cryogenic Liquid	14.2	0.04
Onboard Methanol Reformer	13.9	0.055

2.10.4 Hydrogen Fuel Cell Systems

The problems of onboard hydrogen storage (in terms of both the gravimetric and volumetric storage efficiency), the lack of a hydrogen infrastructure, and the energy costs associated with hydrogen compression or liquefaction mean that onboard fuel reformation is likely to be used in any fuel cell vehicle that is proposed for production in the short term. Figure 2.47 shows a fuel cell system that has been simplified by not showing the internal heat exchangers.

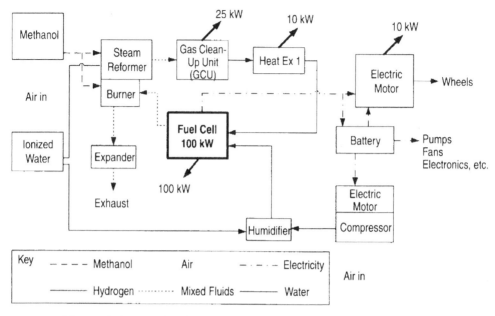

Figure 2.47. *Fuel cell system with a fuel cell and its fuel reformer.*

In Fig. 2.47, the steam reformer reacts the methanol to produce a mixture of mostly hydrogen, carbon dioxide, water vapor, and carbon monoxide. To prevent poisoning of the platinum catalyst within the fuel cell, the carbon monoxide must be selectively oxidized within the gas clean-up unit (GCU). The GCU must be cooled to control the selective oxidation, and further cooling is needed because the fuel cell is likely to operate at 80°C (176°F). This heat is likely to be used for preheating the reactants to the steam reformer, and the air after it is compressed but before it is humidified. The fuel cell is shown with an electrical output of 100 kW and an equal quantity of heat to be rejected. A further 10 kW is dissipated from the electric motor and its control system. Remember that the fuel cell is likely to be operating at 3 bar absolute. Thus, in addition to compressing the air, it is necessary to pump the methanol and water (prior to evaporation) to this pressure.

The anode and cathode off-gases leave the fuel cell and enter a catalytic burner where any unreacted hydrogen from the anode is reacted, so that the exhaust stream is composed solely of carbon dioxide, nitrogen, water vapor, and oxygen. Because this is hot and at a high pressure, it can be expanded in a turbine to produce work. The exhaust gases then are likely to be cooled, so that water can be condensed to eliminate the need for storing (and replenishing) large quantities of water that might freeze.

Finally, it is necessary to consider the overall system efficiency and to make a comparison with internal combustion engines. If the parasitic losses in the fuel cell system (including electrical losses) are taken to be constant at 10% of the maximum output, then the data in Fig. 2.46 for fuel cell performance can be converted to a system efficiency. This has been done in Fig. 2.48, which also utilizes the engine data from Fig. 2.8.

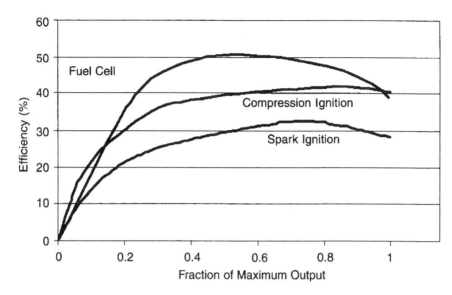

Figure 2.48. *The engine data from Fig. 2.8 have been normalized as a fraction of the maximum output, and the fuel cell data from Fig. 2.46 have been added, assuming a constant parasitic loss that corresponds to 10% of the maximum output.*

To enable a comparison to be made, the efficiency data is plotted against the fraction of maximum output. As with an engine, the fuel cell now has zero efficiency at zero power output. The typical operating point for a car is at 20% of maximum load, whereas a truck or bus has a typical operating point of 40% load.

Figure 2.48 still does not allow a fair comparison to be made because no account has been taken of the energy costs associated with producing and distributing the fuel. These production and distribution costs are shown for different vehicle configurations in Table 2.6, leading to the concept of well-to-wheel efficiency. In the case of gasoline, the production and distribution energy cost amounts to approximately 3 MJ/kg, whereas for diesel, it is approximately 2 MJ/kg, and for fuel cell vehicles, it is very dependent on the source of hydrogen.

When the production and distribution costs are considered, the data in Fig. 2.48 can be plotted again as Fig. 2.49 on a well-to-wheel efficiency basis.

Figure 2.49 shows that a methanol-fueled fuel cell has an efficiency that falls between that of spark ignition and compression ignition engines, whereas the efficiency of a gasoline-fueled fuel cell is comparable to that of a diesel engine. In some ways, this comparison is unfair to the spark ignition and compression ignition engines. These engines are being operated at constant speed, while the power output also can be varied by adjusting the speed. This will tend to flatten the output/efficiency graph around the maximum efficiency. However, the extent of the improvements will depend on the transmission system. Therefore, a sensible comparison can be made only by modeling a complete vehicle over a drive cycle, as discussed

**TABLE 2.6
COMPARISON OF WELL-TO-WHEEL EFFICIENCIES OF
DIFFERENT VEHICLE CONFIGURATIONS FOR A 1300-KG (2866-LB) CAR
OVER THE NEW EUROPEAN DRIVING CYCLE (NEDC)
(ARMSTRONG, 2000)**

	Energy Consumption (MJ/km)			CO_2 Emissions (g/km)		
	Vehicle	Production and Distribution	Total	Vehicle	Production and Distribution	Total
Internal Combustion Engines:						
Gasoline Direct Injection	2.4	0.4	2.8	170	20	190
Diesel Direct Injection	1.9	0.3	2.2	140	20	160
LPG—SI	2.75	0.35	3.1	175	20	195
CNG—SI	2.85	0.2	3.05	160	15	175
Parallel Hybrid Electric Vehicles:						
Gasoline	2.1	0.3	2.4	140	25	165
Diesel	1.7	0.2	1.9	130	10	140
Fuel Cell Electric Vehicles:						
Methanol	1.7	0.8	2.5	120	35	155
Gasoline	1.8	0.4	2.2	130	15	145
Compressed Hydrogen	1.25	0.85	2.1	0	135	135

Notes:
LPG = Liquefied Petroleum Gas
CNG = Compressed Natural Gas
SI = Spark Ignition

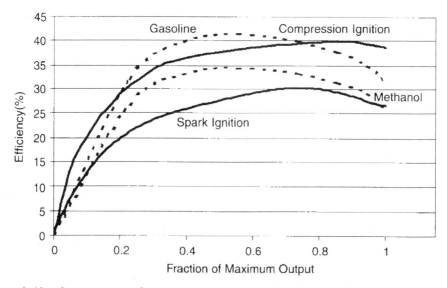

Figure 2.49. *Comparison of engine and fuel cell efficiencies on a well-to-wheel basis. The broken lines are for fuel cells.*

2.11 Concluding Remarks

The efficiency and specific output of gasoline and diesel engines continues to improve, despite the impact of increasingly stringent emissions legislation. Hybrid electric vehicles can utilize existing technology to reduce vehicle energy consumption, but at an increased cost because of the additional technology employed. Fuel cells can achieve an overall energy efficiency approaching that of hybrid electric diesel engined vehicles, but with a potential for a lower cost.

This chapter has concentrated on the fundamentals of gasoline and diesel engines because these dominate vehicle use. The mechanical design aspects have been reviewed, with the underlying thermodynamics and fluid mechanics. The material presented here is common mostly to both types of reciprocating engines, namely, the underlying thermodynamics of their cycles, the operation of two- and four-stroke engines, and their mechanical operation. The discussion of ignition and combustion provides a comparison between the two types of engines and results in the complementary nature of the fuel requirements. Chapters 3 and 4 are devoted to spark ignition and diesel engine technology in greater depth. For convenience, supercharging is discussed in the context of spark ignition engines, whereas turbocharging is treated with diesel engines, although there are, of course, supercharged diesel engines and turbocharged spark ignition engines.

2.12 Problems

2.1 For the ideal air standard diesel cycle with a volumetric compression ratio of 17:1, calculate the efficiencies for cutoff ratios of 1, 2, and 4. Take $\gamma = 1.4$. The answers can be checked with Fig. 2.5.

2.2 Outline the shortcomings of the simple ideal cycles, and explain how the fuel-air cycle model overcomes these problems.

2.3 A 2.5-liter four-stroke naturally aspirated direct injection diesel engine is designed to run at 4500 rpm with a power output of 45 kW; the volumetric efficiency is found to be 80%. The bsfc is 0.071 kg/MJ, and the fuel has a calorific value of 42 MJ/kg. The ambient conditions for the test were 20°C (68°F) and 1 bar. Calculate the bmep, the overall efficiency, and the air-fuel ratio.

2.4 A twin-cylinder two-stroke engine has a swept volume of 500 cm^3. The maximum power output is 60 kW at 9000 rpm. At this condition, the bsfc is 0.11 kg/MJ, and the gravimetric air-fuel ratio is 12:1. If the ambient test conditions were 10°C (50°F) and 1.03 bar, and the fuel has a calorific value of 44 MJ/kg, calculate the bmep, the arbitrary overall efficiency, and the volumetric efficiency.

2.5 A four-stroke 3-liter V-6 spark ignition engine has a maximum power output of 100 kW at 5500 rpm, and a maximum torque of 236 Nm at 3000 rpm. The minimum bsfc is 0.090 kg/MJ at 3000 rpm, and the air flow rate is 0.068 m^3/s. The compression ratio is 8.9:1, and the mechanical efficiency is 90%. The engine was tested under ambient conditions of 20°C (68°F) and 1 bar. Take the calorific value of the fuel to be 44 MJ/kg.

a. Calculate the power output at 3000 rpm and the torque output at 5500 rpm.

b. Calculate for both speeds the bmep and the imep.

c. How does the arbitrary overall efficiency at 3000 rpm compare with the corresponding air standard Otto cycle efficiency?

d. What is the volumetric efficiency and air-fuel ratio at 3000 rpm?

2.6 A four-cylinder, four-stroke gasoline engine is to develop 50 kW at 50 rev/s when designed for a volumetric compression ratio of 10.0:1. The ambient air conditions are 1 bar and 18°C (64°F), and the calorific value of the fuel is 44 MJ/kg.

a. Calculate the specific fuel consumption in grams per kilowatt hour (g/kWh) of brake work if the indicated overall efficiency is 50% of the corresponding air standard Otto cycle, and the mechanical efficiency is 90%. The specific heat capacity ratio for air is 1.4.

b. The required gravimetric air-fuel ratio is 14.5, and the volumetric efficiency is 90%. Estimate the required total swept volume and the cylinder bore if the bore is to be equal to the stroke. Calculate also the brake mean effective pressure.

2.7 A compression ignition engine has a volumetric compression ratio of 16. Find the thermal efficiency of the following air standard cycles having the same volumetric compression ratio as the engine. The specific heat capacity ratio for air is 1.4.

a. An Otto cycle.

b. A diesel cycle in which the temperature at the beginning of compression is 18°C (64°F) and in which the heat supplied per unit mass of air is equal to the energy supplied by the fuel (in terms of its calorific value). The gravimetric air-fuel ratio is 28:1; the calorific value of the fuel is 44 MJ/kg. Assume the specific heat of air at constant pressure is 1.01 kJ/kg-K and is independent of temperature.

2.8 A gasoline engine of volumetric compression ratio 11:1 takes in a mixture of air and fuel in the ratio 14.5:1 by mass; the calorific value of the fuel is 44 MJ/kg. At the start of compression, the temperature of the charge is 50°C (122°F). Assume that compression and expansion are reversible with pv^n constant, and n = 1.325 and 1.240, respectively, and that combustion occurs instantaneously at minimum volume. Combustion can be regarded as adding heat equal to the calorific value to the charge.

However, there is a finite combustion efficiency, and heat transfer from the combustion chamber. Thus, combustion is equivalent to a net heat input that corresponds to 75% of the calorific value of the fuel being burned.

Calculate the temperatures after compression and at the start and end of expansion. Calculate the net work produced by the cycle, and evaluate the indicated efficiency of the engine. Why is it inappropriate to calculate the indicated efficiency in terms of the heat flows?

Use the following thermodynamic data:

	Molar Mass, kg	Specific Heat Capacity at Constant Volume c_v, kJ/kg-K
Air-Fuel Mixture	30	0.95
Combustion Products	28	0.95

2.9 What are the fundamental differences between the ignition and combustion processes in conventional spark ignition and compression ignition engines, and how does this impact their fuel requirements?

2.10 Why does the efficiency of an internal combustion engine fall as the load is reduced, and why is the rate of fall less severe for a compression ignition engine?

2.11 Explain how varying the air-fuel ratio on a spark ignition engine varies the power output, efficiency, and emissions at a constant speed, throttle setting, and ignition timing.

2.12 For a four-stroke five-cylinder engine, show that the primary and secondary forces and moments all can be balanced if the spacing between the cylinder axes 2–3–4 is 1.618 times the spacing of 1–2 or 4–5.

2.13 Using the data in Table 2.3 and Eq. 2.92, reconstruct the data plotted in Fig. 2.46. Then conduct a sensitivity analysis by assuming that the parasitic losses are variously 5%, 10%, and 15% of the maximum power output. Plot an efficiency/output graph similar to Fig. 2.48.

2.14 Discuss the advantages and disadvantages of the different methods of providing onboard hydrogen for a fuel cell.

Chapter 3

Spark Ignition Engines

3.1 Introduction

This review of spark ignition engines starts by looking at how the spark is generated, how it ignites the mixture, the effect of ignition timing on engine performance, and how the engine operating point influences the optimum ignition timing. The control of ignition timing is discussed in Section 3.7, which deals with engine management systems. Mixture preparation (Section 3.3) is concerned solely with fuel injection systems, for both port injected engines and in-cylinder injection. The control of the air-fuel ratio is discussed in the context of engine management systems (Section 3.7).

The most important parameter that influences combustion is turbulence, and this is illustrated, with methods of controlling the in-cylinder motion, in Section 3.4. Also described are combustion system designs for both port injection and direct injection engines. The trend continues for increasing the specific output of engines, as well as the power of engines installed in a given size of vehicle. Section 3.6 illustrates the use of variable valve timing, variable geometry induction systems, and supercharging to increase the power output of engines. (Turbocharging is discussed in Section 4.5 in the context of diesel engines, but data for spark ignition engines are discussed in Section 4.5.3.)

The most significant trend has been the reduction of emissions, and this has been dictated by legislation. Section 3.5 starts with the development of three-way catalysts and includes the issues of light-off, durability, and catalysts that are able to reduce NO in an oxidizing environment. One of the key requirements for proper catalyst operation is close control of the air-fuel ratio, and this is one of the major topics of Section 3.7 (engine management systems). Section 3.7 commences by describing the various sensors used on engines and continues by describing how these sensors are used to control parameters such as the ignition timing, air-fuel ratio, and exhaust gas recirculation (EGR) level.

3.2 Spark Ignition and Ignition Timing

3.2.1 Ignition System Overview

Most engines have a single spark plug per cylinder, a notable exception being in aircraft where the complete ignition system is duplicated to improve reliability. The spark usually is provided by a battery and coil, although until the 1920s, a magneto often was used.

The spark plug (Fig. 3.1) requires its central electrode to operate in the temperature range 350–700°C (572–1292°F) for satisfactory performance. If the electrode is too hot, pre-ignition will occur. If the temperature is too low, carbon deposits will build up on the central insulator, causing electrical breakdown. The heat flows from the central electrode through the ceramic insulator; the shape of this determines the operating temperature of the central electrode. A cool-running engine requires a "hot" or "soft" sparking plug with a long heat flow path in the central electrode (Fig. 3.1a). A hot-running engine, such as a high-performance engine or a high-compression-ratio engine, requires a "cool" or "hard" sparking plug. The much shorter heat flow path for a "cool" spark plug is shown in Fig. 3.1b. The spark plug requires a voltage of 5–15 kV to spark. The larger the electrode gap and the higher the cylinder pressure, the greater the required voltage.

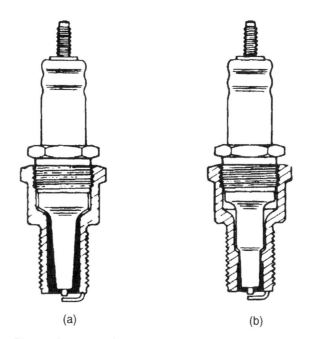

Figure 3.1. *Spark plugs that are (a) hot running and (b) cool running (Stone, 1999).*

Figure 3.2 shows a traditional coil ignition system, in which the coil is in effect a transformer with a primary or low-tension (LT) winding of approximately 200 turns, and a secondary or high-tension (HT) winding of approximately 20,000 turns of fine wire, all wrapped around an iron core. The voltage V induced in the HT winding is

$$V = M \frac{dI}{dt} \qquad (3.1)$$

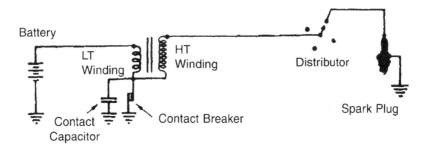

Figure 3.2. *A simple contact breaker-operated ignition system with a contact breaker and distributor for a four-cylinder engine (Stone, 1999).*

where

- I = current flowing in the LT winding
- M = mutual inductance = $k\sqrt{(L_1 \times L_2)}$
- L_1 and L_2 = inductances of the LT and HT windings, respectively (proportional to the number of turns squared)
- k = coupling coefficient (less than unity)

or

- V = k (turns ratio of windings) × (low-tension voltage)

When the contact breaker (or the transistor in an electronic switching circuit) closes to complete the circuit, a voltage will be induced in the HT winding. However, that voltage will be small because dI/dt is limited by the battery voltage and the inductance of the LT winding. Equation 4.9 defines the current flow in the LT winding as

$$I = \left(\frac{V_s}{R}\right)\left[1 - \exp\left(-\frac{Rt}{L_1}\right)\right] \quad (3.2)$$

where

- V_s = supply voltage
- R = resistance of the LT winding
- t = time after the application of V

When the contact breaker (or transistor) opens, dI/dt is much greater, and sufficient voltage is generated in the HT windings to jump the gaps between electrodes. A high voltage (200~300 V) is generated in the LT windings, and this energy is stored in the capacitor. Without the capacitor, severe arcing would occur at the contact breaker. When the spark has ended, the capacitor discharges. In a transistorized ignition system, a diode can be used to short-circuit the voltage generated in the LT winding and thus protect the transistor.

The energy input to the LT side (E_p) of the coil is the integration of the instantaneous current (I) and the supply voltage (V_s) over the period the coil is switched on

$$E_p = \int_0^{t'} I V_s \, dt \qquad (3.3)$$

where t' is the time at which the coil is switched off, the coil-on-time.

The energy stored in the coil (E_s) is less than the energy supplied because of the energy dissipated within the internal resistance of the coil

$$E_s = \frac{1}{2} L_1 I_p^2 \qquad (3.4)$$

where I_p is the LT current at the time when the coil is switched off.

The system shown in Fig. 3.2 requires a mechanical drive, both to operate the contact breaker and to rotate the rotor arm in the distributor. With an engine management system, the time to switch the transistor is determined digitally, and the need for a mechanical drive to the distributor can be eliminated if there is a separate ignition coil for each cylinder. A more elegant solution for four-stroke engines is to use a double-ended coil, as shown in Fig. 3.3, in which the contact breaker also has been replaced by a transistor.

In Fig. 3.3, each end of the HT winding is connected to a spark plug, and the coil will be fired every revolution. Thus, the spark plugs must be connected to cylinders that are 360° out of phase (e.g., in a four-cylinder engine, cylinders 1 and 4, or 2 and 3). This means that a spark

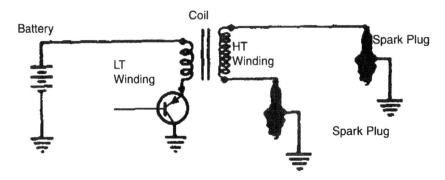

Figure 3.3. *A distributorless ignition system (Stone, 1999).*

also will be generated at each spark plug toward the end of its exhaust stroke, where it should have no effect. Thus, these systems sometimes are referred to as wasted spark systems. There may be slightly increased spark plug wear at the electrodes, but only half the number of HT coils and drive electronics is needed.

Another type of ignition system that can be used is capacitive discharge ignition (CDI, or CD ignition). The battery voltage is used to drive a charging circuit that raises the capacitor voltage to approximately 500 V. At ignition, the energy stored in the capacitor is discharged through an ignition transformer (that is, a coil with primary and secondary windings), the circuit being controlled by a thyristor. The discharge from the capacitor is such that a short-duration (approximately 0.1 ms) spark is generated. The rapid discharge makes this ignition system less susceptible to spark plug fouling. Because the primary voltage is higher than with an inductive system, the CD ignition coils have a lower inductance, and the reduced number of turns also gives a lower resistance. An interesting difference between inductive and capacitive ignition systems is that with inductive systems, the spark occurs when the coil is turned off, whereas for capacitive systems, the spark occurs when the coil is energized. The repetition rate can be much higher with CD systems; thus, these are commonly used in racing applications.

Whether the ignition is by battery and coil (positive or negative earth) or magneto, the HT windings usually are arranged to make the central electrode of the spark plug negative. The electron flow across the electrode gap comes from the negative electrode (the cathode), and the electrons flow more readily from a hot electrode. Because the central electrode is not in direct contact with the cylinder head, this is the hotter electrode. By arranging for the hotter electrode to be the cathode, the breakdown voltage is reduced.

3.2.2 The Ignition Process

The ignition process has been investigated thoroughly by Maly and Vogel (1978) and Maly (1984). The spark that initiates combustion may be considered in the three phases shown in Fig. 3.4:

1. **Pre-breakdown.** Before the discharge occurs, the mixture in the cylinder is a perfect insulator. As the spark pulse occurs, the potential difference across the plug gap increases rapidly (typically 10–100 kV/ms). This causes electrons in the gap to accelerate toward the anode. With a sufficiently high electric field, the accelerated electrons may ionize the molecules with which they collide. This leads to the second phase—avalanche breakdown.

2. **Breakdown.** When enough electrons are produced by the pre-breakdown phase, an over-exponential increase in the discharge current occurs. This can produce currents of the order of 100 A within a few nanoseconds. This is concurrent with a rapid decrease in the potential difference and electric field across the plug gap (typically to 100 V and 1 kV/cm, respectively). Maly suggests that the minimum energy required to initiate breakdown at ambient conditions is approximately 0.3 mJ. The breakdown causes a very rapid

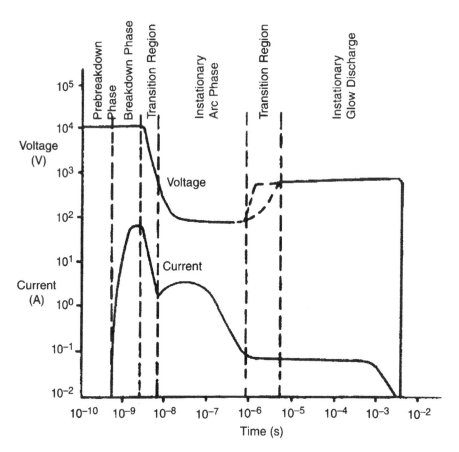

Figure 3.4. *Current and voltage as a function of time during a spark discharge. Adapted from Maly (1984).*

temperature and pressure increase. Temperatures of 60,000 K give rise to pressures of several hundred bars. These high pressures cause an intense shock wave as the spark channel expands at supersonic speed. Expansion of the spark channel allows the conversion of potential energy to thermal energy and facilitates cooling of the plasma. Prolonged high currents lead to thermionic emission from hot spots on the electrodes, and the breakdown phase ends as the arc phase begins.

3. **Arc discharge.** The characteristics of the arc discharge phase are controlled by the external impedances of the ignition circuit. Typically, the arc voltage is approximately 100 V, and the current is greater than 100 mA and is dependent on external impedances. The arc discharge is sustained by electrons emitted from the cathode hot spots. This process causes erosion of the electrodes, with the erosion rate increasing with the plug gap. Depending on the conditions, the efficiency of the energy-transfer process from the arc discharge to the thermal energy of the mixture typically is between 10 and 50%.

With currents of less than 100 mA, this phase becomes a glow discharge, which is distinguished from an arc discharge by the cold cathode. Electrons are liberated by ion impact, a less efficient process than thermionic emission. Although arc discharges are inherently more efficient, glow discharges are more common in practice because of the high electrode erosion rates associated with arc discharges.

3.2.2 Ignition Timing Selection and Control

The determination of the optimum spark type and duration has resulted in disagreement among researchers. Some work concludes that longer arc durations improve the combustion system, whereas other work indicates that short-duration (10–20 ns) high-current arcs (such as those that occur with capacitative discharge ignition [CDI] systems) can be beneficial. The apparent conflict between claims for long-duration and short-duration sparks can be reconciled. The short-duration spark has a better thermal conversion efficiency and can overcome in-cylinder variations by reliable ignition and accelerated flame kernel development. In contrast, the long-duration discharge is successful because it provides a time window long enough to mask the effects of in-cylinder variations. Similarly, a large spark plug gap is beneficial because it increases the likelihood of the existence of a favorable combination of turbulence and mixture between the electrodes.

Ignition timing usually is expressed as degrees before top dead center (°btdc), that is, before the end of the compression stroke. The ignition timing should be varied for different speeds and loads. Section 2.5 of Chapter 2 explained how turbulent flame propagation occupies an approximately constant fraction of the engine cycle because at higher speeds, the increased turbulence gives a nearly corresponding increase in flame propagation rate. However, the initial period of flame growth occupies an approximately constant time (a few milliseconds), and this corresponds to increased crank angles at increased speeds. Consequently, the ignition timing is advanced as speed is increased, so that the main (turbulent) part of combustion is centered around top dead center (tdc).

The ignition tuning must be advanced at part throttle settings because the reduced temperature in the cylinder causes slower combustion. Also, at part throttle settings, there are higher levels of exhaust residuals (because during the valve overlap period, the cylinder pressure will reduce toward the inlet manifold pressure, and the cylinder will tend to fill with exhaust products). Furthermore, higher levels of residuals slow the combustion process. Finally, the rate of combustion also depends on the air-fuel ratio, with the fastest combustion occurring with mixtures that are approximately 10% rich of stoichiometric.

Originally, ignition timings were fixed, and subsequently, they were under direct driver control until the end of the 1920s. Then ignition timing was controlled in response to speed and load by devices within the distributor—spring-loaded rotating masses to control the speed dependence, and a vacuum-operated diaphragm that sensed engine load. With engine management systems, the ignition timing is now controlled digitally (Section 3.7.3).

At any combination of air-fuel ratio and throttle setting, the response to ignition timing variation is illustrated by Fig. 3.5, and the explanation of this response is provided by Fig. 3.6. With an over-advanced ignition timing, the cylinder pressures during combustion will be

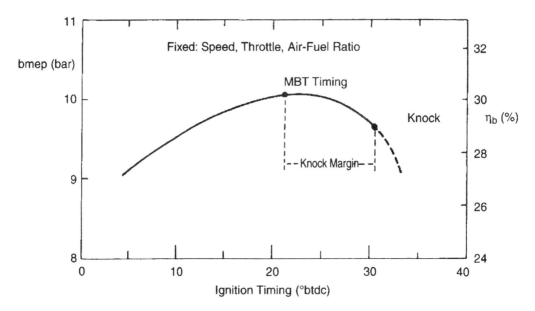

Figure 3.5. *The effect of ignition timing on the output and efficiency of a spark ignition engine (Stone, 1999).*

higher, but the increase in expansion work is less than the extra work done during compression. The converse applies with too retarded an ignition timing.

Advancing the ignition timing raises the cylinder pressure during combustion. As the adiabatic core of the unburned mixture is being compressed isentropically, the unburned gas temperature rises, and it is possible for the unburned mixture to auto-ignite. To maximize the knock margin, the ignition timing is set to the Minimum ignition advance for the Best Torque (MBT)—unless, of course, a timing retarded from MBT is being used to either avoid knock or reduce the NOx emissions.

In summary, the ignition timing must be advanced, as:

a. The mixture is changed from that which gives the maximum burning velocity (essentially the same as the mixture for maximum power)

b. The throttle is closed, because the increased residuals level leads to lower burning velocities

At MBT ignition timing and full throttle operation, approximately 10% of the fuel is burned by top dead center, and the maximum cylinder pressure occurs about 10° after top dead center.

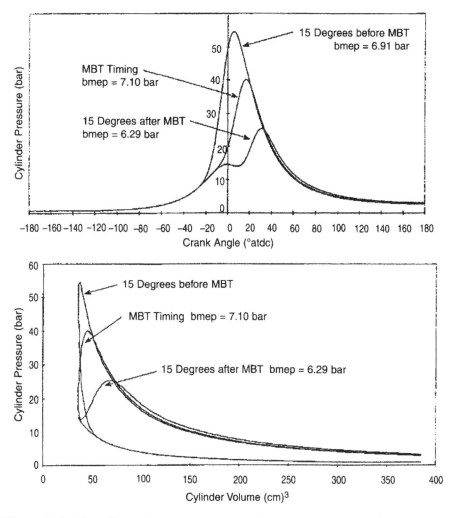

Figure 3.6. *The effect of ignition timing on the pressure–time and pressure–volume diagrams (Stone, 1999).*

3.3 Mixture Preparation

For almost 100 years, carburetors dominated mixture preparation in spark ignition engines. Carburetors rely on the pressure drop of air moving through a venturi to draw fuel through a carefully selected orifice from a reservoir. Numerous refinements were added to carburetors, but by the mid-1980s, emissions legislation was becoming such that only electronically controlled fuel injection systems could provide sufficient control of the air-fuel ratio.

Initially, two types of fuel injection were common—single-point and multi-point. With single-point injection systems (Fig. 3.7), the arrangement was similar to that of a carburetor, and the fuel injector was placed upstream of the throttle plate. Although this gave accurate metering of the fuel, a large part of the induction system would contain fuel as a liquid film, droplets, and vapor. When the throttle is opened, extra fuel is needed for three reasons. First, the air

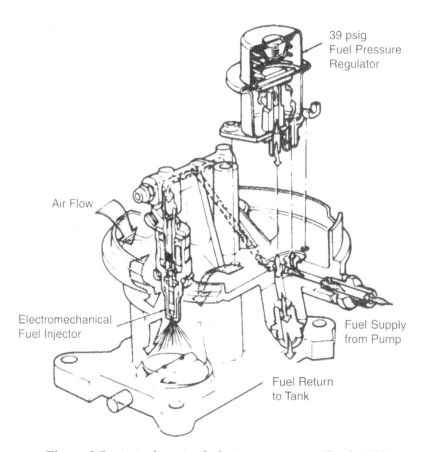

Figure 3.7. *A single-point fuel injection system (Ford, 1982).*

flow into the engine increases more rapidly than the fuel flow because some fuel is in the form of droplets and some is present as a film on the manifold walls. Second, for maximum power, a rich mixture is needed. Finally, when the throttle is opened, the vaporized fuel will tend to condense. When the throttle opens, the pressure in the manifold increases, and the partial pressure of the fuel vapor will increase. (The partial pressure of the fuel vapor depends on the air-fuel ratio.) If the partial pressure of the fuel rises above its saturation pressure, then the fuel will condense; thus, extra fuel is injected to compensate.

When the throttle is suddenly closed, the reduced manifold pressure causes the fuel film to evaporate. This can provide an over-rich mixture and lead to emissions of unburned hydrocarbons. The problem is overcome by a spring-loaded overrun valve on the throttle valve plate that bypasses air into the manifold.

Long inlet manifolds will be particularly bad in these respects because of the large volume in the manifold and the length that the fuel film and droplets must travel. However, inlet manifolds with long pipe lengths give good volumetric efficiency. Thus, the effect of emissions legislation and the desire for increasing the specific output of engines has led to the almost universal use of multi-point injection.

Figure 3.8 shows a solenoid-operated fuel injector. This type of fuel injector usually is mounted close to the inlet valves, so that the fuel spray reaches the back of the inlet valve. In the case of engines with two inlet valves per cylinder, injectors can be used to direct the spray into each inlet port. When the injector is open (which takes approximately 1 ms), the quantity of fuel injected is proportional to the injection duration and the square root of the pressure difference between the fuel supply and the inlet port. The fuel supply is controlled by a regulator (a spring-loaded diaphragm) that maintains a constant pressure difference between the fuel rail and the inlet manifold. The timing and quantity of fuel to be injected must be controlled by the engine management system (Section 3.7.3). With multi-point fuel injection systems, fuel will accumulate on the back of the inlet valve and the adjacent inlet port walls. Therefore, injection strategies must allow for this if the engine performance (especially with respect to emissions) is to be acceptable. However, the quantity of fuel and the time constants are much smaller than with carburetors or single-point injection systems.

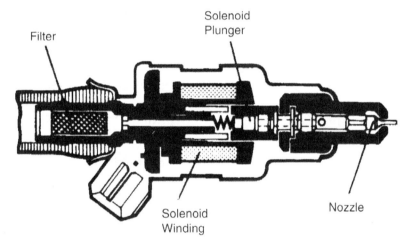

Figure 3.8. *A solenoid-operated fuel injector.*

Initially, fuel injectors were operated in groups to reduce the number of components in the drive electronics. Now, the norm is to have individual injector control and to inject the fuel at the same point in the cycle for every cylinder. This is known as sequential injection. Sequential injection ensures the optimum phasing of injection for each cylinder and helps to minimize the effects of unsteady flows. In both the inlet ports and the high-pressure fuel rail, there will be pressure fluctuations that are in phase with each cylinder. The pressure regulator attempts to maintain a constant mean pressure difference between the inlet manifold and the fuel rail. Thus, with sequential injection, the instantaneous pressure difference across each fuel injector should be similar. (Asymmetry in the fuel rail will prevent complete similarity.) With individual control of the fuel injectors, and a fast-acting lambda sensor (for monitoring the air-fuel ratio in the exhaust), an intelligent control system will be able to fine-tune the injection period for each injector, to allow for injector-to-injector variation as well as the effects of aging or fuels of different compositions.

With multi-point injection systems, the fuel pressure typically is 2 to 3 bar above the manifold pressure, and the breakup of the jet into droplets depends on the relative velocity between the fuel and the air. The maximum size of the droplets will depend on the nozzle diameter; the higher the velocity, the smaller the droplets. An estimate of the maximum possible velocity can be obtained from Bernoulli's equation

$$v = \sqrt{(2\Delta p/\rho)} \tag{3.5}$$

where

v = velocity
Δp = pressure difference across the injector
ρ = density of the fuel

The flow rate will depend on this velocity and the nozzle area. Therefore, it can be seen how reducing the size and increasing the injection pressure (to maintain the same rate of fuel injection) will lead to a smaller diameter jet and more rapid breakup of the jet into smaller droplets. With direct injection spark ignition (DISI) engines, there is little time for fuel evaporation; thus, much higher injection pressures are used (50–200 bar) in association with smaller orifices in the injector nozzle. With multi-point injection systems, the droplets are mostly in the range 100–500 μm. However, by using an "air shroud" around the injector, the droplet sizes can be reduced below 100 μm (Fraidl, 1987). The air shroud works at part throttle, as a result of the pressure drop across the throttle plate. Air is allowed to bypass the throttle plate and enter an annular nozzle around the injector tip. Because the density of air is very low, even a small pressure drop will give a high-velocity air flow. The air flow results in a shear stress being applied to the liquid jet, which promotes the breakup of the fuel jet into fine droplets. Bernoulli's equation (Eq. 3.5) cannot be applied to the air flow because the air is compressible. If the pressure ratio across the nozzle is greater than 1.9, the flow will be choked, and it will have a velocity corresponding to the speed of sound (approximately 300 m/s).

With DISI engines, higher fuel injection pressures are used, with injection times comparable to those of port injectors. Therefore, the injector nozzles have smaller orifices, so that a typical droplet size is smaller than 25 μm. The injectors must be able to withstand in-cylinder pressures and temperatures, and special injector driver hardware is used to provide high-voltage pulses to achieve the higher currents needed to actuate the injectors. Direct injection spark ignition injectors mostly impart swirl to the spray by helical grooves on the needle in the region of the nozzle. This has two benefits. First, it reduces the average droplet size (say, from 20 to 15 μm), and second, it alters the spray characteristics. At low in-cylinder pressures (corresponding to injection during the inlet valve open period for homogeneous operation), the spray tends to be divergent, and this promotes good air and fuel mixing. Conversely, with high in-cylinder pressures (corresponding to injection during the compression stroke for stratified operation with an overall very weak mixture), the spray tends to be less divergent, and the fuel tends to be guided by the piston bowl to form a flammable mixture at the spark plug. These details of mixture preparation are discussed further in Section 3.4.2 and by Stone (1999).

3.4 Combustion System Design

Many designs of combustion chambers have given satisfactory performance. The main considerations are as follows:

a. The distance traveled by the flame should be minimized.

b. The exhaust valve(s) and spark plug(s) should be close together.

c. There should be sufficient turbulence.

d. The end gas should be in a cool part of the combustion chamber.

e. The system should be easy to manufacture and not susceptible to the effects of manufacturing tolerances.

Turbulence is generated during the induction stroke by the high-velocity jet-like flow past the inlet valve, but this turbulence decays during the compression stroke. If barrel swirl (rotation about an axis perpendicular to the cylinder axis) is present, then this ordered flow will degenerate into turbulence as the piston moves toward top dead center (tdc). Turbulence also is generated by the squish areas, which are the areas of the piston that come close to the cylinder head at top dead center. The squish clearance is of the order of 1 mm (0.04 in.).

Figure 3.9 shows the advantages of turbulence, with the increased flammability range and increased knock margin. There is an increased flammability range because combustion is more closely centered around top dead center, when the cylinder pressures and temperatures are higher. This means that a lower temperature rise due to combustion is needed to attain a given self-sustaining combustion temperature. The knock margin is increased because the more rapid combustion with increased turbulence means that combustion will be complete before there is time for unburned mixture to auto-ignite.

There are essentially two different types of spark ignition engine: (1) port injection, which will be discussed first, and (2) direct injection spark ignition (DISI) systems, which are discussed in Section 3.4.2.

3.4.1 Port Injection Combustion Systems

The most common port injected combustion system is the four-valve pent-roof-type combustion system shown in Fig. 3.10. The main characteristic of the four-valve pent-roof combustion chamber is the large flow area provided by the valves. Consequently, there is a high volumetric efficiency, even at high speeds, and this produces an almost constant bmep from mid-speed upward. The inlet tracts tend to be almost horizontal and to converge slightly. During the induction process, barrel swirl (rotation about an axis perpendicular to the cylinder axis) is produced in the cylinder. By disabling one inlet valve, or by having a port throttle, it is possible to increase the flow velocity through the remaining inlet valve at part throttle.

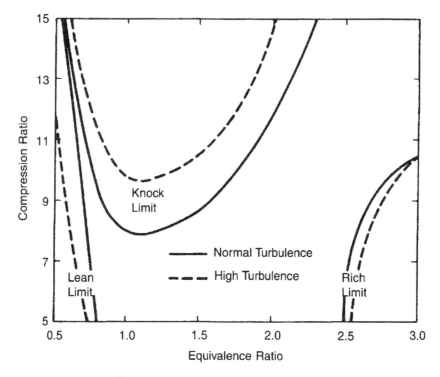

Figure 3.9. *The effect of turbulence on increasing the operating envelope of spark ignition engines (Stone, 1999).*

This generates axial swirl (swirl about the cylinder axis) and a higher level of barrel swirl, both of which lead to faster combustion at part load and a greater tolerance to highly diluted mixtures.

Figure 3.10 shows how the Honda VTEC engine achieves inlet valve disablement (Horie and Nishizawa, 1992). The engine has a central single overhead camshaft, with valve operation through rocker arms with roller followers. There are two separate profiles for the inlet cams, the secondary one having a very low lift (of approximately 0.5 mm [0.2 in.] to prevent liquid fuel accumulating behind the valve). In normal operation, the hydraulically controlled piston locks the two inlet valve rocker arms together. For low-load operation, one inlet valve is controlled by the secondary cam; this can be much narrower because the lower lift requires much lower accelerations with correspondingly reduced contact forces.

Fast-burn systems are tolerant of high levels of exhaust gas recirculation (EGR), whether the EGR is being used for the control of nitrogen oxides (NOx) or to reduce the part-load fuel consumption. The part-load fuel consumption is reduced because EGR leads to a reduction in the throttling loss. To admit a given quantity of air, the throttle must be more fully open, thereby reducing the pressure drop (and loss of work) across the throttle.

Fast-burn systems also are tolerant of very weak mixtures. This is relevant to the development of lean-burn engines that meet emissions legislation without recourse to the use of a

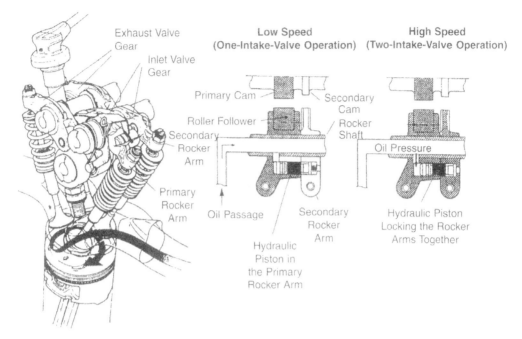

Figure 3.10. *The Honda VTEC combustion system. Adapted from Horie and Nishizawa (1992).*

three-way catalyst. It also is possible for engines fitted with three-way catalysts to be operated in a lean-burn mode prior to the catalyst achieving its light-off temperature, or to operate in a lean-burn mode in selected parts of the operating envelope. A notable example of this is the Honda VTEC engine (Horie and Nishizawa, 1992). When one of the inlet valves is disabled at part load (Fig. 3.10) so that the in-cylinder motion becomes more vigorous, the engine can operate with a weaker mixture. At a part-load operating condition of 1500 rpm and 1.6 bar bmep, the engine can operate with an equivalence ratio of 0.66. This gives a significantly lower brake specific fuel consumption (12% less than at stoichiometric) and less than 6 g/kWh of NOx. The three-way catalyst is still capable of oxidizing any carbon monoxide or unburned hydrocarbons when the engine is in lean-burn mode.

Figure 3.11 shows the different operating regimes of the Honda VTEC engine, with a significant part of the low-speed operation occurring with a weak mixture (an air-fuel ratio of 22 corresponds to an equivalence ratio of 0.66 or lambda of 1.5). A rich mixture is used at full load for higher speeds because this increases the engine power output and prevents overheating of the catalyst. This part of the engine operation is not subject to emissions legislation; therefore, unburned hydrocarbons and carbon monoxide do not need to be oxidized. The rich mixture will have more rapid combustion. Although the combustion temperature is higher than at stoichiometric, more work will be done. Also, because of the increased heat transfer (because of the earlier and higher temperatures), the exhaust temperature will be lower. Furthermore, there is no oxidation within the catalyst; thus, there is no temperature rise (typically 50 K).

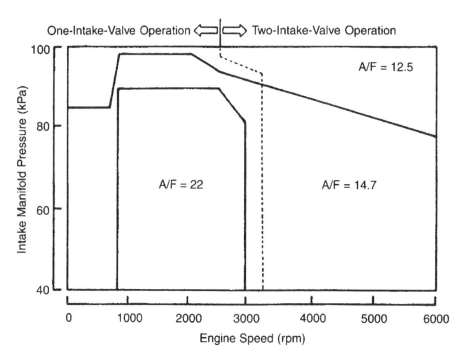

Figure 3.11. *Operating regimes of the Honda VTEC engine. From Horie and Nishizawa (1992). Produced from the Proceedings of the Institution of Mechanical Engineers by permission of the Council of the Institution of Mechanical Engineers.*

3.4.2 Direct Injection Spark Ignition (DISI) Combustion Systems

Direct injection spark ignition (DISI) engines have the potential to achieve the specific output of gasoline engines, but with fuel economy that is said to be comparable to that of diesel engines. Mitsubishi was the first to introduce a DISI engine in a modern car (Ando 1996), and Fig. 3.12 shows some details of the air and fuel handling systems, with the spherically bowled piston being particularly important. Direct injection spark ignition engines operate at stoichiometric near full load, with early injection (during induction) to obtain a nominally homogeneous mixture. This gives a higher volumetric efficiency (by approximately 5%) than a port injected engine because any evaporative cooling is reducing the temperature of only the air, not the inlet port nor the other engine components. Furthermore, the greater cooling of the air means that at the end of compression, the gas temperature will be approximately 30 K lower, and a higher compression ratio can be used (1 or 2 ratios) without the onset of combustion knock. Thus, the engine becomes more efficient.

In contrast, at part load and low speed, direct injection engines operate with injection during the compression stroke. This enables the mixture to be stratified, so that a flammable mixture is formed in the region of the spark plug but the overall air-fuel ratio is weak (and the three-way catalyst operates in an oxidation mode). However, to keep the engine-out NOx emissions low, it is necessary to be careful in the way the mixture is stratified.

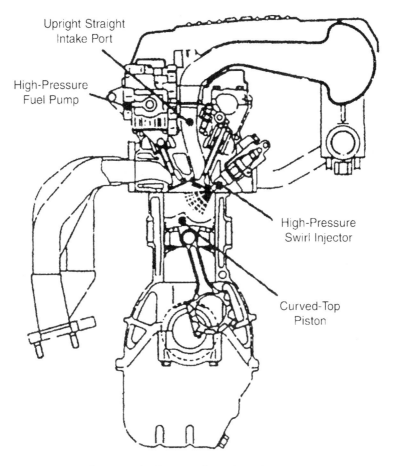

Figure 3.12. *The Mitsubishi GDi (gasoline direct injection) engine. Adapted from Ando (1996).*

The Mitsubishi engine is able to operate in its stratified mode with the air-fuel ratio in the range of 30 to 40, thus reducing the need for throttled operation. In the homogeneous charge mode, it mostly operates at stoichiometric; however, similar to the Honda VTEC engine (Section 3.4.1), it can operate lean at certain load conditions with air-fuel ratios in the range 20–25. With weak mixtures, the air-fuel ratio must be lean enough for the engine-out NOx emissions to need no catalytic reduction. Satisfactory operation of the engine is dependent on careful matching of the in-cylinder air flow to the fuel injection. Reverse tumble (clockwise in Fig. 3.12, the opposite direction to a conventional homogeneous charge engine) must be carefully matched to the fuel injection. The fuel injector is close to the inlet valves (to avoid the exhaust valves and their high temperatures), and the reverse tumble moves the fuel spray toward the spark plug, after impingement on the piston cavity. The injector in the Mitsubishi engine operates at pressures up to 50 bar with a swirl-generating geometry that helps to reduce the droplet size, thereby facilitating evaporation.

For stratified charge operation, the fuel is injected during the start of the compression process, when the cylinder pressure is in the range 3–10 bar. The higher air density makes the spray less divergent than with homogeneous operation, which has injection when the gas pressure is approximately 1 bar. The greater spray divergence with early injection helps to form a homogeneous charge.

In addition to properly controlled air and fuel motion, DISI performance is very sensitive to the timing of injection for stratified charge operation. Jackson et al. (1996) found that cycle-by-cycle variations in combustion are very sensitive to the injection timing. The Ricardo combustion system is similar to the Mitsubishi system and uses an injection pressure of 50–100 bar for stratified charge operation. Figure 3.13 shows that a bowl-in-piston design was needed for stratified charge operation, and that the end of the injection timing window was only approximately 20°ca, if the cycle-by-cycle variations in combustion were to be kept below an upper limit for acceptable driveability (a 10% coefficient of variation for the imep). Furthermore, the injection timing window narrows as the load is reduced, and the end of injection is essentially independent of load. The reason for the sensitivity to injection timing has been explained by Sadler et al. (1998).

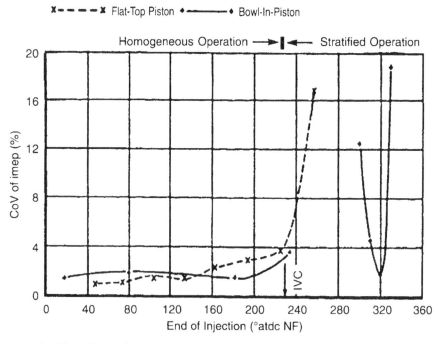

Figure 3.13. *The effect of injection timing on the cycle-by-cycle variations in combustion for homogeneous and stratified charge operation. (Only the bowl-in-piston design would be used for stratified charge mode.) Adapted from Jackson et al. (1996).*

Figure 3.14 shows calculations of the fuel and piston displacements in a DISI engine. Consider the fuel injected with a start of injection (SOI) of 310°ca after top dead center on the non-firing revolution. The fuel strikes the piston (A) and flows to the rim of the piston (B), and it then is swept toward the spark plug by the tumbling flow to arrive at (C). In Fig. 3.14,

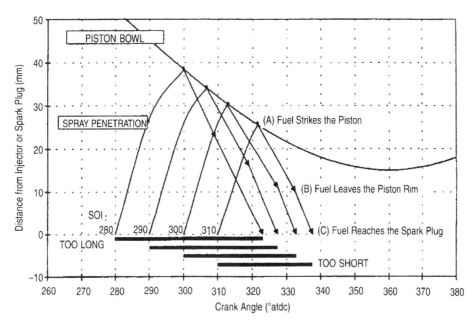

Figure 3.14. *Fuel spray transport calculations for a direct injection spark ignition (DISI) engine, showing the fuel and piston trajectories. Adapted from Sadler et al. (1998).*

the horizontal bars represent the time between the start of injection and some mixture arriving at the spark plug. If the time is too long, the mixture will be over-diluted; if the time is too short, the mixture will be too rich. Results from in-cylinder sampling showed that the best combustion stability coincided with the richest mixture occurring in the region of the spark plug.

Exhaust gas recirculation (EGR) can be usefully applied to direct injection engines, because with lean operation, there is a high level of oxygen and a low level of carbon dioxide in the exhaust gas. Jackson et al. (1996) showed that for a fixed bmep of 1.5 bar at 1500 rpm, applying 40% EGR can lower each of the following: the fuel consumption by 3%, the NOx emissions by 81%, and the unburned hydrocarbons by 35%. Even with this level of EGR, the cycle-by-cycle variations in combustion are negligible. They also point out that at some low-load conditions, it may be advantageous to throttle the engine slightly, because this will have a negligible effect on the fuel consumption but will reduce the unburned hydrocarbon emissions and cycle-by-cycle variations in combustion.

Although DISI engines are being produced commercially, a number of issues might limit their use. First, the Mitsubishi engine has a swept volume of 450 cm^3 per cylinder, and it may be difficult to make this technology work in smaller displacement engines. Second, the combustion stability is very sensitive to the injection timing, as well as the ignition timing relative to the injection timing. Third, although in-cylinder injection should have an inherently good transient response, complex control issues remain, especially when switching between the stratified and homogeneous charge operating modes. Fourth, the operating envelope for unthrottled stratified operation might be quite limited, and this in turn would limit the fuel economy gains. Fifth, there are questions of inlet port cleanliness because in port injected engines, the detergents added to gasoline help to remove combustion deposits from the back of the inlet valve. (These deposits occur as a result of backflow during the valve overlap period when an engine is operating at part throttle.) Finally, even if DISI engines are possible, the higher cost of the fueling system must be justified. Figure 3.15 shows that the different operating regimes for the direct injection engine and its higher compression ratio give a greater high-load efficiency.

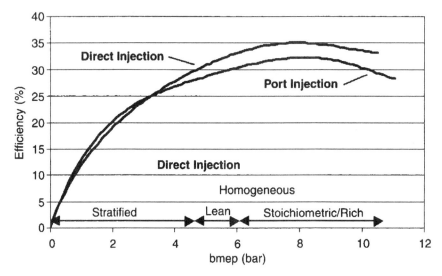

Figure 3.15. *Comparison between the efficiency of the Mitsubishi GDi engine and the port injected spark ignition engine of Fig. 2.8.*

In Fig. 3.15, the port injected engine has a particularly high bmep because of its variable geometry induction system.

3.5 Emissions Control

Whole books have been written about exhaust after-treatment (notably Eastwood [2000]); therefore, the aim here is to review more recent technologies. Three-way catalysts are well established for spark ignition engines, and when such an engine is operating at stoichiometric, the catalyst gives substantial reductions (significantly greater than 90%) in the emissions of carbon monoxide, nitric oxides, and unburned hydrocarbons, when the engine is warmed up. The major developments now are associated with ensuring a faster catalyst light-off and improving the durability with a given catalyst loading.

So-called lean-NOx catalysts also have the potential for application in lean-burn spark ignition engines. Particulate traps might become necessary for spark ignition engines. Direct injection gasoline engines have a limited time for mixing of the air and fuel. Their combustion has some aspects that are similar to combustion in diesel engines. Even with conventional port injected spark ignition engines, there are particulate emissions; however, these particulates are too small to be visible with the naked eye and have not yet been the subject of legislation. The particles smaller than 0.1 μm that are present in both diesel and spark ignition engine exhausts have the greatest deposition efficiency in the lungs (Booker, 2000).

3.5.1 Development of the Three-Way Catalyst

Table 3.1 summarizes the emissions legislation that led to the development of the three-way catalyst.

TABLE 3.1
U.S. FEDERAL EMISSIONS LEGISLATION
(GRAMS OF POLLUTANT PER MILE)

Model Year	CO	HC	NOx	Comments
1966	87	8.8	3.6	Pre-control
1970–1974	34–28	4.1–3.0	4.0–3.1	Retarded ignition, thermal reactors, exhaust gas recirculation (EGR)
1975	155	1.5	3.1	Oxidation catalysts
1977	15	1.5	2.0	Oxidation catalyst and improved EGR
1980	7	0.41	2.0	Improved oxidation catalysts and three-way catalysts
1981	3.4	0.41	1.0	
1994	3.4	0.25	0.4	HC is now non-methane hydrocarbons
2001	3.4	0.125	0.2	Corporate average for hydrocarbons
2004	1.7	0.09	0.07	Phase-in 2004–2007; corporate average for all emissions

The U.S. test is a simulation of urban driving from a cold start in heavy traffic. Vehicles are driven on a chassis dynamometer (rolling road), and the exhaust products are analyzed using a constant-volume sampling (CVS) technique in which the exhaust is collected in plastic bags. The gas then is analyzed for carbon monoxide (CO), unburned hydrocarbons (HC), and nitrogen oxides (NOx) using standard procedures. In 1970, three events—the passing of the American Clean Air Act, the introduction of lead-free gasoline, and the adoption of cold test cycles for engine emissions—led to the development of catalyst systems.

Catalysts in the process industries usually work under carefully controlled steady-state conditions, but this is obviously not the case for engines, especially after a cold start. While catalyst systems were being developed, engine emissions were controlled by retarding the ignition and using exhaust gas recirculation (both to control NOx) and a thermal reactor to complete oxidation of the fuel. These methods of NOx control led to poor fuel economy and poor driveability (that is, poor transient engine response). Furthermore, the methods used to reduce NOx emissions tend to increase CO and HC emissions, and vice versa (Fig. 2.11). The use of EGR and retarding the ignition also reduce the power output and fuel economy of engines.

Catalysts were able to overcome these disadvantages and meet the 1975 emissions requirements. Figure 3.16 shows the operating regimes of the different catalyst systems.

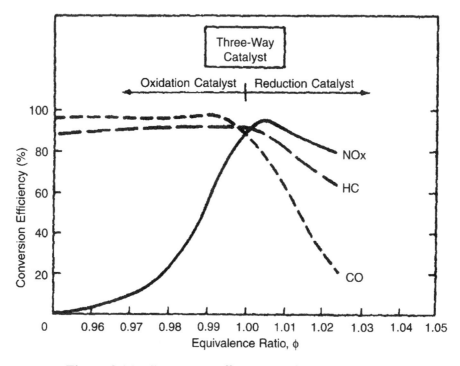

Figure 3.16. *Conversion efficiencies of catalyst systems. Courtesy of Johnson Matthey.*

With rich-mixture running, the catalyst promotes the reduction of NOx by reactions involving HC and CO:

$$4HC + 10NO \rightarrow 4CO_2 + 2H_2O + 5N_2 \quad \text{and} \quad 2CO + 2NO \rightarrow 2CO_2 + N_2$$

Because there is insufficient oxygen for complete combustion, some hydrocarbons and carbon monoxide will remain unreacted. With lean-mixture conditions, the catalyst promotes the complete oxidation of the hydrocarbons and carbon monoxide

$$4HC + 5NO_2 \rightarrow 4CO_2 + 2H_2O \quad \text{and} \quad 2CO + O_2 \rightarrow 2CO_2$$

With oxygen present, nitric oxides will not be reacted.

Oxidation catalyst systems were the first to be introduced, but NOx emissions still had to be controlled by EGR. Excess oxygen was added to the exhaust (by an air pump) to ensure the catalyst could always oxidize the CO and HC.

Dual catalyst systems control NOx emissions without resorting to EGR or retarded ignition timings. A feedback system incorporating an exhaust oxygen sensor is used with a carburetor or fuel injection system to control the air-fuel ratio. The first catalyst is a reduction catalyst, and by maintaining a rich mixture, the NO is reduced. Before entering the second catalyst, air is injected into the exhaust to enable oxidation of the CO and HC to take place in the oxidation catalyst.

Conventional reduction catalysts are liable to reduce NOx but produce significant quantities of ammonia (NH_3). This would be oxidized in the second catalyst to produce NOx. However, by using a platinum/rhodium system, the selectivity of the reduction catalyst is improved, and a negligible quantity of ammonia is produced.

Three-way catalyst systems control CO, HC, and NO emissions as a result of developments to the platinum/rhodium catalysts. As Fig. 3.16 shows, close control is needed on the air-fuel ratio. This normally is achieved by electronic fuel injection, with a lambda sensor to provide feedback by measuring the oxygen concentration in the exhaust. A typical air-fuel ratio perturbation for such a system is +0.25 (or ±0.02φ). When a three-way catalyst is used, it requires first an engine management system capable of accurate air-fuel ratio control, and second a catalyst. Both of these requirements add considerably to the cost of an engine. Because a three-way catalyst must always operate with a stoichiometric air-fuel ratio, at part load this means the maximum economy cannot be achieved (Fig. 2.9). The fuel consumption penalty (and the increase in carbon dioxide emissions) associated with stoichiometric operation is approximately 10%.

Fortunately, the research into lean-burn combustion systems can be exploited, to improve the part-load fuel economy for engines operating with stoichiometric air-fuel ratios, by using high levels of EGR. A combustion system designed for lean mixtures also can operate satisfactorily when a stoichiometric air-fuel mixture is diluted by exhaust gas residuals. At part load up to approximately 30%, EGR can be used. This reduces the volume of flammable mixture induced, and consequently, the throttle must be opened slightly to restore the power output. With a more open throttle, the depression across the throttle plate is reduced, and the pumping work (or pmep) is lower. Nakajima et al. (1979) showed that for a bmep of 3.24 bar with stoichiometric operation at a speed of 1400 rpm, they were able to reduce fuel consumption by approximately 5% through the use of 20% EGR on an engine with a fast-burn combustion system.

Figure 3.17 shows the construction of a typical three-way catalyst, and the technology used in the catalyst has been reviewed by Milton (1998). The catalyst is supported on either a metal or ceramic substrate (alumina Al_2O_3, aluminum oxide), which typically has an approximately square cross-section passage of width 1 mm (0.02 in). Ceramic substrates are used more commonly and have a flow area of approximately 70% of the cross section, compared to

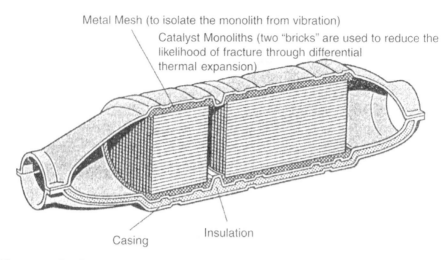

Figure 3.17. *Construction of a three-way catalyst. Adapted from Gunther (1988).*

approximately 90% for metal substrates. The substrate is covered with a washcoat based on $\gamma\text{-}Al_2O_3$ (chosen for its porosity), but also containing the catalyst materials and stabilizers such as cerium oxide (CeO_2) and barium oxide (BaO). Unfortunately, $\gamma\text{-}Al_2O_3$ turns to $\alpha\text{-}Al_2O_3$ at 900°C (1652°F), and the porosity is lost; therefore, stabilizers are added to enable higher operating temperatures. The catalyst material is mostly platinum (approximately 85–95%), but there are also significant quantities of rhodium.

The cerium oxide (ceria) has the ability to store oxygen under weak operating conditions and release it during rich-mixture operation. This is important for the operation of the three-way catalyst because the simple lambda sensor (Section 3.7.2) can determine only whether a mixture is weak or rich. Thus, the engine management system seeks to oscillate the air-fuel ratio around stoichiometric with a frequency of approximately 1 Hz. This means that the ability of the catalyst to store oxygen and pollutants is important. By installing a second lambda sensor downstream of the catalyst and looking at the amplitude of its response compared to the upstream sensor, it is possible to assess the oxygen storage performance of the catalyst. The better the condition of the catalyst, the greater the attenuation of the oxygen fluctuations at the catalyst exit (Rieck et al., 1998).

3.5.2 Durability

As a catalyst ages, its performance changes in several ways. First, its light-off temperature rises, then its conversion efficiency falls, and finally, its response to different components in the exhaust changes. The catalyst performance deteriorates through a number of mechanisms: poisoning of the catalyst, failure of the substrate, and sintering (a process by which the catalytic material agglomerates and its effective area is reduced).

Poisoning is deactivation of the catalytic material through deposits. Operating an engine on leaded fuel (0.125 gPb/L) would degrade the performance of the catalyst by approximately

25% after 60 hours of use. Sulfur also degrades the catalyst through sulfate deposits. European legislation limits sulfur content to 500 ppm by mass, but 200 ppm is typical. The catalyst will convert the sulfur dioxide to sulfur trioxide (and then sulfuric acid) in an oxidizing environment. In a rich mixture, hydrogen sulfide (H_2S, which has the odor of rotten eggs) is formed, and this often is noticeable during engine warm-up, especially with the removal of nickel oxide from the catalyst on health grounds. Nickel oxide promotes the oxidation of hydrogen sulfide.

3.5.3 Catalyst Light-Off

With increasingly demanding emissions legislation, it is even more important for the catalyst to start working as soon as possible. The thermal inertia of the catalyst can be reduced by using a metal matrix because the foil thickness is approximately 0.05 mm (0.002 in.). Ceramic matrices usually have a wall thickness of approximately 0.3 mm (0.01 in.), but this can be halved (Yamamoto et al., 1991) to give a slight improvement in light-off performance. Systems to promote catalyst light-off might usefully be classified as passive or active, with active being when an external energy input is used. Two active systems are electrically heated catalysts (using metal substrates) and exhaust gas ignition (EGI). Exhaust gas ignition requires the engine to be run very rich of stoichiometric and then adds air to the exhaust stream, so that an approximately stoichiometric mixture then can be ignited in the catalyst (Eade et al., 1996). The mixture is ignited by a glow-plug situated in the chamber formed between two catalyst bricks.

Electrically heated catalysts (EHCs) are placed between the close-coupled catalyst and the downstream catalyst. Electrically heated catalysts have a power input of approximately 5.5 kW and are energized for 15–30 seconds before engine cranking, to raise their temperature to approximately 300°C (572°F). When the engine is firing, the electrical power input is reduced by a controller that responds to the catalyst temperature. Table 3.2 shows results from a study of two vehicles fitted with an EHC (Heimrich et al., 1991).

TABLE 3.2
FEDERAL TEST PROCEDURE PERFORMANCE FOR TWO VEHICLES FITTED WITH ELECTRICALLY HEATED CATALYSTS (EHC) (HEIMRICH ET AL., 1991)

Configuration	NMOG, g/mi	CO, g/mi	NOx, g/mi	Fuel Economy, mpg
Vehicle 1, No EHC	0.15	1.36	0.18	20.2
Vehicle 1, With EHC+	0.02	0.25	0.18	19.7
Vehicle 2, No EHC	0.08	0.66	0.09	25.4
Vehicle 2, With EHC*	0.02	0.30	0.05	24.3

+ With injection of 300 L/min of air; 75 s for cold start, 30 s for hot start.
* With injection of 170 L/min of air for 50 s for cold start.

The performance of an EHC probably would be better than this when it is incorporated into the engine management strategy by the vehicle manufacturer. However, with EHC systems, there are questions about durability, and indeed any active system is likely to be used when it is the only solution.

Passive systems rely on thermal management. Typically, a small catalyst is placed close to the engine, so that its reduced mass and higher inlet temperatures give quicker light-off. Its small volume limits the maximum conversion efficiency; thus, a second larger catalyst is placed farther downstream, under the car body. Another strategy is to operate with a retarded ignition timing. Because there is less in-cylinder heat and work transfer, the exhaust temperatures are higher. Proposals also have been made for storing the unburned hydrocarbons prior to catalyst light-off, and then re-introducing them into the exhaust stream after light-off.

A recent development from Johnson Matthey is a catalyst with light-off temperatures in the range 100–150°C (212–302°F) for carbon monoxide and hydrogen. The engine is initially operated very rich (thereby reducing NO emissions and increasing the levels of carbon monoxide and hydrogen). Air is added after the engine to make the mixture stoichiometric, and the exothermic oxidation of the carbon monoxide and hydrogen heats the catalyst. Initially, unburned hydrocarbons must be stored in a trap, for release after the catalyst is fully warmed up.

3.5.4 Lean-Burn NOx-Reducing Catalysts, "DENOx"

It has already been reported how stoichiometric operation compromises the efficiency of engines, but that for control of NOx, it is necessary to operate either at stoichiometric or sufficiently weak (say, an equivalence ratio of 0.6), such that there is no need for NOx reduction in the catalyst. If a system can be devised for NOx to be reduced in an oxidizing environment, this gives scope to operate the engine at a higher efficiency.

A number of technologies are being developed for "DENOx," some of which are more suitable for diesel engines than spark ignition engines. The different systems are designated active or passive (passive being when nothing must be added to the exhaust gases). The systems are as follows:

a. **Selective Catalytic Reduction (SCR).** In this technique, ammonia (NH_3) or urea ($CO(NH_2)_2$) is added to the exhaust stream. This is likely to be more suited to stationary engine applications. Conversion efficiencies of up to 80% are quoted, but the NO level must be known, because if too much reductant is added, ammonia would be emitted.

b. **Passive DENOx.** These use the hydrocarbons present in the exhaust to chemically reduce the NO. There is a narrow temperature window (in the range 160–220°C [320–428°F] for platinum catalysts) within which the competition for HC between oxygen and nitric oxide leads to a reduction in the NOx (Joccheim et al., 1996). The temperature range is a limitation and is more suited to diesel engine operation. More recent work with copper-exchanged zeolite catalysts has shown them to be effective at higher temperatures.

By modifying the zeolite chemistry, a peak NOx conversion efficiency of 40% has been achieved at 400°C (752°F) (Brogan et al., 1998).

c. **Active DENOx Catalysts.** These use the injection of fuel to reduce the NOx, and a reduction in NOx of approximately 20% is achievable with diesel-engined vehicles on typical drive cycles, but with a 1.5% increase in fuel consumption (Pouille et al., 1998). Current systems inject fuel into the exhaust system, but there is the possibility of late in-cylinder injection with future diesel engines.

d. **NOx Trap Catalysts.** In this technology (first developed by Toyota), a three-way catalyst is combined with a NOx-absorbing material to store the NOx when the engine is operating in lean-burn mode. When the engine operates under rich conditions, the NOx is released from the storage media and reduced in the three-way catalyst.

NOx trap catalysts have barium carbonate deposits between the platinum and the alumina base. During lean operation, the nitric oxide and oxygen convert the barium carbonate to barium nitrate. A rich transient (approximately 5 s at an equivalence ratio of 1.4) is needed every five minutes or so, such that the carbon monoxide, unburned hydrocarbons, and hydrogen regenerate the barium nitrate to barium carbonate. The NOx that is released is then reduced by the partial products of combustion over the rhodium in the catalyst. Sulfur in the fuel causes the NOx trap to lose its effectiveness because of the formation of barium sulfate. However, operating the engine at high load to give an inlet temperature of 600°C (1112°F), with an equivalence ratio of 1.05, for 600 s can be used to remove the sulfate deposits (Brogan et al., 1998).

3.6 Power Boosting

The way to increase the power output of an engine is to increase the air flow into the engine. This can be done passively by exploiting the unsteady nature of the inlet and exhaust flows (Section 3.6.1) or by using an external compressor, which is a technique known as supercharging. When the compressor is driven by an exhaust gas turbine, the device is known as a turbocharger, and these are most effective on diesel engines. In spark ignition engines, there is a wider speed range and an even wider variation in the air flow (because of throttling); therefore, this exacerbates the problem of poor transient response, known as turbo-lag. Turbochargers will be discussed fully in the context of diesel engines (Section 4.5); supercharging will be restricted in meaning here to a compressor driven from the engine crankshaft and is the subject of Section 3.6.2.

3.6.1 Variable Valve Timing and Induction Tuning

The intermittent flow in reciprocating engines leads to pressure pulsations, the amplitude and frequency of which depend on the geometry of the induction and exhaust systems. This subject has been treated comprehensively in two books by Winterbone and Pearson (1999 and 2000). The volumetric efficiency of an engine can be improved if a pressure pulsation arrives at the inlet valve between bdc and its closure (approximately 40°). With a fixed geometry induction or exhaust system, there will be a sequence of speeds at which resonance occurs; however, between each resonance, there also will be a minima in the volumetric efficiency. The minima can be avoided either by using a variable geometry induction system or by changing the valve timing.

An example of a variable valve timing (VVT) mechanism in production that can control the duration and phasing of a camshaft is provided by the Rover VVC (variable valve control) system, which is used on the inlet camshafts of some K16 engines (Whiteley, 1995; Parker, 2000). The mechanism is essentially a non-constant velocity drive system interposed between the cam drive pulley and the cam lobes. The VVC mechanism controls the valve duration from 220 to 295°ca, with a corresponding change in valve overlap from 21 to 58°ca. Thus, with the maximum valve period, inlet valve closure has also been delayed by 38°ca. Table 3.3 compares the performance of the 1.8-liter VVC and conventional engines.

TABLE 3.3
COMPARISON OF ROVER K16 1.8-LITER IN-LINE FOUR-CYLINDER ENGINES WITH AND WITHOUT VVC

Engine	Maximum bmep (bar)	Speed (rpm) for Maximum bmep	Maximum Power (kW)	Speed (rpm) for Maximum Power
Standard	11.6	3000	90	5500
VVC	12.2	4500	108	7000

The VVC engine also had larger valves, with the inlet valves increased from 27.3 to 31.5 mm (1.1 to 1.24 in.), and the exhaust valves increased from 24.0 to 27.3 mm (0.9 to 1.1 in.).

A variable geometry induction system has been used on the Rover KV6 engine (the V-6 version of the engine discussed in Table 3.3), for which the full-load operating modes and torque curve are shown in Fig. 3.18. Below approximately 2750 rpm, both the connect valve and balance valve are closed, so that the engine behaves similarly to two three-cylinder engines with long secondary pipes. Between 2750 and 3500 rpm, the connect valve is opened, and above 3500 rpm, the balance valve also is opened to give the maximum torque at 4000 rpm, with a bmep of 12.1 bar. This compares with a maximum bmep of 12.2 bar obtained at 4500 rpm in the K16 engine with VVC. At full load, the system operates to maximize the volumetric efficiency, whereas at part load, the two valves are closed. This reduces the volumetric efficiency, so that wider open throttle settings are needed, thereby reducing the pumping losses and reducing the fuel consumption.

3.6.2 Supercharging

Supercharging is restricted in meaning here to compressors driven from the crankshaft. When the compressor is driven by an exhaust gas turbine, it is known as turbocharging, and

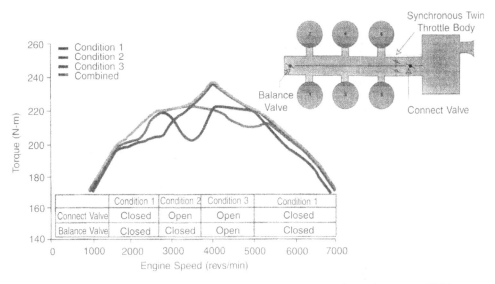

Figure 3.18. *The variable geometry induction system used on the Rover KV6 engine, with its torque curve and operating modes. Adapted from Whiteley (1995).*

turbocharged spark ignition engines are discussed in Section 4.5.3. Supercharging usually is achieved by driving a positive displacement compressor from the crankshaft. Radial flow compressors are not used because they would have to be driven at least an order of magnitude faster than the engine, and with a variable ratio drive. In contrast, because the engine is a positive displacement device, a positive displacement compressor will be well matched to an engine with a simple constant ratio drive.

Figure 3.19 shows two types of positive displacement compressor, the Roots blower and a vane compressor, with their pressure volume diagrams. The operation of the Roots blower is described by Fig. 3.19, and it depends on there being a lower volumetric flow rate on the high-pressure side than at its inlet. A volume of air is trapped at ambient pressure, and the pressure is unchanged until this trapped volume is transported to the high-pressure side. At the instant that the high-pressure port is uncovered, the air in the rotor is compressed by the irreversible reverse flow of the high-pressure air from downstream of the Roots blower. Thus, the work done is the rectangular area abcd in Fig. 3.19. This work is executed as the rotor turns, to displace the volume of high-pressure gas (corresponding to the trapped volume of the rotor) out of the Roots blower. This model assumes the downstream volume is sufficiently large compared to the trapped volume, for the delivery pressure to remain constant. The efficiency of the Roots blower falls rapidly with an increase in its pressure ratio because of the absence of any internal compression. The performance of a Roots compressor is further compromised by its mechanical losses and its volumetric efficiency. Because of losses (including internal leakage), the volume flow rate into the compressor is reduced, but the work is still area abcd in Fig. 3.19 (Stone, 1988).

The vane compressor has internal compression, and it will have a high efficiency when the system requires operation at this pressure ratio. If this internal pressure rise is below the

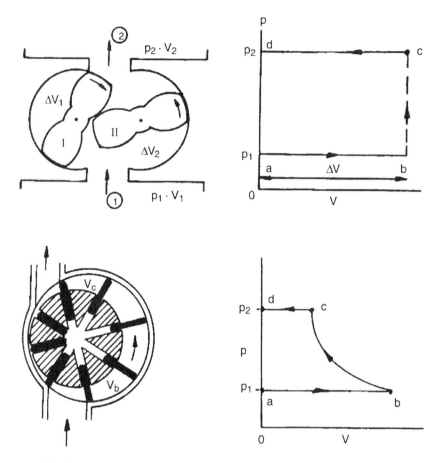

Figure 3.19. *Positive displacement compressors, a Roots blower (upper) and vane compressor (lower), and their pressure–volume diagrams. Adapted from Stone (1988).*

required pressure ratio, then there is external irreversible compression as with the Roots blower. If the required pressure ratio is lower than the internal pressure ratio, there will be irreversible expansion and a rapid fall in efficiency (Stone, 1988).

With a supercharged spark ignition engine, the compression ratio will have to be reduced to avoid knock, and this will lower the engine efficiency. However, a small displacement supercharged engine can still be more efficient than a large displacement naturally aspirated engine because there is scope for reducing the part-load throttling losses. This is best illustrated by using a case study, such as the Jaguar four-liter engine (Joyce, 1994). In normal use, it will be operating under comparatively low load and speed conditions, the exact conditions from which turbo-lag would be most significant. Obviously, a supercharger will lead to worse fuel economy than a turbocharger, because there is no recovery of the exhaust gas expansion work. However, because of the limit to boost pressure imposed by fuel quality and knock, the pressure ratio will be comparatively low, and the fuel economy penalty will not be unacceptable. In any case, people who want a supercharged four-liter engine probably will be more concerned with performance than economy. Jaguar adopted a Roots compressor with a pressure ratio of 1.5 and an air/coolant/air intercooler. The other main loss is leakage past the rotors, which depends solely on the pressure ratio and seal clearances, and not on the flow rate through the compressor. Leakage losses are most significant at low speeds because the flow rates are low and the running clearances are greatest (due to the comparatively low temperature of the compressor). Jaguar found that inlet system deposits on the Roots blower rotors led to an in-service reduction in the leakage loss.

At part-load operation, the supercharger is unnecessary; therefore, one option is to use a clutch to control its use. However, to avoid a loss of refinement through engaging and disengaging the supercharger, Jaguar adopted a permanent drive with a step-up speed ratio of 2.5. (The drive must meet a requirement of 34 kW. If a continuously variable ratio drive is available, then throttling losses can be reduced at part-load operation by running the supercharger more slowly and using it as an expander.) To avoid unnecessary compression in the supercharger, a bypass valve opens at part load (in response to manifold pressure), and there is no compression (Fig. 3.20). The main throttle is upstream of the supercharger.

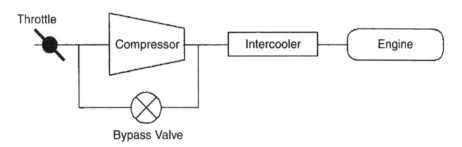

Figure 3.20. *Arrangement of the supercharger and bypass valve to avoid unnecessary compressor work.*

An intercooler is used; otherwise, the performance gains from supercharging could be reduced by approximately 50%. Without an intercooler, the supercharger outlet temperature would be limited to approximately 80°C (176°F), compared to 120°C (248°F) with the intercooler. After the intercooler, the temperature might be approximately 50°C (122°F). Thus, the intercooler increases the output of the engine in two ways. First, the higher pressure ratio and the cooler air temperature both increase the air density. Second, the lower temperature allows a higher boost pressure or compression ratio for knock-free operation with a given quality fuel.

Table 3.4 compares the naturally aspirated and the supercharged Jaguar AJ6 engines. Both have an aluminum cylinder head with four valves per cylinder, a bore of 91 mm, and a stroke of 102 mm.

The supercharged engine uses the same camshafts (242°ca valve open period) as the naturally aspirated engine, but with zero overlap at top dead center to avoid short-circuiting loss of the mixture at high load conditions and to give good idle stability.

TABLE 3.4
COMPARISON OF NATURALLY ASPIRATED AND SUPERCHARGED JAGUAR AJ6 ENGINES

	Naturally Aspirated	Supercharged
Compression Ratio	9.5	8.5
Maximum Torque (Nm)	370 at 3500 rpm	510 Nm at 3000 rpm
bmep (bar)	11.6	16.0
Maximum Power (kW)	166 at 5000 rpm	240 at 5000 rpm
bmep (bar)	10.0	14.4

The supercharged engine has a higher power and torque output than the naturally aspirated 6-liter V-12 engine. Table 3.5 compares the brake specific fuel consumptions of the engines at both the same bmep and the same torque (64 Nm, corresponding to a bmep of 2 bar for the 4-liter engine and 1.33 bar for the 6-liter engine).

TABLE 3.5
COMPARISON OF BRAKE SPECIFIC FUEL CONSUMPTIONS (bsfc) (kg/kWh) FOR NATURALLY ASPIRATED AND SUPERCHARGED JAGUAR ENGINES AT 2000 RPM, FOR A FIXED BMEP AND A FIXED TORQUE

Engine	2 bar bmep	64 Nm Torque
6.0-Liter V-12	0.414	0.54
4.0-Liter Supercharged	0.412	0.41
4.0-Liter Naturally Aspirated	0.390	0.39

The supercharged engine shows a slight loss of fuel economy compared with the naturally aspirated engine, but a substantial fuel economy benefit compared to the V-12 engine, especially when the comparison is made on the basis of a specified torque. The greater output of the supercharged engine allows a higher gearing compared to the naturally aspirated engine, which partially compensates for the lower fuel economy of the engine.

3.7 Engine Management Systems

3.7.1 Introduction

Figure 3.21 shows an engine management system. The purpose of all the sensors and actuators will be discussed in this section, with some control strategies. Turner and Austin (2000) provide a useful summary of sensor technologies, with details of the signal conditioning and processing. They also indicate how the information is used, and the likely developments in sensors, especially in terms of intelligent sensors (in which some of the signal conditioning is

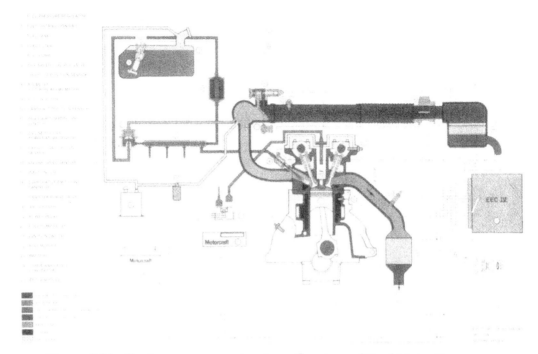

Figure 3.21. *Engine management system. Courtesy of Ford Motor Company.*

integral to the transducer). The engine management system must control the engine in response to both engine and vehicle inputs, but initially, only engine inputs will be considered. The engine management system also might be responsible for additional functions, such as transmission ratio selection. The most important parameters to be controlled are the ignition timing and air-fuel ratio, but it also is necessary to control some or all of the following:

- EGR valve
- Air purge to the fuel vapor storage canister
- Throttle position (or idle speed adjustment)
- Variable geometry induction system settings
- Variable valve timing system settings

On a test bed, it is usual to define the engine operating point in terms of speed and bmep (or torque). However, because torque is difficult to measure in a vehicle installation, it is usual to deduce the torque from measurements such as throttle setting, manifold pressure, and air flow rate.

Sensors that will be found in a typical installation are listed here, with their usual technology type and whether the signal is analog or digital. The technology of each type will be discussed, in addition to how the information is used.

Crankshaft speed/position—Inductive pickup (digital)

Camshaft position—Inductive pickup (digital)

Throttle position—Potentiometer (analog)

Air flow rate—Moving vane with a potentiometer or a hot wire probe (analog)

Inlet manifold absolute pressure—Strain-gauged diaphragm (analog)

Air temperature—Thermistor (analog)

Coolant temperature—Thermistor (analog)

Air-fuel ratio—Lambda sensor (digital or analog)

Knock detector—Accelerometer (analog)

There are two types of control systems: open loop, and closed loop. Open loop control systems rely on a parameter (e.g., ignition timing) being set on the basis of stored information (usually in some form of look-up table), with the particular ignition timing being selected on the basis of measurements such as engine speed, manifold pressure, and coolant temperature. In contrast, closed loop control systems rely on measuring the effect of a parameter that is being varied, to control the parameter to a target value. The best example of closed loop is probably air-fuel ratio control, which uses a lambda sensor to deduce the injection duration needed for stoichiometric operation. The closed loop system will make use of stored information to make an estimate of the necessary injection duration, and it is possible for this stored information to be updated on the basis of feedback, to allow for changes in the injector flow characteristics. Also, there will be no feedback before the lambda sensor has warmed up; therefore, during warm-up, it is necessary to use open loop control. The operation of the principal sensors will now be described.

3.7.2 Sensor Types

3.7.2.1 Crankshaft Speed/Position and Camshaft Position

Figure 3.22 shows the construction of an inductive pickup and its associated analog output. A magnetic field originating from the magnet passes through the soft iron core of the coil. The strength of the magnetic field in the coil will depend on the "magnetic conductance" in the remainder of the magnetic circuit. When a tooth on the toothed wheel (typical spacing 10°ca) aligns with the pickup core (as depicted in Fig. 3.22), the magnetic field will be strong, as the magnetic path is completed through the engine structure. Conversely, when there is an air gap opposite the core, the field will be weaker, and when there is a big gap (due to the missing tooth), the field will be very weak. The voltage generated by the coil is proportional to the

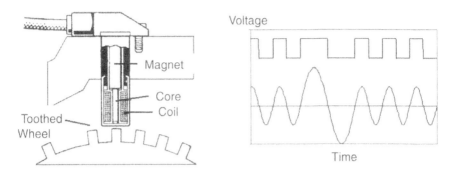

Figure 3.22. *An inductive pickup, and its waveform before and after signal conditioning. Adapted from Lembke (1988).*

rate of change in the strength of the magnetic field; thus, when the missing tooth passes the pickup, there will be an increase in both the amplitude and period. The analog signal can be signal-conditioned (by a high-gain amplifier that saturates at a fixed voltage) to provide pulses of constant voltage. Otherwise, the faster the engine, the higher the amplitude of the pulses.

The reference pulse provides the engine management system with the crankshaft datum position, and the shorter duration pulses provide engine position and speed information. The pulses can be fed to a phase lock loop circuit that generates pulses at a higher frequency that is a multiple of the base frequency (ignoring the missing tooth). The pulses can be generated every degree of crank angle, which is finer than the spacing of the teeth on the toothed wheel. The camshaft pickup needs only to establish whether a reference cylinder is on the gas exchange or compression/expansion strokes, because accurate positional information comes from the crankshaft, unless it also is being used as part of a variable valve timing control system.

The crankshaft timing information can be used for controlling the following:

- Ignition timing
- Ignition coil-on time
- Start of injection
- Switching of any cam profiles (e.g., in the Honda VTEC system)

In the Rover VVC system (which introduces a non-constant velocity in the camshaft drive to lengthen or shorten the cam period), the cam flag can be used to provide feedback on the VVC system setting. The cam flag usually is arranged so that its output is high for half a revolution and low for the other half. With a constant velocity, this will lead to a square wave. When the VVC system is being used (which produces a non-constant velocity ratio drive, as discussed in Section 3.6.1), the ratio of the high period to the low period will enable the engine management system to deduce the cam period (Parker, 2000).

Finally, because of the intermittent nature of combustion and the kinematics of the reciprocating engine, the torque produced by an engine varies within its cycle. This in turn leads to speed fluctuations, from which it is theoretically possible to deduce the torque being contributed by each cylinder. In practice, the speed fluctuations are quite small (especially at high speeds); therefore, the resulting information has a poor signal-to-noise ratio. Nonetheless, crankshaft speed fluctuations deduced from the period variations of the toothed wheel pulses can be used to detect misfire (e.g., Connolly, 1994). This is important for two reasons. First, a misfire will lead to a high level of unburned hydrocarbons, even after a catalyst. Second, if there is persistent misfire, then oxidation of fuel within the catalyst can lead to overheating and failure of the catalyst.

3.7.2.2 Throttle Position

The throttle position information is provided by a potentiometer, and this provides an indication to the engine of the load being demanded by the driver. In the case of a direct connection between the accelerator pedal and the throttle, the potentiometer provides immediate information about a change in demand, even before there is a change in air flow rate. In the case of drive-by-wire systems, the accelerator pedal is connected to a potentiometer, and this informs the engine management system of the driver's demands. The throttle potentiometer then provides feedback of its position to the throttle actuator and engine management system.

Immediate knowledge of throttle transients is important because it enables strategies to be implemented for controlling fueling changes during transients. Because small changes in throttle position have a large effect on the air flow when the throttle starts to open, an idle speed control valve allows an air flow to bypass the throttle plate, to provide idle speed control.

3.7.2.3 Air Flow Rate

The air flow rate initially was measured by a pivoting vane (or flap) with a potentiometer, but now a hot wire probe is more common because it has a faster response (say, 5 ms, compared with 35 ms for the vane). The hot wire is part of an electrical bridge circuit, and by determining the resistance of the wire, its temperature can be deduced. By keeping the wire at a constant temperature above the air temperature, and measuring the power that is dissipated by the wire, the air mass flow passing it can be calculated. The 70-mm platinum wire is placed in a venturi section because this speeds the air flow to give a greater cooling effect and thus a stronger signal. Because dirt on the wire would change its performance, the hot wire can be heated briefly when the engine is started or switched off, to burn off any contamination; the platinum has a catalytic effect that promotes oxidation of any dirt on the wire. The circuitry for controlling the wire temperature and amplifying the mass flow signal is integral to the air flow meter. Thus, a signal of a few volts is sent to the electronic control unit—high-level signals are used to make them less susceptible to the effects of electrical noise. Further details are provided by Felger (1988).

Unfortunately, the hot wire anemometer is insensitive to the flow direction, and as wide open throttle (WOT) is approached, the flow becomes more strongly pulsating, and there can be a backflow of air out of the engine (especially with a tuned induction system). If a hot wire anemometer is used, it is necessary to deduce the flow direction (by means of a pressure drop, or by looking for when the flow rate falls to zero), or alternatively to rely on the engine management system to correct the hot wire anemometer results for when backflow is present.

The air flow rate into the engine is a key input for determining the quantity of fuel to be injected and defining the engine operating point. Deducing the air flow rate from the engine speed and throttle position might seem reasonable in theory. However, the throttle response to air flow is very nonlinear, and such a system cannot take into account changes in ambient pressure—at 3000 m, atmospheric pressure is only 70% of the sea-level value.

3.7.2.4 Inlet Manifold Absolute Pressure

The inlet manifold pressure can be measured by a piezo-resistive transducer comprising a silicon diaphragm that has strain gauges etched onto its surface. The signal conditioning (based on a Wheatstone bridge) is integrated into the transducer, so that the signal voltage is large compared to any background noise. One side of the transducer must be hermetically sealed to ensure that the pressure measured is indeed an absolute pressure (i.e., it is not relative to ambient pressure). The engine speed, air temperature, and manifold absolute pressure enable an estimate to be made of the mass flow rate into the engine. When this is compared with the air mass flow rate, it is possible to deduce the level of exhaust gas recirculation. An alternative to a piezo-resistive transducer is when the displacement of the diaphragm is deduced by the change in capacitance relative to a stationary electrode.

3.7.2.5 Air Temperature and Coolant Temperature

Temperatures are conveniently measured by thermistors. These are semiconductor devices that have a resistance that falls rapidly with increasing temperature. They are cheap, and the high-temperature coefficient of resistance enables a signal to be generated with a good ratio of signal to noise. They have a nonlinear temperature dependence, but this is easily linearized by the high computing power in the electronic control unit.

3.7.2.6 Air-Fuel Ratio

Stoichiometric operation can be determined from a lambda sensor (Fig. 3.23); however, a more sophisticated oxygen sensor is needed to measure the air-fuel ratio. When a three-way catalyst is to be used, it is essential to use a feedback system incorporating such a sensor to maintain an air-fuel ratio that is within approximately 1% of stoichiometric. The oxygen or lambda sensor has been described by Wiedenmann et al. (1984). One electrode is exposed to air, and the other electrode is exposed to the exhaust gas. The difference in the partial pressures of oxygen leads to a voltage difference that is related to the difference in partial pressures. Because the platinum electrode also acts as a catalyst for the exhaust gases, then for rich or stoichiometric air-fuel ratios, there is a high output from the lambda sensor (just below 1 V) because the partial pressure of the oxygen will be many orders of magnitude lower than for air. With weak mixtures, the voltage falls to the order of 0.1 V. The platinum

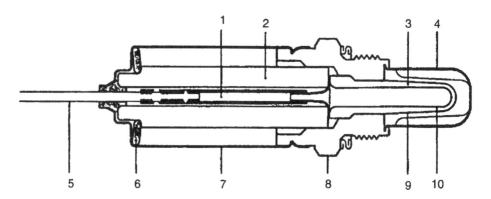

Figure 3.23. *Cross-section drawing of an exhaust oxygen sensor: (1) Contact element. (2) Protective ceramic element. (3) Sensor ceramic. (4) Protective tube (exhaust end). (5) Electrical connection. (6) Disc spring. (7) Protective sleeve (atmosphere end). (8) Housing (−). (9) Electrode (−). (10) Electrode (+). Reproduced with permission from Robert Bosch Ltd.*

electrodes are deposited on zirconia (ZrO_2) stabilized with yttrium oxide. At high temperatures, this acts as a solid electrolyte for oxygen ions (O^{2-}), so that a voltage difference is established between the two electrodes. Measuring the voltage will lead to a flow of electrons in the measuring circuit to and from the electrodes, and this circuit is completed by the flow of oxygen ions through the zirconia.

Because the sensor is used in a way that decides whether the mixture is rich or weak, a control system is needed that makes the air-fuel ratio perturbate around stoichiometric. Because the lambda sensor will work only when it has reached a temperature of approximately 300°C (572°F), this feedback control system can be used only after the engine has started to warm up. The warm-up time can be reduced by having an electrical heating element in the center of the sensor, and this reduces the warm-up time (to 20–30 s). The response time falls with increasing temperature—from the order of seconds at 300°C (572°F) to below 50 ms at 600°C (1112°F). This is another reason for having electrical heating.

For lean-burn engine control, it is necessary to evaluate the partial pressure of the oxygen (as opposed to deciding whether or not oxygen is present), and this has led to the development of the universal exhaust gas oxygen sensor (UEGO—sometimes prefixed with H because the sensor is heated).

Figure 3.24 shows the UEGO sensor, which in effect is a pair of lambda sensors, constructed from three layers of zirconia. All of the layers are heated, but only the top two layers have platinum electrodes and electrical connections. The UEGO sensor operates by seeking to maintain a constant low oxygen partial pressure in the cavity between the sensing cell and the pumping cell. The whole assembly is heated to maintain a constant temperature because the voltage-oxygen partial pressure response is temperature dependent. The sensing cell measures the oxygen partial pressure, and the current imposed on the pumping cell seeks to

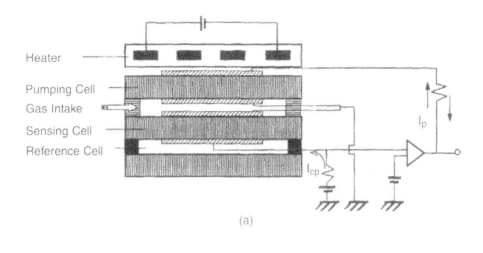

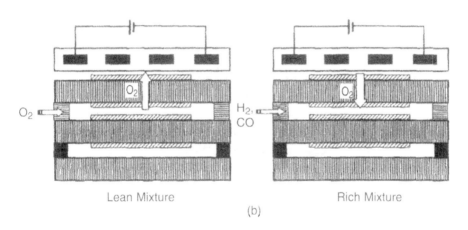

Figure 3.24. *The universal exhaust gas oxygen (UEGO) sensor: (a) structure, and (b) operation (Zhao and Ladommatos, 2001).*

maintain a constant oxygen partial pressure in the measurement cavity between the sensing cell and the pumping cell.

When a weak mixture is present, the concentration gradient between the exhaust gas and the measurement cavity will cause oxygen to diffuse through the porous gas intake, and the electrical current in the pumping cell (which is removing oxygen) will be proportional to the oxygen concentration in the exhaust gas. With rich mixtures, the partial products of combustion (CO, H_2, and hydrocarbons) that have diffused into the measurement cavity will be oxidized, thus causing more to diffuse through the porous gas intake. The richer the mixture, the greater the flow rate of reactants diffusing into the measuring cavity, and the greater the current flow in the pumping cell that is adding oxygen (to maintain the constant oxygen level), so the current will be proportional to the exhaust gas richness.

3.7.2.7 Knock Detector

The knock detector is an accelerometer that detects engine structural vibrations (Fig. 3.25). If a knock detector is fitted, it can retard the ignition at the onset of knock, thereby preventing damage to the engine. The cylinder pressure oscillations associated with combustion knock cause the engine structure to vibrate, and this is why knock can sometimes be heard. Tests must be conducted on the engine to find a location that gives a good vibrational signal, regardless of which cylinder is knocking. Forlani and Ferrati (1987) report that the signal typically is filtered with a pass band of 6–10 kHz and examined in a window from top dead center to 70° after top dead center. Because the signal is examined for a particular time window, the knocking cylinder can be identified and the ignition timing can be retarded selectively. In the case of a V engine, there is likely to be an accelerometer on each bank. The accelerometer consists of a mass mounted on a spring, with a method of detecting the movement of the mass or the force in the spring.

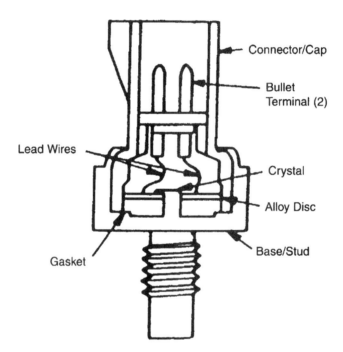

Figure 3.25. *A piezo-electric knock detector. From Turner and Austin (2000). Produced from the Proceedings of the Institution of Mechanical Engineers by permission of the Council of the Institution of Mechanical Engineers.*

It is convenient to combine the spring and the sensing element, and this can be done by mounting a mass on a piece of silicon that is etched with strain gauges, or on a piezo-electric crystal. The piezo-electric crystal has the advantage of high stiffness, so that it is easier to design a transducer with a high natural frequency. The disadvantage is that the piezo-electric crystal produces an electrical charge proportional to the acceleration, and this must be amplified by a high-input impedance amplifier to give a voltage signal. Normally, accelerometers are designed to have a much higher natural frequency than any component present in the signal being measured, so that the transducer has a response that is independent of frequency. However, with the knock detector, structural vibrations at a particular frequency are being detected. Therefore, if this corresponds to the natural frequency of the transducer, the resonance provides "dynamic amplification," and the transducer will give a better signal-to-noise ratio. As with the pressure transducer, it is expedient to incorporate some signal conditioning into the transducer.

The knock sensor provides a safety margin that otherwise would be obtained by having a lower compression ratio or permanently retarded ignition. A striking example of this is provided by Meyer et al. (1984), in which an engine fitted with a knock sensor was run with fuels of a lower octane rating than the octane rating for which it had been designed. Figure 3.26 shows the results, and it can be seen that there is no significant change in the fuel consumption for a range of driving conditions. Indeed, only a slight deterioration was found in the full-load fuel economy.

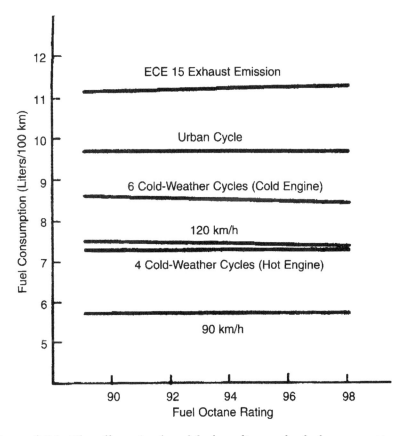

Figure 3.26. *The effect of reduced fuel quality on the fuel consumption of a vehicle fitted with a knock detector. Adapted from Meyer et al. (1984).*

3.8 Engine Management System Functions

The key parameters to be controlled are the ignition timing and air-fuel ratio, with possibly the following:

- EGR valve
- Air purge to the fuel vapor storage canister
- Throttle position (or idle speed adjustment)
- Air conditioning compressor
- Variable geometry induction system settings
- Variable valve timing system settings
- Variable geometry turbocharger settings

3.8.1 Ignition Timing

The ignition timing is determined from a look-up map, in which the values might be recorded as functions of the engine speed and manifold pressure. Figure 3.27 shows that the ignition timing variation is quite complex and would be difficult to obtain from a purely mechanical system.

The ignition timing settings would be modified by knock (as already described in the context of the knock detector) and by coolant and air temperatures. For example, to promote rapid catalyst warm-up, a very retarded ignition timing can be used. This gives a hotter exhaust

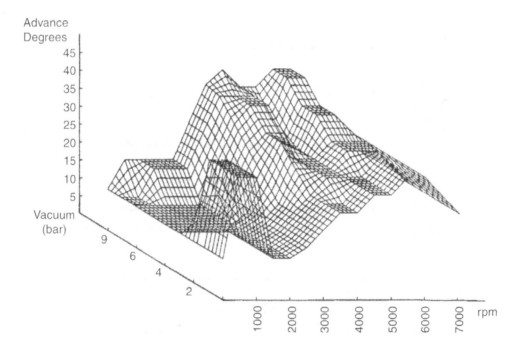

Figure 3.27. *The ignition timing map (as a function of engine speed and inlet manifold pressure), as used in an engine management system (Forlani and Ferrati, 1987).*

because there will have been less time for in-cylinder heat transfer, and the retarded ignition timing will have reduced the work output.

The coil-on-time also might be controlled in response to battery voltage (because the current achieved when the coil is switched off determines the spark energy) and in response to the operating point. When the engine is operating at full load with a rich mixture, the ignition energy requirements are very low. In contrast, at a light load with a weak mixture (or one that is diluted by a high level of exhaust gas recirculation), increasing the spark energy can improve the combustion stability. However, high spark energies can reduce the life of the spark plugs.

3.8.2 Air-Fuel Ratio Control

Steady-state air-fuel ratio control is comparatively straightforward because the requirements can be specified in terms of speed and load (Fig. 3.11). When the engine is warmed up, the lambda sensor provides feedback so that the nominal injection durations stored in memory can be modified to account for different quality fuels (e.g., a fuel containing oxygenates will require an increased injection duration because a lower air-fuel ratio is needed for a given stoichiometry) or the buildup of deposits in the injectors. These corrections can be stored, so that they are used before the lambda sensor is operational the next time the engine is run. There also will be upper and lower bounds on the possible injection durations. Furthermore, the injection duration will be sensitive to the battery voltage because this influences the time it takes for the solenoid valve to open and has a slight effect on the release time—the effective injection period is reduced as the battery voltage decreases.

Fuel will not be injected when the engine is on overrun (i.e., the throttle is closed, and the engine is being used to brake the vehicle), and the engine speed is above a threshold value (say, 1500 rpm). In engines with six or more cylinders, there may even be selective cylinder deactivation (by not injecting fuel) when the engine is at part load. If only half the air entering the engine is being used for combustion, the throttle will have to be open wider, which will reduce the throttling loss. However, precautions are needed to avoid the disabled cylinders cooling down. Thus, it is necessary either to circulate exhaust gas through the cylinders or to disable the valves in such a way as to trap exhaust gases in the cylinder. Fuel also will be turned off when the engine speed exceeds its designed maximum by approximately 1%, and the fueling will not be resumed until the speed has fallen below the rated maximum by approximately 1%. Excess fuel sometimes is supplied through a supplementary injector for cold starting.

The transients that cause the greatest difficulty for port injected engines are those associated with opening and closing the throttle. When the throttle is opened, the air flow increases to its new steady-state value in approximately one engine cycle, whereas the fuel film and droplets respond much more slowly over a few cycles. Also, the increase in manifold pressure is likely to raise the partial pressure of the fuel vapor above its saturation pressure, so that fuel will condense on the walls and droplets. Thus, when the throttle is opened suddenly, there will be an instantaneous reduction in the fuel that is being inducted into the cylinder while the air flow is increasing! As a result, it is necessary to inject extra fuel to achieve a mixture strength close to stoichiometric (or slightly rich if the management strategy is aiming for maximum torque output and not the lowest emissions). Numerous attempts have been made to model the extra fueling requirements, but establishing the control parameters remains a largely empirical exercise during calibration of the engine management system in the vehicle. Conversely, when the throttle is suddenly closed, fuel will evaporate from the droplets and liquid films, and the mixture will tend to richen. Thus, on throttle closing transients, it is necessary to inject less fuel than for the corresponding steady-state air flow.

3.8.3 Exhaust Gas Recirculation (EGR) Control

Exhaust gas recirculation (EGR) is used for reducing the levels of nitrogen oxides (NOx) and also the reduction of part-load throttling losses, for which up to approximately 30% of the mass entering the engine might be from EGR. However, too much EGR interferes with combustion because it reduces the burn rates and increases the level of cycle-by-cycle variations that lead to high hydrocarbon emissions and poor driveability. Thus, it is important to know the level of EGR that is being employed, to ensure compliance with NOx emissions legislation and to be able to set the correct ignition timing for the slower combustion.

An EGR system is not shown as part of Fig. 3.21 but is shown separately in Fig. 3.28. The EGR is drawn into the inlet manifold, as the throttle reduces the pressure below atmospheric pressure. The EGR valve is a very simple spring-loaded needle valve or poppet valve that moves axially; EGR tends to cause deposits, so it is important that the valve is immune to sticking. The actuator for the valve can be a high-power solenoid that is pulsed on and off (with different on-to-off ratios), to hold the valve against the spring at a specified position. A more usual arrangement is to have a diaphragm actuator, on which the actuation pressure is controlled by a solenoid valve that is pulsed on and off. The solenoid valve switches between atmospheric pressure and the inlet manifold pressure to provide a variable pressure on the actuator diaphragm. There usually is no position feedback for the valve, because its flow performance will be dependent on deposits; therefore, feedback is needed from other information.

The level of EGR in the inlet manifold could be established if a UEGO sensor was used to compare the oxygen level in the inlet manifold with that in the exhaust; however, this would require an extra sensor. Instead, the level of EGR is inferred by computing the volumetric flow rate of air into the inlet manifold (based on the air mass flow rate, temperature, and pressure), and comparing that with the volumetric flow rate into the engine, based on the engine speed and inlet manifold pressure.

3.8.4 Additional Functions

The engine management system must control the vapor that is produced by the fuel tank. The major source of evaporative emissions from a passenger car is a consequence of the fuel tank being subject to diurnal temperature variations. The fuel tank must be vented to atmosphere to avoid a pressure buildup as the fuel tank warms during the day. By venting the fuel tank

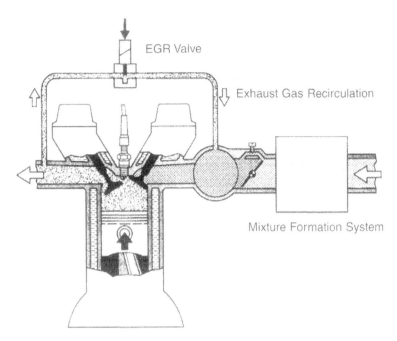

Figure 3.28. *An EGR system (Gunther, 1988).*

through a canister containing active charcoal (Fig. 3.21), the fuel vapor is absorbed. When the fuel tank cools at night, air is drawn in through the carbon canister, and this removes some of the hydrocarbons from the active charcoal. However, to ensure adequate purging of the active charcoal, it is necessary to draw additional air through the active charcoal. This is achieved by drawing some of the air flowing into the engine through the carbon canister. Clearly, the air-fuel ratio of this stream will be unpredictable, but a closed loop engine management system will enable the engine to always adopt the intended air-fuel ratio.

The engine management system also is likely to control the air conditioning pump. The air conditioning pump can absorb about 5 kW, and it usually is driven from a belt drive via an electromagnetic clutch. The engine management system can decouple the pump when the driver demands maximum power (e.g., during overtaking) because 5 kW can represent a significant fraction of the output of a European engine. Also, when the engine is idling, if the air conditioning pump is switched by the engine management system, it can anticipate the load being applied or removed and thus provide better idle speed control, or indeed increase the idle speed if there is a high cooling demand. In the case of turbocharged or supercharged engines, boost pressure control and any variable geometry settings can be controlled by the engine management system.

Inputs from the vehicle will define the vehicle speed and gear ratio selected, and, in the case of a vehicle with an automatic transmission, the engine management system can determine the gear ratio. Different modes can be selected, either from explicit driver instructions (via a switch to select either "sport" or "economy"), or by detecting the driver's use of the throttle and brakes. Vehicle inputs also will be used by the engine management system to provide average speed and fuel consumption display information. With drive-by-wire throttle systems, the engine management system can provide cruise control; obviously, it is necessary to have an additional input to detect when the brakes are being applied. When an antilock brake system (ABS) is installed, the information about incipient wheel slip can be used to regulate the power output of the engine to provide a traction control system (TCS). The engine torque output control can be provided by throttle control or very rapidly by retarding the ignition timing. Apart from saving tire wear, this also provides maximum torque transmission. The vehicle speed information can be used to adjust the pressure in a power steering system so that there is more assistance at low speeds and conversely more "feel" or feedback through the steering wheel to the driver at high speeds.

3.8.5 Concluding Remarks on Engine Management Systems

It is difficult to provide optimum engine control because of the interdependence of many engine parameters and the absence of some direct measurements of what is being controlled (e.g., direct measurements are not made of either the engine torque or its emissions). With the multitude of control loops, the hierarchy must be carefully defined; otherwise, one loop might be working directly against another.

Furthermore, even supposedly identical engines have a different performance when new, and, of course, they will age differently, too. What is needed is some form of adaptive control system, in which the operating point would be found by optimizing the fuel economy or emissions. An example of an adaptive engine control system is provided by Wakeman et al. (1987) and Holmes et al. (1988). Perturbations are applied to the ignition timing, and the slope of the ignition timing/torque curve is inferred from the response of the engine speed. In this way, the control system can find the MBT ignition timing for a particular operating condition and then store this as a correction value from the ignition timing map. A similar system can be used for controlling EGR to minimize NOx emissions. Holmes et al. also describe an approach to knock control, in which the timing retard from the ignition map is stored as a function of operating point. The system is arranged so that when there is a change in fuel quality, the ignition timing will converge to the new optimum timing schedule. The fuel economy benefits of an adaptive ignition system are shown by Wakeman et al. Four nominally identical vehicles were tested on the ECE 15 urban driving cycle, and it was found that the adapted ignition timing map was different for each vehicle (and different from the manufacturer's calibration). Fuel savings of up to 9% were achieved.

Improvements to control strategies can be made when there are mathematical models to describe the processes being controlled. Hendricks and co-workers (e.g., Hendricks et al., 1993 and 1996) have modeled intake filling dynamics and fuel transport for throttle transients. A related idea (again for air-fuel ratio control) is the use of neural networks that are able to "learn" about the dynamic characteristics of throttle transients and mixture preparation. Again with neural networks, mathematical models to describe sub-processes enable a more refined approach to be developed. Examples of this for mixture control are provided by Won et al. (1998) and Manzie et al. (1998).

3.9 Conclusions

The combination of sophisticated engine management systems and catalysts has enabled spark ignition engines to be developed with remarkably low emissions. Over the 30-year period from 1970 to 2000, U.S. emissions legislation has required more than a thirty-fold decrease in unburned hydrocarbons, a ten-fold reduction in carbon monoxide, and a twenty-five-fold reduction in nitric oxides. During that same time, there has been an increase in the specific output, fuel economy, and design life. Technologies such as variable geometry induction systems, variable valve timing, and gasoline direct injection have permitted the simultaneous increase of output and efficiency. Supercharging also is being used as a means of increasing the specific output and giving a better part-load fuel economy than a larger-capacity naturally aspirated engine of comparable output.

3.10 Questions

3.1 Why does the optimum ignition timing change with engine-operating conditions? What are the advantages of electronic ignition with an electronic control system?

3.2 Two spark ignition gasoline engines having the same swept volume and compression ratio are running at the same speed with wide open throttles. One engine operates on the two-stroke cycle, and the other operates on the four-stroke cycle. State with reasons for the following:

   a. Which engine has the greater power output?
   b. Which engine has the higher efficiency?

3.3 The Rover M16 spark ignition engine has a swept volume of 2.0 liters and operates on the four-stroke cycle. When installed in the Rover 800, the operating point for a vehicle speed of 120 km/h (74.5 mph) corresponds to 3669 rpm and a torque of 71.85 Nm, for which the specific fuel consumption is 298 g/kWh.

Calculate the bmep at this operating point, the arbitrary overall efficiency, and the fuel consumption (liters per 100 km). If the gravimetric air-fuel ratio is 20:1, calculate the volumetric efficiency of the engine, and comment on the value.

The calorific value of the fuel is 43 MJ/kg, and its density is 795 kg/m^3. Ambient conditions are 27°C (81°F) and 1.05 bar.

Explain how both lean-burn engines and engines fitted with three-way catalysts obtain low exhaust emissions. What are the advantages and disadvantages of lean-burn operation?

3.4 The Rover K16 engine is a four-stroke spark ignition engine with a swept volume of 1.397 liters and a compression ratio of 9.5. The maximum torque occurs at a speed of 4000 rpm, at which point the power output is 52 kW. The maximum power output of the engine is 70 kW at 6250 rpm. Suppose the brake specific fuel consumption is 261.7 g/kWh using 95 RON lead-free fuel with a calorific value of 43 MJ/kg.

Calculate the corresponding Otto cycle efficiency, the maximum brake efficiency, and the maximum brake mean effective pressure. Give reasons why the brake efficiency is less than the Otto cycle efficiency. Show the Otto cycle on the p-V state diagram, and contrast this with an engine indicator diagram—identify the principal features.

Explain why, when the load is reduced, the part-load efficiency of a diesel engine falls less rapidly than the part-load efficiency of a spark ignition engine.

3.5 Because the primary winding of an ignition coil has a finite resistance, energy is dissipated while the magnetic field is being established after switching on the coil. If the primary winding has an inductance of 5.5 mH and a resistance of 1.9 ohms, and the supply voltage is 11.6 V, show that:

a. The energy stored after 2 ms would be 26 mJ, increasing to 79 mJ after 6 ms.

b. The theoretical efficiency of energy storage is approximately 65% for a 2-ms coil-on-time, falling to 32% for a 6-ms coil-on-time.

What are the other sources of loss in an ignition coil?

Define MBT ignition timing, and describe how it varies with engine speed and load. Under what circumstances are ignition timings other than MBT used?

3.6 Redraw Fig. 3.9 using a scale of lambda (the relative air-fuel ratio, the reciprocal of equivalence ratio), and comment regarding under what circumstances is this a more appropriate form for presenting the data.

3.7 A 4-liter swept volume spark ignition engine is supercharged using a Roots blower to give a pressure ratio of 1.5. At maximum power, the engine output is 240 kW at 5000 rpm.

If the air mass flow rate is 243 g/s at ambient conditions of 1 bar and 25°C (15.5°F), calculate the power requirement of a Roots compressor with a pressure ratio of 1.5, for mechanical and volumetric efficiencies both of 80%. What would the exit air temperature be from the Roots blower?

After intercooling, the air temperature is reduced to 50°C (31°F). Neglecting any pressure drop in the intercooler, calculate the volumetric efficiency of the engine.

CHAPTER 4

Diesel Engines

4.1 Introduction

The key characteristics of diesel engines have already been introduced in Chapter 2. The diesel cycle analysis (Section 2.4.2) shows that the efficiency increases as the compression ratio is increased, but reduces slightly as the load ratio is increased. However, it must be appreciated that increasing the compression ratio has a diminishing effect on increasing the ideal cycle efficiency, and that for a practical engine, there are no benefits in increasing the compression ratio above approximately 20:1. This is true because increases in frictional losses (associated with the higher pressures) and increases in heat transfer (because of the higher temperatures and pressures, and the adverse surface-area-to-volume ratio) outweigh any gains in the ideal cycle efficiency. Combustion is initiated by self-ignition of the fuel (hence the term compression ignition engines), and the main requirement of the fuel is that it is susceptible to self-ignition (Section 2.7).

The compression ratio is likely to be selected on the basis of the lowest compression ratio that will provide satisfactory cold-starting performance, and even this compression ratio can be higher than that for optimum efficiency. For direct injection diesel engines with a displacement of 0.5 L/cylinder, the compression ratio is approximately 18:1, whereas for larger engines (say, 1 L/cylinder and greater), the compression ratio will be approximately 15:1 or slightly lower. The larger displacement engines have a better volume-to-surface-area ratio; thus, less heat transfer occurs, and satisfactory cold-starting can be achieved with a lower compression ratio. In very high-output diesel engines (i.e., those with a high boost pressure and a bmep of 20 bar and higher), the compression ratio may be lowered to limit the peak pressure, even if this results in difficulties in starting the engine without starting aids.

The performance of a diesel engine is critically dependent on its combustion system, and this means both the combustion chamber and the fuel injection system (Section 4.3). There are two types of diesel combustion systems: direct injection, and indirect injection. Although development of indirect injection engines has all but ended, many remain in use; thus, both types are described in Section 4.2. As explained in Section 2.4, diesel engine combustion includes diffusion combustion, so that the air-fuel ratio is always weaker than stoichiometric (to avoid hydrocarbon and particulate emissions). In consequence, the specific output of a naturally aspirated diesel engine is much lower than that of a spark ignition engine. Fortunately, turbocharging can increase both the efficiency and output of diesel engines, and this is treated in Section 4.5.

4.2 Direct and Indirect Injection Combustion Chambers

Two types of combustion chambers are used in diesel engines. These are direct injection (DI) and indirect injection (IDI), as shown in Fig. 4.1. The IDI engine offers faster combustion and thus the potential for higher engine speeds. When combustion speed is not the limiting factor, there is nonetheless a limitation on the maximum mean piston speed (v_p), which limits the engine speed (N)

$$N = \frac{v_p \times 60}{2 \times L} \tag{4.1}$$

The maximum mean piston speed is a result of mechanical considerations and the flow through the inlet valve, and is limited to approximately 12 m/s. For an engine speed of 3000 rpm, this corresponds to a stroke of 0.12 m, or a swept volume of approximately 1 L/cylinder. Thus, if the combustion speed limits direct injection engines to 3000 rpm, then for engines of less than 1 L/cylinder swept volume, higher speeds are obtainable only by using the faster indirect injection combustion system. With careful development of the combustion system, direct injection engines can now operate at speeds of up to 5000 rpm.

The pre-chamber of the IDI engine in Fig. 4.1 constitutes approximately half of the clearance volume, and the lower part of the swirl chamber is an insert made from a heat-resisting material that is thermally isolated from the cylinder head. The insert heats when the engine has started, thereby reducing heat transfer losses, and its high temperature helps fuel evaporation, ignition, and combustion. Indirect injection engines invariably have a heater plug projecting into the pre-chamber. This is switched on prior to cranking the engine, so that when the fuel is injected, its ignition is facilitated. Small DI engines (say, 500 cm^3/cylinder and smaller)

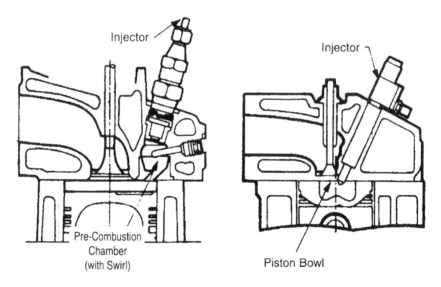

Figure 4.1. *Indirect injection (IDI) (left) and direct injection (DI) (right) combustion chambers.*

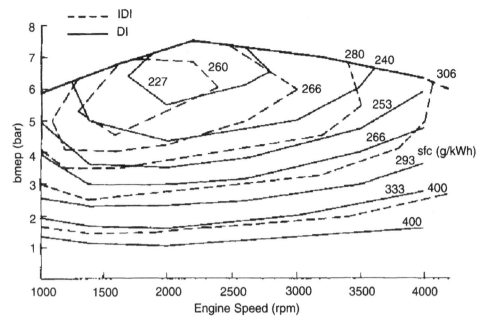

Figure 4.2. *Comparison of specific fuel consumption (g/kWh) maps for four-cylinder 2.5-liter naturally aspirated DI and IDI engines (Hahn, 1986).*

also can use heater plugs, and these are sited so that their tips are close to the injector. Sometimes the heater plugs remain switched on (possibly at a reduced power setting) so that cold running is quieter.

A direct comparison between DI and IDI engines has been reported by Hahn (1986), and the differences in brake specific fuel consumption (bsfc) are shown by Fig. 4.2. The IDI engine has a higher bsfc than the DI engine because of the following:

a. The compression ratio (typically greater than 20) has been selected on the basis of starting the engine, not for the maximum brake efficiency.

b. There is substantial heat transfer and a pressure drop in the pre-chamber throat.

Direct injection engines demand rigorous matching of the fuel spray and air motion. Initially, this was achieved in high-speed DI engines by having a modest injection pressure (say, 600 bar) and swirl (a rotating flow with its axis parallel to the cylinder axis). However, the kinetic energy associated with swirl comes from the pressure drop in the induction process; thus, there is a tradeoff between swirl and volumetric efficiency (and power output). In addition to promoting good mixing of the fuel and air, swirl increases heat transfer, which, of course, lowers the engine efficiency. Therefore, the current trend is toward lower levels of swirl, in which case four-valves-per-cylinder layouts can be used, with benefits for the volumetric efficiency and power output (Pischinger, 1998). The good air-fuel mixing now comes from more advanced fuel injection equipment.

4.3 Fuel Injection Equipment

To obtain small droplets that will evaporate quickly, the fuel injector nozzle holes can be as small as 0.15 mm (0.006 in.). To inject sufficient fuel in the short time available, injection pressures must be 1500 bar or higher. These high pressures also ensure that the fuel jet disperses well within the combustion chamber. Likewise, these high pressures have led to the use of electronic unit injectors (EUI) and common rail (CR) injection systems in preference to the traditional pump–line–injector (PLI) systems.

Unit injectors have the pumping element and injector packaged together, with the pumping element operated from a camshaft in the cylinder head. This eliminates the high-pressure fuel line and its associated pressure propagation delays and elasticity. Common rail fuel injection systems have a high-pressure fuel pump that produces a controlled and steady pressure, and the injector controls the start and end of injection.

Common rail and EUI systems have scope for pilot injection (to control the amount of fuel injected during the ignition delay period, thereby controlling combustion noise) and more desirable injection pressure characteristics. Figure 4.3 shows how injection pressure varies significantly with engine speed for pump–line–injector (PLI) systems, and that for low speeds, only low injection pressures are possible. The low injection pressures limit the quantity of fuel that can be injected because of poor air utilization, thereby limiting the low-speed torque of the engine. With CR injection systems, there is independent control of the injection pressure across a wide range of operating speed.

Each of these systems will now be discussed in more detail.

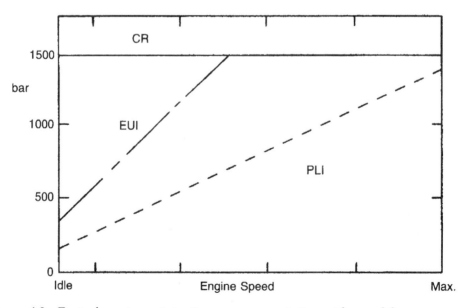

Figure 4.3. *Typical maximum injection pressure variation with speed for common rail (CR), electronic unit injector (EUI), and pump–line–injector (PLI) systems (Stone, 1999).*

4.3.1 Pump–Line–Injector (PLI) Systems

A comprehensive treatment of diesel fuel injection equipment (both in-line pumps and rotary pumps) is provided by Adler (1994). The discussion here will be limited to rotary or distributor pumps, because the use of in-line pumps has been displaced by unit injectors. The pump is driven from the crankshaft by gears, chains, or nowadays a timing belt, at half crankshaft speed for a four-stroke engine. Because of the high pressures (of the order of 1000 bar) and although only a small quantity of fuel might be injected (say, 50 mm^3) because that fuel is injected very quickly (of the order of 1 ms), the instantaneous power is quite high ($1000 \times 10^5 \times 50 \times 10^{-9}/10^{-3} = 5$ kW/cylinder) and the instantaneous torque is very high for a timing belt (20 Nm at 2500 rpm pump speed). The pump is connected by a line (a thick-walled tube) to the injector. Figure 4.4 shows a distributor-type fuel injection pump.

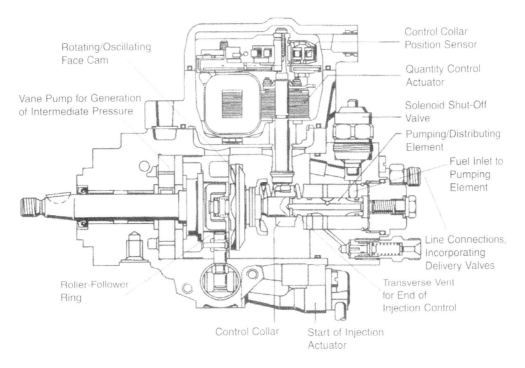

Figure 4.4. *Distributor-type fuel injection pump. Adapted from Tschöke (1994).*

The key to the distributor fuel pump is the face cam that is rotating. Because it reacts against roller-followers in a ring that is fixed axially, the face cam is made to oscillate axially. Thus, the face cam causes the pumping/distributing element to oscillate and rotate, thereby both pumping the fuel to a high pressure and directing it to the appropriate delivery valve and fuel line. The roller-follower ring can be rotated a small amount by the start of injection actuator, and this changes the phasing of the cam lift relative to the crankshaft, thereby controlling the start of injection timing.

The quantity of fuel injected is determined by the axial position of the control collar, which allows fuel to exhaust from the pumping/distributing element through the transverse vent in the pumping/distributing element. The vane pump raises the fuel pressure to an intermediate pressure, which is kept constant by a pressure relief valve (not shown). The intermediate pressure fuel is fed to the pumping/distributing element in the position shown, but this inlet is subsequently sealed as the pumping/distributing element rotates. The axial position of the control collar (and hence the quantity of fuel injected) is determined by the quantity control actuator. The solenoid shut-off valve stops injection by preventing admission of fuel to the pumping/distributing element.

The pump is connected to the injectors by thick-walled pipes. These all should be of the same length to ensure the same pressure propagation delay for all cylinders (approximately 1 ms/m) and that any pressure pulsation effects in the fuel pipes affect all cylinders equally. The injector comprises two parts: the nozzle, and the nozzle holder, as shown in Fig. 4.5.

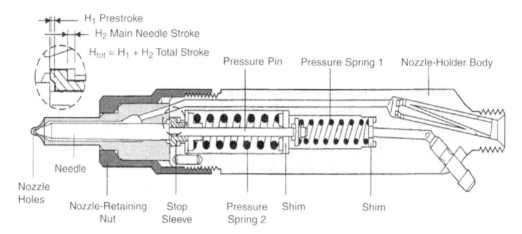

Figure 4.5. *Injector and nozzle assembly. Adapted from Warga (1994).*

The most important part of the fuel injector is the nozzle. The nozzle has a needle that closes under a spring load when it is not spraying. Although less prone to blockage, open (needle-less) nozzles are not used because they dribble. When an injector dribbles, combustion deposits build up on the injector, and the engine exhaust is likely to become smoky. The needle-opening and needle-closing pressures are determined by the spring load and the projected area of the needle. The pressure to open the needle is greater than that required to maintain it in an open position, because in the closed position, the projected area of the needle is reduced by the seat contact area. The differential pressures are controlled by the relative needle diameter and seat diameter. A high needle-closing pressure is desirable because it maintains a high seat pressure, thereby giving a better seal. This also is desirable because it keeps the nozzle holes free from blockages caused by decomposition of leaked fuel. In automotive applications, the nozzles typically are approximately 20 mm (0.8 in.) in diameter and 45 mm (1.8 in.) long, with 4-mm (0.16-in.) diameter needles. Figure 4.5 illustrates a valve covered orifice (VCO) arrangement, in which the needle seats across the nozzle holes. After the end of injection, any fuel in the nozzle holes or downstream of the seating area will be heated and can enter the combustion chamber. This fuel does not necessarily have time to be properly burned; thus, it can lead to deposits, hydrocarbon emissions, and particulate emissions. The VCO design minimizes the amount of fuel that can enter the combustion system after injection.

The injector shown in Fig. 4.5 has a two-stage spring arrangement that provides pilot injection. Pilot injection, in which a small amount of fuel is injected during the ignition delay period, is a means of reducing diesel knock, by limiting the amount of mixture that is burned rapidly. The weak spring (1), acting through the central pressure pin, allows the nozzle to open a small amount (H_1, usually less than 0.1 mm [0.004 in.]) at low pressures, during the early part of the injection pump pumping plunger stroke. (The cam profile in the pump obviously must be matched to the injector characteristics.) During pilot injection, the nozzle needle is held against the stop sleeve, which is held against the strong spring (2). As the pressure rises further, the nozzle needle and strong spring lift further ($H_1 + H_2$) for the main part of the injection. The pre-loads from the two springs are controlled by shims.

The spray pattern from the injector is very important, and high-speed combustion photography and laser-based techniques for measuring droplet size distributions and velocities can be very informative. The testing can be either in an engine or in a special combustion rig that replicates the end of compression temperatures and pressures.

4.3.2 Electronic Unit Injectors (EUI)

In the Delphi Diesel Systems electronic unit injector (EUI) (Fig. 4.6), both the quantity and the timing of injection are controlled electronically through a Colenoid actuator. The Colenoid is a solenoid of patented construction that can respond very quickly (injection periods are of

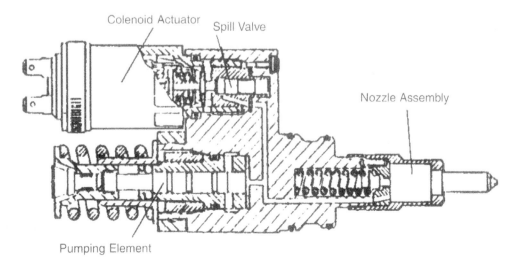

Figure 4.6. *A Delphi electronic unit injector (EUI). Adapted from Frankl et al. (1989).*

the order 1 ms), to control very high injection pressures (up to 1600 bar or so). The Colenoid controls a spill valve, which in turn controls the injection process. The pumping element is operated directly from a camshaft (or indirectly via a rocker), and the whole assembly is contained within the cylinder head.

An alternative approach to the EUI is the Caterpillar Hydraulic Electronic Unit Injector (HEUI, also supplied by other manufacturers). Figure 4.7 shows the HEUI, which uses a hydraulic pressure intensifier system with a 7:1 pressure ratio to generate the injection pressures. The hydraulic pressure is generated by pumping engine lubricant to a controllable high pressure. Similar to CR injection systems, there is control of the injection pressure. The HEUI uses a two-stage valve to control the oil pressure, and this is able to control the rate at which the fuel pressure rises, thereby controlling the rate of injection, because a lower injection rate can help control NOx emissions.

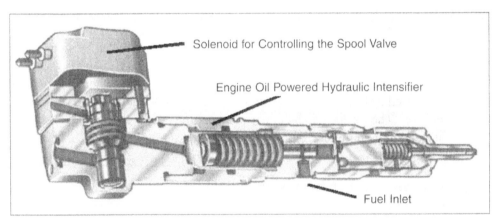

Figure 4.7. *A hydraulic electronic unit injector (HEUI).*
Adapted from Walker (1997).

4.3.3 Common Rail (CR) Fuel Injection Systems

Common rail (CR) fuel injection systems decouple the pressure generation from the injection process and have become popular because of the possibilities offered by electronic control. The key elements of a CR fuel injection system are as follows:

- A (controllable) high-pressure pump
- The fuel rail with a pressure sensor
- Electronically controlled injectors
- An engine management system (EMS)

The injector is an electro-hydraulic device, in which a control valve determines whether or not the injector needle lifts from its seat. The engine management system can divide the injection process into four phases: two pilot injections, main injection, and post-injection (for supplying a controlled quantity of hydrocarbons as a reducing agent for NOx catalysts).

Common rail injection also enables a high output to be achieved at a comparatively low engine speed (Piccone and Rinolfi, 1998).

4.4 Diesel Engine Emissions and Their Control

4.4.1 *Diesel Engine Emissions*

Diesel engine emissions have already been previewed in Section 2.6.3. Figure 4.8 shows a comparison of the emissions performance of DI and IDI engines.

The NOx emissions increase with load because of the increase in combustion temperature, and this increase in combustion temperature (or rather the later time in the expansion stroke when the temperature falls to a level at which reactions cease) is why the hydrocarbon emissions fall. Similarly, retarding the injection timing in all cases leads to lower combustion temperatures. This lowers the NOx emissions but increases the hydrocarbon emissions. The DI diesel has lower NOx emissions than the IDI engine because the lower compression ratio gives lower in-cylinder temperatures. The hydrocarbon emissions are higher from the DI engine because there is a longer ignition delay period, resulting in more over-dilution at the fringes of the spray. The ignition delay period increases with a reduced load in the DI engine because the combustion chamber surface temperatures are lower, and there will be more cooling of the charge during compression. In contrast, the pre-chamber insert in the IDI

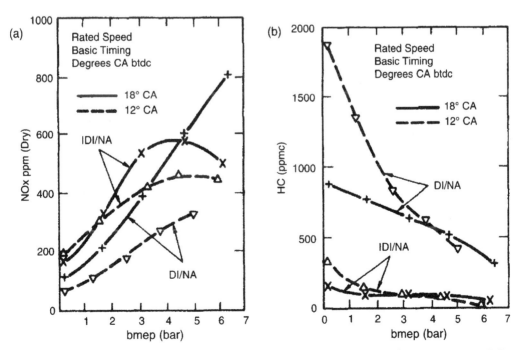

Figure 4.8. *A comparison of emissions from naturally aspirated direct injection (DI) and indirect injection (IDI) diesel engines: (a) nitrogen oxide emissions and (b) hydrocarbon emissions (Pischinger and Cartellieri, 1972).*

engine is always hot enough to ensure a short ignition delay period. Particulate emissions have already been discussed in Section 2.6.3

4.4.2 Diesel Engine Emissions Control

Catalysts have already been discussed in the context of spark ignition engines (Section 3.5), and this includes lean-burn NOx reducing catalysts (Section 3.5.3). However, a discussion is needed here of exhaust gas recirculation (EGR) (because its effect on diesel engine combustion is somewhat different than on spark ignition engines) and particulate traps.

4.4.2.1 Exhaust Gas Recirculation (EGR)

In a diesel engine, the exhaust gas recirculation (EGR) is displacing oxygen. Therefore, only a limited amount of EGR can be used before there is insufficient oxygen for thorough combustion, with consequential rises in the emissions of carbon monoxide, particulates, and unburned hydrocarbons. When these emissions rise, because they are products of partial combustion, it is inevitable that the fuel consumption also will rise. The highest levels of EGR are used at low speeds and low loads, and the EGR level must be decreased as either the load or the speed are increased. There is a tradeoff between particulate emissions and NOx, because increasing the level of EGR reduces the amount of oxygen that is necessary for soot oxidation. However, a higher level of swirl (produced by closing one inlet port of a four-valves-per-cylinder combustion system) gives a more advantageous tradeoff between NOx and particulates because the swirl ensures better oxygen utilization. High injection pressure also can be used to obtain a better NOx/particulates tradeoff, but Herzog (1998) argues that this leads to noisier combustion than if EGR is used.

Suzuki (1997) points out that EGR in diesel engines has only approximately half the effect that occurs with stoichiometrically operated spark ignition engines. In the spark ignition engine, the stoichiometry is unaffected, and there is a significant increase in the heat capacity of the mixture. In contrast, EGR richens the mixture in a diesel engine and has less effect on the heat capacity. A higher heat capacity mixture, of course, means lower combustion temperatures, which in turn means lower NOx emissions. The mechanism for NOx reduction by EGR is not obvious but has been elucidated by some experiments in which the dilution, chemical, and thermal effects have been isolated for the carbon dioxide present in EGR (Ladommatos et al., 1998):

a. The "dilution" effect was examined by replacing oxygen with a nitrogen/argon mixture with the same heat capacity as the oxygen.

b. The "chemical" effects were examined by replacing nitrogen with a carbon dioxide/argon mixture with the same heat capacity as nitrogen, so that the carbon dioxide could lower the combustion temperature through dissociation.

c. The "thermal" effects were examined by replacing nitrogen with a nitrogen/helium mixture with the same heat capacity as carbon dioxide, so that there would be the same cooling effect.

After making due allowance for changes in the ignition delay, Ladommatos et al. (1998) concluded that the main effect of the carbon dioxide was through reducing the oxygen availability and that a small effect was attributable to the change in heat capacity. The dissociation of carbon dioxide was found to have a small contribution to the NOx reduction, but it did contribute toward reducing particulate emissions. Similar arguments would apply to the water vapor because it has a similar heat capacity to carbon dioxide, and it too can dissociate. A disadvantage of EGR is its tendency to increase engine wear rates, but low-sulfur-level fuels reduce the differences between wear rates with and without EGR.

Cooled EGR also is used, because if it displaces the same amount of air, it will represent a larger fraction of the charge and will increase the NOx reductions due to the "thermal" and "chemical" effects. In practice, this means that a lower level of EGR can be used for a specified level of NOx emissions, with a consequentially reduced increase in the particulate and other emissions. This also means that in a turbocharged engine at high loads, the necessary levels of EGR can be achieved without recourse to devices such as inlet throttles. There are many different schemes for implementing EGR in a turbocharged engine, and several of these are reviewed by Suzuki (1997). Cooling EGR has its disadvantages—most notably a tendency to increase the ignition delay period and thereby increase the combustion noise.

When EGR is used, it is necessary to have some form of feedback system. The level of EGR achieved for a given EGR valve position will depend on the condition of many components, not the least deposits in the EGR system. Feedback can be provided by measuring the oxygen level in the inlet manifold, or by measuring the air mass flow rate, the inlet manifold temperature, and the absolute pressure. Diesel engine management systems are discussed further in Section 4.6.

4.4.2.2 Particulate Traps

Hawker (1995) points out that for diesel engines, conventional platinum-based oxidation catalysts give useful reductions in the gaseous unburned hydrocarbons (and, indeed, any carbon monoxide) but have little effect on the soot. However, before catalyst systems can be considered, the levels of sulfur in the diesel must be 0.05% by mass or less. This is because an oxidation catalyst would lead to the formation of sulfur trioxide and then sulfuric acid. In turn, this would lead to sulfate deposits that would block the catalyst. An additional advantage of using a catalyst is that it should lead to a reduction in the odor of diesel exhaust. Particulates can be oxidized by a catalyst incorporated into the exhaust manifold, in the manner described by Enga et al. (1982). However, for a catalyst to perform satisfactorily, it must be operating above its light-off temperature. Because diesel engines have comparatively cool exhausts, catalysts do not necessarily attain their light-off temperatures.

Particulate traps are usually filters that require temperatures of approximately 550–600°C (868–1112°F) for soot oxidation. This led to the development of electrically heated regenerative particulate traps, examples of which are described by Arai and Miyashita (1990) and Garrett (1990). The regeneration process does not occur with the exhaust flowing through the trap. Either the exhaust flow is diverted, or the regeneration occurs when the engine is inoperative. Air is drawn into the trap, and electrical or other heating is used to obtain a temperature high enough for oxidation of the trapped particulate matter. Pischinger (1998) describes how additives in the diesel fuel can be used to lower the ignition temperature, so that electrical ignition is needed only under very cold ambient conditions or when the driving pattern is exclusively short-distance journeys. Particulate traps have trapping efficiencies of 80% and higher, but it is important to ensure that the back-pressure in the exhaust is not too high. An alternative to a filter is the use of a cyclone. To make the particulates large enough to be separated by the centripetal acceleration in a cyclone, the particles must be given an electrical charge so that they agglomerate before entering the cyclone (Polach and Leonard, 1994).

An oxidation catalyst and soot filter can be combined in a single enclosure, as shown in Fig. 4.9 (Walker, 1998). Hawker (1995) details the design of such a system. The platinum catalyst is loaded at 1.8 g/L onto a conventional substrate with 62 cells/cm^2, and this oxidizes not only the carbon monoxide and unburned hydrocarbons but also the NOx to nitrogen dioxide (NO_2). The nitrogen dioxide (rather than the oxygen) is responsible for oxidizing the particulates in the soot filter. The soot filter is an alumina matrix with 15.5 cells/cm^2, but with adjacent channels blocked at alternate ends. As the exhaust gas enters a channel, it then flows through the wall to an adjacent channel—hence, the name of "wall flow" filter. With the presence of a platinum catalyst, the processes of soot trapping and destruction are continuous above temperatures of 275°C (527°F), and the system is known as a continuously regenerating trap

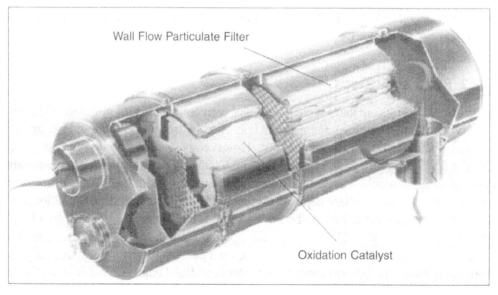

Figure 4.9. *An oxidation catalyst and soot filter assembly for use in diesel engines. Adapted from Walker (1998).*

(CRT). The system introduces a back-pressure of approximately 50 mbar, and the duty cycle of the vehicle must ensure that a temperature of 275°C (527°F) is regularly exceeded. Such an assembly also can be incorporated into a silencer (muffler), so that existing vehicles can be retrofitted (Walker, 1998).

4.5 Turbocharging

4.5.1 Introduction

Automotive engines invariably use turbochargers with radial flow compressors and turbines as shown in Fig 4.10; larger engines can use axial flow turbines. Watson and Janota (1982) provide a comprehensive treatment of turbochargers and turbocharging, whereas the treatment here is derived from Stone (1999).

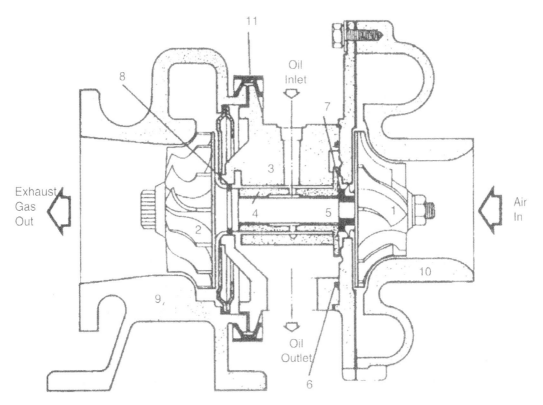

Figure 4.10. *An automotive turbocharger with a radial compressor and a radial turbine: (1) Compressor wheel. (2) Turbine wheel. (3) Bearing housing. (4) Bearing. (5) Shaft. (6) Seal. (7) Mechanical face seal. (8) Piston ring seal. (9) Turbine housing. (10) Compressor housing. (11) "V" band clamp. (Allard, 1982).*

Radial flow compressors work well without stator blades because of the conservation of the moment of momentum in the diffuser, and the increase of the flow area as the radius increases; this can be explained with reference to Fig. 4.11.

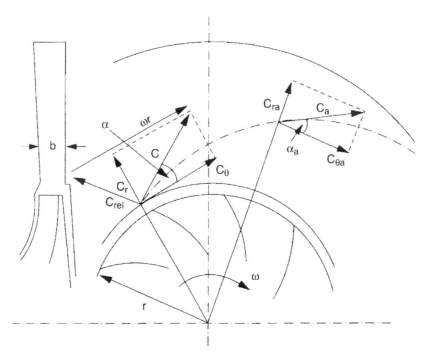

Figure 4.11. *Velocity triangles for a radial flow compressor with a vaneless diffuser (Stone 1999).*

In Fig. 4.11, the flow has a velocity relative to the blade of C_{rel}, and this can be resolved into two components: the radial component (C_r), and a tangential component. The tangential component must be converted to an absolute velocity by subtracting the blade tip velocity (ωr) to give the absolute tangential (or whirl) velocity (C_θ). The flow in the stator immediately after leaving the rotor thus has absolute velocity components comprising the radial component (C_r) and the tangential velocity (C_θ). These absolute velocities also can be defined as the velocity C with a flow angle α, and this is tangential to the particle path, which is shown as the chain dotted line. However, the action of the diffuser is best explained separately in terms of the radial and tangential velocity components.

For convenience, the flow will be considered as frictionless. First, even with a constant depth (b) diffuser, as the radius is increased there is a larger circumferential area, so that the radial velocity component will be reduced. (The mass flow is the product of the density, the circumferential area, and the radial velocity component.) Second, the tangential velocity component will reduce because of the conservation of the moment of momentum, which states that the product of the radius and the tangential velocity will be a constant. Thus, it can be seen how the flow is diffused (decelerated), which will lead to a rise in pressure. Finally, this rise in pressure will increase the density, and this too contributes to the lower velocities.

Analysis of turbocharged engines is similar to open-circuit gas turbines, except that the combustion chamber is replaced by an engine, as illustrated in Fig. 4.12. By applying the steady

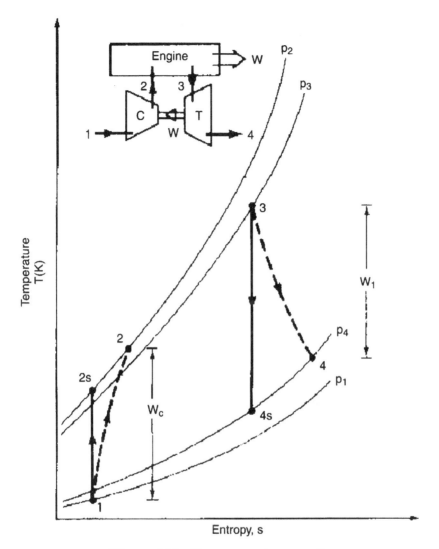

Figure 4.12. *Temperature/entropy diagram for a turbocharger (Stone, 1999).*

flow energy equation and neglecting heat transfer, changes in kinetic energy (unless stagnation temperatures are used), and changes in potential energy, it is possible to write equations for the turbine and compressor specific work, noting that the compressor work will be treated as a negative quantity:

Turbine work: $\quad w_t = h_3 - h_4$

(4.2)

Compressor work: $\quad w_c = h_2 - h_1$

By treating the gases as perfect (or alternatively as semi-perfect gases with an appropriate mean value of the heat capacity), the specific work equations become

Turbine work: $$w_t = c_{p34}(T_3 - T_4)$$

(4.3)

Compressor work: $$w_c = c_{p12}(T_2 - T_1)$$

The compressor and turbine isentropic efficiencies are defined in the usual way as

Turbine isentropic efficiency: $$\eta_t = \frac{(h_3 - h_4)}{(h_3 - h_{4s})} = \frac{(T_3 - T_4)}{(T_3 - T_{4s})}$$

(4.4)

Compressor isentropic efficiency: $$\eta_t = \frac{(h_{2s} - h_1)}{(h_2 - h_1)} = \frac{(T_{2s} - T_1)}{(T_2 - T_1)}$$

for which the isentropic temperatures are defined by

$$T_{4s} = T_3 (p_4/p_3)^{(\gamma-1)/\gamma} \quad \text{and} \quad T_{2s} = T_1 (p_2/p_1)^{(\gamma-1)/\gamma}$$

(4.5)

Finally, it is necessary to define a mechanical efficiency, noting that the mass flow rates through the compressor and the turbine are likely to differ

$$\eta_m = \frac{w_c}{w_t} = \frac{m_{12} c_{p12}(T_2 - T_1)}{m_{34} c_{p34}(T_3 - T_4)}$$

(4.6)

Figure 4.13 illustrates the significance of the turbocharger efficiencies.

4.5.2 Turbocharger Performance

The purpose of turbocharging is to increase the density in the inlet manifold. This is achieved by raising the pressure; however, in doing so, the temperature also rises, and this tends to reduce the density. By substituting for T_{2s} from Eq. 4.5 into Eq. 4.4

$$\eta_c = \frac{(T_{2s} - T_1)}{(T_2 - T_1)}$$

(4.4)

$$T_{2s} = T_1 (p_2/p_1)^{(\gamma-1/\gamma)}$$

(4.5)

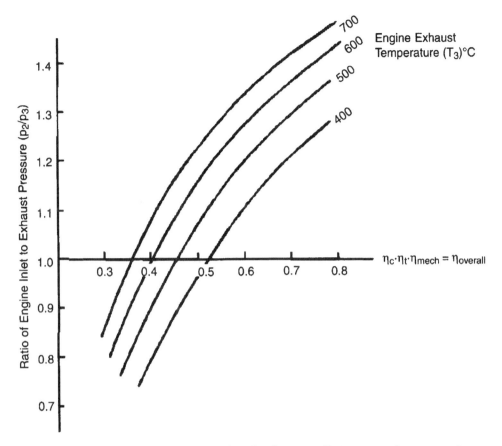

Figure 4.13. *Effect of the overall turbocharger efficiency on the scavenging pressure ratio, for a compressor pressure ratio of 2 (Stone, 1999).*

and rearranging gives

$$T_2 = T_1\left[1 + \frac{(p_2/p_1)^{(\gamma-1)} - 1}{\eta_c}\right] \qquad (4.6)$$

The density ratio now can be found by applying the equation of state ($\rho = p/RT$)

$$\frac{\rho_2}{\rho_1} = \frac{p_2}{p_1}\left[1 + \frac{(p_2/p_1)^{(\gamma-1)} - 1}{\eta_c}\right]^{-1} \qquad (4.7)$$

Figure 4.14 shows the density ratio as a function of the isentropic compressor efficiency.

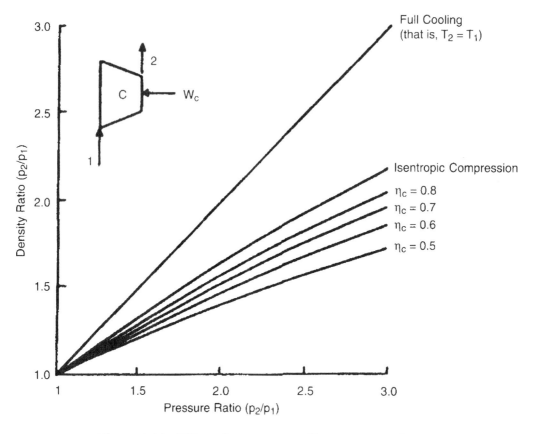

Figure 4.14. *Effect of compressor efficiency on air density in the inlet manifold (Stone, 1999).*

As the pressure ratio increases or the compressor isentropic efficiency falls, the density ratio departs further from the ideal of isothermal compression. However, the density ratio can be increased by cooling the compressed gas. This is known as both intercooling and aftercooling, for which it is necessary to define an effectiveness (ε):

$$\varepsilon = \text{actual heat transfer/maximum possible heat transfer}$$

If the cooling media is assumed to be available at the ambient temperature (T_1), then the limiting case is when the temperature of the flow leaving the intercooler (T_3) is at the same temperature as the ambient temperature (T_1):

$$\varepsilon = (T_2 - T_3)/(T_2 - T_1) \quad \text{or} \quad T_3 = T_2(1-\varepsilon) + \varepsilon T_1 \tag{4.8}$$

Substituting from Eq. 4.6 for T_2, then Eq. 4.8 becomes

$$T_3 = T_1 \left\{ \left[1 + \frac{(p_2/p_1)^{(\gamma-1)/\gamma}}{\eta_c} \right](1-\varepsilon) + \varepsilon \right\}$$

$$= T_1 \left[1 + (1-\varepsilon)\frac{(p_2/p_1)^{(\gamma-1)/\gamma} - 1}{\eta_c} \right] \quad (4.9)$$

Neglecting the pressure drop in the intercooler and applying the ideal gas equation of state, the equation for the density ratio across the compressor and intercooler is

$$\frac{\rho_3}{\rho_1} = \frac{p_2}{p_1} \left[1 + (1-\varepsilon)\frac{(p_2/p_1)^{(\gamma-1)/\gamma} - 1}{\eta_c} \right]^{-1} \quad (4.10)$$

Figure 4.15 illustrates the benefits of intercooling on the density ratio.

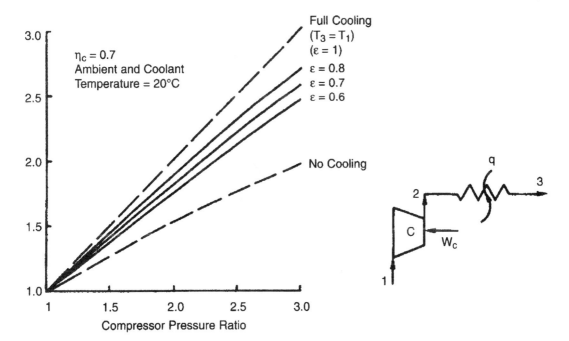

Figure 4.15. *Effect of charge cooling on inlet air density (Stone, 1999).*

Additional advantages of intercooling arise because all the temperatures in the subsequent processes within the engine are lower, which leads to the following:

a. A lower thermal loading for a given fueling rate
b. Lower NOx emissions
c. Less heat transfer and improved efficiency

If the fueling level is increased to maintain the same thermal loading, then a significant increase in output can be achieved, which in turn leads to a further improvement in the brake efficiency. Intercooling increases turbo-lag and the ignition delay, and this can cause problems in diesel engines with low compression ratios that were selected to limit peak pressures.

The compressor performance is characterized by maps such as Fig. 4.16. These plots are not dimensionless, so great care is needed in their use. The safest approach is to convert to the units used on the map, and convert back if necessary at the end. The air flow parameter would be made dimensionless if multiplied by $R/A\sqrt{c_p}$, but because this group is a constant for a given compressor, it can be omitted. Similarly, the speed parameter $\left(N/\sqrt{T}\right)$ would be made dimensionless by dividing by $\sqrt{c_p}$.

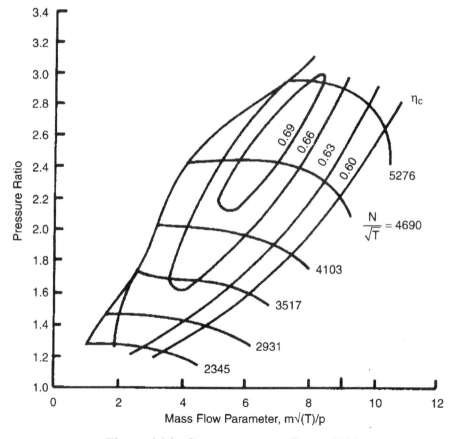

Figure 4.16. *Compressor map (Stone, 1999).*

Radial compressors invariably employ a vaneless diffuser. Although this leads to a lower peak isentropic efficiency, there can be no mismatch between the flow angle and the stator blade angle; thus, the isentropic efficiency is less sensitive to flow rate variations. As explained (Fig. 4.11), the pressure rise occurs in the diffuser because of the following:

a. The radial component of velocity decreases as first the flow area increases with the circumference, and second the pressure is rising.

b. The tangential component of the velocity (whirl) reduces because of the conservation of the moment of momentum.

4.5.3 Turbocharged Engine Performance

Figure 4.17 shows how the compressor operating point changes with load and speed. When the load is increased, the exhaust temperature rises, more turbine work is done, and the compressor pressure ratio rises. Because the engine speed is constant (fixing the volumetric flow rate into the engine), only a slight increase in the mass flow rate results from the increased pressure ratio. Conversely, when the engine speed is increased at constant load, a significant increase occurs in the mass flow rate, with only a slight increase in the pressure ratio. As the

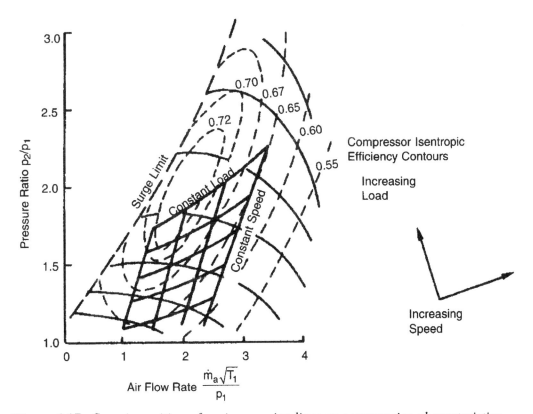

Figure 4.17. *Superimposition of engine running lines on compression characteristics—constant engine load and speed lines (Stone, 1999).*

exit from the turbine is invariably choked, the increase in mass flow can occur only with an increased pressure ratio across the turbine. This leads to an increase in the specific turbine work and, in turn, an increase in the compressor pressure ratio.

Figure 4.18 shows that turbocharging a diesel engine leads to a significant increase in output. Also, because the mechanical losses do not rise in proportion, there is a reduction in fuel consumption. In the case of spark ignition engines, the compression ratio must be lowered to avoid knock. Figure 4.19 shows that turbocharging a spark ignition engine does not necessarily lead to a reduction in fuel consumption.

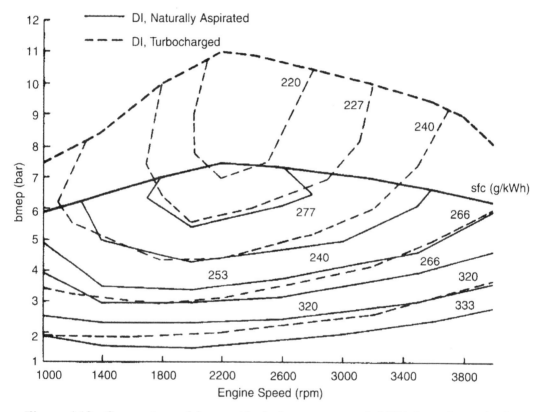

Figure 4.18. *Comparison of the specific fuel consumption (g/kWh) for a four-cylinder naturally aspirated and turbocharged direct injection diesel engine (the same DI engine as in Fig. 4.2). Adapted from Hahn (1986).*

The main disadvantage of turbocharging is turbo-lag, due primarily to the inertia of the turbocharger. Turbo-lag can be minimized by the following:

a. Using ceramic rotors

b. Undersizing the turbine and using a waste-gate

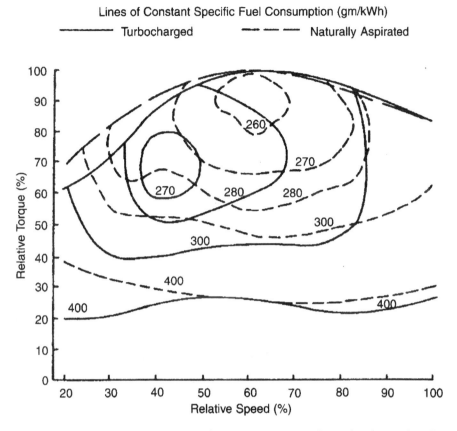

Figure 4.19. *Comparative specific fuel consumption of a turbocharged and naturally aspirated engine scaled for the same maximum torque (Stone, 1999).*

c. Replacing a single turbocharger with two smaller units

d. Employing an additional energy input: Hyperbar, with an auxiliary combustion chamber, or a Pelton wheel driven by a lubricating oil jet, or electrically

e. Minimizing the manifold volumes

Turbo-lag is more significant with spark ignition engines because they have the following characteristics:

a. A lower inertia and an intrinsically faster response
b. A wider flow range because of their wider speed range and throttling

Undersizing the turbocharger and using a waste-gate is a popular means for reducing turbo-lag, with the additional benefit that it helps to limit the maximum cylinder pressure. Figure 4.20 shows how the boost pressure is sensed, and when it rises above a value (as determined by the spring loading and diaphragm area), some of the exhaust flow is permitted to bypass the

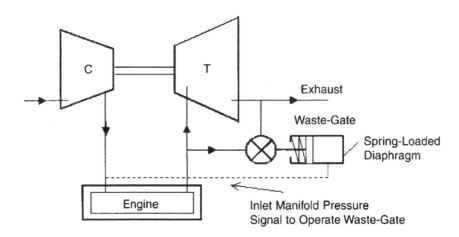

Figure 4.20. *Waste-gate control of the compressor boost pressure.*

turbine. This stops the turbine overspeeding and limits the boost pressure from the compressor. The waste-gate is a simple flap valve incorporated into the turbine casing. (It must operate red hot. It does not matter if it leaks a bit, but it must not jam.)

4.6 Diesel Engine Management Systems

The engine management system shown in Fig. 4.21 is for a turbocharged engine using the Delphi EPIC system, which is based around an electronically controlled rotary fuel pump (Bostock and Cooper, 1992). The sensors and actuators are as already discussed in Section 3.7.2 for spark ignition engines.

For every engine operating condition (starting, idling, speed limited by the governor, providing engine braking, and so forth), sets of rules determine the required fueling (quantity and timing) and the EGR level, as a function of load and speed. Figure 4.21 shows that the EGR valve and the air throttle are operated by vacuum actuators, with a feedback signal telling the ECU their positions. The vacuum is generated by the vacuum pump for the braking servo, and the vacuum signal level is adjusted by controlling the on-to-off ratio of a pulsed solenoid valve that switches between the vacuum and atmospheric pressure. This system ensures accurate positional control of the EGR valve. The EGR valve position correlates well with the EGR level; thus, it is not necessary to have a direct measurement of the EGR. However, it is necessary to ensure that the EGR calibration does not drift with engine condition. The effect of faults such as a blocked air filter, wear on the engine (reducing its volumetric efficiency), a restricted exhaust system, and deposits in the EGR system all should be reviewed.

Such a study was undertaken by Lancefield et al. (1996) using a system modeling approach to model the dynamic response. When the model had been verified with experimental data, it was used to investigate different engine management system strategies. The favored engine management system strategies were then tested on the engine.

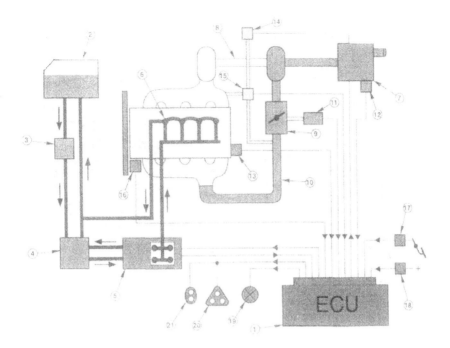

1	Electronic control unit (ECU)	11	Manifold absolute pressure (MAP) sensor
2	Fuel tank	12	Air charge temperature (ACT) sensor
3	Fuel lift pump	13	Engine coolant temperature (EC) sensor
4	Fuel filter	14	EGR vacuum flow regulator
5	Fuel injection pump	15	Exhaust gas recirculation (EGR) valve
6	Injection nozzles	16	Engine speed sensor
7	Air cleaner	17	Accelerator pedal position sensor
8	Turbocharger	18	Power hold relay
9	Throttle plate	19	Engine management system lamp
10	Crossover duct	20, 21	Self test connectors

Figure 4.21. *A diesel engine management system.
Courtesy of Ford Motor Company.*

Figure 4.22a shows a typical start of injection timing map. In general, the injection is more advanced as the speed and load are increased. However, there are additional rules, so that when the coolant is cold, the injection is more advanced. There are also five different maps for ambient temperatures in the range −20 to +60°C (−4 to 140°F). In the normal driving mode, the fuel injected is the minimum of the following:

1. That signaled by the driver demand

2. The maximum allowed by the preset torque curve, or

3 The maximum allowed by the density of the air in the inlet manifold

The air density is calculated from the manifold absolute pressure and temperature, and it was found always to be known to an accuracy of better than 3%, thus not needing an air mass

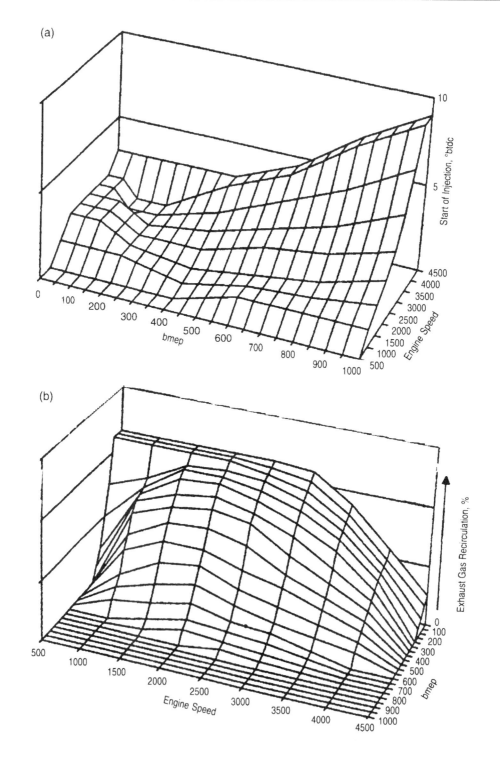

Figure 4.22. *Typical (a) start of injection timing and (b) EGR level maps for the Ford 2.5 HSDI turbocharged diesel engine. Courtesy of Ford Motor Company.*

flowmeter. The fueling is limited when there is a lower than expected boost pressure, such as high-altitude operation, a turbocharger fault, or rapid engine acceleration. If the waste-gate should fail in a closed position, the maximum fueling is limited by the pre-set torque curve, to avoid the engine combustion pressures rising too high.

Exhaust gas recirculation is a well-established method of controlling NOx emissions (Section 4.2.2), and the Ford HSDI engine was the first application in which cooled EGR was used. Figure 4.22b shows the EGR schedule, with up to half of the air being replaced by EGR at light loads. The intake throttle valve (after the compressor) is needed to generate a sufficient pressure differential at low loads to achieve sufficient flow of the exhaust gases. At speeds above 1500 rpm with zero fuel delivery (corresponding to engine braking), the EGR level is set to zero. This fills the intake system with air because engine braking frequently is followed by a need for acceleration. If EGR had been applied, there is a possibility of over-fueling because it takes a finite time for the EGR to be eliminated from the inlet manifold.

Two features not shown in this engine management system are the start of injection sensors and the control of heater or glow plugs. As explained in Section 4.2, some engines use glow plugs or heater plugs to facilitate starting and to reduce the ignition delay in a cold engine (to give quieter combustion). The engine management system thus can tell the driver when it is time to start cranking the engine after turning on the heaters, and it also can control their use during engine warm-up.

A start of injection sensor can be used with a distributor-type fuel pump (Section 4.3.1 and Fig. 4.4) to provide a feedback signal to the start of injection actuator. Start of injection sensors have been made using Hall effect sensors (in which a magnetic field is used to switch a semiconductor), and only one cylinder may be sensed. A more fundamental measurement would be the start of combustion, because this would allow for changes in the fuel quality (e.g., its cetane rating) or engine condition or operation (e.g., ambient or coolant temperature). Toyota used an optical start of combustion sensor in an IDI engine (Takata et al., 1987), but these sensors have not been generally used because of either cost or durability issues.

4.7 Concluding Remarks

Despite increasingly stringent emissions legislation, diesel engines have increased their specific output significantly, and there have been useful improvements in fuel economy. Another trend with diesel engines has been for them to be made smaller. Because it is difficult to make injectors smaller than a certain size, this limits the minimum bore diameter. The 1.2-liter direct injection diesel used by Volkswagen in the Lupo (which has achieved a fuel consumption below 3 L/100 km on the European MVEG drive cycle) has three cylinders of 88-mm bore (Ermisch et al., 2000). The performance map in Fig. 4.23 shows that a bmep of greater than 16 bar is achieved in the speed range 1750 to 2750 rpm, and there is a minimum bsfc of 205 g/kWh (corresponding to an efficiency of 42%). These fuel consumption data have already been used in plotting Fig. 4.23.

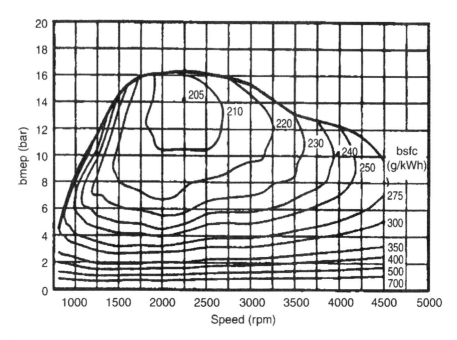

Figure 4.23. *Brake specific fuel consumption (bsfc) map for the Volkswagen Lupo diesel engine. Adapted from Ermisch et al. (2000).*

A comparison of Fig. 4.23 with Fig. 4.18 shows an increase in bmep from 11 to 16 bar with a flatter torque curve, and a reduction in fuel consumption from 220 to 205 g/kWh. Figure 4.24 shows how the unit injector (with injection pressures up to 2000 bar) is driven from the camshaft within the cylinder head. Also, a counterbalance shaft driven at engine speed eliminates the primary out-of-balance moments (Example 2.2). There are no out-of-balance primary or secondary forces in a three-cylinder engine.

In large diesel engines, there is a trend to use variable geometry turbochargers. Fixed geometry turbochargers have an efficiency that falls quite rapidly when they are operating away from their design points. This adverse effect can be reduced by using variable geometry devices. These either control the flow area of the turbine or change the orientation of the stator blades. The Holset moving sidewall variable geometry turbine increases the low-speed torque in a truck engine application; the maximum torque engine speed range is extended to lower speeds by 40%, and there is a 43% improvement in the torque at 1000 rpm (Stone, 1999).

Truck diesel engines also have increased in specific output, with bmeps exceeding 20 bar. Pfluger (1997) describes how the performance of a 2-liter/cylinder truck engine was improved by two-stage turbocharging. The bmep was raised to 22 bar (constant for 50–75% of the rated speed), and the minimum specific fuel consumption was reduced to 185 g/kWh. The equivalent single-stage turbocharged engine had an output of slightly below 22 bar bmep, with a very narrow maxima, and a minimum brake specific fuel consumption of 190 g/kWh. The two-stage turbocharged engine used a waste-gate on the high-pressure turbine to control the overall boost pressure, with coolers after each compressor.

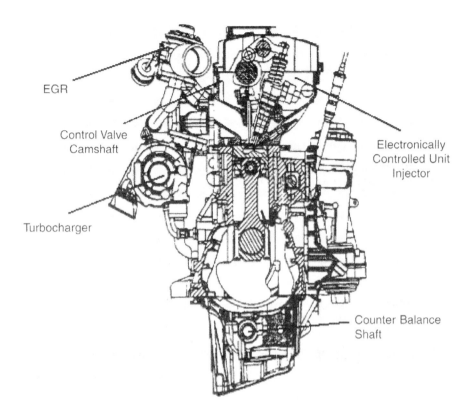

Figure 4.24. *The Volkswagen Lupo 1.2-liter direct injection diesel engine. Adapted from Hilbig et al. (1999).*

4.8 Examples

Example 4.1 A turbocharged and intercooled twelve-cylinder diesel engine has a swept volume of 39 liters. The inlet manifold conditions are 2.0 bar and 53°C (127°F). The volumetric efficiency of the engine is 95%, and it is operating at a load of 16.1 bar bmep, at 1200 rpm with an air-fuel ratio of 21.4. The power delivered to the compressor is 100 kW, with entry conditions of 25°C (77°F) and 0.95 bar. The fuel has a calorific value of 42 MJ/kg.

Stating any assumptions, calculate the following:

a   The power output of the engine
b. The brake efficiency of the engine
c. The compressor isentropic efficiency
d. The effectiveness of the intercooler

Estimate the effect of removing the intercooler on the power output and emissions of the engine, and the operating point of the turbocharger.

Solution: To find the brake power (W_b), we need to use Eq 2.4

$$\text{Power,} \quad W_b = p_b \times V_s \times N^*$$

where

p_b = the bmep (N/m²)
V_s = the engine swept volume (m³)
N^* = rpm/120 for a four-stroke engine (s⁻¹)

Thus,

$$W_b = 1.61 \times 10^5 \times 39 \times 10^{-3} \times 1200/120 = 628 \text{ kW}$$

To determine the brake efficiency (η_b), we need to find the fuel mass flow rate (m_f) from the air mass flow rate (m_a), and the air-fuel ratio (AFR).

The air mass flow rate (m_a) can be found from the volumetric efficiency (Eq. 2.11)

$$\eta_v = \frac{V_a}{\left(V_s \times N^*\right)}$$

where V_a is the volumetric flow rate of the air (m³/s) and

$$m_a = \rho V_a = \eta_v \rho V_s N^*$$

where $\rho = p/(RT)$.

Thus,

$$\begin{aligned} m_a &= \eta_v V_s N^* p/(RT) \\ &= 0.95 \times 39 \times 10^3 \times (1200/120) \times 2 \times 10^5 / (287 \times [273+53]) \\ &= 0.792 \text{ kg/s} \\ m_f &= m_a/\text{AFR} = 0.792/21.4 = 0.037 \text{ kg/s} \end{aligned}$$

The brake efficiency of the engine is the ratio of the work output to the energy released by the combustion of the fuel

$$\eta_b = W_b/(m_f \times CV) = 628 \times 10^3 / (0.037 \times 42 \times 10^6) = 0.404, \quad \text{or } 40.4\%$$

Assuming perfect gas behavior, with no pressure drop in the intercooler, and no heat transfer in the compressor, the compressor work input (W_c) is

$$W_c = m_a c_{p,a}(T_2 - T_1)$$

From this equation, we can determine the temperature rise, for comparison with the temperature rise in an isentropic process:

$$(T_2 - T_1) = W_c / (m_a c_{p,a}) = 100 \times 10^3 / (0.792 \times 1.01 \times 10^3) = 125 \text{ K}$$

We need to determine the isentropic compression temperature (T_{2s}) from the pressure ratio across the compressor

$$T_{2s} = T_1 (p_2/p_1)^{([\gamma-1]/\gamma)} = (2.73 + 25) \times (2/0.95)^{([1.4-1]/1.4)} = 369 \text{ K}$$

The isentropic compressor efficiency (η_c) is (refer to Fig. 4.12 and Eq. 4.4)

$$\eta_c = (T_{2s} - T_1)/(T_2 - T_1) = (369 - 298)/125 = 0.57, \text{ or } 57\%$$

The intercooler effectiveness (ε) (Eq. 4.8) is defined as

$$\varepsilon = (T_2 - T_3)/(T_2 - T_1)$$

where T_3 is the temperature after the intercooler.

Substitution of numerical values gives

$$\varepsilon = (423 - 326)/(23 - 298) = 0.78$$

Removing the intercooler will reduce the density of the air at the inlet to the engine and thus the air mass flow rate. The air-fuel ratio of 21.4 is probably smoke limited, so that as the air flow is reduced, a corresponding reduction in the fuel flow rate must occur. A pessimistic assumption would be to assume that the pressure ratio across the compressor remained the same. If the brake efficiency of the engine is assumed to remain constant (a slightly optimistic assumption, because the frictional losses will become more significant), then the power output of the engine is directly proportional to the air mass flow rate.

The air mass flow rate would become

$$m_a T_3/T_2 = 0.792 \times 326/423 = 0.61 \text{ kg/s}$$

and the power output would be reduced by

$$T_3/T_2 (326/423) = 0.77, \text{ or } 77\%$$

to

$$(0.77 \times 628 =) 484 \text{ kW}$$

The constant air-fuel ratio and the higher inlet temperature will mean that all processes within the engine will occur at higher temperatures. This means the following are true:

1. The increased turbine entry temperature will cause the turbine specific work output to increase, and this in turn means a higher compressor delivery pressure (unless this is limited by a waste-gate), so that the mass flow rate of air into the engine will not be decreased by as much as was assumed in the preceding discussion.

2. The NOx emissions will be increased, because the formation of NOx is strongly temperature dependent. A 10% increase in the in-cylinder temperature at the inlet valve closure will cause something similar to a 30% increase in the NOx emissions.

3. The combustion noise will be reduced, because higher in-cylinder temperatures will reduce the ignition delay period and the mass of flammable mixture formed prior to ignition.

Example 4.2. A turbocharged 2-liter direct injection diesel engine operates on a four-stroke cycle. At 2900 rpm and a bmep of 9.7 bar (full load), it is operating with a 22:1 gravimetric air-fuel ratio and a brake specific fuel consumption of 230 g/kWh. The turbocharger is fitted with a waste-gate to regulate the pressure ratio to 2.0. Use the compressor map in Fig. 4.16, for which the pressure units are kN/m², the mass flow is grams per second, and the temperature units are Kelvin. The turbine entry temperature is 850 K, its pressure ratio is also 2.0, and the turbine isentropic efficiency is 0.75.

The compressor entry conditions are 1 bar and 298 K, and you should assume the following thermodynamic properties:

	Air	Exhaust
Specific heat capacity at constant pressure, c_p (kJ/kgK)	1.01	1.12
Ratio of heat capacities, γ	1.4	1.33

a. Stating any assumptions you make, calculate the brake power output of the engine, its volumetric efficiency (based on inlet manifold conditions), and the fraction of the exhaust gas that passes through the turbine.

b. Suggest ways of increasing the bmep of this engine in order of increasing complexity, with an indication of the likely increase in output (and how this would be calculated). Comment on how other aspects of the engine performance would be affected.

Diesel Engines

Solution:

a. To calculate the brake power output, we can use the definition of bmep (Eq 2.4), where

p_b = the bmep (N/m²)
V_s = the engine swept volume (m³)
N^* = rpm/120 for a four-stroke engine (s⁻¹)
W_b = $p_b \times V_s \times N^* = 9.7 \times 10^5 \times 2.0 \times 10^{-3} \times (2900/120) = 46.9$ kW

To determine the volumetric efficiency of the engine, we need to know the air mass flow rate into the engine, as well as the compressor delivery conditions. The compressor delivery conditions depend on the inlet conditions, the pressure ratio (all of which are known), and the compressor isentropic efficiency, which we must deduce from the operating point on the compressor map. In turn, this depends on the pressure ratio and the air flow rate. Thus, a good way to proceed is to calculate the air mass flow rate next.

Because we know the power output and the brake specific fuel consumption (bsfc) of the engine, we can calculate the fuel mass flow rate (m_f), and then use the air-fuel ratio (AFR) to calculate the air mass flow rate (m_a)

$$m_f = \text{bsfc} \times W_b = 230 \text{ (g/kWh)} \times 46.9 \text{ (kW)} = 10{,}787 \text{ g/h}$$

or

$$10{,}787/3600 = 2.996 \text{ g/s}$$

$$m_a = \text{AFR} \times m_f = 22 \times 2.996 = 65.9 \text{ g/s air}$$

Next, we need to find the value of the compressor mass flow parameter (m^*), taking careful note of the units because the mass flow parameter is not dimensionless

$$m^* = m_a \text{(g/s)} \times \sqrt{T(K)} / p(kN/m^2) = 6.59 \times \sqrt{298}/100 = 11.38$$

The intersection of the 11.38 mass flow parameter and the pressure ratio of 2.0 gives (by interpolation) a compressor isentropic efficiency of 0.67. We can now find the compressor delivery temperature, but first we need the isentropic compression temperature (T_{2s}) from the pressure ratio across the compressor

$$T_{2s} = T_1 (p_2/p_1)^{([\gamma-1]/\gamma)} = 298 \times (2.0)^{([1.4-1]/1.4)} = 363 \text{ K}$$

The isentropic compressor efficiency (η_c) is (Fig 4.16)

$$\eta_c = (T_{2s} - T_1)/(T_2 - T_1)$$

or

$$T_2 = (T_2 - T_1)/\eta_c + T_1 = (363 - 298)/0.67 + 298 = 395 \text{ K}$$

The volumetric efficiency (Eq. 2.11) is

$$\eta_v = \frac{V_a}{(V_s \times N^*)}$$

where V_a is the volumetric flow rate of the air (m³/s), and

$$V_a = m_a/\rho$$

where $\rho = p/(RT)$. Thus,

$$\eta_v = m_a(RT/p)/(V_s N^*)$$
$$= 0.0659\left(287 \times 395/2 \times 10^{-5}\right)/\left(2.0 \times 10^3 \times 2900/120\right) = 0.77$$

To determine what fraction of the exhaust flow passes through the turbine, we need to calculate the power absorbed by the compressor. Assuming it to be adiabatic, the compressor work input (W_c) is

$$W_c = m_a c_{p,a}(T_2 - T_1) = 0.0659 \times 1.01 \times 10^3 (395 - 298) = 6.46 \text{ kW}$$

Assuming that the turbine is also adiabatic and that there are no mechanical losses, a power balance can be written to define the mass flow through the turbine (m_t)

$$W_c = m_t c_{p,ex}(T_3 - T_4)$$

where 3 is the turbine entry, and 4 is its exit.

We have been given the turbine entry temperature (T_3), but we will have to calculate T_4 from the turbine pressure ratio and its isentropic efficiency

$$(T_3 - T_4) = \eta_t (T_3 - T_{4s})$$

where

$$T_{4s} = T_3 (p_4/p_3)^{([\gamma-1]/\gamma)} = 850 \times 2.0^{(1.33-1)/1.33} = 716 \text{ K}$$

Therefore,

$$(T_3 - T_4) = 0.75(850 - 716) = 100.5 \text{ K}$$

and

$$W_c = m_t c_{p,ex} 100.5$$

or

$$m_t = W_c / (c_{p,ex} 100.5) = 6460/(1120 \times 100.5) = 57.4 \text{ g/s}$$

We must compare this with the mass flow rate of the exhaust (m_{ex})

$$m_{ex} = m_a + m_f = m_a(1 + 1/\text{AFR}) = 65.9(1 + 1/22) = 68.9 \text{ g/s}$$

The fraction of the exhaust flowing through the turbine is

$$m_t/m_{ex} = 57.4/68.9 = 0.833, \text{ or } 83.3\%$$

b. Two options will be considered: (1) raising the boost pressure, or (2) adding an intercooler

1. Increasing the boost pressure (and increasing the fueling rate to maintain the same air-fuel ratio) will lead to an increase in output. Because the fuel injection pump should be fitted with a boost control unit to limit the maximum fueling rate according to the inlet manifold pressure (and possibly temperature), then for small increases in boost pressure, there should be a corresponding increase in the fueling rate. However, there will be a corresponding increase in the maximum combustion pressures and the thermal loading. We can estimate the possible increase in output by estimating the increase in the turbine output. We will need to make three assumptions:

   - The turbine isentropic efficiency is unchanged. (There might be a slight increase.)

   - The turbine entry temperature does not increase. (It will, so this is a conservative assumption.)

   - The volumetric efficiency of the engine is unchanged.

The calculation procedure will be iterative:

a. Estimate by how much the pressure ratio across the compressor will increase.

b. Assuming the same compressor isentropic efficiency, calculate the air mass flow rate into the compressor and engine.

c. Identify the operating point on the compressor map, and if the estimate of the compressor isentropic efficiency was wrong, then recalculate the air flow rate. Repeat until satisfactory.

d. Calculate the power input required to drive the compressor.

e. Assuming the same pressure ratio across the turbine and compressor, calculate the power output from the turbine.

f. Compare the power output of the turbine, with the power requirement of the compressor, and make a better estimate of the compressor pressure ratio.

g. Repeat steps a to f until satisfactory convergence is achieved.

In this case, if all the exhaust flows through the turbine, then there is an increase from 57.4 to 68.9 g/s (an increase of 20%). This suggests that the air mass flow might be increased by approximately 20%, with a similar increase in the engine output. However, some practical considerations might limit the increase in air flow: (1) the maximum pressure and temperature already referred to, and (2) the compressor operating point will move closer to the surge line. An increase of 10% might be a more realistic limit to the increase in output.

2. Intercooling lowers the temperature of the air, thereby increasing its density. The mass flow rate of air is increased for a given pressure ratio, so this gives the option of increasing the fuel flow rate, thereby increasing the power output and efficiency. (The efficiency will increase because the mechanical losses do not rise in proportion to the brake power output.) To a first order, the air mass flow rate will be increased in inverse proportion to the absolute temperature (because we can ignore the pressure drop in the intercooler).

Temperature after the compressor	395 K
Ambient temperature	298 K

With an effectiveness of 1.0, the temperature after the intercooler would be 298 K. More realistically, with an effectiveness of 0.7, the temperature after the intercooler would be

$$395 - 0.7(395 - 298) = 327 \text{ K}$$

and the increase in output would be

$$(1/327 - 1/395)/(1/395) = (395 - 327)/327 = 0.21, \text{ or } 21\%$$

Because the mass flow rate through the compressor and the turbine both will have increased, and the pressure ratio across the compressor is limited to 2.0 (by the waste-gate), then the compressor operating point will move to the right. The reduction in the compressor isentropic efficiency means an increase in the compressor specific work, so the fraction of the flow through the waste-gate might be reduced slightly.

The cooler air at the start of the compression process will lower the temperatures in all subsequent processes, and thus lower the heat transfer and thermal loading on the engine. The lower combustion temperatures will lead to lower NOx emissions, because NOx formation is highly temperature dependent.

The disadvantages of intercooling are as follows:

a. The space needed by the intercooler and its associated ductwork (for both the engine air and the cooling medium).

b. The increased volume between the compressor and engine will worsen the transient response.

c. The lower temperatures during compression will increase the ignition delay. This will lead to an increase in the mass of flammable air-fuel mixture formed before self-ignition, and thus increase the initial rate of pressure rise and then the combustion noise. This will be most significant at low loads, but the ignition delay can be reduced by using engine coolant as the cooling medium for the intercooler. This will heat the air under very low load conditions, thereby reducing the ignition delay period. The disadvantage is that the air will not be cooled as much by the intercooler at higher loads. Alternatively, hot EGR at part load will help to maintain a higher inlet manifold temperature.

4.9 Problems

4.1 A diesel engine is fitted with a turbocharger that comprises a radial flow compressor driven by a radial flow exhaust gas turbine. The gravimetric air-fuel ratio is 20:1; the air is drawn into the compressor at a pressure of 1 bar and at a temperature of 15°C (59°F). The compressor delivery pressure is 2.0 bar. The exhaust gases from the engine enter the turbine at a temperature of 510°C (950°F); the gases leave the turbine at a pressure of 1.05 bar. The isentropic efficiencies of the compressor and turbine are 65% and 75%, respectively.

Treating the exhaust gases as perfect gas with the same properties as air, and assuming a mechanical efficiency of 100%, calculate the following:

a. The temperature of the gases leaving the compressor
b. The pressure ratio across the turbine

4.2 Why is it more difficult and less appropriate to turbocharge spark ignition engines than compression ignition engines? Under what circumstances might a supercharger be more appropriate?

4.3 Show that the density ratio across a compressor and intercooler is given by

$$\frac{\rho_3}{\rho_1} = \frac{p_2}{p_1}\left[1+(1-\varepsilon)\frac{(p_2/p_1)^{(\gamma-1)}-1}{\eta_c}\right]^{-1}$$

where

1 = compressor entry
2 = compressor delivery
3 = intercooler exit
η_c = compressor isentropic efficiency
ε = intercooler effectiveness = $(T_2 - T_3)/(T_2 - T_1)$

Neglect the pressure drop in the intercooler, and state any assumptions that you make.

Plot a graph of the density ratio against effectiveness for pressure ratios of 2 and 3, for ambient conditions of 1 bar, 300 K, if the compressor isentropic efficiency is 70%.

What are the advantages and disadvantages in using an intercooler? Explain under what circumstances should it be used?

4.4 A turbocharged diesel engine has an exhaust gas flow rate of 0.75 kg/s. The turbine entry conditions are 500°C (932°F) at 2.5 bar, and the exit conditions are 400°C (752°F) at 1.1 bar.

a. Calculate the turbine isentropic efficiency and power output.

The engine design is changed to reduce the heat transfer from the combustion chamber, and for the same operating conditions, the exhaust temperature becomes 550°C (1022°F). Assume that the pressure ratio remains the same, and assume the same turbine isentropic efficiency.

b. Calculate the increase in power output from the turbine.

How will the performance of the engine be changed by reducing the heat transfer, in terms of economy, power output, and emissions? Assume ratio of specific heat capacities = 1.33, and c_p = 1.15 kJ/kgK.

4.5 A turbocharged diesel engine has a compressor operating point that is marked by a cross on Fig 4.25. If the compressor entry conditions are a pressure of 1 bar and a temperature of 20°C (68°F), determine the volume flow rate out of the compressor, and the power absorbed by the compressor. Assume the following properties for air: $\gamma = 1.4$, and $c_p = 1.01$ kJ/kgK.

The fueling rate to the engine is increased and the air mass flow rate into the engine are both increased by 50%, and the volume flow rate out of the compressor increases by 22%. State any assumptions, and establish approximately the new operating point for the compressor, its rotor speed, and the power that it is absorbing.

List briefly:

The advantages and disadvantages of turbocharging a diesel engine
and
How the disadvantages can be ameliorated.

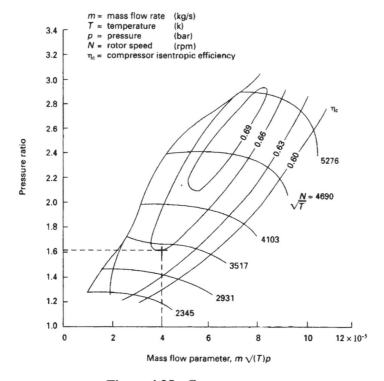

Figure 4.25. *Compressor map.*

Chapter 5

Ancillaries

5.1 Introduction

This chapter on ancillaries starts inside the engine, with lubrication systems and bearing types, and continues with cooling systems (Section 5.3) that are, of course, both inside and outside the engine. The coolant pump is invariably belt driven (external to the engine), and belt drives are the subject of Section 5.4. The belt drive also is used for the air conditioning system, which is the subject of Section 5.5. Finally, Section 5.6 deals with the major electrical machines used in a vehicle, namely, the starter motor and alternator.

5.2 Lubrication System

The lubrication system can justifiably be credited with advancements in high-performance engines, for without adequate lubricants and a system for employing them, the engine would quickly cease to function. The most critical parts of the lubrication system are the bearings, and regardless of the quality of the lubricant, a poor bearing surface will be destroyed quickly when operated under the conditions present in all modern engines. Thus, the first part of this section will describe the types of bearing found in a modern engine. The section will then discuss lubricants and introduce the subject of tribology with fundamental calculations for the friction created (and hence the power lost) in bearings. Finally, a description of the components of a typical lubrication system will be described.

5.2.1 Bearings

Bearings may be broadly defined as surfaces between which there is relative motion. This motion can be rotating motion or linear, but in either case, the bearing must allow the motion with a minimum of friction, as well as supporting any loads present. For the purposes of this work, four main types of bearings will be discussed:

a. Anti-friction bearings (e.g., ball bearings, roller bearings)
b. Guide bearings
c. Thrust bearings
d. Journal bearings

5.2.1.1 Anti-Friction Bearings

Anti-friction bearings, also called rolling contact bearings, are not widely used in the engine proper. They are used more prevalently in the transmission and drivetrain. Nonetheless, they are found in several engine components such as alternators and water pumps. Their primary function is to support a rotating shaft. Figure 5.1 shows a selection of types of anti-friction bearings.

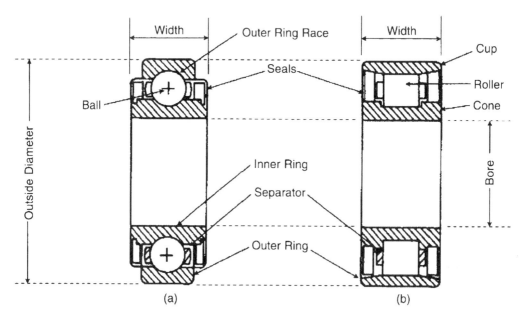

Figure 5.1. *Anti-friction bearings: (a) ball bearing, and (b) roller bearing. Adapted from Krutz et al. (1994).*

Anti-friction bearings offer very low coefficients of friction. Furthermore, some types are able to support axial (thrust) loads in addition to the radial, or shaft, loads. Deep groove ball bearings can support small thrust loads. Tapered roller bearings are specifically designed to support significant thrust loads; hence, they are used as wheel bearings. The rollers or balls are made of hardened, high carbon chromium alloy steels, and their construction is somewhat complicated. Furthermore, these bearings do not handle shock loads very well, and their performance is greatly reduced by the presence of dirt. Thus, anti-friction bearings must be well sealed to keep in the lubricant and keep out dirt and other contaminants.

5.2.1.2 Guide Bearings

Guide bearings, as their name implies, exist to guide a machine component undergoing lengthwise motion. Thus, the cylinder bore is a guide bearing, as are the valve guides shown in Fig. 5.2. Guide bearings may be lubricated by either a pressure system or splash lubrication, depending on the manufacturer. In the case of valve guides, some means of controlling the

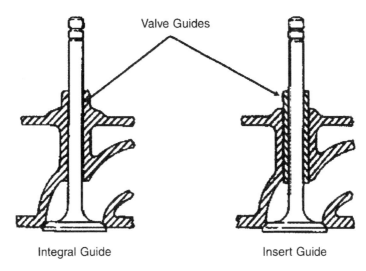

Figure 5.2. *Valve guides. Adapted from TM 9-8000 (1985).*

amount of oil used for lubrication must be employed. Too little lubrication results in high heat buildup and eventual failure of the valve stem; too much causes excessive oil consumption in the combustion chamber.

As shown in Fig. 5.2, valve guides may be integral or insert. The integral guides are formed by the material of the cylinder head itself, whereas the insert guide is made of a sleeve that is press-fitted into a bore in the head. Alloy cylinder heads require the use of valve guide inserts, and these often are made from cast iron. To allow for differential thermal expansion between the valve stem and the guide, there is a large clearance (usually approximately 1%). The advantage of insert guides is that if the guide becomes worn, it can be removed and a new guide inserted. If an integral guide becomes worn, it can be over-bored to accept an insert guide. If the wear is within specified limits, the integral guide can be knurled. This builds up a series of ridges in the guide that are then rebored to the diameter of the valve stem. The ridges provide a bearing surface for lubrication while maintaining the appropriate level of sealing. This repair is not long-lasting, but it is quick and inexpensive.

5.2.1.3 Thrust Bearings

Thrust bearings must support both rotating loads and longitudinal loads. Several examples exist in the engine proper, most notably as camshaft supports. Camshaft lobes can be ground with a small amount of taper, as shown in Fig. 5.3. As the crowned tappet acts on the tapered lobe, it tends to rotate the tappet, ensuring even wear of the tappet surface.

However, this taper also induces a longitudinal load on the camshaft, which must be absorbed by a thrust bearing. Several strategies incorporate a thrust bearing into the camshaft, including a combination journal and thrust bearing around the camshaft, and an end thrust bearing as shown in Fig. 5.4.

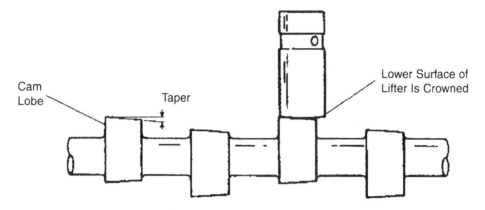

Figure 5.3. *Generation of thrust load on a camshaft. Adapted from TM 9-8000 (1985).*

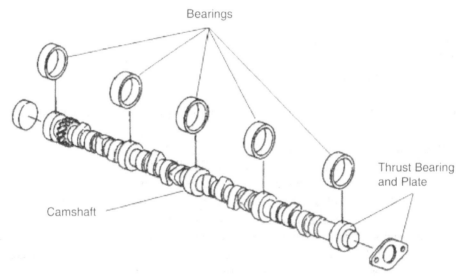

Figure 5.4. *Camshaft thrust bearing. Adapted from TM 9-8000 (1985).*

5.2.1.4 Journal Bearings

Journal bearings also carry a rotating shaft, and they are the most prevalent bearing in the engine proper. Journal bearings are used on the connecting rods at both ends, at the main bearings that support the crankshaft, and at the camshaft supports. They generally are composed of two semi-circular inserts made to close tolerances. The inserts allow for ease of manufacturing and repair of damaged bearings. Figure 5.5 shows an example of the journal bearings on a crankshaft.

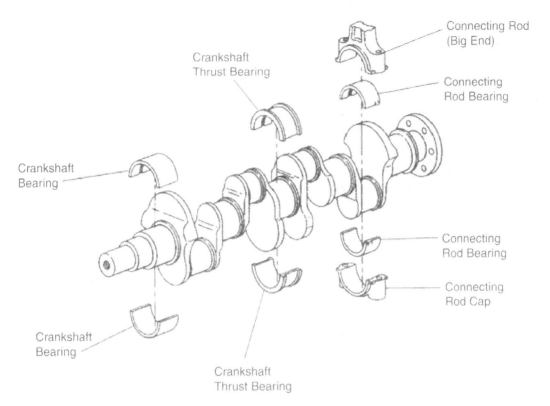

Figure 5.5. *Typical journal bearing installation. Adapted from TM 9-8000 (1985).*

There are two conflicting sets of requirements for a good bearing material:

1. The material should have a satisfactory compressive and fatigue strength.

2. The material should be soft, with a low modulus of elasticity and a low melting point.

Soft materials allow foreign particles to be absorbed without damaging the journal. A low modulus of elasticity enables the bearing to conform readily to the journal. The low melting point reduces the risk of seizure that could occur in the boundary regime—all bearings at some stage during startup will operate in the boundary regime. These conflicting requirements can be met by the steel-backed bearings discussed next.

Initially, Babbit metal, which is a tin-antimony-copper alloy, was widely used as a bearing material in engines. The original composition of Babbit metal or white metal was 83% tin, 11% antimony, and 6% copper. The hard copper-antimony particles were suspended in a soft copper-tin matrix to provide good wear resistance, as well as conformability and the ability to embed foreign particles. The disadvantages were the expense of the tin and the poor high-temperature performance; consequently, lead was substituted for tin. The white metal bearings originally were cast in their housings or made as thick shells, sometimes with a thick bronze or steel backing. In all cases, it was necessary to fit the bearings to the engine and then hand-scrape the bearing surfaces. This technique made the manufacture and repair of engines a difficult and skilled task.

These problems were overcome in the 1930s by the development of thin-wall or shell-type bearings. Thin-wall bearings are made by casting a thin layer of white metal, typically 0.4 mm (0.015 in.) thick, onto a steel strip backing approximately 1.5 mm (0.06 in.) thick. The manufacture is precise enough to allow the strip to be formed into bearings, which then are placed in accurately machined housings. These bearings kept all the good properties of white metal, but they gained in strength and fatigue life from the steel backing. To provide bearings for higher loads, a lead-bronze alloy was used; however, this required hardening of the journals, which was an expensive process. To overcome this difficulty, a three-layer bearing was developed. A thin layer of white metal was cast on top of the lead-bronze lining. To prevent diffusion of this tin into the lead-bronze layer, a plated-nickel barrier was necessary. The expense of three-layer bearings led to the development of single-layer aluminum-tin bearings with up to 10% tin. More recently, an 11% silicon-aluminum alloy has been developed for heavily loaded bearings.

The manufacture of such bearings is a specialist task carried out by firms such as Glacier Vandervell. There are essentially three different types of bearing materials. In decreasing order of strength, these are copper-based, aluminum-based, and white metals (tin or lead-based). In their simplest form, thin-wall bearings are of bimetal construction—a steel backing with a bearing material on top. In trimetal bearings, there is an overlay material above the bearing material.

Copper-based bearings are mostly lead bronzes, in which the lead is dispersed in a bronze matrix, which is either cast or sintered onto the steel backing. Copper-based materials are relatively hard and are overplated with a soft-phase material (such as lead-indium) for crankshaft applications. The soft-phase material provides conformability to the journal and embedability, so that wear debris from elsewhere can be absorbed into the bearing without harm. Copper-based bearings are used in high-performance racing engines and heavy-duty diesel engines.

Aluminum bimetal bearings are widely used for medium-duty engine applications and consist of tin or tin and silicon in an aluminum matrix. Tin gives the bearing good soft-phase properties (for conformability and embedability). These bearings are economical and have good corrosion resistance; they are manufactured by roll bonding onto a steel backing with an aluminum foil interlay.

White metals are either lead- or tin-based and traditionally have been used because of their excellent soft-phase properties. They have a comparatively low strength and are rarely used now; they are made by casting onto the steel backing.

In trimetal bearings, the overlay is very thin (14–33 μm) and originally was applied by electroplating. The overlay leads to improvements in corrosion resistance, reduced friction, and wear and seizure resistance (especially in the early life of the engine). Thicker overlays provide greater conformability and embedability; thinner overlays promote a higher load-carrying capacity. The overlay is applied only after the bimetal bearing has been finish bored.

During electroplating, it is possible to incorporate inert particles that have been in suspension in the plating bath. Alumina particles of 1 μm or smaller can be incorporated in the metal matrix this way (up to 2.0% by mass), and this leads to reduced wear rates. Adding 1% alumina can halve the wear rate. An alternative to electroplating is to sputter the overlay onto the bearing. In the case of aluminum-tin overlays, this leads to a greater mechanical strength (Eastham et al., 1995).

5.2.2 Engine Lubricants

Engine lubricants, generally oils, are derived from crude oil through several refining processes, which are outlined in SAE J357 (1999). The engine oil performs several important functions in the engine, including lubrication, cooling, sealing, cleaning, and protection against wear and corrosion. The base oil stock is not capable of performing all these functions; thus, additive agents are used to enhance their performance. The additives are used at concentrations ranging from several parts per million to greater than 10% by volume (SAE J357). Although some of the additives are naturally occurring materials, the majority must be chemically synthesized to produce the desired performance. Additives fall into three primary functional groups: engine protectors, oil property modifiers, and base stock protectors (SAE J357). The engine protectors include seal swell inhibitors, detergents, dispersants, and friction modifiers. Oil modifiers include pour point depressors, antifoam agents, and viscosity index improvers. Finally, base stock protectors include antioxidants and metal deactivators.

The primary oil classification system is the SAE viscosity, outlined in SAE J300 (1999). The oil is tested for viscosity using procedures outlined in SAE J30 (1999). Four categories are defined by viscosity measurements at 0°C (32°F) (5W, 10W, 15W, and 20W), and another four categories are defined by viscosity measurements at 99°C (210°F) (20, 30, 40, and 50). Multi-grade oils have been developed to satisfy both requirements by adding polymeric additives that thicken the oil at high temperatures, and their designations are 10W30, 10W40, and so forth. Multi-grade oils give better cold-start fuel economy because the viscosity of an SAE 10W40 oil will be less than that of an SAE 40 oil at ambient conditions. Table 5.1 shows the viscosity ranges for each classification.

One common question from the average motorist concerns the frequency with which he or she should change the oil in the engine. Most popular quick-change oil establishments recommend oil changes every 3,000 miles or three months. This is necessary only when the vehicle has been driven under extreme conditions. Operation under extreme conditions includes dusty or dirty environments, towing heavy loads, poor mechanical condition of the engine, an inoperative positive crankcase ventilation (PCV) system, or short trips in cold conditions (where the engine does not reach normal operating temperature). Most manufacturers recommend a change interval of 7,500 miles, and this applies to normal driving conditions. However, the definition of normal driving conditions is not standardized. Thus, 7,500 miles is a good upper bound for changing the engine oil.

TABLE 5.1
SAE VISCOSITY GRADES FOR ENGINE OILS (SAE J300)

SAE Viscosity Grade	Low-Temp. (°C) Cranking Viscosity cP, Max.	Low-Temp. (°C) Pumping Viscosity cP, Max. with No Yield Stress	Low Shear Rate Kinematic Viscosity (cSt) at 100°C Min.	Low Shear Rate Kinematic Viscosity (cSt) at 100°C Max.	High Shear Rate Viscosity (cP) at 150°C Min.
0W	3250 at −30	60,000 at −40	3.8		
5W	3500 at −25	60,000 at −35	3.8		
10W	3500 at −20	60,000 at −30	4.1		
15W	3500 at −15	60,000 at −25	5.6		
20W	4500 at −10	60,000 at −20	5.6		
25W	6000 at −5	60,000 at −15	9.3		
20			5.6	<9.3	2.6
30			9.3	<12.5	2.9
40			12.5	<16.3	2.9 (0W40, 5W40, and 10W40)
40			12.5	<16.3	3.7 (15W40, 20W40, and 25W40)
50			16.3	<21.9	3.7
60			21.9	<26.1	3.7

Figure 5.6 shows the results of viscosity tests on used 10W30 oil compared to new oil of the same grade. The tests were performed as part of an undergraduate laboratory experiment, and although there certainly is some absolute experimental error, the results nonetheless illustrate the effects of use on the viscosity of engine oil. The used oil had been in the owner's car for 10,000 miles. However, the owner did not know when the last oil change was performed by the previous owner; hence, the used oil could have many more miles on it than stated. As the figure indicates, at low temperatures, the used oil behaves similarly to a heavier (higher viscosity) oil; at high temperatures, the used oil behaves similarly to a lighter oil. Obviously, this is an undesirable situation and should motivate regular oil changes.

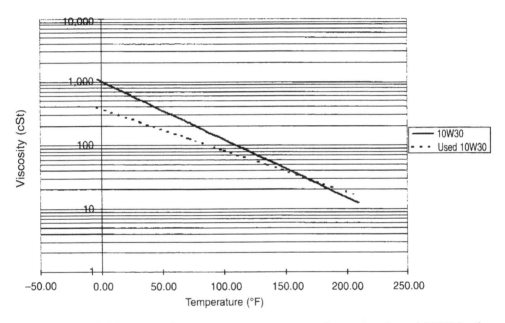

Figure 5.6. *Viscosity change versus temperature for new and used 10W30 oil.*

5.2.3 Lubrication of Journal Bearings

Before delving into the specifics of journal bearing lubrication, it is prudent to discuss the basics of fluid film lubrication in general. Suppose a plate is supported on a film of fluid of height, h, and is moving with velocity, V, as shown in Fig. 5.7.

The fluid in contact with the stationary surface has zero velocity, whereas the fluid in contact with the moving plate has the same velocity as the plate, v = V. This induces a shear stress in the fluid, and Newton's viscous effect states that the shear stress is proportional to the rate of change of fluid velocity with respect to y, or

$$\tau = \mu \frac{du}{dy} \tag{5.1}$$

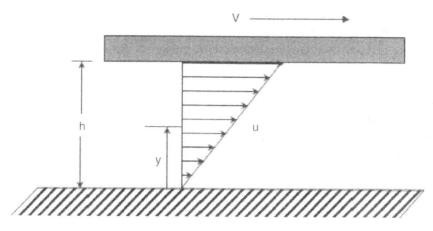

Figure 5.7. *Velocity gradient in lubricating fluid.*

where μ is the dynamic viscosity of the fluid, and $\frac{du}{dy}$ is the velocity gradient. If the rate of change of velocity is assumed to be constant, then

$$\tau = \mu \frac{V}{h} \tag{5.2}$$

Consider a horizontal shaft supported by a bearing as shown in Fig. 5.8. Initially, it is assumed that there is no vertical load on the shaft (W = 0) and that the shaft remains concentric with the bearing. The shaft has a radius r, a radial clearance c, and a length l. If the shaft is rotating at a speed of N_s rev/sec, the velocity of the shaft surface is given by

$$V = 2\pi r N_s \tag{5.3}$$

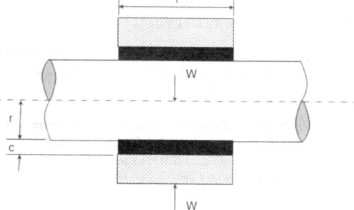

Figure 5.8. *A lightly loaded shaft in bearing.*

Because the shear stress at the shaft surface is equal to the velocity gradient times the viscosity (Eq. 5.2), the shear stress in the lubricant is equal to

$$\tau = \mu \frac{V}{h} = \frac{2\pi r \mu N_s}{c} \tag{5.4}$$

where the height of the lubricant, h, has been replaced by the radial clearance, c. The torque on the shaft induced by the shear stress in the lubricant is equal to the stress times the shear surface area times the shaft radius, or

$$T = \tau A r = \left(\frac{2\pi r \mu N_s}{c}\right)(2\pi r l) r = \frac{4\pi^2 r^3 l \mu N_s}{c} \tag{5.5}$$

The horsepower loss due to the lubricant shear is the torque times the shaft speed. In English units, this becomes

$$HP = \frac{T(\text{in}-\text{lbf}) N_s(\text{rev/sec})}{1050} \tag{5.6}$$

Now, allow the radial shaft load, W, to become some non-zero force, and assume the shaft remains concentric in the bearing. This induces an average pressure in the lubricant, which is equal to the radial load times the projection of the shaft area on the bearing

$$P = \frac{W}{2rl} \tag{5.7}$$

or

$$W = 2Prl \tag{5.8}$$

The torque due to surface friction is the coefficient of friction, f, times the normal force, W, times the radius, r,

$$T_f = fWr \tag{5.9}$$

Substituting Eq. 5.8 into Eq. 5.9 and solving for the coefficient of friction gives

$$f = 2\pi^2 \left(\frac{r}{c}\right)\left(\frac{\mu N_s}{P}\right) \tag{5.10}$$

Equation 5.10 is Petroff's equation, first published in 1883. The bearing characteristic number, or Sommerfeld number, is defined as

$$S = \left(\frac{r}{c}\right)^2 \left(\frac{\mu N_s}{P}\right) \tag{5.11}$$

The Sommerfeld number contains many of the design parameters for the bearing. Of particular importance is the term $\frac{\mu N_s}{P}$, which can be plotted on a Stribeck diagram as shown in Fig. 5.9. This diagram confirms Petroff's equation, in that to the right of point A on the figure, the coefficient of friction is proportional to $\frac{\mu N_s}{P}$.

Figure 5.9 shows the three lubrication regimes that are important for engine components. Hydrodynamic lubrication is when the load-carrying surfaces of the bearing are separated by

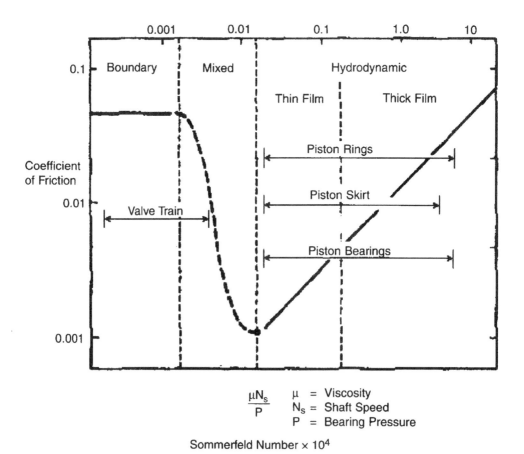

Figure 5.9. *Engine lubrication regimes on a Stribeck diagram. Adapted from Stone (1999).*

a film of lubricant of sufficient thickness to prevent metal-to-metal contact. The flow of oil and its pressure between the bearing surfaces are governed by their motion and the laws of fluid mechanics. The oil film pressure is produced by the moving surface drawing oil into a wedge-shaped zone, at a velocity high enough to create a film pressure that is sufficient to separate the surfaces. In the case of a journal and a bearing, the wedge shape is provided by the journal running with a slight eccentricity in the bearing. Hydrodynamic lubrication does not require a supply of lubricant under pressure to separate the surfaces (unlike hydrostatic lubrication), but it does require an adequate supply of oil. It is convenient to use a pressurized oil supply, but because the film pressures are much greater, the oil must be introduced in a way that does not disturb the film pressure.

As the bearing pressure is increased and either the viscosity or the sliding velocity is reduced, the separation between the bearing surfaces reduces until contact occurs between the asperities of the two surfaces—point A on Fig. 5.9. As the bearing separation reduces, the solid-to-solid contact increases and the coefficient of friction rises rapidly, leading ultimately to the boundary lubrication mode shown in Fig. 5.10. The transition to boundary lubrication is controlled by the surface finish of the bearing surfaces, and the chemical composition of the lubricant becomes more important than its viscosity. The real area of contact is governed by the geometry of the asperities and the strength of the contacting surfaces. In choosing bearing materials that have boundary lubrication, it is essential to choose combinations of material that will not cold weld or "pick up" when the solid-to-solid contact occurs. The lubricant also convects heat from the bearing surfaces, and there will be additives to neutralize the effect of acidic combustion products.

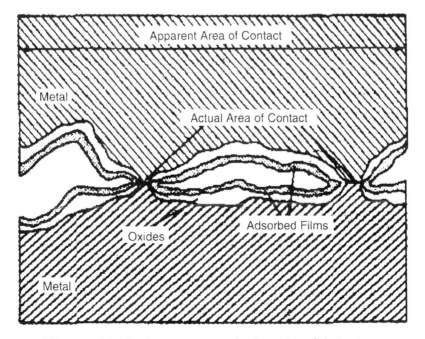

Figure 5.10. *Surface contact under boundary lubrication. Adapted from Stone (1999).*

In reality, the shaft does not remain concentric in the bearing, nor is the pressure uniform around the bearing. The eccentricity and the pressure distribution are shown schematically in Fig. 5.11. An early and reliable solution to this problem was published by Raimondi and Boyd (1958). This technique uses tabulated or graphical data to calculate the coefficient of friction, film pressure, point of maximum film pressure, side flow, and bearing temperature rise on the basis of the Sommerfeld number for the bearing under analysis. A thorough treatment of this method is found in Shigley and Mischke (2001).

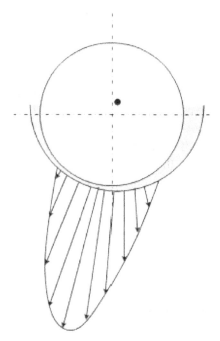

Figure 5.11. *A schematic of journal bearing film pressure distribution.*

5.3 Vehicle Cooling Systems

Gruden and Kuper (1987) conducted a systematic study of the energy balance in a spark ignition engine and presented a series of contour plots for the different energy flows (i.e., fuel in, brake power, coolant, oil, and exhaust) as functions of bmep and engine speed for a 2.5-liter engine. They also presented contour plots of the brake, mechanical, and indicated efficiencies. The brake efficiency results imply that the engine has been tuned for maximum economy at part load, whereas at full throttle, the mixture has been richened to give the maximum power. The mechanical efficiency is directly affected by the load (with zero mechanical efficiency by definition at no load). Also, the mechanical efficiency at full load falls from approximately 90% at 1000 rpm to 70% at 6000 rpm. At 6000 rpm, the frictional losses represent approximately 34 kW. Friction dissipates useful work as heat, some of which appears in the coolant and some in the oil. The heat loss recorded to the oil is almost solely a function of speed, with approximately 5 kW dissipated at 3000 rpm, and 15 kW dissipated at 6000 rpm.

Figure 5.12 shows the contours of the energy flow to the coolant as a function of the load and speed. For convenience, the brake power output hyperbolas (calculated from the bmep and speed) also have been added. At a bmep of approximately 1 bar, the energy flow to the coolant is approximately twice the brake power output, whereas at a load of 3 bar bmep, the energy flow to the coolant is comparable to the brake power output. In the load range 8–10 bar bmep, the energy flow to the coolant is approximately half the brake power output. However, of greater importance to the vehicle cooling system are the absolute values of the heat rejection. Figure 5.12 shows that heat rejected to the coolant is a stronger function of speed than load.

The heat rejection to a direct injection diesel engine is approximately a third lower, and this can lead to the need for a supplementary heater for passenger compartment heating. In both diesel engines and spark ignition engines, approximately half the heat flow to the coolant

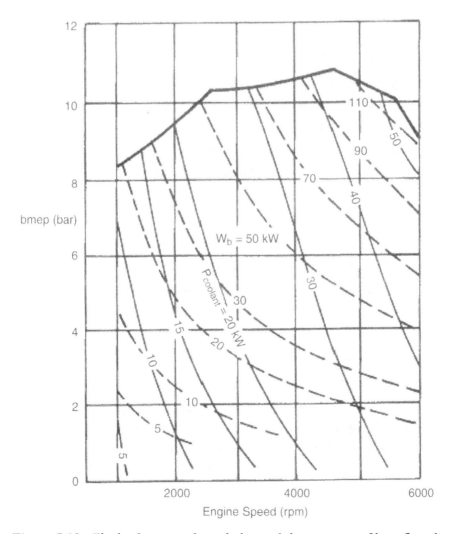

Figure 5.12. *The brake power hyperbolas and the contours of heat flow dissipated in the coolant system for a 2.5-liter spark ignition engine. Adapted from Gruden and Kuper (1987).*

comes from in-cylinder heat transfer. The remainder comes from heat transfer via the exhaust valve and port, and the geometry of the port and its extent within the cylinder head have a significant influence on the heat transfer to the coolant.

Originally, a coolant pump was not used; instead, natural convection led to a thermosiphon effect. Figure 5.13 shows the cooling circuit from a Rover 800 (a transverse-mounted engine with front-wheel drive), as a typical example of a cooling system. Initially, with a cold engine, the thermostat is closed and the pump circulates the coolant within the primary circuit, which is completed by the internal passages within the engine. The interior heater matrix is part of the primary coolant circuit, with the inlet manifold (when it is coolant heated), so that these items reach their proper working temperature as quickly as possible.

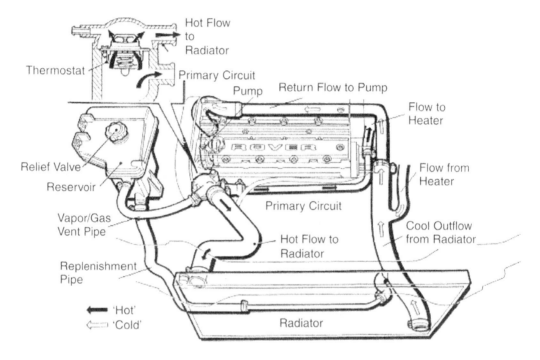

Figure 5.13. *The Rover 800 cooling system. Adapted from the Rover workshop manual.*

Some typical results recorded during engine warm-up are plotted in Fig. 5.14. These results were obtained from a Rover 800 engine installed on a dynamometer and operated at fixed throttle and speed. There was no interior heater matrix, but an oil cooler was installed in the primary coolant circuit. This has a beneficial effect in causing the engine oil to warm up more rapidly. Note that the exhaust temperature rises very rapidly, and this is important for catalyst light-off. The engine load or torque (bmep) rises quickly immediately after starting, although it took longer than 20 minutes for the engine to achieve its steady-state output under these conditions with a bmep of 2 bar.

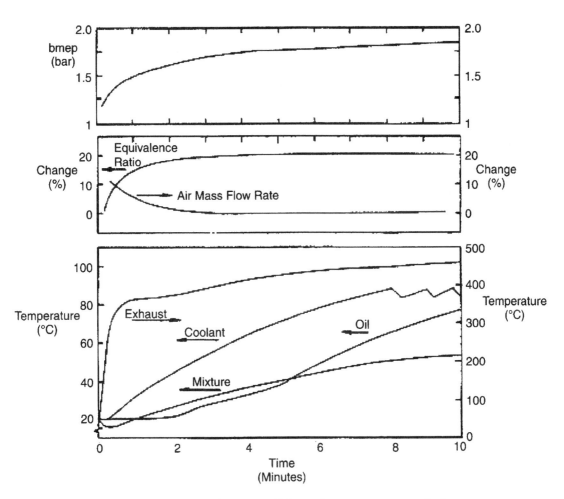

Figure 5.14. *The performance of a Rover 800 spark ignition engine during warm-up, at a speed of 1500 rpm, with a fixed throttle setting that gives a steady-state bmep of 2 bar.*

This particular version of the engine had a single-point injection system with a coolant-heated inlet manifold. The mixture temperature is seen to track the coolant temperature but with an initial fall that is attributable to the evaporative cooling of the fuel. The engine management system was arranged to give a fixed fuel flow rate into the engine. Soon after starting, there is a 10% fall in the air flow rate, but this only partly accounts for the 24% richening of the burned air-fuel ratio. The remaining difference probably is due to changes in the liquid fuel film on the wall of the inlet manifold.

When the engine has reached its operating temperature (usually around 90°C [194°F]), the thermostat opens so that some coolant flows through the main circuit and is cooled by the radiator. The radiator is a tube-and-fin heat exchanger, with coolant flowing through round tubes, and air flowing between the fins that zig-zag between the flat surfaces of the adjacent tubes. Originally, radiators were made of brass and copper, with nickel silver on the exposed surfaces. Current practice is to use aluminum (which is cheaper than copper but has a lower mass and higher thermal conductivity) for the tubes and fins, with plastic header tanks (often at each side rather than at the top and bottom). Figure 5.14 shows that when the thermostat opens, fluctuations occur in the temperature of the coolant in the primary circuit. This is due to cold coolant entering from the main circuit. These oscillations would be smaller if the thermostat opened more slowly, but there would be a risk of overheating if the thermostat did not open quickly enough at a high engine load. Even when the thermostat is fully open, there will still be a significant flow through the primary circuit.

The trend now is to reduce the volume of coolant in the primary circuit, so that rapid engine warm-up is achieved. This often is combined with precision cooling, in which drilled passages are used in the cylinder head to provide closer control on the dimensions and positioning of cooling passages. Today, it is common to have drilled passages between the valve seats because these are subject to high levels of heat transfer.

The only remaining element to discuss in Fig. 5.13 is the expansion tank, which has the highest elevation. The vent pipe allows any gas to leave the cooling system and to be separated from the coolant. Such gas comes from the degassing of dissolved gases in the coolant, and under some circumstances, there might be a leakage of combustion gases through the cylinder head gasket. (This can be identified by using an exhaust gas analyzer.) The expansion tank has a pressure relief valve to limit the system pressure (usually approximately 1 bar gauge pressure), and there also is an outlet into the main circuit so that makeup liquid can enter the engine. The expansion tank should be large enough to accommodate expansion of the coolant and have a large enough gas/vapor space to ensure that the gas/vapor is not compressed to a high enough pressure to open the pressure relief valve. Otherwise, when the system cools, the pressure would fall below atmospheric pressure.

The coolant pump usually is located at a low part of the cooling system with a cooled inlet flow; this is done to minimize the risk of cavitation. Frequently, the pump is partly formed by the cylinder block because this simplifies assembly and reduces the component costs. Traditionally, the coolant pump is of a simple design with a correspondingly low efficiency. The rotor may be stamped from sheet metal or be a simple casting.

5.3.1 Coolant

The composition of the coolant also is important, and an important part of any antifreeze mixture is its corrosion-inhibiting properties.

Water is a very effective cooling media, with a high enthalpy of vaporization, a high specific heat capacity, and a high thermal conductivity. Its saturation temperature of 99.6°C (211.3°F) at 1 bar also is convenient. Less convenient are its freezing point of 0°C (32°F) and its contribution toward corrosion, especially when there are different metals in the cooling system. Removal of heat from metal into a liquid coolant can be achieved by forced convection, sub-cooled boiling (where the bubbles detach when small and collapse into the bulk fluid that is below its saturation temperature), and saturated boiling (which has large bubbles and no condensation in the bulk fluid). With saturated boiling, there is the risk of vapor blankets and film boiling. This can cause overheating that leads to thermal fatigue, fracture of the component, or some form of abnormal combustion.

An aqueous ethylene glycol solution provides a sound basis for automotive engine coolant. However, to prevent corrosion and other shortcomings, it is necessary to have a range of additives in the ethylene glycol concentrate. Typically, these are the following:

a. Inhibitors to prevent metal corrosion
b. Alkaline substances to provide a buffering action against acids
c. An antifoam additive
d. A dye for ready identification
e. A small amount of water to dissolve certain additives

Aqueous mixtures of ethylene glycol are now well established as engine coolant, with the percentage of ethylene glycol typically ranging from 25 to 60% on a volumetric basis (i.e., 26.9 to 61.9% by weight, or 9.7 to 32.1% on a molar basis). Table 5.2 shows that in addition to lowering the freezing point, antifreeze raises the boiling point. The boiling points also will be increased because cooling systems are invariably pressurized; a 1 bar gauge system will raise the boiling point by approximately 20 K. At 2000 m (6560 ft), atmospheric pressure will fall to 0.8 bar, and in an unpressurized system, the boiling point will fall by approximately 6 K. However, for a system pressurized by 1 bar, the boiling point will fall by only approximately 3 K. Higher coolant temperatures enable smaller radiators to be used, which saves money and facilitates their installation.

TABLE 5.2
COOLANT PROPERTIES

Property	Water	Ethylene Glycol and Water (50/50)	Ethylene Glycol
Boiling point, 1 bar (°C)	100	111	197
Freezing point (°C)	0	−37	−9
Enthalpy of vaporization (MJ/kmol)	44.0	41.2	52.6
Specific heat capacity (kJ/kg-K)	4.25	3.74	2.38
Thermal conductivity (W/m-K)	0.69	0.47	0.33
Density, 20°C (kg/m^3)	998	1057	1117
Viscosity, 20°C (cS, 10^{-6} m^2/s)	0.89	4.0	20

5.4 Drive Belts

5.4.1 Flat Belt Drives

Belt drives are used to transmit rotary motion in much the same way as gears. They are cheaper than gears to manufacture, and they can run more quietly than gears. They require no lubrication, but they do require regular maintenance, which may include routine replacement. Size for size, they can transmit less power than a gear drive. Nevertheless, they are used widely and can be found in every car manufactured. Toothed belts (also known as timing belts or synchronous belts) can carry a greater power than friction belts, and because there is no slip, they have a higher efficiency. Manufacturers' data sheets provide guidance on how to rate the power transmission capabilities of these belts. The analysis here is restricted to friction belt drives. Consider the drive pulley and belt in Fig. 5.15. Note that the coefficient of friction is f to distinguish it from viscosity.

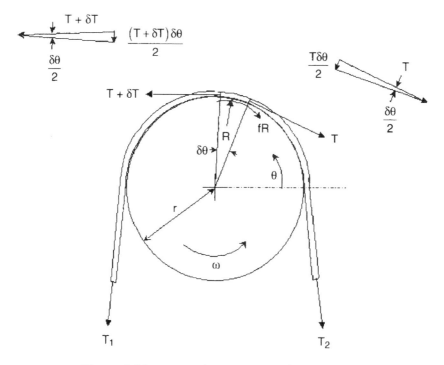

Figure 5.15. *Ratio of tensions in a flat belt pulley.*

Taking a differential mass element of the belt defined by the arc $\delta\theta$, and employing the small angle assumption ($\sin\theta \approx \theta$), the sum of the forces in the radial direction is

$$\sum F_r = -\delta m r \omega^2 = R - T\frac{\delta\theta}{2} - (T + \delta T)\frac{\delta\theta}{2}$$

or

$$R + \delta m r \omega^2 = T \delta \theta \tag{5.12}$$

Likewise, summing forces in the tangential direction and assuming small angles ($\cos \theta \approx 1$),

$$\sum F_t = 0 = -\mu R - T + (T + \delta T)$$

or

$$\mu R = \delta T \tag{5.13}$$

If the mass per unit length of the belt is denoted by m, and its average velocity is v (where $v = r\omega$), then

$$\frac{\delta T}{\mu} + m r \delta \theta \frac{v^2}{r} = T \delta \theta$$

or

$$\frac{\delta T}{(T - mv^2)} = \mu \delta \theta \tag{5.14}$$

Equation 5.14 then can be integrated over the contact arc length of the belt to give

$$\left[\log(T - mv^2) \right]_2^1 = [T\theta]_0^\theta \tag{5.15}$$

Expanding the limits yields

$$\frac{T_1 - mv^2}{T_2 - mv^2} = e^{\mu \theta} = \frac{T'_1}{T'_2} \tag{5.16}$$

where $T'_1 = T_1 - mv^2$, and $T'_2 = T_2 - mv^2$. If $v = 0$, $T'_1 = T_1$ and $T'_2 = T_2$. Thus, the torque being transmitted is

$$T = r(T'_1 - T'_2) = rT'_1 \left(1 - \frac{1}{e^{\mu \theta}}\right) \tag{5.17}$$

When this value of torque is reached, gross slip will occur. Prior to this value of torque being reached, creep will take place due to the slight differences in density between the "slack" side and the "tight" side of the belt. The presence of creep makes "smooth belt" drives unsuitable for drives where timing (i.e., synchronization) is important. Creep is inevitable because of the elasticity of the belt.

If the initial tension in the belt when at rest is T_0, then applying a torque, T, to the pulley would increase and decrease the tensions by δT (only when it is tight, and slack sides are of equal length), as shown in Fig. 5.16.

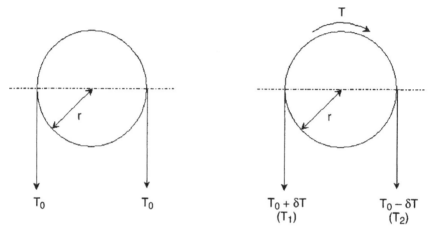

Figure 5.16. *The influence of initial tension (T_0).*

Thus,

$$T_1 = T_0 + \delta T$$
$$T_2 = T_0 - \delta T \qquad (5.18)$$

and

$$T_1 + T_2 = 2T_0 = T_1' + T_2' + 2mv^2 \qquad (5.19)$$

Therefore, from Eq. 5.16,

$$T_1'\left(1 + \frac{1}{e^{\mu\theta}}\right) = 2T_0 - 2mv^2 \qquad (5.20)$$

Finally,

$$T_1' = \frac{2(T_0 - mv^2)}{(1 + 1/e^{\mu\theta})} = T_1 - mv^2 \qquad (5.21)$$

It can be seen that the value of T_1 (and therefore the maximum torque) is limited by the initial tension (T_0) and the "centripetal" tension (mv^2). In other words,

$$T_{1(max)} = \frac{2(T_0 - mv^2)}{(1 + e^{-\mu\theta})} + mv^2 \qquad (5.22)$$

The maximum power that can be transmitted can be calculated by noting that power is equal to force times velocity, or

$$P = (T_1 - T_2)r\omega = (T_1 - T_2)v \qquad (5.23)$$

Therefore,

$$P = (T_1' - T_2')v = T_1'(1 - e^{-\mu\theta}) \qquad (5.24)$$

Assuming the belt had not reached its breaking load, then at a particular speed

$$P_{max} = \frac{2(T_0 - mv^2)}{(1 + e^{-\mu\theta})}(1 - e^{-\mu\theta})v$$

$$P_{max} = 2v(T_0 - mv^2)\tan\left(\frac{\mu\theta}{2}\right) \qquad (5.25)$$

If T_1 is at its maximum due to the belt material, then

$$T_1 = T_{max} \quad \text{and} \quad P_{max} = (T_{max} - mv^2)(1 - e^{-\mu\theta})v \qquad (5.26)$$

If there is a choice of speed for the belt drive, then to find the maximum value of P_{max}, differentiate Eq. 5.25 or Eq. 5.26 with respect to v.

5.4.2 V-Belts

One way of increasing the ratio T_1/T_2 is by using a belt that has a cross section in the shape of a truncated "Vee." Figure 5.17 shows wedge action with a V-belt.

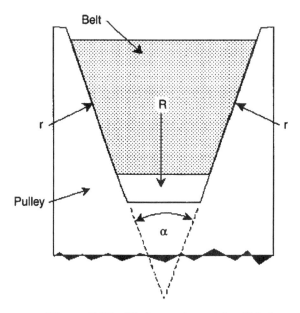

Figure 5.17. *Wedge action with a V-belt.*

Resolving radially gives

$$R = 2r\sin\left(\frac{\alpha}{2}\right) \tag{5.27}$$

Therefore,

$$r = \frac{R}{2\sin\left(\dfrac{\alpha}{2}\right)} \tag{5.28}$$

The friction force between belt and pulley is therefore

$$2\mu r = \frac{\mu R}{\sin\left(\dfrac{\alpha}{2}\right)} \tag{5.29}$$

Thus, by inspection from Eq. 5.13,

$$\frac{\mu R}{\sin\left(\dfrac{\alpha}{2}\right)} = \delta T \tag{5.30}$$

or

$$R = \frac{\sin\left(\dfrac{\alpha}{2}\right)\delta T}{\mu} \tag{5.31}$$

and from the previous analysis (Eq. 5.16),

$$\frac{T_1 - mv^2}{T_2 - mv^2} = e^{\dfrac{\mu\theta}{\sin\left(\dfrac{\alpha}{2}\right)}} = e^{\mu'\theta} \tag{5.32}$$

where μ' is the "effective coefficient" of friction,

$$\mu' = \frac{\mu}{\sin\left(\dfrac{\alpha}{2}\right)} \tag{5.33}$$

Typically, $\alpha = 30°$ and $\mu' = 3.86\mu$.

5.5 Air Conditioning Systems

5.5.1 Overview

Air conditioning invariably is obtained in vehicles by using a vapor compression refrigeration cycle (a reversed power vapor cycle). During steady-state operation, a vehicle might require 2–3 kW of cooling. However, to cool a vehicle after it has been parked might be three times this value. The Peltier effect could be used, but this would demand electrical energy that is "expensive" because of the comparatively low efficiency of vehicle alternators and the need to increase the alternator capacity. Utilizing the exhaust gas thermal energy may appear an attractive alternative, and different systems have been reviewed by Boatto et al. (2000). They conclude that an absorption system is the most attractive alternative to a vapor compression refrigeration system, but that an auxiliary firing system might be needed to provide heat when the engine power is less than approximately 10 kW. Furthermore, remember that absorption systems have a low coefficient of performance. (COP = refrigeration effect/power input; the power input is either high-temperature heat or mechanical power.) The COP may be only approximately 0.5, and this places a significant load on the vehicle cooling system, as shown in Fig. 5.18.

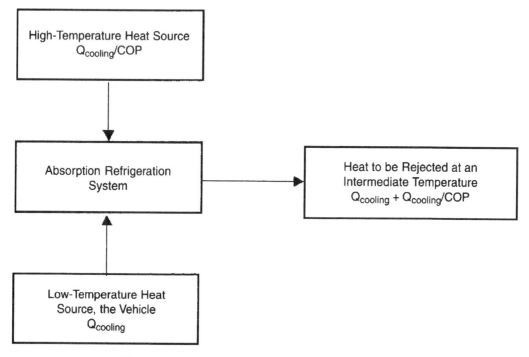

Figure 5.18. *An absorption refrigeration system.*

If the cooling requirement is 5 kW at 8°C (46°F) and the COP is 0.5 when the high-temperature heat source is 120°C (248°F), it will be necessary to reject 15 kW at the intermediate temperature (say, 60°C [140°F]). With ambient temperatures that might be 30°C (86°F) or higher, rejecting 15 kW with a temperature difference of 30 K or less implies a large air-cooled heat exchanger.

Figure 5.19 shows a vehicle air conditioning system based around a reversed power vapor cycle. The purpose of the key components is explained within the figure. The compressor usually is of a swash-plate design, in which an inclined rotating disc causes pistons (with a stationary axis) to reciprocate. The pumping capacity of the compressor could be varied by adjusting the inclination of the disc, but the usual method is to have a fixed displacement compressor driven through an electromagnetically controlled clutch. The receiver/drier provides storage for the liquid refrigerant, and the drier (a water-absorbing material) is needed to prevent any freezing of moisture. This would be especially troublesome if it occurred within the expansion valve.

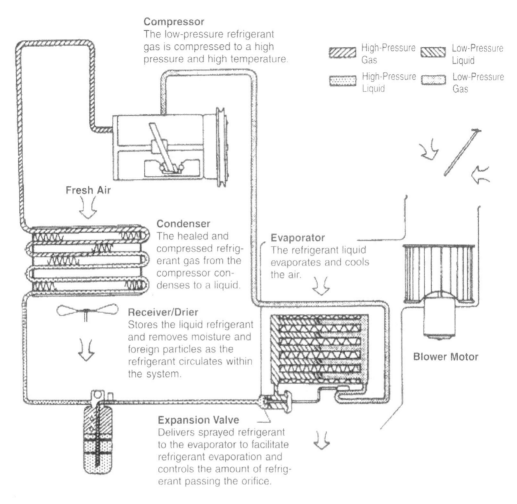

Figure 5.19. *A system diagram for a vapor compression air conditioning system.*

5.5.2 Thermodynamic Performance and Operation

The reversed power vapor cycle can be analyzed by using a few simplifying assumptions:

a. The only pressure changes occur in the expander and compressor,

b. There is no extraneous heat transfer (this is a common assumption for refrigeration compressors, but in this particular case, the high temperature in the engine compartment is likely to make it valid), and

c. Changes in the potential energy and kinetic energy can be neglected.

The steady flow energy equation (SFEE) then can be written in simplified (and specific—i.e., per unit mass) form as

$$h_1 + q = h_2 + w \qquad (5.34)$$

The SFEE can be applied in turn to the four components in the cycle:

$$\begin{aligned}
\text{Compressor:} &\quad w_{in} = h_2 - h_1 \\
\text{Condenser:} &\quad q_{out} = h_3 - h_2 \\
\text{Expander:} &\quad w_{out} = h_3 - h_{4s} \\
\text{Evaporator:} &\quad q_{in} = h_1 - h_{4s}
\end{aligned} \qquad (5.35)$$

The four processes, using the numbering from Fig. 5.19, can be conveniently represented on the temperature/entropy and pressure enthalpy planes of the refrigerant, as shown in Fig. 5.20.

State 4 is within the saturation region (where liquid and vapor coexist in equilibrium), so pressure and temperature are not independent. However, because we are neglecting changes in kinetic energy and potential energy, assuming there is no extraneous heat transfer and clearly there is no work output from the throttle, then

$$h_3 = h_4 \qquad (5.36)$$

The First Law is always valid; therefore, the difference between the heat flow out and heat flow in will correspond to the net work input (W_{net}):

$$W_{net} = q_{out} - q_{in} \qquad (5.37)$$

The net work input, of course, equates to the area within the cycle.

5.5.3 Coefficient of Performance (CoP)

$$\text{Refrigerator:} \quad CoP_{Ref} = Q_{in}/W_{net} = Q_{in}/(Q_{out} - Q_{in})$$

Clearly, it would be undesirable to expand vapor from a high pressure to a low pressure, because the irreversibilities would mean that more work would be input to the compressor than might be produced by the expander. The usual practice is to make the refrigerant slightly sub-cooled at exit from the condenser (as shown in Fig. 5.20). The low specific volume of the liquid and very wet vapor (i.e., they are dense) means that little work is produced by the expander. Furthermore, it is difficult to design an expander for liquids or wet vapors; instead, a simple expansion throttle is used. The irreversibilities associated with the throttle are, in

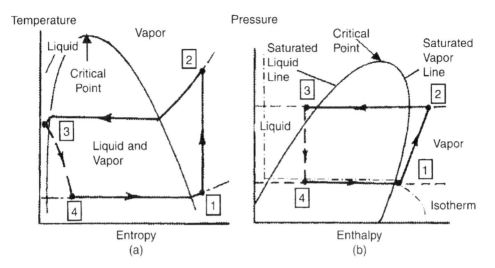

Figure 5.20. *(a) Temperature/entropy and (b) pressure/enthalpy diagrams for a practical reversed power cycle.*

Notes on the Pressure Enthalpy Plot:

1. Changes in enthalpy will correspond to the horizontal distances on the p–h chart (Fig. 5.20b).

2. Isotherms are plotted, so that measurements of temperature and pressure can define the operating points of the refrigeration cycle. The exception is in the saturation region (where the liquid and vapor coexist) as the temperature and pressure are no longer independent of one another. (They are related by the saturation pressure/temperature characteristics of the refrigerant.)

3. Isotherms obviously are horizontal in the saturation region (because both pressure and temperature are constant), and are almost vertical in the liquid region (because the enthalpy of a liquid is essentially only a function of temperature; there is negligible pressure dependency). In the vapor region, the isotherm would be vertical if the vapor behaved as an ideal gas. As the pressure is reduced and the isotherm moves away from the saturation line, the vapor behavior becomes closer to ideal, and the isotherm slope comes closer to vertical.

4. What is meant by the critical point? At pressures above the critical point, there is no longer any distinction between liquid and vapor. One way of thinking about this is to consider a point vertically below the critical point (in the saturation region), and then raise the pressure. As the pressure is raised, the density of the vapor will increase significantly, while the density of the liquid will hardly increase at all. When the critical pressure has been reached, there no longer will be any difference between the density of the liquid and the vapor. Therefore, there no longer can be any distinction between liquid and vapor. It is merely a dense fluid.

fact, reduced by sub-cooling the refrigerant before expansion and are significant only with low-temperature systems, such as those used for gas liquefaction.

It would be undesirable for liquid droplets to enter most compressors; therefore, the vapor entering the compressor is superheated slightly. In vehicle air conditioning systems, the small flow rates mean that reciprocating compressors are used, and liquid droplets would tend to transfer lubricant from the compressor to the condenser. The thermostatically controlled expansion valve (Fig. 5.21) controls the level of superheat; it is designed to give a constant level of vapor superheat at entry to the compressor. The temperature-sensing bulb contains a fluid that expands as the temperature rises, thereby tending to open the valve to increase both the refrigerant mass flow rate and the evaporator pressure. The increased evaporator pressure has a higher saturation temperature, which results in a lower superheat. The increased evaporator temperature also reduces the amount of heat transferred into the evaporator. When this is coupled with an increased mass flow rate, the negative feedback provided by the thermostatically controlled expansion valve leads to a stable control system.

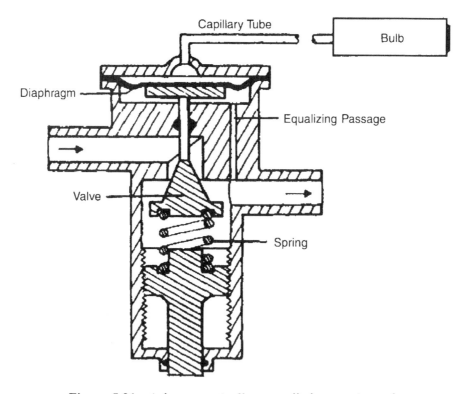

Figure 5.21. *A thermostatically controlled expansion valve.*

The compressor has a comparatively small displacement (say, 150 cm^3/rev) and is driven via the magnetically operated clutch at approximately the engine speed. The volume flow rate into the compressor essentially is constant; therefore, as the evaporator pressure rises, the higher density leads to the increased flow rate. To summarize, the greater the temperature difference between the evaporator and the condenser, the lower the amount of heat that can be "pumped" from the evaporator. Also, the power input increases with the temperature differential, because the pressure differential increases and the volume flow rate into the compressor is approximately constant.

Dentis et al. (1999) report on tests with different refrigerants in an automotive air conditioning system. When using refrigerant R134a in a compressor with a displacement of 150 cm³ at 1800 rpm, cabin air with a flow rate of 400 m³/h was cooled from 45 to 15°C (113 to 59°F). Assuming an air density of 1.15 kg/m³ and a specific heat capacity of 1.01 kJ/kgK, then the cooling load (Q_{cool}) is

$$Q_{cool} = 1.15 \times (400/3600) \times 1.01(45-15) = 3.87 \text{ kW}$$

Because the coefficient of performance is 2, the power input to the compressor is 1.94 kW.

R134a (CH_2FCF_3, tetrafluoroethane) has replaced R12 because of environmental concerns about refrigerants containing chlorine. Table 5.3 presents thermodynamic data for R134a. Refrigerant tables frequently are in a more compact form (e.g., Table 5.3) than steam tables, which means that more interpolation is required. The data are a combination of saturation and superheat values on a single line. Each line corresponds to a particular pressure, the increments of which are chosen to given uniform steps in the saturation temperature.

The superheat data do not refer to absolute temperatures but are for fixed superheat temperature increments relative to the saturation temperature. In Table 5.3, the superheat increment is 15 K, and as an example, the values of vapor enthalpy that have been underlined all refer to 25°C (77°F). The three values of enthalpy are not the same because the enthalpy of a real gas increases slightly with reducing pressure.

It is possible to analyze the thermodynamic performance of the refrigeration cycle if the air conditioning system being considered here has the following operating points:

- Evaporator exit (state 1 in Figs. 5.19 and 5.20) –2°C (28°F) at 2.012 bar (T_{sat} of –10°C [14°F])

- Compressor exit (state 2 in Figs. 5.19 and 5.20) 14.837 bar (T_{sat} of 55°C [131°F])

- Condenser exit (state 3 in Figs. 5.19 and 5.20) 14.837 bar at 45°C [113°F])

First, we must find the enthalpy and entropy at state 1. Because this has a superheat of 8 K, this will require linear interpolation between the saturation data and 15 K superheat data in Table 5.3 at a pressure of 2.012 bar:

$$h_1 = h_g + 8/15(h_{sat+15\,K} - h_g) = 290.99 + 8/15(303.96 - 290.99) = 297.9 \text{ kJ/kg}$$

and

$$s_1 = s_g + 8/15(s_{sat+15\,K} - s_g) = 1.727 + 8/15(1.775 - 1.727) = 1.753 \text{ kJ/kgK}$$

TABLE 5.3
THERMODYNAMIC PROPERTIES OF REFRIGERANT R134a, DERIVED FROM THE ICI KLEA CALC® SOFTWARE

		Saturation Data					Superheat Data			
							T_{sat} + 15 [Deg C]		T_{sat} + 30 [Deg C]	
T_{sat} [Deg C]	P_{sat} [bar]	ρ_g [kg/m³]	h_f [kJ/kg]	h_g [kJ/kg]	s_f [kJ/kg-K]	s_g [kJ/kg-K]	h [kJ/kg]	s [kJ/kg-K]	h [kJ/kg]	s [kJ/kg-K]
−50.00	0.297	1.665	33.66	266.01	0.733	1.774	276.99	1.821	288.35	1.868
−45.00	0.395	2.174	40.32	269.17	0.762	1.765	280.37	1.813	291.92	1.859
−40.00	0.517	2.798	46.96	272.32	0.791	1.757	283.75	1.805	295.51	1.851
−35.00	0.667	3.556	53.58	275.47	0.819	1.751	287.14	1.798	299.10	1.844
−30.00	0.850	4.467	60.21	278.61	0.846	1.745	290.52	1.792	302.70	1.838
−25.00	1.071	5.551	66.82	281.74	0.873	1.739	293.89	1.787	306.30	1.833
−20.00	1.335	6.833	73.44	284.84	0.900	1.735	297.26	1.782	309.90	1.828
−15.00	1.646	8.334	80.06	287.93	0.925	1.731	300.62	1.778	313.49	1.824
−10.00	2.012	10.084	86.69	290.99	0.951	1.727	303.96	1.775	317.07	1.821
−5.00	2.437	12.109	93.33	294.01	0.975	1.724	307.28	1.772	320.64	1.818
0.00	2.929	14.444	100.00	297.00	1.000	1.721	310.57	1.770	324.19	1.816
5.00	3.493	17.122	106.69	299.95	1.024	1.719	313.83	1.768	327.71	1.814
10.00	4.137	20.185	113.43	302.84	1.048	1.717	317.06	1.766	331.22	1.812
15.00	4.868	23.676	120.20	305.67	1.071	1.715	320.24	1.764	334.68	1.811
20.00	5.693	27.648	127.03	308.43	1.095	1.713	323.37	1.763	338.12	1.810
25.00	6.621	32.158	133.93	311.11	1.118	1.712	326.45	1.762	341.51	1.809
30.00	7.659	37.277	140.91	313.70	1.141	1.711	329.46	1.761	344.85	1.809
35.00	8.818	43.083	147.98	316.18	1.163	1.709	332.39	1.761	348.13	1.808
40.00	10.104	49.675	155.17	318.52	1.186	1.708	335.24	1.760	351.35	1.808
45.00	11.529	57.171	162.50	320.72	1.209	1.706	337.99	1.760	354.49	1.808
50.00	13.103	65.714	169.98	322.74	1.232	1.705	340.64	1.759	357.56	1.808
55.00	14.837	75.489	177.66	324.56	1.255	1.703	343.15	1.758	360.54	1.808
60.00	16.741	86.731	185.57	326.12	1.278	1.700	345.53	1.757	363.42	1.808
65.00	18.828	99.755	193.76	327.37	1.302	1.697	347.75	1.756	366.20	1.808
70.00	21.111	114.994	202.29	328.25	1.327	1.694	349.79	1.755	368.85	1.807

The compression process (1→2) in an ideal refrigeration process is isentropic (as shown by the vertical line in Fig. 5.20), and inspection of the entropy data in Table 5.3 for the 14.837 bar isobar shows that an entropy of 1.753 kJ/kgK corresponds to a superheat of slighly below 15 K. Textbooks usually employ two stages of linear interpolation, in which entropy is interpolated with temperature and then temperature is interpolated with enthalpy. However, because we are assuming linear interpolation, it is possible to interpolate directly between enthalpy and entropy, as shown in Fig. 5.22.

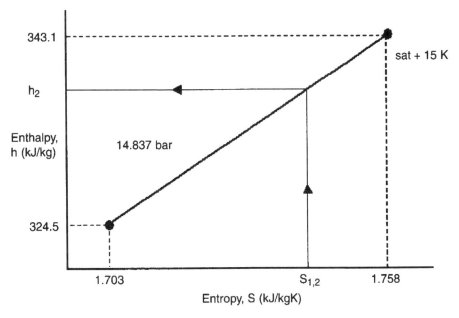

Figure 5.22. *Interpolation between enthalpy and entropy in the superheated vapor region.*

Inspection of Fig 5.22 gives

$$\frac{h_2 - h_g}{s_2 - s_g} = \frac{h_{sat+15} - h_g}{s_{sat+15} - s_g}$$

Rearrangement gives

$$h_2 = h_g + \left(h_{sat+15} - h_g\right) \times \left(s_2 - s_g\right) / \left(s_{sat+15} - s_g\right)$$

Substitution of the numerical values for 14.837 bar gives

$$h_2 = 324.56 + (343.15 - 324.56) \times (1.753 - 1.703)/(1.758 - 1.703) = 341.46 \text{ kJ/kg}$$

At exit from the condenser, the enthalpy of the liquid refrigerant is assumed to be dependent on only its temperature. Thus, the enthalpy at state 3 can be assumed to be equal to that of saturated liquid at 11.529 bar, for which the saturation temperature is 45°C (13°F).

$$h_3 = 162.50 \text{ kJ/kg}$$

We have already calculated the heat flow into the evaporator (Q_{cool} = 3.87 kW); thus, we can apply an energy balance to the evaporator to determine the refrigerant mass flow rate (m). Because for the throttle $h_3 = h_4$, then

$$Q_{cool} = m(h_1 - h_4) = 3.87 \text{ kW}$$

or

$$m = 3.87/(297.9 - 162.50) = 27.0 \text{ g/s}$$

The thermodynamic power required for the compression process is

$$m(h_2 - h_1) = 0.027(341.46 - 297.9) = 1.18 \text{ kW}$$

A comparison with the power input to the compressor (1.94 kW) shows that the compressor has an overall efficiency $(1.18/1.94)$ of 61%.

5.5.4 Air Conditioning System Performance

The air conditioning operating point will depend on several parameters. There is almost a linear relationship between the air flow rate through the evaporator and the cooling effect. Table 5.4 show that as ambient temperature rises, there are significant increases in the system pressures. This is a consequence of the almost exponential rise in saturation pressure with temperature.

TABLE 5.4
INFLUENCE OF AMBIENT AIR TEMPERATURE ON SYSTEM PRESSURES

Temperature (°C)	Condenser Pressure (bar)	Evaporator Pressure (bar)
20	9.2	0.8
25	11.5	1.5
30	13.9	2.1
35	16.2	2.8
40	18.6	3.5

Löhle et al. (1999) show how the evaporator performance and the coefficient of performance are related to the air flow through the condenser and its temperature (Fig. 5.23).

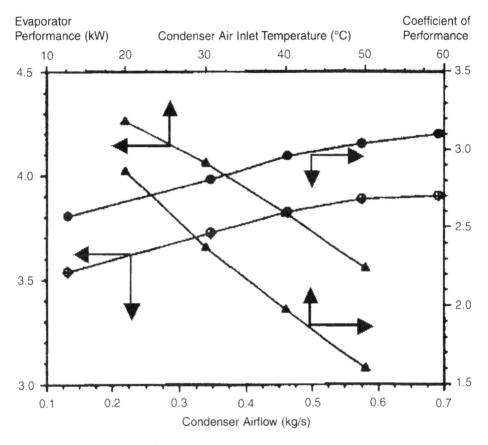

Figure 5.23. *The dependence of the evaporator performance and the coefficient of performance on the air flow through the condenser and its temperature. Adapted from Löhle et al. (1999).*

As would be expected, increasing the air flow through the condenser increases both the cooling effect in the evaporator and the coefficient of performance. In contrast, increasing the air temperature flowing through the condenser decreases both the cooling effect in the evaporator and the coefficient of performance.

5.6 Generators, Motors, and Alternators

5.6.1 Fundamentals

Figure 5.24 shows the simplest possible DC electric motor or generator, using permanent magnets to provide the stator magnetic field. Initially, we will ignore any currents flowing in the rotor and consider only the voltage being generated. The voltage (or EMF—electromotive force) generated by a conductor moving in a magnetic field is

EMF = Conductor velocity normal to the magnetic flux lines
　　　× Length of the conductor (in a direction normal to the velocity vector)
　　　× Magnetic flux density

If the magnetic field is assumed to be uniform in Fig. 5.24, then a sinusoidally varying voltage will be generated in the conductors, and the action of the brushes in the commutator is to rectify the voltage to give the half sine wave shown in Fig 5.24. Whether a DC electrical machine is acting as a motor or generator, this EMF is generated. In the case of an electric motor, it is termed a back EMF because it opposes the voltage being supplied to the motor. When a DC electric motor is connected to a battery, it will turn at a certain speed and draw current from the battery. However, if the motor is made to turn faster (by some external mechanical work input), the back EMF will become greater than the battery voltage, and current will be delivered to the battery.

The simple generator or motor shown in Fig. 5.24 is not practical because its output varies widely and at times is zero. Thus, practical generators and motors have three or more windings,

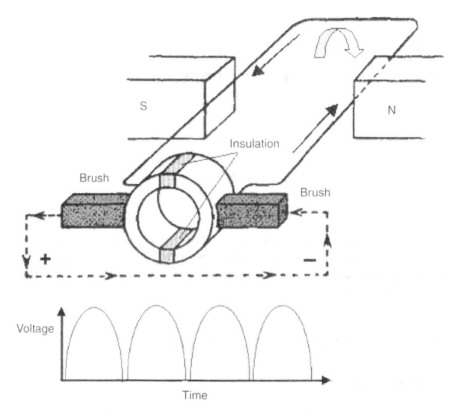

Figure 5.24. *A simple DC generator and its output voltage.*

and the greater the number of windings (and thus contacts on the commutator), the smoother the voltage.

If we ignore resistive losses in the armature windings and any frictional losses, then it can be seen that the motor speed is proportional to the battery voltage. Because the power produced by the motor is equal to both the product of speed and torque, and the voltage and current, then

Torque is proportional to current

This result also can be derived from magnetic arguments. The main simplifying assumption here has been to assume no resistance in the armature windings. When current (I) flows, the resistive voltage drop (IR) will cause the speed to drop. Figure 5.25 shows the resulting motor response at two different supply voltages (V) and its equivalent electrical model.

The torque produced by the motor will be less than the "magnetic torque" because of frictional losses.

Thus,

$$\text{Power output} = V \times (I - I_o) - IR \tag{5.38}$$

where I_o is the no-load current at the particular speed, and

$$\text{Power output} = V \times I$$

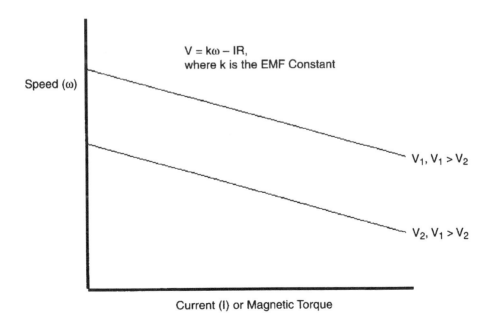

Figure 5.25. *Simple voltage, current, speed, and torque characteristics for a DC motor.*

This simple model for the voltage/speed response ignores the way that currents flowing in the rotor modify the static magnetic field, a phenomenon known as secondary armature reaction. However, that is outside the scope of this section.

Electric motors primarily use coils wrapped around cores of carefully selected iron alloys because this increases the strength of the magnetic field compared to air by approximately 500 to 5000 at the operating flux densities. Similarly, these materials will be used to construct a magnetic circuit, so that the magnetic field from permanent magnets or electromagnets is brought into the correct geometrical location relative to the rotor. The air gaps, of course, are minimized. Most of the electric motors in cars are small (e.g., windshield wipers, interior blower, seat adjustments) and use permanent magnets. Speed control traditionally is achieved either by switching in resistances (e.g., blower motor) or by switching the energization to a pair of brushes that are approximately 120° apart, so that a lower back EMF is generated. However, large motors, used for either propulsion or as starter motors, have an electromagnet energized by field windings. The two basic configurations are to have the field either in series or in parallel (referred to as a shunt winding). Figure 5.26 shows the two different configurations and their torque speed characteristics.

The characteristic of the shunt motor is essentially the same as the permanent magnet motor. However, if the resistance of the field coils is reduced (but maintaining the same number of turns) or the current is increased by some other means, then the magnetic field will strengthen and the back EMF constant will increase. Thus, with a fixed voltage operation and a specified speed, the current flowing through the rotor will be reduced. Therefore, less power will be produced, and the torque is reduced. However, the gradient of the torque/speed curve is increased. It might be thought that the efficiency of a shunt wound motor would be significantly

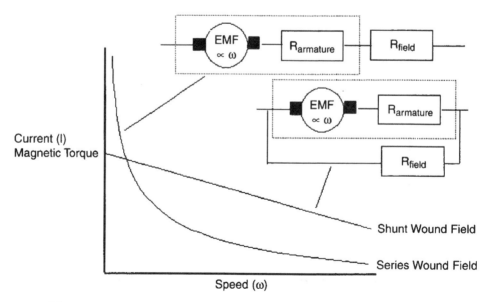

Figure 5.26. *Configuration and torque speed characteristics of series and shunt DC motors; the rotor is contained within the broken line.*

less than that of a permanent magnet motor. However, the current flowing through the field windings is very small compared to that of the armature, because the strength of the magnetic field is proportional to the product of the current and the number of turns in the coil. Thus, the field windings have a large number of turns and a high resistance.

Historically, the main automotive use of shunt "motors" was as a generator. At low rotational speeds, a high field excitation enabled a high enough voltage to recharge the battery. As the speed was increased, the field excitation could be reduced. This voltage control and current control (as well as protection to ensure that the battery would not discharge through the generator at very low speeds) were provided by ingenious electromechanical devices. The DC generator has a worse low-speed generating capability than an alternator, and all the current being generated must flow through the commutator and brushes. Therefore, the alternator has been in almost universal use since around 1970.

The series configuration motor is used in starter motors and is characterized by having very low resistance armature and field windings. This ensures a high starting torque (but also a high current), which reduces as the speed increases because the back EMF generated in the rotor reduces the current flowing through both the field and armature windings.

With traditional DC motors used in propulsion, there was scope for using combinations of both shunt and series field windings. However, with solid-state electronics that are capable of switching high currents, it is now common to have separate excitation of the field and the armature. By rapidly switching a fixed voltage supply with different ratios of on and off, it is possible to supply the field and armature windings with variable DC voltages, such that there is complete control of the motor speed and torque. However, DC motors still have the major drawback, which is that the current flow leading to the power output must flow through the brushes.

If a permanent magnet is rotated inside a coil with opposed windings or using the arrangement shown in Fig. 5.27, a sinusoidal alternating voltage will be produced. This is the basic principle of an alternator. The permanent magnet can be replaced by an electromagnet; however, in both cases, the voltage will be essentially proportional to the product of the rotor speed and field strength.

The alternator will be run at different speeds, and its voltage output will depend on the current being drawn. Thus, the rotating magnetic field is provided by an electromagnet because its energization can be varied. The alternating voltage must be rectified.

5.6.2 Practical Alternators

A practical alternator uses three sets of coils 120° apart to generate a three-phase alternating current. When this is rectified, it will generate an almost constant DC voltage. By using multiple poles in the rotating magnetic field, a higher frequency AC voltage is generated. This too helps to ensure a smooth DC voltage after rectification.

228 | *Automotive Engineering Fundamentals*

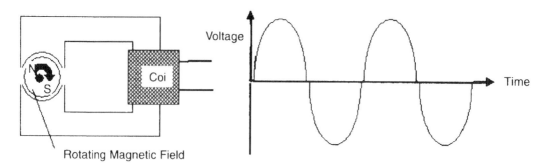

Figure 5.27. *A basic alternator.*

Figure 5.28 shows the construction of a Bosch claw-pole alternator. The axial construction of the rotor electromagnet gives good high-speed strength, and the rotor has twelve poles. The energization current (a maximum of approximately 3 A for a 35-amp-output alternator) passes through the slip rings. This is much lower than the current in a dynamo, and it is, of course, continuous.

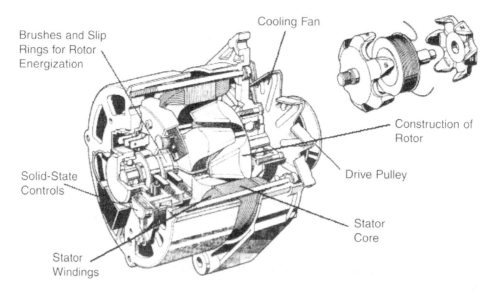

Figure 5.28. *Construction of a Bosch claw-pole alternator. Adapted from Meyer and Gerhard (1988).*

Figure 5.29 shows the basic electrical configuration of the alternator. The three stator windings are 120° apart physically, and they generate an alternating voltage that is 120° out of phase, as shown in Fig. 5.30. The power diodes allow only the highest and lowest voltages from the three phases to pass through. Because there is a voltage drop (nominally 0.7 V but rising to 1 V when the rated current is flowing) at each diode, the resulting envelope for the

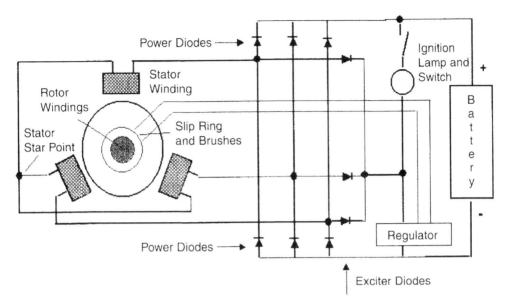

Figure 5.29. *The basic electrical configuration of a three-phase alternator.*

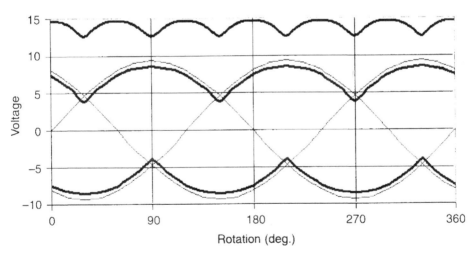

Figure 5.30. *The voltage characteristics of the alternator.*

positive and negative voltage variation is as shown in Fig. 5.30. Because the negative going voltage is connected to the negative terminal (0 V) of the battery, the DC voltage being generated is the difference of these two envelopes. This is the upper trace in Fig. 5.30. The amplitude of the AC voltage and thus the current (because the current depends on the voltage and the external load) generated in each stator are determined by the energization level of the rotor and its speed. As the alternator speed is reduced, the energization current must be increased until the maximum permissible rotor current is reached. Below this speed, the maximum current output of the alternator is below its rating.

There will be some residual magnetism in the rotor, but this is insufficient to make the alternator self-energizing. To be self-energizing, a voltage greater than the voltage drop across the diodes must be generated. Thus, energization initially comes from the battery via the ignition warning lamp (usually red). When the engine is stationary and the ignition is turned on, current flows through the ignition warning lamp (a maximum of approximately 0.15 A) via the regulator to energize the rotor. As soon as the alternator is at a high enough speed to generate a voltage to match the battery, there is no potential difference across the ignition lamp, and it will extinguish. The regulator now controls the current drawn from the exciter diodes to energize the rotor, in order to maintain a suitable voltage for recharging the battery. As a minimum, the regulator contains a Zener diode to provide a reference voltage, as well as power transistors (usually as a Darlington pair) to control the energization current supplied to the rotor.

The use of an increasing number of electrically operated or actuated accessories is increasing the power demand from the alternator. The loads include electrically heated windows, seats, and catalysts; motors for cooling fans (and possibly coolant pumps); and electric actuators for power steering, seat adjustment, and possibly brakes. In 1990, a typical requirement was 500 W. Today, it is 1 to 1.5 kW, and by 2005, it is forecast to be approximately 3 kW (Smith, 2001). With a 12 V system, 3 kW would require a current of 250 amps. Therefore, vehicle manufacturers have agreed to implement 42 V electrical systems, this being the highest voltage considered safe with the possibility of exposed connections in a vehicle. This is three times the present voltage. (Although the battery nominally is 12 V, the alternator is designed around an output of 14 V, to fully charge the battery.) The 42 V alternator will require stator coils with three times as many turns. Therefore, for a given power alternator using the same size conductors, the resistive losses inside the alternator will be reduced by two-thirds. Initially, alternators had an efficiency of approximately 30%, but current alternators with an output of 80 A can achieve efficiencies of approximately 60%. In the 14 V alternator, there is always a diode voltage drop of 2 V, a 14% loss, whereas for a 42 V alternator, this represents only a 5% loss.

The increase in electrical demand also increases the attraction of high-output integrated starter alternator (ISA) systems. These devices can replace the flywheel, and the control system speed of response is such that they can be used to smooth crankshaft torque fluctuations within each revolution. With a power rating of the order of 5 kW, they can start the engine in 0.2 seconds, and this provides the opportunity to turn off the engine during stationary parts of the drive cycle. The elimination of idle periods provides substantial reductions in both emissions and fuel consumption. An ISA also provides some hybrid capability, because electrical assist can be used for acceleration, and there can be regenerative braking. The ISA provides up to 20% lower fuel consumption for small vehicles on the European driving cycle. Toyota is introducing an ISA that replaces the existing alternator and uses a drive belt connection to the crankshaft, thereby facilitating its introduction into existing product ranges.

The higher voltage will not affect lead-acid battery design; however, if "hybrid" type operation is being planned, the battery will need to be designed for many more discharge cycles. It also will need to have a greater energy capacity because when it is being recharged after starting, it must not be fully charged. Otherwise, there would be no capacity for regenerative braking.

5.6.3 Practical Starter Motors

Starter motors usually have four poles because, all things being equal, this will cut in half the rotational speed and provide twice the torque of a two-pole motor at a given power. Figure 5.31 shows one arrangement for the stator windings and brushes. By having two parallel paths for the current flow (each through a pair of brushes and two stator windings), a higher current will flow (compared to having all the windings in series), and the torque also will be higher.

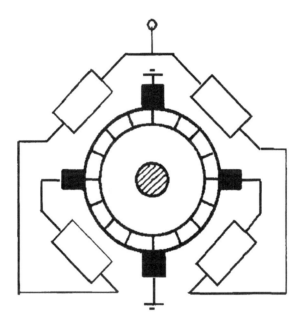

Figure 5.31. *Arrangement of the stator windings and brushes in a four-pole starter motor.*

A pinion on the starter motor engages with teeth on the "starter ring," which usually is shrunk-fit onto the flywheel; the reduction ratio typically is approximately 14:1. The pinion slides axially into engagement, and this is facilitated by chamfers on the leading edges of the pinion teeth. The most common arrangement originally was the Bendix system, which relied on a coarse pitch helical spline between the pinion and the starter motor shaft. When the starter motor was switched on, the inertia of the pinion led to relative rotation between the pinion and its shaft, so that it would slide along the helix and into engagement. When the engine had started, it would attempt to drive the starter motor. However, the pinion would now rotate in the opposite direction relative to its shaft, and it would slide along the helix out of engagement with the starter ring. The Bendix system was superseded by the so-called "pre-engaged" starter, of which there are several variations. Figure 5.32 shows the basic electrical and mechanical arrangements of a pre-engaged starter.

The key to the pre-engaged starter is the solenoid, which is a solenoid with two windings, and a pair of contacts that controls the main current flow to the starter motor. When the starter

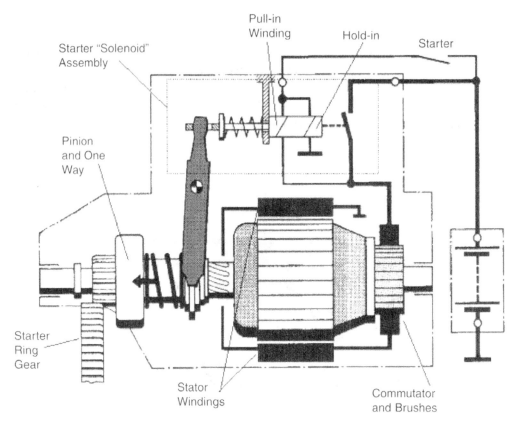

Figure 5.32. *The basic electrical and mechanical arrangements of a pre-engaged starter. Adapted from Adler (1988).*

switch is closed, current flows through both the pull-in and hold-in windings of the solenoid. The solenoid acts through a lever to slide the pinion into mesh with the starter ring gear. Meshing is facilitated by the helical spline on the motor shaft and by the low-torque rotation of the starter motor. Figure 5.32 shows how the current through the pull-in winding also flows through the armature and stator. This current is limited to a few amps by the pull-in winding, but it is sufficient to turn the starter motor. When the pinion is fully meshed, the main current contacts are closed by the solenoid, and the full battery voltage is applied across the starter motor. This also stops the current flow through the pull-in winding on the solenoid. A starter motor typically takes 250 amps when cranking the engine, and double this amount when it is stalled. When the engine fires, the one-way clutch prevents any torque being used to drive the starter motor. When the starter switch is opened, the solenoid hold-in winding is de-energized, and the pinion is retracted by the spring.

Starter motors also are made with permanent magnets for small engines (spark ignition engines below approximately 2 liters displacement). When an epicyclic reduction gearbox is incorporated into a permanent magnet starter motor, they can be used for spark ignition engines up to 6 liters displacement. The starter motor obviously is not rated for continuous operation, and this also might apply to the main conductor from the battery to the solenoid. However, for cold starting, the engine will require the greatest starting torque, and the battery will have an increased internal resistance and lower open-circuit voltage. Thus, it is essential that the voltage drop in the connections between the battery and motor are not excessive. Additional facilities that can be included in the starter circuit are timers to prevent excessive use of the starter motor, as well as interlocks to prevent the starter from being used when the engine is firing.

5.7 Conclusions

The various topics within this chapter may not seem to share much in common. However, one important common factor is that the efficiency of all the ancillary items has a major impact on vehicle fuel economy. This includes even the starter motor because some designs are suitable for management strategies that turn off the engine when the vehicle is stationary. Effective lubrication and bearing systems allow the mechanical power to be produced with minimal mechanical losses. It is then necessary to transmit and use this power as effectively as possible. The power required by a car at a steady 60 mph may be only 20 kW, and at a typical corresponding engine speed, the frictional losses in the engine might be 4 kW. Thus, it can be seen that the power used by the ancillaries (e.g., 2–3 kW for air conditioning, or 1–2 kW for the alternator) can be highly significant.

CHAPTER 6

Transmissions and Driveline

6.1 Introduction

The most powerful engine in the world is of little use unless the power from the engine can be safely and effectively transmitted to the ground. This, then, is the primary function of the transmission and driveline. In addition to being able to transmit the torque and power from the engine, the transmission and driveline also must allow the vehicle to operate over a wide range of speeds—from a standstill to the maximum speed of the vehicle. This implies that the system must inherently have some method of disconnecting the engine from the remainder of the driveline to allow the vehicle to remain stationary. Furthermore, the transmission also must be designed to satisfy the conflicting requirements of quick acceleration, high speed, and adequate fuel economy.

Figure 6.1 illustrates the need for a transmission. This figure shows the tractive force at the wheels plotted against vehicle speed. The individual curves have been generated by taking the basic engine torque curve and multiplying it by the appropriate gear ratio.

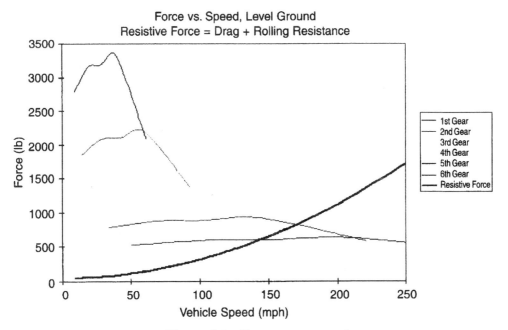

Figure 6.1. *Force versus speed.*

As shown in Fig. 6.1, a high torque multiplication is desired to accelerate the vehicle from a standstill. However, if first gear were all that were available, the maximum speed of the vehicle would be limited to 62 mph by the engine red-line. Although this might be an acceptable top speed for some, the penalty in fuel economy would be excessive. Figure 6.1 also plots the resistive force acting on the car, which is the sum of the aerodynamic drag and rolling resistance. As shown in Fig. 6.1, the car is capable of exceeding 62 mph by a large margin, but only by having more gear ratios available. At the other extreme, if the car had only sixth gear, any attempt to accelerate the car from a standstill inevitably would stall the engine or, at best, result in extreme clutch wear.

This chapter takes a logical progression from the engine flywheel to the drive axles. The first section deals with manual transmission systems and covers friction clutches, gear theory, and manual transmission operation and analysis. The next section discusses automatic transmission systems, including hydrodynamic torque converters, planetary gear analysis, general transmission operation, and gear ratio analysis. The chapter includes a brief discussion of continuously variable transmissions (CVTs). The power flow then is traced through the driveline and the differential to the drive axles. Although the case can be made for including tires and wheels in the driveline, for the purposes of this work, they will be treated as part of the vehicle control system. Thus, tires will receive mention under steering, vehicle dynamics, and braking, with a full discussion of tire nomenclature and design in Chapter 9. Chapter 6 concludes with a case study of a modern, five-speed, electronically controlled automatic transaxle.

6.2 Friction Clutches

The friction clutch is the link between the engine and the transmission, and it exists to provide the operator with the ability to engage and disengage the engine from the transmission. The clutch consists of a cover, a pressure plate, and a disc with friction facings, as shown in Fig. 6.2. The cover is bolted to the engine flywheel; thus, it rotates with the engine at all times. Inside the cover is a pressure plate, which also rotates with the cover and the flywheel. Sandwiched between the pressure plate and the flywheel is the friction disc. This disc is connected to the transmission input shaft by means of splines.

Torque is transmitted to the transmission when the pressure plate is forced against the friction disc, thus squeezing the friction disc between the pressure plate and the flywheel. The axial force required to squeeze the disc is supplied by a series of springs arranged circumferentially around the pressure plate. Some clutches use coil springs, as shown in Fig. 6.2, although some manufacturers use a diaphragm spring (Belleville spring) for this function (Fig. 6.3).

The operation of a clutch may be explained by reference to Figs. 6.3 through 6.5. When the driver wishes to disengage the clutch, he or she does so by depressing the clutch pedal. Through a series of either mechanical linkages or hydraulics, this action causes the clutch fork to move against a release, or thrust, bearing. This bearing allows the axial force to be transmitted from the clutch fork to the diaphragm spring, while minimizing the wear that would exist if a stationary, solid piece of steel were used to contact the rotating spring. This action causes the

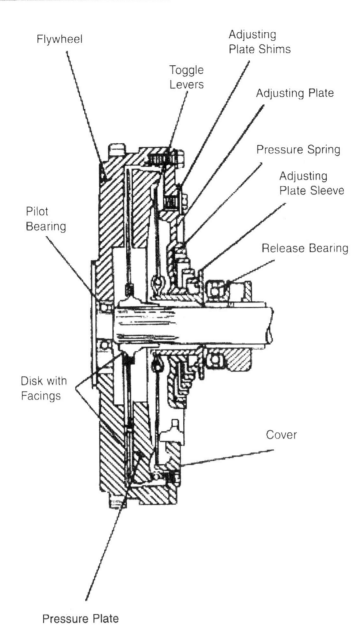

Figure 6.2. *Cross-sectional view of a coil spring clutch (TM 9-8000, 1985).*

diaphragm spring to pull the pressure plate away from the friction disc. At this point, no torque can be transmitted to the transmission shaft, and the clutch is disengaged. Engagement of the clutch is produced when the driver releases the clutch pedal. The diaphragm spring now returns to its unloaded position, thus applying the necessary axial force to the friction disc and allowing torque to be transmitted

238 *Automotive Engineering Fundamentals*

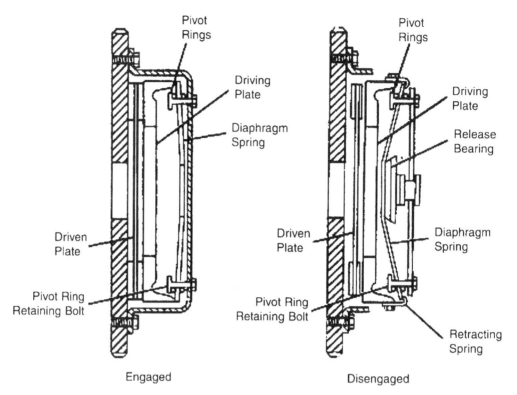

Figure 6.3. *Operation of a diaphragm spring. Adapted from TM 9-8000 (1985).*

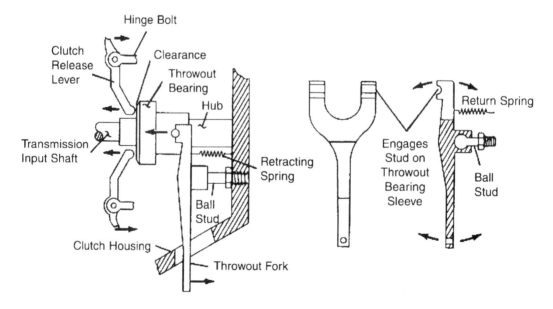

Figure 6.4. *Clutch activation (TM 9-8000, 1985).*

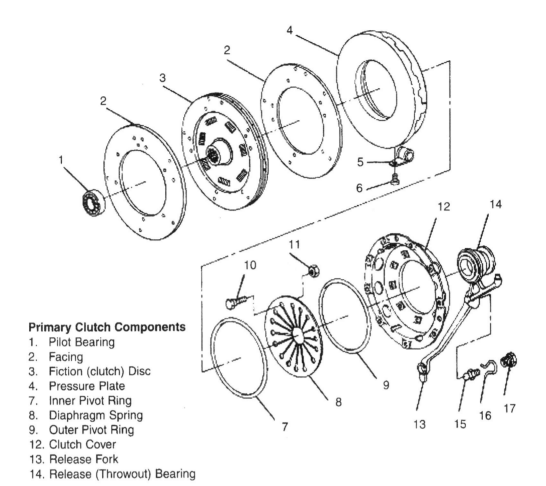

Primary Clutch Components
1. Pilot Bearing
2. Facing
3. Fiction (clutch) Disc
4. Pressure Plate
7. Inner Pivot Ring
8. Diaphragm Spring
9. Outer Pivot Ring
12. Clutch Cover
13. Release Fork
14. Release (Throwout) Bearing

Figure 6.5. *Exploded view of a diaphragm clutch. Adapted from TM 9-8000 (1985).*

6.2.1 Torque Capability of an Axial Clutch

Figure 6.6 shows a typical friction disc. The disc normally has an annulus of friction material riveted or glued to both sides of the plate. The coil springs shown in Fig. 6.6 provide some measure of torsional flexibility to the plate, the better to extend its life when the operator is either learning to operate a clutch or is an aggressive driver.

The clutch designer must consider activating force, torque delivery, energy loss, temperature rise, and wear. Thus, the friction material must be selected to provide all of the following:

- A uniform coefficient of friction over the surface
- A coefficient of friction that remains stable with temperature changes
- Good thermal conductivity
- Resistance to wear

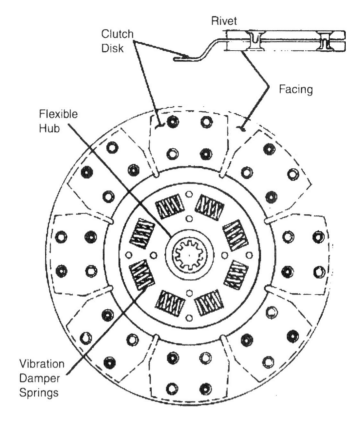

Figure 6.6. *Clutch disc with flexible hub (TM 9-8000, 1985).*

- Resistance to thermal fatigue
- Good high-temperature strength

The torque capability of a clutch may be analyzed with reference to Fig. 6.7. There is some pressure operating over the face of the clutch; therefore, in general, the force on the clutch face is given by $F = pA$. Looking at a differential area on the face, this becomes $dF = p r dr d\theta$. The total force on the clutch face may be obtained by integrating over the face of the friction material. However, some assumptions must be made regarding the pressure.

6.2.1.1 Uniform Pressure: $p = p_a$

This assumption is valid if the surface of the friction material is truly planar. Thus, it is a good assumption in the analysis of a new clutch. The integral used to find the force is

$$F = p_a \int_0^{2\pi} \int_{d/2}^{D/2} r \, dr \, d\theta \tag{6.1}$$

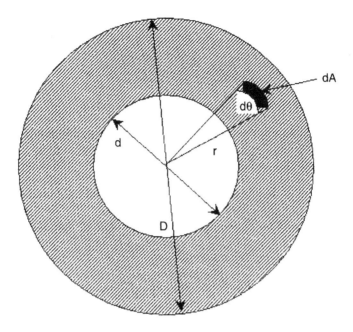

Figure 6.7. *Schematic of the friction material on a clutch surface.*

which upon evaluation yields

$$F = \frac{\pi p_a}{4}\left(D^2 - d^2\right) \tag{6.2}$$

The frictional force on the face of the clutch is simply the normal force (F) times the coefficient of friction for the material (μ). The torque transmitted by the clutch is found by integrating the product of the frictional force and the radius

$$T = \mu p_a \int_0^{2\pi} \int_{d/2}^{D/2} r^2 \, dr \, d\theta = \frac{\pi \mu p_a}{12}\left(D^3 - d^3\right) \tag{6.3}$$

Equation 6.3 gives the torque capacity of the clutch as a function of pressure. A more convenient form for the designer results by solving Eq. 6.2 for the pressure (p_a), and substituting this expression into Eq. 6.3. This results in

$$T = \frac{F\mu}{3}\frac{D^3 - d^3}{D^2 - d^2} \tag{6.4}$$

which is the torque capacity of the clutch as a function of activation force (F). Note that this expression gives the torque capacity for a single friction surface. For multiple friction surfaces, Eqs. 6.3 or 6.4 must be multiplied by the number of friction surfaces (N).

6.2.1.2 Uniform Wear

Obviously, a clutch does not remain new for long. When the clutch has been broken in, the pressure distribution changes and permits uniform wear. The wear is a function of normal force and linear velocity, or

$$FVdt = p\mu r\omega dA dt = \text{constant}$$

The implication of this is that the pressure is not constant, and the maximum pressure (p_a) occurs along the inner radius $\left(\dfrac{d}{2}\right)$. Furthermore, the pressure at any other radius must be inversely proportional to the maximum, or $p = p_a \dfrac{d}{2r}$. The same procedure is followed to determine the activation force on the clutch face

$$F = \frac{p_a d}{2} \int_0^{2\pi} \int_{d/2}^{D/2} dr d\theta = \frac{\pi p_a d}{2}(D-d) \tag{6.5}$$

Again, the torque is the integral of the frictional force times the radius, and the torque capacity as a function of pressure is given by

$$T = \frac{\mu p_a d}{2} \int_0^{2\pi} \int_{d/2}^{D/2} r dr d\theta = \frac{\pi \mu p_a d}{8}\left(D^2 - d^2\right) \tag{6.6}$$

Again, solving Eq. 6.5 for p_a and substituting into Eq. 6.6 gives torque capacity as a function of activation force (F)

$$T = \frac{F\mu}{4}(D+d) \text{ for a single surface and}$$

$$T = N\frac{F\mu}{4}(D+d) \text{ for multiple friction surfaces} \tag{6.7}$$

Now, it appears by Eq. 6.7 that the torque capacity of the clutch would be maximized by allowing the inner radius to equal the outer radius (d = D). It would indeed generate maximum torque because this gives the largest radius for the clutch. However, to do this would require infinite pressure to achieve any torque transfer. The proper way to maximize torque capacity is to take the appropriate partial differential of Eq. 6.6 (see problem 6.3 at the end of this chapter).

Another consideration in designing the clutch is the fact that the activation force provided by the springs must be overcome by the driver to disengage the clutch. Although the clutch pedal and linkages provide a mechanical advantage, another solution is to add centrifugal weights to the clutch assembly. Then, as the cover rotates, these weights can be used to increase the force on the pressure plate. This can be analyzed with reference to Fig. 6.8.

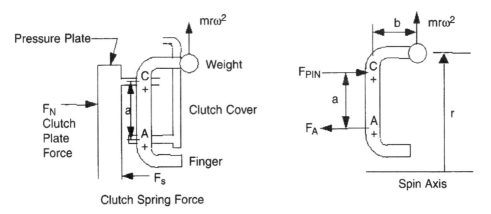

Figure 6.8. *Schematic of a clutch with centrifugal weights and free-body diagram of a finger.*

Summing moments about point A in Fig. 6.7 yields $F_{PIN}a = mr\omega^2 b$, or

$$F_{PIN} = \frac{mr\omega^2 b}{a} \qquad (6.8)$$

which is the additional activation force on the clutch face.

6.3 Gear Theory

Before explaining the operation of a manual transmission, the fundamentals of gear operation will be examined. Gears are most often used to transmit power, as well as to change angular velocity and torque, because the teeth provide a positive driving action (i.e., no slippage). The nomenclature, design, fabrication, and analysis of gears are governed in the United States by standards bodies such as the American Gear Manufacturers Association (AGMA). For this information, the student is directed to any good machine design textbook such as Shigley and Mischke (2001). This book is concerned with the application of gears and their use in automobiles.

Almost every type of gear can be found in an automobile. The most common varieties are discussed next.

6.3.1 Straight-Tooth Spur Gears

Figure 6.9 shows an example of this type of gear. Straight-tooth spur gears have straight teeth parallel to the axis of rotation. When the teeth engage, they do so instantaneously along the tooth face. This sudden meshing results in high impact stresses and noise. Thus, these gears have been replaced with helical gears in most transmissions. However, these gears do not generate axial (or thrust) loads along the shaft axis. Furthermore, they are easier to manufacture and can transmit high torque loads. For these reasons, many transmissions use spur gears for first and reverse gears. This accounts for the distinctive "whine" when a car is reversed rapidly.

Figure 6.9. *Spur gears.*

6.3.2 Helical Spur Gears

Figure 6.10 shows an example of a helical gear. Helical gears have teeth that are cut in the form of a helix on a cylindrical surface. As the teeth begin to mesh, contact begins at the leading edge of the tooth and progresses across the tooth face. Although this greatly reduces the impact load and noise, it generates a thrust load that must be absorbed at the end of the shaft by a suitable bearing.

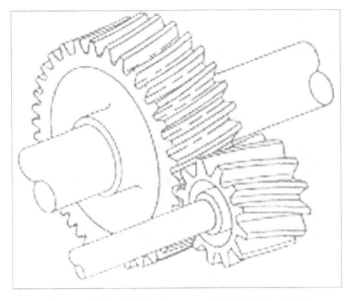

Figure 6.10. *Helical gears (TM 9-8000, 1985).*

6.3.3 Straight-Tooth Bevel Gears

These gears, shown in Fig. 6.11, have straight teeth cut on a conical surface. They are used to transmit power between shafts that intersect but are not parallel. They are used in differentials. Similar to straight-tooth spur gears, they will be noisy. However, in the differential, they rotate only when the axles are rotating at different speeds.

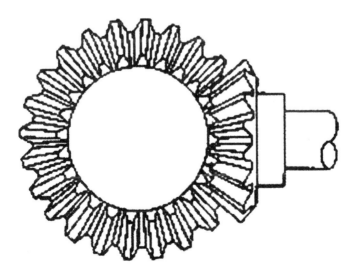

Figure 6.11. *Straight-tooth bevel gears (TM 9-8000, 1985).*

6.3.4 Spiral Bevel Gears

These gears have teeth cut in the shape of a helix on a conical surface. They can be used for final drives to connect intersecting shafts (Fig. 6.12).

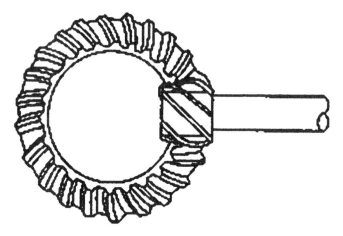

Figure 6.12. *Spiral bevel gears (TM 9-8000, 1985).*

6.3.5 Hypoid Gears

These gears have helical teeth cut on a hyperbolic surface (Fig. 6.13). They are used in final drives to connect shafts that are neither parallel nor intersecting. These gears have high tooth loads and must be lubricated with special heavy-duty hypoid gear oil because greater sliding occurs between the teeth. The sliding increases with the amount of offset between the shaft axes. With zero offset, a spiral bevel gear results, whereas the maximum offset corresponds to a worm/wheel configuration. Despite having a lower efficiency than spiral bevel gears,

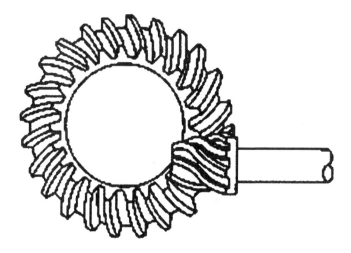

Figure 6.13. *Hypoid gears (TM 9-8000, 1985).*

hypoid gears allow the driveshaft to be lowered, thereby requiring a smaller "transmission tunnel" in the body.

The relationship between two meshed gears can be derived from Fig. 6.14, which shows a schematic of two meshed gears and a free-body diagram of both the gear and the pinion. Summing moments about the center of the input gear yields

$$F_t = \frac{T_{in}}{\left(\frac{d_{in}}{2}\right)}$$

and by the same process on the output gear

$$F_t = \frac{T_{out}}{\left(\frac{d_{out}}{2}\right)}$$

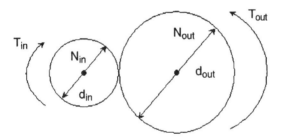

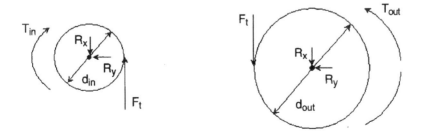

Figure 6.14. *Schematic of the gear mesh and free-body diagram of each gear.*

Because the tangential force at the mesh must be equal and opposite at the point of meshing,

$$\frac{T_{out}}{T_{in}} = -\frac{d_{out}}{d_{in}}$$

Furthermore, for any gear, the pitch diameter is proportional to the number of teeth (N), and an analysis of angular velocity also shows that the speeds of the gears are inversely related to the diameter. Combining this leads to what is commonly known as the gear law

$$\frac{T_{out}}{T_{in}} = -\frac{d_{out}}{d_{in}} = -\frac{N_{out}}{N_{in}} = -\frac{\omega_{in}}{\omega_{out}} \quad (6.9)$$

The negative sign in Eq. 6.9 accounts for the reversal of rotation in a single mesh. Equation 6.9 also assumes no friction loss in the mesh. In reality, there is a slight loss on the order of 1 to 2%.

The gear law can be extended to multiple gear meshes such as the one shown in Fig. 6.15. In this case, the gear train is analyzed by

$$\frac{T_3}{T_2}\frac{T_4}{T_3} = \left(-\frac{N_3}{N_2}\right)\left(-\frac{N_4}{N_3}\right) \Rightarrow \frac{T_4}{T_2} = \frac{N_4}{N_3}\left(-1^n\right) \quad (6.10)$$

where N is the number of gear meshes.

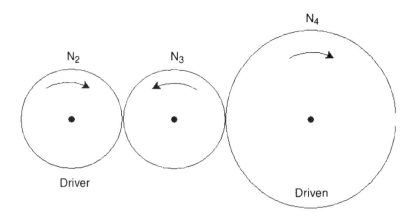

Figure 6.15. *Gear train.*

In the case presented in Fig. 6.15, the center gear has no effect on the overall torque ratio. However, it causes the output shaft to rotate in the same direction as the input shaft. This gear is called an idler. Thus, it can be seen that stringing several gears together is no way to gain large increases in torque. The final torque ratio is governed solely by the first and last gears in the train. The only way to gain a larger torque ratio is to go to a compound gear train, in which two of the gears are connected by a common shaft, as shown in Fig. 6.16. In this case, the torque relationship is given by

$$\frac{T_6}{T_2} = -\left(\frac{N_3}{N_2}\right)\left(\frac{N_4}{N_3}\right)\left(\frac{N_6}{N_5}\right) \tag{6.11}$$

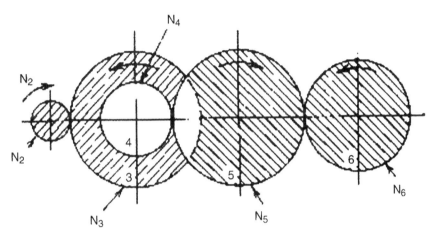

Figure 6.16. *Compound gear train (Krutz et al., 1994).*

Note that gears 2, 4, and 5 in Fig. 6.16 are driving gears, whereas gears 3, 5, and 6 are driven gears. This leads to the definition of a train value, e, as

$$e = \frac{\text{Product of driving tooth numbers}}{\text{Product of driven tooth numbers}} \tag{6.12}$$

By observation, the train value is the inverse of the gear (or torque) ratio and is equal to the speed ratio for a gear train. With this knowledge, it is now possible to analyze the manual transmission.

6.4 Manual Transmissions

The earliest transmissions were sliding gear types, in which the gears were splined to the appropriate shafts and were engaged and disengaged by the driver. They were used universally through the late 1920s but have since been replaced in passenger cars by synchronized, constant-mesh transmissions. Figure 6.17 shows a photograph of a four forward-speed plus reverse transmission, with a schematic of the transmission. At this point, the discussion is

250 | *Automotive Engineering Fundamentals*

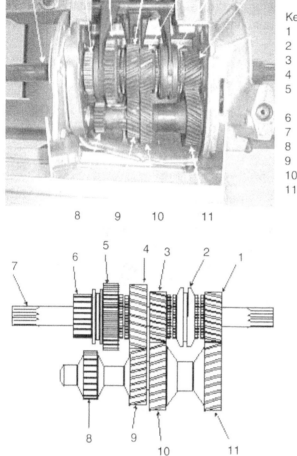

Key
1   Clutch Gear
2   3–4 Synchronizer
3   Third Gear (27 teeth)
4   Second Gear (33 teeth)
5   Second Synchronizer Sleeve and First Reverse Driven Gear (40 teeth)
6   Second Gear Synchronizer Hub
7   Mainshaft
8   Countergear First Gear (18 teeth)
9   Countergear Second Gear (25 teeth)
10 Countergear Third Gear (31 teeth)
11 Countergear Shaft Gear (35 teeth)

Figure 6.17. *A four-speed transmission.*

limited to analyzing gear ratios and will not delve into the operation of the linkages and safety devices. Furthermore, the operation of the synchronizer will be discussed in the next section.

The transmission has three shafts: the input shaft, the countershaft, and the output shaft (or mainshaft). The clutch gear (1) is an integral part of the transmission input shaft and always rotates with that shaft. The countershaft gears normally are machined from a single piece of steel and sometimes are referred to as the cluster gears. These are mounted to the countershaft on roller bearings, and the countershaft is fixed in place so it does not rotate. The gears on the output shaft, called speed gears, also are mounted on roller bearings. They are always meshed with the cluster gears and continuously rotate around the main shaft. The speed gears are locked onto the main shaft by the action of the synchronizers and, when locked, transmit torque to the output shaft. (Synchronizer operation will be discussed in the next section.)

6.4.1 Transmission Power Flows

Now it is possible to outline the flow of power through the transmission shown in Fig. 6.17, as well as to calculate the gear ratio for each gear.

6.4.1.1 First Gear

Torque flows through the input shaft to the clutch gear (23 teeth) and then to the countershaft via the cluster gear (35 teeth) (Fig. 6.18). The countershaft torque then is transmitted through the countershaft first gear (18 teeth) to the second gear synchronizer sleeve (40 teeth). The second gear synchronizer sleeve is splined to the main shaft through its hub, and torque is transmitted to the main shaft. Applying the gear law to this gear train, the torque ratio (or gear ratio) is

$$GR_1 = \frac{T_{out}}{T_{in}} = \frac{35}{23}\frac{40}{18} = 3.38 \qquad (6.13)$$

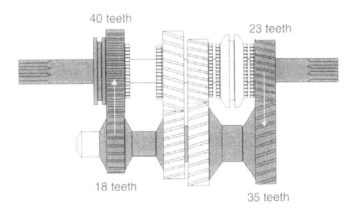

Figure 6.18. *Power flow in first gear.*

6.4.1.2 Second Gear

Torque flow again is from the input shaft to the clutch gear and countershaft. In second gear, the rear synchronizer has moved forward, locking the second speed gear onto the main shaft (Fig. 6.19). Torque flows through the countershaft second gear (25 teeth) to the second speed gear (33 teeth). The gear ratio is thus

$$GR_2 = \frac{T_{out}}{T_{in}} = \frac{35}{23}\frac{33}{25} = 2.01 \qquad (6.14)$$

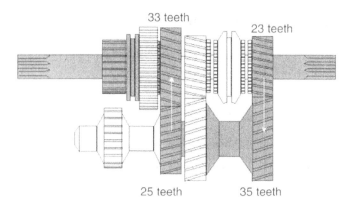

Figure 6.19. *Power flow in second gear.*

6.4.1.3 Third Gear

The torque flow again is input shaft, clutch gear, and countershaft. In this case, the forward synchronizer has moved to lock the third speed gear onto the main shaft (Fig. 6.20). Torque flow from the countershaft is to the countershaft third gear (31 teeth) to the third speed gear (27 teeth). The gear ratio is

$$GR_3 = \frac{T_{out}}{T_{in}} = \frac{35}{23}\frac{27}{31} = 1.33 \tag{6.15}$$

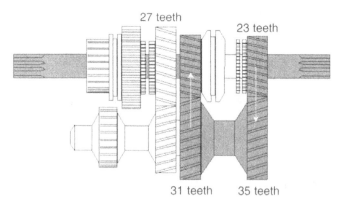

Figure 6.20. *Power flow in third gear.*

6.4.1.4 Fourth Gear

In fourth gear, the forward synchronizer has moved forward and locked the clutch gear onto the main shaft (Fig. 6.21). This results in power flowing directly from the input shaft to the main shaft, and the gear ratio thus is 1:1. In other words, in fourth gear, there is no torque

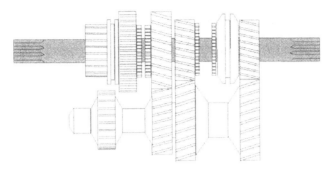

Figure 6.21. *Power flow in fourth gear.*

multiplication or speed reduction. Fourth gear is used for steady cruising to gain maximum fuel economy. To accelerate or pass, the driver must downshift to a lower gear. Most modern transmissions incorporate a fifth (or sixth) gear that is an overdrive. This has a gear ratio of less than unity and further increases fuel economy.

6.4.1.5 Reverse

With reverse gear selected, it is necessary to reverse the direction of rotation of the main shaft (Fig. 6.22). This is accomplished by introducing an idler gear into the train between the countershaft first gear and the second gear synchronizer sleeve. When reverse is selected, both synchronizers are in the neutral position, and the linkages move the reverse idler gear into mesh with the appropriate gears. The torque flow is thus: input shaft, clutch gear, countershaft, reverse idler, reverse speed gear, and main shaft. The corresponding gear ratio is

$$\text{GR}_{\text{rev}} = \frac{T_{\text{out}}}{T_{\text{in}}} = -\frac{35}{23}\frac{20}{18}\frac{40}{20} = -3.38 \tag{6.16}$$

which is the same ratio as first gear, except with reversed direction of rotation.

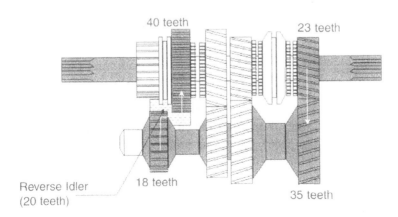

Figure 6.22. *Power flow in reverse.*

6.4.2 Synchronizer Operation

The primary function of the synchronizer is to ensure that the speed gear and the main shaft are rotating at the same speed prior to engagement. The synchronizer also locks the speed gear onto the main shaft so that torque may be transmitted. It accomplishes the first function through the blocking ring and gear shoulder (Fig. 6.24). These components function as a cone clutch. A schematic of a cone clutch is shown in Fig. 6.23.

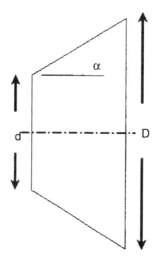

Figure 6.23. *Cone clutch schematic.*

The analysis of a cone clutch is identical to that of an axial clutch, except that the friction surface is now at some angle, α. Following the same procedure and assuming uniform wear, the activation force and the torque capacity of a cone clutch are found to be

$$F = \frac{\pi p_a d}{2}(D-d)$$

$$T = \frac{Ff}{4\sin\alpha}(D+d) \tag{6.17}$$

Several important conclusions can be derived from Eq. 6.9. First, the activation force for a cone clutch is identical to that of an axial clutch for a given diameter (inner and outer). Second, for a given activation force, the torque transmission is much larger due to the $\sin\alpha$ term in the denominator. Thus, cone clutches are used when spatial constraints limit the diameter of the clutch and would lead to a required activation force that would be too high for an axial clutch (synchronizers). If $\alpha < 8°$, the cone clutch is difficult to disengage. The normal range is $10° < \alpha < 15°$ (Shigley and Mischke, 2001).

Figure 6.24 shows the operation of the synchronizer. As the driver moves the gearshift lever to select the desired gear, the synchronizer sleeve is moved horizontally toward the desired gear by means of shift forks. The blocking ring (which is splined to the synchronizer sleeve, which in turn is splined to the shaft) makes contact with the gear shoulder, and although it initially slips, it begins to transmit torque. As the synchronizer sleeve continues to move, the activating force on the blocking ring increases until there is no more slip. At this point, the speed gear (which is rotating about the main shaft) and the synchronizer (which is fixed to the main shaft through the hub) are rotating at the same speed. Final engagement occurs when the synchronizer sleeve engages the dog teeth on the driven gear. This locks the speed gear to the main shaft, and the gear change is complete.

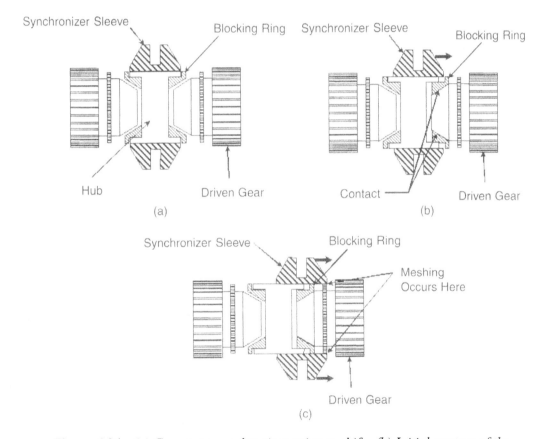

Figure 6.24. *(a) Cone-type synchronizer prior to shift. (b) Initial contact of the blocking ring with the gear shoulder. (c) Final engagement with the sleeve locking the driven gear onto the main shaft.*

6.5 Automatic Transmissions

Figure 6.25 shows a typical three-speed automatic transmission. For the purposes of this book, the automatic transmission will be considered to have the following four components:

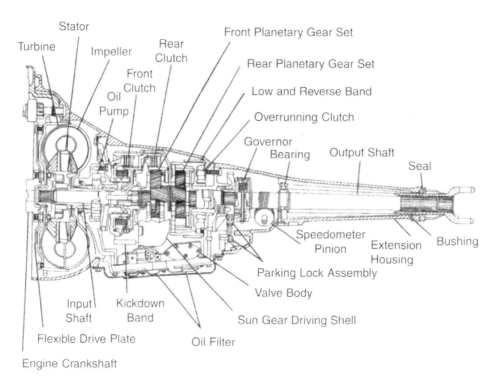

Figure 6.25. *Three-speed automatic transmission (TM 9-8000, 1985).*

1. Hydrodynamic torque converter

2. A planetary gear-set torque converter

3. A control system (The hydraulic control system will be described first, followed by electronic control.)

4. Working fluid

6.5.1 Fluid Couplings and Torque Converters

The friction clutch has the advantages of being lightweight, reliable, simple, easy to repair, and efficient. However, its major drawback is that it requires driver input to operate. With the advent of automatic transmissions, a means of transferring torque from the engine to the transmission was required, which eliminated driver operation while allowing the engine to idle when the vehicle was stopped. This led to the development of the fluid coupling. Figure 6.26 shows a simple fluid coupling, which consists of a housing, a pump, and a turbine.

The pump is the driving member of the converter. It is one-half of a torus, and the blades usually are streamlined with rounded leading edges. The pump normally is the rear member

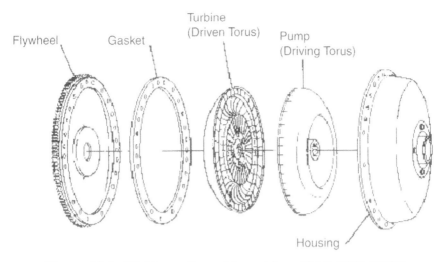

Figure 6.26. *Fluid coupling. Adapted from TM 9-8000 (1985).*

of the converter and is connected to the engine crankshaft. The pump is filled with transmission fluid, as are all components of the torque converter.

The turbine is the driven member of the converter. In shape, it is similar to the pump, and the blades often have an airfoil shape to enhance non-turbulent fluid flow under conditions of high fluid velocity. The turbine is the forward member of the converter and is splined to the transmission input shaft. The only connection between the pump and the turbine is the working fluid.

The principle of a fluid coupling is best illustrated by taking two electric fans and placing them face to face. One fan is turned on, and as its speed increases, the air flow causes the motionless fan to rotate. The second fan accelerates until it is rotating almost as rapidly as the powered fan. The same principle is used in the fluid coupling, except that the working medium is hydraulic fluid rather than air.

Figure 6.27 shows the flow of fluid in a fluid coupling. As the pump is rotated by the engine, it forces fluid against the faces of the turbine blades, causing the turbine to rotate in the same direction as the pump. If the rotation speed of the pump is much higher than that of the turbine, the fluid enters the turbine with great force. However, in this case, not all of the fluid energy is transmitted to the turbine, and the fluid leaves the turbine with high velocity in a direction opposite to the pump rotation. This produces drag on the pump and results in a horsepower drain on the engine.

The solution to this problem is found in the hydrodynamic torque converter. The simplest torque converter used in passenger cars is the three-member, single-stage device shown in Fig. 6.28. The torque converter contains a pump and a turbine but adds a stator. The stator is mounted on a stationary shaft between the pump and the turbine. It also contains an

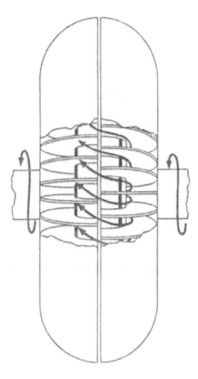

Figure 6.27. *Fluid flow in a fluid coupling (TM 9-8000, 1985).*

overrunning clutch; with one direction of rotation, it is locked to the shaft, whereas in the opposite direction, it can rotate freely. The reasons for this will become clear with a qualitative analysis of the fluid flow in the converter.

In the case of a high pump velocity relative to the turbine, the stator is locked onto its stationary shaft. It then reverses the direction of the fluid, and the fluid re-enters the pump in the same direction as the pump rotation (Fig. 6.29). This eliminates the braking effect of the fluid and its corresponding large loss of engine power. When the turbine is rotating at nearly the same velocity as the pump (e.g., at steady highway speeds), the fluid leaves the turbine with little residual velocity. In this case, the fluid strikes the back of the stator blades, causing them to unlock (or overrun) the stationary shaft. The stator is then free-wheeling, and the converter becomes a simple hydrodynamic coupling.

As mentioned, the transmission fluid completely fills the converter. Through the dynamics of the fluid, torque is transmitted from the pump to the turbine. In addition, the fluid must have good lubricating properties, resist oxidation, have good antifoaming characteristics, and have a low enough pour point to allow operation in cold climates. Because of the high demands placed on automatic transmission fluid, most oil companies have developed specialized fluids for use in automatic transmissions, and regular engine lubricant should never be used in an automatic transmission.

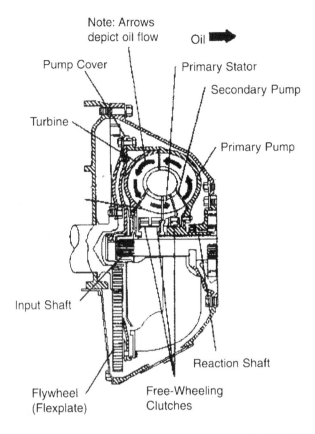

Figure 6.28. *Cross section of a torque converter (TM 9-8000, 1985).*

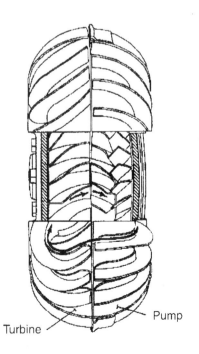

Figure 6.29. *Fluid flow in a torque converter (TM 9-8000, 1985).*

It should be apparent that there is never a time when the pump and turbine are rotating at the same speed. If this were the case, no torque would be transmitted. The implication of this is that there is always some slippage, and hence loss, in a torque converter. Thus, a torque converter never has 100% efficiency, although modern converters attain efficiencies of 96 to 98%. Furthermore, this power dissipation appears as heat in the fluid and must be removed. This can be done by air-cooling the converter housing, an acceptable method for many vehicles. However, for vehicles that are subjected to high loads at low speeds, such as pickup trucks and sport utility vehicles used for heavy towing, the addition of a transmission cooler is required. This usually takes the form of a secondary radiator through which the transmission fluid is passed to cool it. The other feature of many modern torque converters is a lockup device, as shown in Fig. 6.30. The lockup unit consists of a friction disc that is locked to the turbine housing by a hydraulic piston. The hydraulic piston usually is controlled by the automatic transmission control system. Because this feature eliminates slippage, the primary benefit of this device is increased fuel economy, but it has the added advantage of eliminating heat buildup caused by the slippage.

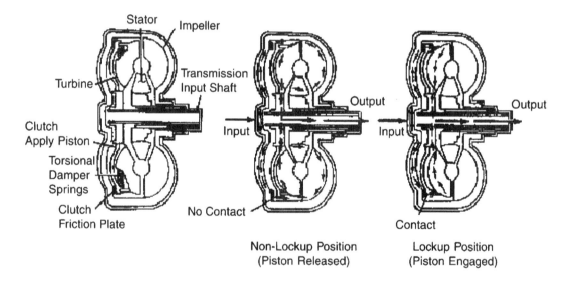

Figure 6.30. *Lockup torque converter (TM 9-8000, 1985).*

Figure 6.31 shows performance curves for a fluid coupling and a torque converter. Several conclusions may be drawn from these curves.

1. Converter efficiency increases up to the design point, after which it decreases. Coupling efficiency increases linearly with speed ratio and is less than converter efficiency until the coupling point is reached. When the two efficiency curves intersect, it is desirable to allow the stator to free-wheel, allowing the converter to operate as a fluid coupling. As discussed previously, the stator free-wheels when fluid impinges on the back of the blade. Thus, the speed at which the stator free-wheels is determined by the stator blade angle.

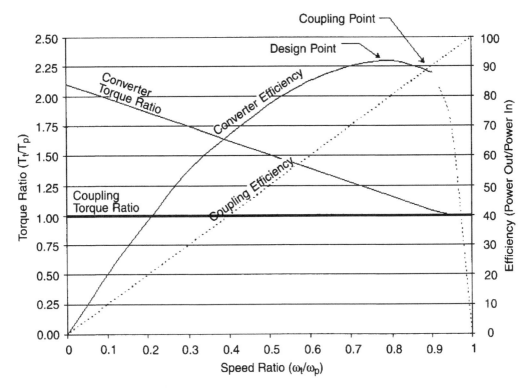

Figure 6.31. *Performance curves for a torque converter and a fluid coupling. Data taken from* Automotive Engineering *(1982).*

2. Converter torque ratio is highest at the stall point, or where the turbine has zero revolutions per minute. Thus, the converter provides an infinite number of torque ratios between stall and maximum vehicle speed. At the coupling point, the torque ratio has a value of one and remains at one because the converter acts as a simple fluid coupling.

3. The torque converter can be made to approach the efficiency of an axial clutch by the addition of stators, whose blades are set at different angles. This causes the individual stators to free-wheel at different speeds and reduces turbulence. Ideally, a single stator with infinitely variable blades could be used, but this would add greatly to the complexity of the converter.

6.5.2 Planetary Gears

The heart of all automatic transmissions is the planetary gear set, an example of which is shown in Fig. 6.32.

A planetary gear set consists of a sun gear in the center, a planet carrier with the planetary pinions (or planets) mounted to it, and a ring gear with internal teeth. While the planets spin as they rotate around the sun gear, the planets and the carrier act as a single unit. Any member

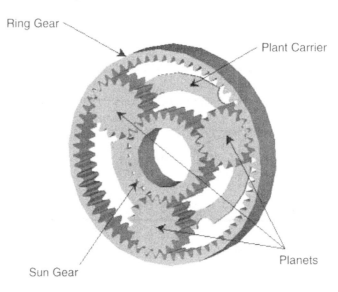

Figure 6.32. *Planetary gear set.*

of the gear set can be fixed or can spin or revolve. To transmit torque, one of the elements must be fixed. If all elements are allowed to rotate, no torque can be transmitted, and the gear set is in neutral. Planetary gear sets have several advantages over compound gear trains, as follows:

1. They are strong and compact. Because the load is distributed over many teeth, individual tooth loading is less than that experienced by a conventional gear train. Also, as will be shown, a single planetary gear set provides six gear ratios.

2. The gears are in constant mesh, which eliminates the risk of damage due to engagement or disengagement.

3. All elements rotate around the same central axis. This provides advantages in packaging, choice of output element, lubrication, and control.

4. Gear ratios can be changed with no interruption in torque transfer.

Analysis of planetary gear ratios is more complex than the analysis of simple gear trains, but it proceeds straightforwardly from a basic kinematic analysis of the elements.

Figure 6.33 shows a schematic of a planetary gear set that consists of four elements. (The additional planets are not shown because their presence does not affect the analysis.) The sun gear, ring gear, and carrier all rotate around the center of the device, O. The planet is pinned to the end of the carrier and rolls between the sun and the ring. The three circular elements have radii r_s, r_p, and r_r, and the carrier has radius r_c. Note that

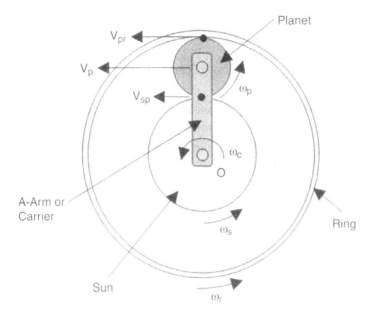

Figure 6.33. *Schematic of a planetary gear set.*

$$r_r = r_s + 2r_p \qquad (6.18)$$

Denoted in Fig. 6.33 are the absolute angular velocities of the four elements ω_s, ω_c, ω_p, and ω_r, and the linear velocities at three points: the sun-planet interface, V_{sp}; the center of the planet, V_p; and the planet-ring interface, V_{pr}. To relate the motions of these elements and hence to discover possible transmission ratios, two basic facts must be recalled from dynamics. The first relates to the relative velocities between two points and states that the velocity of one point is equal to the velocity of the other point plus the relative velocity between the two, $V_a = V_b + V_{a/b}$. The second fact is that the relative velocity between two points fixed on a rotating body is the angular velocity of the body crossed with the relative position vector between the points, $V_{a/b} = \omega \times R_{a/b}$.

Applying this to the velocity of the planet, the angular velocities of the sun, planet, and carrier are related by

$$V_p = r_s\omega_s + r_p\omega_p = r_c\omega_c \qquad (6.19)$$

Applying this to the planet-ring interface, the angular velocities of the sun, planet, and ring are related by

$$V_{pr} = r_s\omega_s + 2r_p\omega_p = r_r\omega_r \qquad (6.20)$$

Most often, it is desired to know the relationship among the sun, carrier, and ring. Thus, Eqs. 6.19 and 6.20 can be combined, eliminating ω_p, to give

$$r_s\omega_s + r_r\omega_r = 2r_c\omega_c \qquad (6.21)$$

Although Eq. 6.21 is valid for analyzing a planetary gear set, it usually is more convenient to analyze gear ratios based on the number of gear teeth in the train. Whereas the gear radius is proportional to the number of teeth, the presence of the carrier radius in Eq. 6.21 makes its use somewhat awkward. This can be eliminated by noting that

$$r_c = r_s + r_p \qquad (6.22)$$

Substitution of Eq. 6.22 into Eq. 6.21 yields

$$r_s\omega_s + r_r\omega_r = 2r_s\omega_c + 2r_p\omega_c \qquad (6.23)$$

Although Eq. 6.23 has eliminated the carrier radius, it still requires calculation of linear velocities to analyze the gear set. Equation 6.23 can be put into a more workable form by using a rearranged Eq. 6.18

$$2r_p = r_r - r_s \qquad (6.24)$$

Substitution of Eq. 6.24 into Eq. 6.23 and collecting terms gives

$$-\frac{r_s}{r_r} = \frac{\omega_r - \omega_c}{\omega_s - \omega_c} \qquad (6.25)$$

Equation 6.25 finally relates gear size to angular velocity. Furthermore, because gear radius is proportional to the number of teeth, Eq. 6.18 can be written in the usual form for planetary analysis

$$-\frac{N_s}{N_r} = \frac{\omega_r - \omega_c}{\omega_s - \omega_c} \qquad (6.26)$$

Equation 6.26 is a general relationship and may be used to analyze all cases of planetary gearset operation. The left side of the equation is simply the train value of the set when the planet carrier is fixed. The right side of the equation contains the absolute angular velocity of the elements with respect to earth. Thus, shaft speeds (in revolutions per minute) can be substituted directly into Eq. 6.26 instead of converting them into angular velocities (in radians per second).

Two inferences can be drawn from Eq. 6.26. First, any time the planet carrier is fixed ($\omega_c = 0$), the gear set produces a reverse. Second, if any two elements are locked together, the entire set rotates as a whole, and a direct drive is produced.

Table 6.1 summarizes the remaining six gear ratios available with a planetary gear set.

**TABLE 6.1
SIMPLE PLANETARY GEAR OPERATION**

Sun Gear	Carrier	Ring Gear	Speed	Torque	Direction
Input	Output	Held	Maximum Reduction	Increase	Same as input
Held	Output	Input	Minimum Reduction	Increase	Same as input
Output	Input	Held	Maximum Increase	Reduction	Same as input
Held	Input	Output	Maximum Increase	Reduction	Same as input
Input	Held	Output	Reduction	Increase	Reverse
Output	Held	Input	Increase	Reduction	Reverse

6.5.3 Planetary Gear-Set Torque Converter

As mentioned in the previous section, a planetary gear set requires that one element must be held for torque transfer to occur. Within an automatic transmission, there are three primary means of fixing planetary elements. The first is with a multiple disc clutch, or clutch pack, as shown in Fig. 6.34. The clutch pack is hydraulically operated and is used to lock a planetary element to either an input or an output shaft, or to fix an element so that it remains stationary. These also may be used to lock two elements together, which results in a direct drive (gear ratio of 1:1).

The second means is with a brake band as shown in Fig. 6.35. These elements also are hydraulically operated and generally are used to fix an element so that it remains stationary.

The third means of fixing planetary elements in an automatic transmission is through an overrunning (or one-way) clutch. Figure 6.36 shows two examples. One overrunning clutch uses a series of spring-loaded ball bearings to lock the inner hub onto the shaft; the other uses a series of sprags. Either way, when the torque flow is in one direction, the clutch locks the hub onto the shaft and allows torque transfer. If the torque flow reverses, the overrunning clutch releases, and the hub free-wheels around the shaft.

266 *Automotive Engineering Fundamentals*

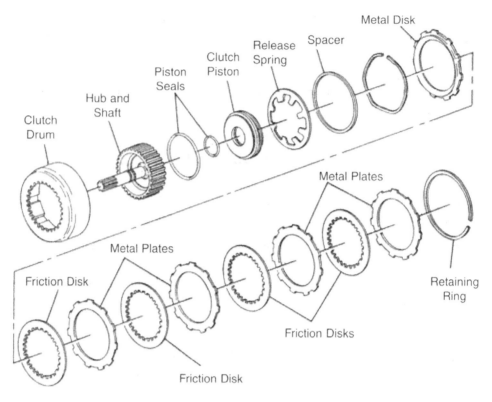

Figure 6.34. *Exploded view of a multiple disc clutch pack (TM 9-8000, 1985).*

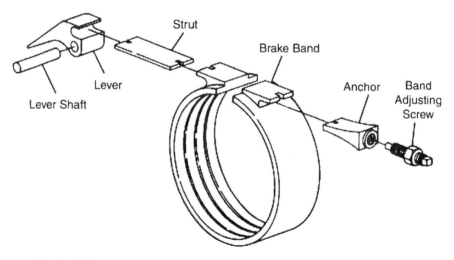

Figure 6.35. *Hydraulically actuated brake band (TM 9-8000, 1985).*

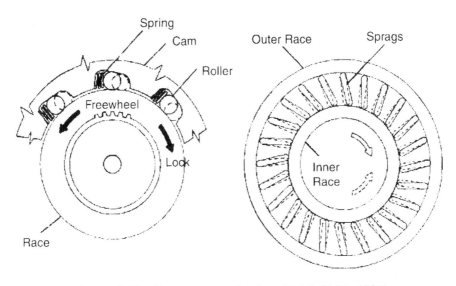

Figure 6.36. *Overrunning clutches (TM 9-8000, 1985).*

6.5.4 Simpson Drive

The most common planetary configuration for years has been the Simpson drive, a schematic of which is shown in Fig. 6.37. The Simpson drive provides three forward speeds, reverse, and neutral. The drive consists of two clutch packs (front and rear), two brake bands (front and rear), an overrunning clutch, and two planetary gear sets that share a common sun gear. The ring gear of the rear set and the planetary carrier of the front set both are splined to the output shaft and always rotate at the output shaft speed. The sun gear is free to rotate around

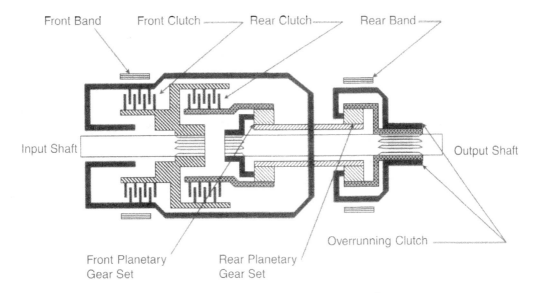

Figure 6.37. *A schematic of a Simpson drive.*

the output shaft and is not connected to it directly, but it is connected to the casing that can be braked by the front band. Power flow and control actuation are outlined in Table 6.2 and in the following discussion and figures. For illustrative purposes, it is assumed that the sun gear has 31 teeth and that the ring gear has 65 teeth. The numbers among individual manufacturers will differ, but these result in realistic gear ratios.

TABLE 6.2
ELEMENT ACTUATION FOR A SIMPSON DRIVE

Selector	Gear	Front Clutch	Rear Clutch	Front Band	Rear Band	Over-Running Clutch
D—Drive	First	Off	On	Off	Off	Holds
	Second	Off	On	On	Off	Overruns
	Third	On	On	Off	Off	Overruns
2—Drive	Second	Off	On	On	Off	Overruns
	First	Off	On	Off	Off	Holds
1—Low	First	Off	On	Off	On	Holds
Reverse	—	On	Off	Off	On	No Movement
Neutral	—	Off	Off	Off	Off	No Movement

6.5.4.1 Power Flow in First Gear

Figure 6.38 shows the power flow in first gear. In first gear, the rear clutch is applied, which connects the front ring gear to the input shaft. The ring gear drives the planetary pinions (whose carrier is splined to the output shaft), and the pinions in turn rotate the sun gear. The sun gear drives the rear planetary gear set. The rear planetary carrier is held by the overrunning clutch; thus, the rear planetary pinions drive the ring gear.

Analysis of the gear ratio begins with the equation derived previously, noting that the two sets must be analyzed in sequence. Beginning with Eq. 6.26 for the front set, the speed of the front ring gear is equal to the transmission input shaft speed ($\omega_r = \omega_{in}$). Furthermore, because the front carrier is splined to the output shaft, ω_c in Eq. 6.26 must equal the output speed (ω_{out}). Making these substitutions in Eq. 6.26 produces

$$-\frac{N_s}{N_r} = \frac{\omega_{in} - \omega_{out}}{\omega_s - \omega_{out}} \qquad (6.27)$$

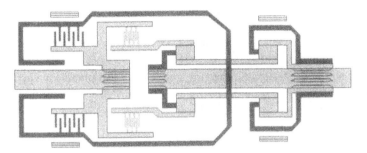

Figure 6.38. *Power flow in first gear, Simpson drive.*

Equation 6.27 may be solved for the speed of the sun gear

$$\omega_s = -\frac{N_r}{N_s}\omega_{in} + \omega_{out}\left(\frac{N_r}{N_s}+1\right) \qquad (6.28)$$

Moving now to the rear gear set, the sun gear is the input, the planet carrier is held by the overrunning clutch, and the ring gear is the output. Thus, Eq. 6.26 for the rear set becomes

$$-\frac{N_s}{N_r} = \frac{\omega_{out}}{\omega_s} \qquad (6.29)$$

Equation 6.28 gives the speed of the sun gear; hence, it may be substituted into Eq. 6.29 for ω_s. After some algebraic manipulation, the ratio of input speed to output speed may be found, which is the gear ratio (GR)

$$\frac{\omega_{in}}{\omega_{out}} = \left(\frac{N_s}{N_r}+2\right) = \frac{T_{out}}{T_{in}} = GR$$

Inserting our hypothetical (but nonetheless realistic) tooth count into Eq. 6.29 gives the gear ratio (GR)

$$GR = \left(\frac{31}{65}+2\right) = 2.48:1 \qquad (6.30)$$

Note that with the selector in drive and the transmission in first gear, the rear band is not applied. If the driver lifts off the throttle, the transmission is driven by the rear wheels. With torque flowing in this direction, the overrunning clutch releases and overruns. Thus, engine braking is not available, and the vehicle coasts. If the driver selects low, the rear band is applied, fixing the rear planet carrier and providing engine braking. Thus, only the rear brake band ever has to resist the engine braking torque.

6.5.4.2 Power Flow in Second Gear

Figure 6.39 shows the power flow in second gear. In second gear, the rear clutch remains applied. The front band engages and fixes the sun gear by preventing rotation of the sun gear case. With the sun gear stationary, the ring gear drives the planetary carrier around the sun. The front planetary carrier thus is the output element and drives the output shaft. Note that in this case, the rear planetary set is in a neutral condition and is not used.

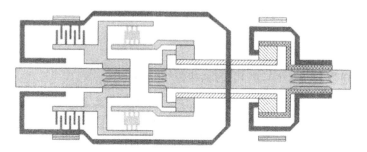

Figure 6.39. *Power flow in second gear, Simpson drive.*

In this case, only the front planetary set must be analyzed to determine the gear ratio. The sun gear is held, the ring gear is the input, and the carrier is the output. Equation 6.26 then becomes

$$-\frac{N_s}{N_r} = \frac{\omega_{in} - \omega_{out}}{0 - \omega_{out}} \qquad (6.31)$$

Solving for the gear ratio (GR) yields

$$\frac{\omega_{in}}{\omega_{out}} = \left(\frac{N_s}{N_r} + 1\right) = \frac{T_{out}}{T_{in}} = GR \qquad (6.32)$$

which, upon insertion of the tooth numbers, gives

$$GR = \left(\frac{31}{65} + 1\right) = 1.48:1 \qquad (6.33)$$

6.5.4.3 Power Flow in Third Gear

Figure 6.40 shows the power flow in third gear. In third gear, the front band is released while the front and rear clutches are engaged. This locks the front ring and the sun gear together. As

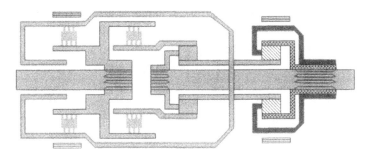

Figure 6.40. *Power flow in third gear, Simpson drive.*

noted in Section 6.5.2. on general planetary analysis, whenever two elements are locked together, the entire gear set rotates as a unit, providing a direct drive or a gear ratio of 1:1. Again, the rear gear set has no effect on the gear ratio.

6.5.4.4 Power Flow in Reverse

Figure 6.41 shows the power flow in reverse. In reverse, the front clutch and the rear band are engaged. Engagement of the front clutch results in the sun gear being driven at input speed, while the rear band fixes the rear planetary carrier. Thus, the sun is the input, the rear ring is the output, and the rear carrier does not rotate. Equation 6.26 becomes

$$-\frac{N_s}{N_r} = \frac{\omega_{out}}{\omega_{in}} \tag{6.34}$$

and the gear ratio (GR) is

$$\frac{\omega_{in}}{\omega_{out}} = -\left(\frac{N_r}{N_s}\right) = \frac{T_{out}}{T_{in}} = GR \tag{6.35}$$

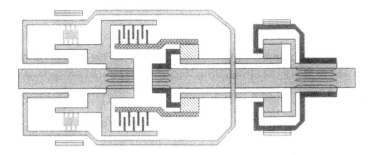

Figure 6.41. *Power flow in reverse, Simpson drive.*

Continuing with the previous tooth count, the reverse gear ratio (GR) is

$$\text{GR} = -\left(\frac{65}{31}\right) = -2.10:1 \qquad (6.36)$$

In this case, the front planetary set is in a neutral state and does not affect the gear ratio.

6.5.5 Hydraulic Control System

Until recently, the automation of the automatic transmission was provided by a hydraulic control system. The system is basically an analog computer that uses hydraulic fluid under pressure rather than electrons to operate the logic. The driver manually selects a desired range with the shift lever. From there, the control system automatically upshifts and downshifts through the available gear ratios. The control system actuates and releases the clutch packs and bands by the automatic opening and closing of valves in the hydraulic lines. Figure 6.42 shows a diagram of a typical hydraulic control system.

The control system contains two pumps. The front pump is driven by the torque converter, and the pressure it develops is a function of engine speed. It provides operating pressure to the control unit at all engine speeds above idle. The pressure is regulated to a specific value when the transmission is in reverse. The rear pump is driven by the transmission output shaft; thus, its pressure is a function of vehicle speed. It provides fluid under pressure at shaft speeds above some predetermined speed. When the pressure of the rear pump exceeds that of the front pump, the output from the front pump is bypassed back to the input side of the pump, minimizing power losses. When the vehicle is in reverse, the rear pump provides no output.

There are six broad categories of control valves in the control system. These valves are as follows:

1. The regulator valve
2. The manual valve
3. The throttle valve
4. The kickdown valve
5. The governor valve
6. The shift valve

All of these valves are basically spool valves. The spool valves are used to regulate line pressure (valve type 1 in the preceding list) and to control the direction of the flow (valve types 2 through 6 inclusive in the preceding list). Pressure regulation may be explained by reference to Fig. 6.43.

The valve consists of a piston with two lands, three ports, a reaction chamber, and a spring-loaded plunger. Port 1 is connected to the suction side of the pump; port 2 is connected to the pressure side. When the pump pressure rises above the regulation pressure, the pressure in

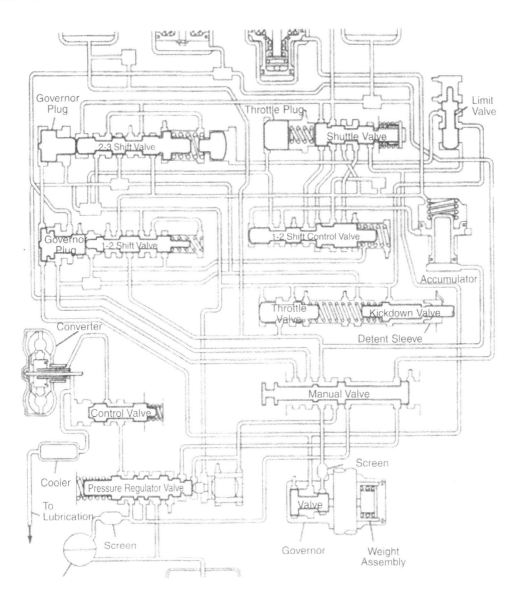

Figure 6.42. *Schematic of a typical three-speed automatic transmission control system (TM 9-8000, 1985).*

the reaction chamber causes the piston to move to the left. Land 1 uncovers port 1, and the fluid is bypassed back to the suction side of the pump. The pressure in the reaction chamber quickly reduces, causing port 1 to be closed, and the fluid now flows out of port 3 at the regulation pressure. The output pressure value may be altered by mechanically moving the plunger to the right. This induces a higher pre-load in the spring, which results in a higher output pressure through port 3 because it requires a higher pressure to move the piston to the left.

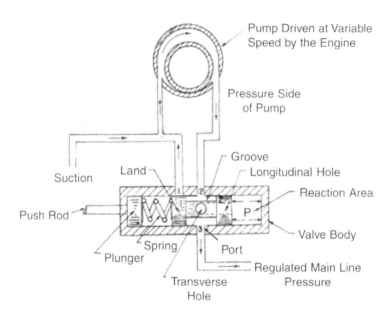

Figure 6.43. *Schematic of a pressure-regulating spool valve* (Automotive Engineering, *1982).*

Figure 6.44 shows the way the spool valves direct the pressure. In this case, regulated mainline pressure enters port 2. When the variable pressure at port 1 increases to a specified point (as determined by the pre-load on the spring), the piston moves to the right, allowing fluid to flow through port 3. The variable pressure can be applied either hydraulically by the pumps or mechanically through the shift lever, throttle linkage, or kickdown linkage.

The function of the regulator valve is to provide fluid at constant pressure to operate the control system. The manual valve directs the pressure and is controlled by the operator through the gear selector lever. The throttle valve is operated by linkages connected to the throttle plate. The kickdown valve is basically a spring-loaded ball check valve. It is closed for all throttle positions except wide open throttle (WOT). When the throttle is opened fully, the valve opens and causes the transmission to downshift. The governor valve contains a centrifugal governor that is driven by the output shaft. Pressure in the valve is a function of the square of the vehicle speed and is used to prevent engine overspeed when manually downshifting the transmission. Finally, the shift valve uses variable pressure as an input, and its output is sent to the actuating servos on the clutch packs and/or brake bands.

In a modern automatic transmission, control logic and band/clutch control are provided by a computer. Although the bands and clutches still require hydraulic pressure to operate, they are actuated by electronically controlled hydraulic solenoids. A case study of a modern, four-speed automatic transaxle is given at the conclusion of this chapter.

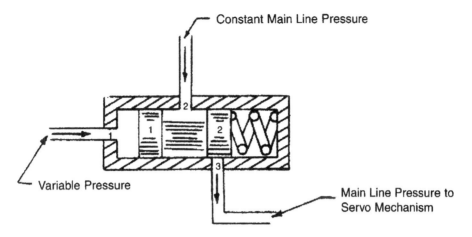

Figure 6.44. *Directional spool valve (*Automotive Engineering, *1982).*

6.6 Continuously Variable Transmissions (CVT)

6.6.1 Introduction

For a continuously variable transmission (CVT) to produce an improvement in fuel economy, it must be efficient and have a wide span (range of ratios). There are two significant types of CVT—the Van Doorne belt system, and the Torotrak system—both of which have been the subject of much development work. The belt system is most suited to low-power applications and has particular advantages for front-wheel-drive vehicles. The Torotrak system has been built in much larger sizes and lends itself to conventional in-line engine gearbox installations; prototypes have been installed for testing in numerous vehicles. In both cases, the control strategy is of great importance if maximum fuel economy, maximum acceleration, and engine braking all are to be obtained. Indeed, a microprocessor-based control system is almost inevitable.

6.6.2 Van Doorne Continuously Variable Transmission (CVT)

The Van Doorne CVT has been used by Daf (now owned by Ford) on cars since 1955. The essential part of this CVT, shown in Fig. 6.45, is a pair of conically faced driving pulleys, in which the separation between the two sides of each pulley can be adjusted. Because the belt is incompressible, the effective radius of the pulleys is varied by adjusting the separation between the pulley sides. Conventionally, the separation between the sides of the driving pulley is varied, and the driven pulley adjusts automatically because the axial loading between the driven pulley sides is provided by springs or hydraulic pressure. This axial loading also controls the belt tension.

On the original system, the ratio control was by centrifugal masses on the driving pulley and an engine vacuum actuator. The ratio range was approximately 4, and the belt transmitted the

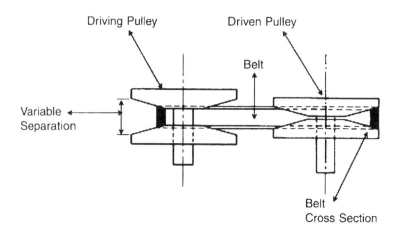

Figure 6.45. *The Van Doorne continuously variable transmission (CVT) (Stone, 1989).*

power in tension. The belt-type CVTs that are under current development have a wider ratio range (approaching 6) and improved efficiency (in the range 86 to 90%). This is the result of a radically different belt design (Fig. 6.46). The steel bands are used to carry the tensile forces, but power is transmitted by the compressive forces between the belt elements.

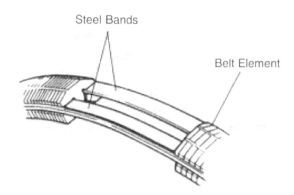

Figure 6.46. *Continuously variable transmission (CVT) belt construction (Stone, 1989).*

Figure 6.47 shows a complete Ford CTX (continuously variable transaxle), and a description of some of the development work is given by Hahne (1984). The belt drive is particularly suited to compact front-wheel-drive vehicles because it provides a convenient means of connecting the engine and differential. For the CTX shown in Fig. 6.47, the input drives the planet carrier and the inner part of the forward drive clutch. The outer part of the forward clutch is connected directly to the driving pulley, providing the path for power transmission.

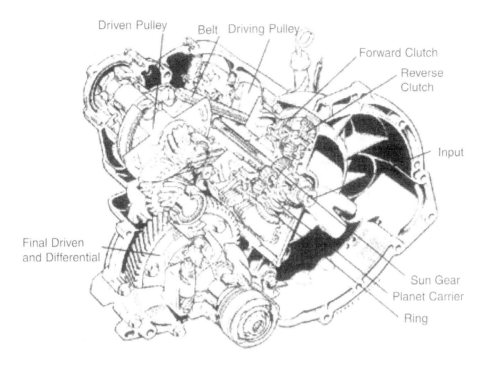

Figure 6.47. *Ford continuously variable transmission (CVT) (Stone, 1989).*

The sun gear also is connected directly to the driving pulley, and if the reverse clutch brakes the annulus, the planet carrier will drive the sun gear in reverse. These clutches are of the wet type with hydraulic actuation and have more predictable performance during their lifetime than dry friction clutches. A torque converter could have been incorporated, but the associated losses would remove the benefits of a CVT. The Ford CTX is controlled hydraulically and responds to five inputs: shift lever position, accelerator pedal position, pulley ratio, engine speed, and primary pulley speed. The transmission is designed for a torque of 125 Nm.

Recent developments are an increase in the torque capacity to 165 Nm and the use of electronic control systems. Electronic control provides much more accurate and sophisticated control than a purely hydraulic system. Options for either sport or economy modes are available, and the transmission also can imitate a manual transmission by operating at a series of discrete ratios. Such ratios can be selected by a "gear lever" with a gate, or by switches such as the buttons used on Formula 1 cars, with one button for changing up, and the other button for changing down.

6.6.3 Torotrak Continuously Variable Transmission (CVT)

A key part of the Torotrak CVT is the variator, a tilting roller assembly as shown in Fig. 6.48; this had already been used on cars in the 1930s by Hayes. The variator consists of three discs, with the outer pair connected. The discs have a part toroidal surface on their inner faces, upon which the spherically faced rollers roll. The rollers can rotate around their own axes, and the

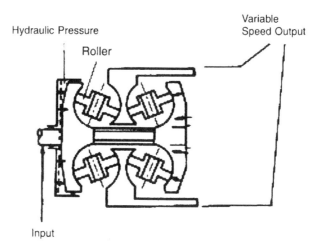

Figure 6.48. *Variator assembly (Stone, 1989).*

inclination of these axes is varied; however, the carrier for the rollers is fixed. The inclination of these rollers is varied by a control sleeve acting through rockers, which leads to the continuously variable ratio changes in the variator.

To prevent wear, lubrication is needed, and an elasto-hydrodynamic oil film exists between the rollers and the discs. Relative slip (typically 1 to 2%) is inevitable because it produces the shear in the oil film, which transmits the tractive forces that are tangential to the variator surfaces (hence the term "traction drive"). Even with specially formulated oils, the coefficient of friction (or in this case, traction) is very low; thus, high contact forces are needed between the discs and the rollers. This is achieved in a system with internally balanced forces, by applying hydraulic pressure to the outer face of the driving disc that is splined to the input shaft. Because the pressure can be varied, the unit life is improved by reducing the contact forces when the variator is lightly loaded.

The control of the rollers is critical if the three rollers on each side are to transmit equal loads and operate at the same ratio. This is achieved automatically in the Torotrak CVT by ingenious design of the roller inclination actuation mechanism. Three disadvantages that the basic variator shares with the Van Doorne belt system are that the ratio range is less than approximately 5, there is no ratio that will give zero output (to eliminate the need for a clutch), and there is no reverse ratio. All these shortcomings are overcome in the Torotrak transmission by using a shunt transmission system with an epicyclic gearbox.

Figure 6.49 shows the key elements of the Torotrak transmission. There are two clutches, one of which is engaged for the high-speed regime, and the other engaged for the low-speed regime. The various modes will be considered now, assuming a constant input speed.

a. **High regime**. When the high-regime clutch is engaged, the epicyclic gear train is not used, and the variator output is coupled directly to the output shaft via the high-regime

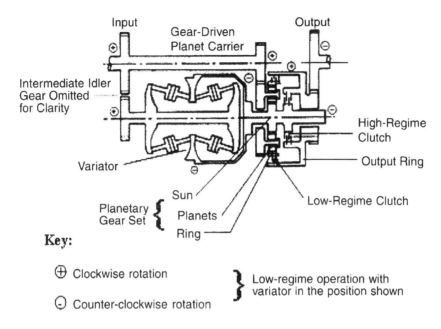

Figure 6.49. *Elements of a Torotrak continuously variable transmission (CVT) (Stone, 1989).*

clutch, the output annulus, and the output gear. With the rollers in the position shown in Fig. 6.49, the output speed is at its greatest. As the roller inclination is reduced, the output speed also will decrease.

b. **Regime changeover.** The gear ratios have been chosen so that when the output from the variator is at its lowest speed (the opposite to that illustrated in Fig. 6.49), the gear-driven planet cage and the sun gear are traveling at the same speed. Thus, the epicyclic annulus and the output annulus are also traveling at the same speed. This is called the synchronous ratio, and it enables a smooth transition between the regimes to occur by engaging the low-regime clutch and disengaging the high-regime clutch, or vice versa.

c. **Low regime.** The epicyclic gear train can act as a differential. If the input speed to the gearbox is constant, then the gear-driven planet carrier will rotate the planets around the sun gear at constant speed. If the output speed of the variator is now increased, the differential action of the epicyclic gear train will reduce the speed of the epicyclic annulus and the gearbox output. As the variator output is increased even further, the gearbox output is reduced, and a point will be reached at which the gearbox output speed is zero. This is termed "geared neutral." If the variator output is increased further, then the gearbox output will reverse in direction; the maximum reverse speed will occur in the variator position shown in Fig. 6.49.

Figure 6.50 shows the overall Torotrak transmission characteristics. In the high regime, the drive is occurring solely through the variator, whereas in the low regime, the drive occurs

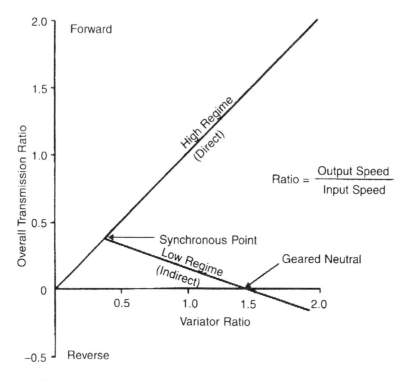

Figure 6.50. *Operating regimes of a Torotrak continuously variable transmission (CVT) (Stone, 1989).*

through the variator and a direct connection into the epicyclic gear train. Thus, the low regime is a shunt transmission system, with the epicyclic gearing acting as a differential to combine the two inputs. Because the geared neutral provides zero speed output or infinite reduction, there is theoretically infinite torque multiplication; in practice, the only limitation on the torque output will be the torque capacity of the variator. Figure 6.51 shows a complete Torotrak transmission.

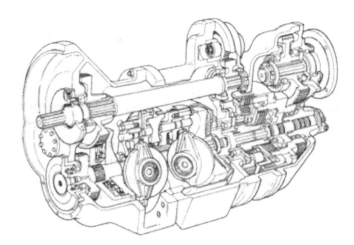

Figure 6.51. *The Torotrak continuously variable transmission (CVT) (Stone, 1989).*

The control of the CVT is of paramount importance; in particular, the geared neutral must be held accurately if unwanted vehicle motion is to be prevented. This is achieved by using a hydro-mechanical feedback control for the roller inclination, with the input hydraulic pressure representing the demand from the accelerator pedal for the torque to be transmitted. Any difference between the actual torque and the demand torque causes the inclination of the rollers to adjust until the demand is met. If the transmission was controlled by a ratio demand and the vehicle was restrained, then very high torques would be produced as soon as the ratio changed from the geared neutral position.

The Torotrak CVT is actuated hydraulically. Two on/off signals are needed for the high- and low-regime clutches, and a proportional control signal is required for the torque demand. The control signals are generated by a microprocessor, in response to driver inputs of drive selector position and accelerator position. The microprocessor can be readily programmed for different optimum economy torque/speed curves, and it can provide additional facilities. For example, the torque can be limited at low vehicle speeds to prevent wheel spin.

6.7 Driveshafts

Upon exiting the transmission, the next link in the driveline is the driveshaft itself. Of course, several options exist: the power may go to the rear wheels, the front wheels, or both. In the case of a front-wheel-drive vehicle, the transmission is called a transaxle. The difference in terminology is due to the fact that a transaxle combines the functions of a transmission, a driveshaft, and a differential into one unit. Details of such a configuration are contained in the case study at the conclusion of this chapter. For now, this chapter will consider the case of a rear-wheel-drive vehicle and will examine each component in sequence. Normally, it is not possible to mount the transmission such that the differential is in the same plane as the transmission output shaft. Thus, the driveshaft must transmit the torque through an angle. This is accomplished through the use of Hooke's joints or constant velocity (CV) joints, and these will be analyzed here before delving into the driveshaft.

6.7.1 Hooke's Joints

Due to suspension travel at the rear axle, it is impossible to connect the transmission to the differential with a single, rigid shaft. The torque must be capable of being transferred through some angle. Thus came the development and implementation of Hooke's joints, or universal joints (U-joints). Figure 6.52 shows an example of a U-joint. Furthermore, as the rear suspension deflects, it generally does not move in an arc centered on the transmission output. As a result, the distance between the transmission output and the differential case changes with suspension motion. To accommodate the changing length, most driveshafts also incorporate a slip yoke, as shown in Fig. 6.53.

The kinematics of a U-joint may be analyzed by reference to Fig. 6.54, which shows two shafts connected by a U-joint with an operating angle of ε. The left portion of the figure shows a schematic of the joint looking along the axis of the input shaft. The datum point for

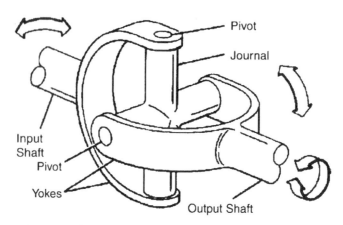

Figure 6.52. *Schematic of a universal joint (U-joint). Adapted from TM 9-8000 (1985).*

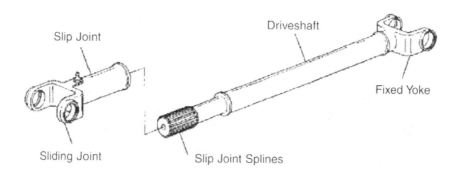

Figure 6.53. *Details of a slip joint on a driveshaft. Adapted from TM 9-8000 (1985).*

analysis is with the input (driving) yoke in the vertical position (A–C) and the output (driven) yoke horizontal (b–d). Point A on the input yoke follows a circular path ABCD, as shown. The projection of the path of point b on the output yoke describes an ellipse, bCdA. Suppose the input shaft is rotated through an arbitrary angle, θ, as shown in Fig. 6.54. The output yoke moves to point b′, and because the yoke is rigid, angle nOb′ = θ. However, this angle is the projection of the follower motion onto the plane perpendicular to the input shaft. Thus, the actual angle through which the output yoke rotates is angle mOB′ and is defined here as ϕ. Also, the line Eb′ is the projection of line EB′ onto the plane of reference, which is at angle ε to the output yoke. Making use of the dot product, the length of Eb′ is thus

$$Eb' = EB'\cos\varepsilon \tag{6.37}$$

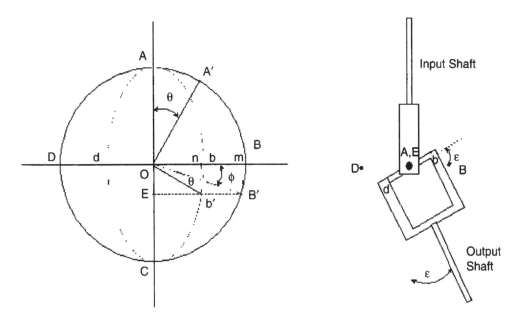

Figure 6.54. *The angular relationships of a universal joint (U-joint).*

Referring to Fig. 6.54,

$$\tan \phi = \frac{mB'}{Om} \tag{6.38}$$

Noting that $mB' = nb'$, $Om = EB'$, and making use of Eq. 6.37,

$$\tan \phi = \frac{mB'}{Om} = \frac{nb'}{EB'} = \frac{nb'}{Eb'} \cos \varepsilon \tag{6.39}$$

Because $\tan \theta = \dfrac{nb'}{Eb'}$, Eq. 6.29 becomes

$$\tan \phi = \cos \varepsilon \tan \theta \tag{6.40}$$

which relates the angular motion of the output shaft to the input shaft. Equation 6.30 may be differentiated with respect to time, assuming the operating angle (ε) is constant, to yield

$$\dot{\phi} \sec^2 \phi = \dot{\theta} \sec^2 \theta \cos \varepsilon \tag{6.41}$$

Making use of the trigonometric identity $\sec^2(x) = 1 + \tan^2(x)$, the ratio of output angular velocity to input angular velocity is then given by

$$\frac{\dot{\phi}}{\dot{\theta}} = \frac{\sec^2 \theta \cos \varepsilon}{1 + \tan^2 \phi} \tag{6.42}$$

Substituting Eq. 6.40 into Eq. 6.42 yields

$$\frac{\dot{\phi}}{\dot{\theta}} = \frac{\sec^2 \theta \cos \varepsilon}{1 + \tan^2 \theta \cos^2 \varepsilon} = \frac{\sec^2 \theta \cos \varepsilon}{1 + \tan^2 \theta (1 - \sin^2 \varepsilon)} = \frac{\cos \varepsilon}{\cos^2 \theta \left[1 + \frac{\sin^2 \theta}{\cos^2 \theta}(1 - \sin^2 \varepsilon)\right]} \tag{6.43}$$

which can be reduced to

$$\frac{\dot{\phi}}{\dot{\theta}} = \frac{\cos \varepsilon}{1 - \sin^2 \theta \sin^2 \varepsilon} \tag{6.44}$$

Figure 6.55 shows a plot of Eq. 6.44 for a range of shaft operating angles.

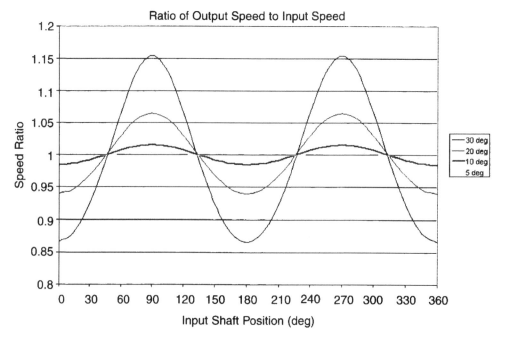

Figure 6.55. *Ratio of output to input shaft speed for shafts with various operating angles.*

Equation 6.44 and Fig. 6.55 lead to several conclusions. First, the only way to eliminate the speed variation with U-joints is to have an operating angle of 0. If this were possible, the need for the U-joint would be removed. Second, when using U-joints, the operating angle should be kept small—ideally less than 5°. Finally, what may not be obvious from Eq. 6.44 is that if the driveshaft is constructed with a U-joint at each end, and the operating angles of each end are equal, the second U-joint will cancel the effects of the first, as long as the yokes are properly phased. As a result, the entire shaft system produces a uniform angular velocity. Although this eliminates variations in the differential, the driveshaft itself is still subject to speed variation. If there is any imbalance in the shaft, it is possible to excite a resonance in the shaft with corresponding shaft whirl. This becomes an issue in the case of installing a lift kit on a four-wheel-drive vehicle. The stock vehicle usually is designed with equal operating angles at each end of the driveshaft. If the user lifts the body substantially, this usually is no longer the case. The user then has two options: (1) to live with the ensuing speed variations and vibration, or (2) to attempt to tilt the differential to again achieve equal operating angles at each end.

One method of eliminating the speed variations inherent in the U-joint is to use double U-joints, or double-cardan joints. Figure 6.56 shows an example of such a joint. This joint uses two cross-yokes linked to a center ball socket. The center ball and socket divide the U-joint operating angle into two equal halves. Because the two angles are equal, speed variations are cancelled within the joint, and the driveshaft exhibits constant velocity over large total operating angles.

The other difficulty with limiting the operating angles is in the case of front-wheel-drive cars. In this case, the joint must transfer torque through the angle caused by both suspension travel

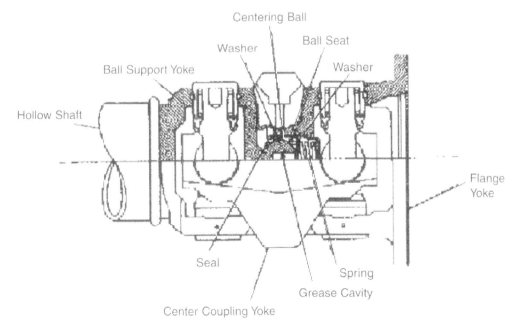

Figure 6.56. *A double-cardan joint (TM 9-8000, 1985).*

and steering. In the case of steering angles, they can be quite large, rendering the U-joint unusable. The solution here is to use the constant velocity (CV) joint shown in Fig. 6.57. Within this joint, the ends of the shafts are ground to precision fit to a central ball, thus locating the joint axially. Furthermore, the ball cages have an internal spherical profile that also is a precision fit to the ball tracks. When the joint articulates, the cage steers the balls into the homokinetic plane. This plane bisects the operating angle and provides a uniform velocity transfer.

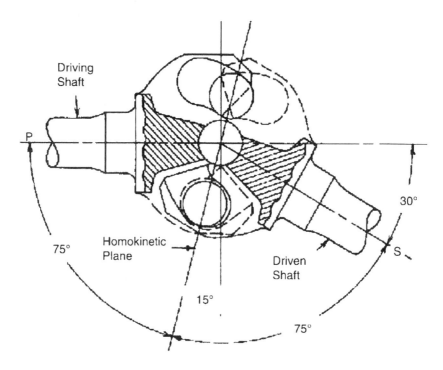

Figure 6.57. *A constant velocity (CV) joint (Krutz et al., 1994).*

6.7.2 Shaft Whirl

The driveshaft, being a rotating shaft, will tend to bow outward at certain speeds. This phenomenon is known as "shaft whirl" and may be analyzed with reference to Fig. 6.58.

The driveshaft is modeled here as a disc of mass m attached to a slender shaft. The shaft passes through the geometric center of the disc, s. The center of mass of the disc is located at point G. The distance from the disc center to point G (e) is known as the eccentricity. The disc rotates at a constant angular velocity ω, while the line r = OS rotates at some angular velocity $\dot{\theta}$ that is not equal to ω. The acceleration of point G is then

$$\bar{a}_G = \bar{a}_s + \bar{a}_{G/s} \tag{6.45}$$

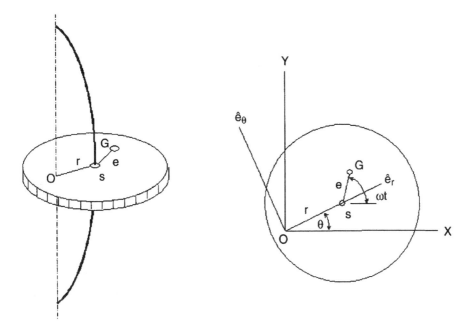

Figure 6.58. *A rotating shaft with a dynamic imbalance.*

Because ω is constant, the terms on the right side of Eq. 6.45 become

$$\bar{a}_s = \left(\ddot{r} - r\dot{\theta}^2\right)\hat{e}_r + \left(r\ddot{\theta} + 2\dot{r}\dot{\theta}\right)\hat{e}_\theta$$

$$\bar{a}_{G/s} = -e\omega^2\left(\cos(\omega t - \theta)\hat{e}_r + \sin(\omega t - \theta)\hat{e}_\theta\right)$$

and the acceleration of the center of mass of the disc is

$$\bar{a}_G = \left[\left(\ddot{r} - r\dot{\theta}^2\right) - e\omega^2\cos(\omega t - \theta)\right]\hat{e}_r + \left[\left(r\ddot{\theta} + 2\dot{r}\dot{\theta}\right) - e\omega^2\sin(\omega t - \theta)\right]\hat{e}_\theta \qquad (6.46)$$

The shaft provides a restoring force due to its flexural stiffness, and it will be assumed that it also provides equivalent viscous damping. If the shaft is modeled as a simply supported shaft, and it is assumed that the eccentricity is located at the midpoint of the shaft, the first mode flexural stiffness is found from mechanics of materials to be

$$k = \frac{48EI}{L^3} \qquad (6.47)$$

where L is the shaft length, E is the modulus of elasticity, and I for a hollow shaft is given by

$$I = \frac{\pi}{64}\left(D^4 - d^4\right)$$

where D and d are the outer and inner diameters of the shaft, respectively. The equations of motion resolved in the radial and tangential directions thus are

$$-kr - c\dot{r} = m\left[\ddot{r} - r\dot{\theta}^2 - e\omega^2 \cos(\omega t - \theta)\right]$$
$$-cr\dot{\theta} = m\left[r\ddot{\theta} + 2\dot{r}\dot{\theta} - e\omega^2 \sin(\omega t - \theta)\right] \quad (6.48)$$

Generally, a driveshaft will exhibit synchronous whirl, which is defined as the whirl speed being equal to the rotational speed, or $\dot{\theta} = \omega$. This relation can be integrated to produce

$$\theta = \omega t - \phi \quad (6.49)$$

where ϕ is the phase angle between e and r. Furthermore, with synchronous whirl, $\ddot{\theta} = \ddot{r} = \dot{r} = 0$. Thus, Eq. 6.48 reduces to

$$\left(\frac{k}{m} - \omega^2\right)r = e\omega^2 \cos\phi$$
$$\frac{c}{m}\omega r = e\omega^2 \sin\phi \quad (6.50)$$

From basic vibration analysis, the critical speed ω_n, damping ratio ζ, and critical damping c_{cr}, are defined as

$$\omega_n = \sqrt{\frac{k}{m}}$$

$$\zeta = \frac{c}{c_{cr}}$$

$$c_{cr} = 2m\omega_n$$

With these definitions, Eq. 6.46 may be solved for the phase angle and amplitude

$$\tan\phi = \frac{2\zeta\dfrac{\omega}{\omega_n}}{1-\left(\dfrac{\omega}{\omega_n}\right)^2}$$

$$r = \frac{e\left(\dfrac{\omega}{\omega_n}\right)^2}{\sqrt{\left[1-\left(\dfrac{\omega}{\omega_n}\right)^2\right]^2+\left[2\zeta\dfrac{\omega}{\omega_n}\right]^2}} \tag{6.51}$$

An analysis of Eq. 6.51 indicates that the line of eccentricity, sG, leads the displacement line, Os, by the angle ϕ. At low speeds ($\omega \ll \omega_n$), ϕ is less than 90°. At resonance ($\omega = \omega_n$), $\phi = 90°$. Finally, at high speeds ($\omega \gg \omega_n$), point G tends to approach the fixed point O, and the shaft center rotates around point O in a circle of radius e.

In addition to considering shaft whirl, the driveshaft must be designed to withstand the loads applied to it. Although properly oriented Hooke's joints provide an overall constant torque transfer, it should be apparent that the driveshaft is still subject to a fluctuating torque. Assuming no frictional losses in the Hooke's joint, the power must remain constant from the transmission output to the driveshaft through the joint, or

$$T_{in}\dot{\theta} = T_s\dot{\phi} \tag{6.52}$$

where $\dot{\theta}$ and $\dot{\phi}$ are the input and output angular velocities of the Hooke's joint, respectively, and T_s is the torque in the driveshaft. Making use of Eq. 6.28, Eq. 6.36 may be rewritten as

$$T_s = T_{in}\left(\frac{1-\sin^2\varepsilon\sin^2\theta}{\cos\varepsilon}\right) \tag{6.53}$$

From basic strength of materials, the shear stress in the driveshaft is then

$$\tau = \frac{T_s r}{J} \tag{6.54}$$

Obviously, the driveshaft is subjected to a fluctuating shear stress, and fatigue should immediately come to mind. The details of a fatigue analysis are beyond the scope of this work, and the interested engineer is referred to any good machine design text or the American Society of Mechanical Engineers (ASME) shaft design standards for guidance.

6.8 Differentials

If cars could travel continually in a straight line, there would be no need for a differential. Because this is not the case, one is required. Figure 6.59 illustrates the reason for this.

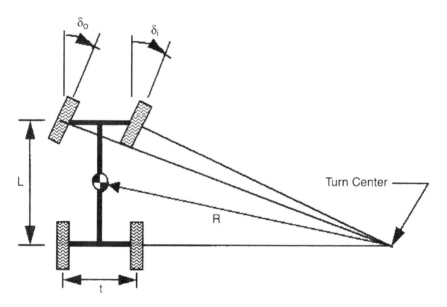

Figure 6.59. *A vehicle in a low-speed turn.*

As the vehicle in Fig. 6.59 negotiates the turn, the two rear (driven) wheels proscribe arcs of different radii. Thus, the outer wheel would tend to rotate more times than the inner wheel. If the vehicle had a solid drive axle, this would result in twisting of the axle, a phenomenon referred to as "wind-up." The solution is to allow each wheel to rotate independently, each at its own speed. This is accomplished with a differential.

Figure 6.60 shows a schematic of a differential. Before analyzing the planetary motion of the differential, note that the pinion gear and ring gear provide another gear ratio in the driveline. This "final drive" ratio follows from the basic gear law and is used to classify "rear ends." A "tall" rear end has a low torque multiplication (3.07:1) and provides better cruising fuel economy. Conversely, a "low" rear end (4.10:1) provides quicker acceleration but poorer fuel economy.

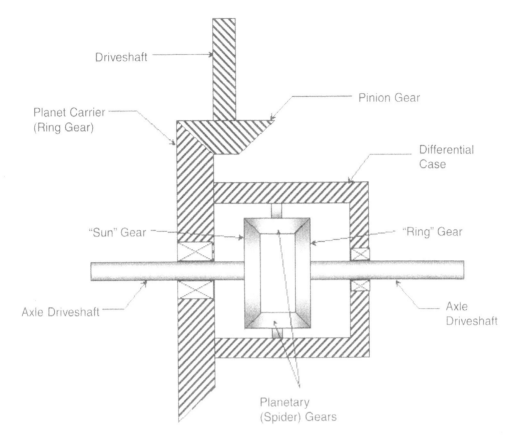

Figure 6.60. *Schematic of a differential.*

The differential is really a set of planetary gears, although the terminology applied to it confuses the issue. Referring to Fig. 6.60, the "ring gear" of a differential actually functions as the planetary carrier. Furthermore, the gears driving the two drive axles are identical. Figure 6.60 refers to one as the "sun" and the other as the "ring" to clarify the ensuing analysis. The planetary gears, sometimes called spider gears, enable the axles to rotate at different speeds. Because the differential is merely a planetary gear set, and because Eq. 6.25 applies to all planetary sets, it can be used to analyze the differential. Beginning with Eq. 6.25,

$$-\frac{N_s}{N_r} = \frac{\omega_r - \omega_c}{\omega_s - \omega_c} \tag{6.55}$$

Note that in the differential, the "sun" gear and the "ring" gear (the two attached to the driving axles) both have the same number of teeth. Thus, the left side of Eq. 6.18 is always −1 for the differential. With this fact, Eq. 6.18 becomes

$$\omega_c - \omega_s = \omega_r - \omega_c \tag{6.56}$$

which, upon further manipulation, yields

$$\omega_c = \frac{\omega_r + \omega_s}{2} \tag{6.57}$$

Thus, the angular velocity of the differential case is equal to the average angular velocity of the axles. This allows the vehicle to corner without wind-up, but it also has a few less desirable consequences. If one of the drive wheels is positioned on a patch of ice while the other is on dry pavement, there is no resistance to rotation at the wheel on the ice. As a result, the wheel on the ice spins freely at twice the speed of the differential case. Such a differential is called an "open" differential. Anyone who has lived in a northern climate has surely seen numerous examples of this problem as someone spins one wheel on snow or ice in a vain attempt to get his or her vehicle moving.

The solution to the problem of the open differential is the limited slip differential. Such devices are marketed under various names, most notably the "posi-track." These differentials contain some sort of clutch inside the differential case. When wheel spin occurs, the clutch engages on the spinning wheel. This provides a resistance to rotation and transfers torque to the wheel that grips, thus allowing the vehicle to move.

One option is the use of a clutch pack, as shown in Fig. 6.61. This design uses two sets of clutch packs. Each pack contains friction discs that are splined to the hub of the side gears, and steel plates between the friction discs that are tanged to the differential case. Pressure is applied to the clutch packs by springs, and as long as the torque remains below the capacity of the clutch packs (as determined by the spring force), the axles are in effect locked together.

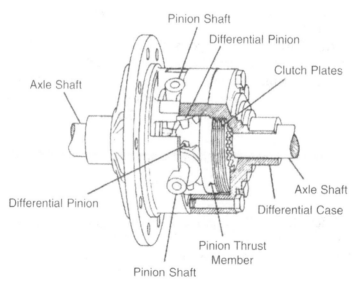

Figure 6.61. *Clutch-pack-style limited slip differential (TM 9-8000, 1985).*

Thus, if one wheel loses traction, torque is transmitted to the opposite wheel. The high torque generated by differential wheel speed while turning causes the clutch packs to slip and allows normal differential motion.

The other popular option is the use of cone clutches as shown in Fig. 6.62. This unit contains cone clutches that have spiral grooves cut into the outer edge. As one wheel rotates faster than the other, it causes the cone to be drawn into the case. This engages the friction surface and allows torque transfer to the other wheel.

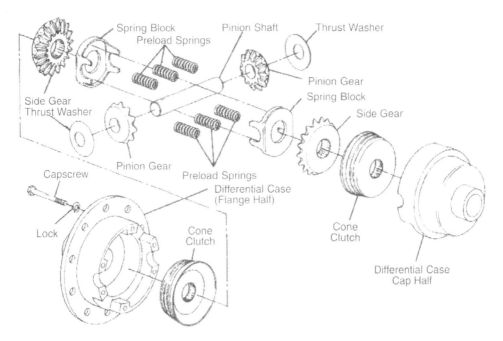

Figure 6.62. *Cone-clutch-type limited slip differential (TM 9-8000, 1985).*

6.9 Four-Wheel Drive (4WD) and All-Wheel Drive (AWD)

A vehicle that provides power to all four wheels has some key advantages in slippery or rough terrain. First and foremost, four-wheel drive (4WD) enables the vehicle to move under conditions of reduced traction. What is lost on many drivers is that four-wheel drive does not enable the vehicle to stop more rapidly, evidence of this being commonly seen in the mountainous states of the western United States. Nevertheless, the U.S. market has seen an explosion in the sale of four-wheel-drive sport utility vehicles (SUVs). Several automakers also have successfully marketed all-wheel-drive (AWD) vehicles, most notably Subaru and Audi. Furthermore, there is a vast array of adjectives used to define the systems, including part-time four-wheel drive, full-time four-wheel drive, all-wheel drive, and so forth. The differences among these systems often owe more to marketing than engineering. Adding to the confusion is the fact that the automakers themselves use various terms for their systems, often meaning something quite different to a competitor. In short, any attempt to classify four- or all-wheel drive systems invariably will meet with exceptions to the classification. This work will classify these systems into three broad categories: part-time four-wheel drive, full-time four-wheel drive, and all-wheel drive.

6.9.1 Part-Time Four-Wheel Drive (4WD)

The key feature of a part-time four-wheel-drive system is the inclusion of a separate transfer case aft of the transmission, an example of which is shown in Fig. 6.63. This is the lowest-cost option and can be considered the first-generation option. It is called part-time because it can be used only in conditions that will allow for wheel slip, such as dirt roads, full snow coverage, and so forth. The reason for this is that there is no mechanism to eliminate driveline wind-up. Recall that a differential is used at the rear axle to allow differences in wheel rotation while the vehicle is cornering. With four-wheel drive, the same thing is happening with the front axle and the rear axle. One is traveling faster than the other; therefore, something must allow for the speed difference. In the absence of a center differential, the only mechanism allowing wheel speed variation is for the wheel to break free at the contact patch. Because this requires large forces on dry pavement, the part-time system cannot be used on dry pavement without serious drivetrain damage.

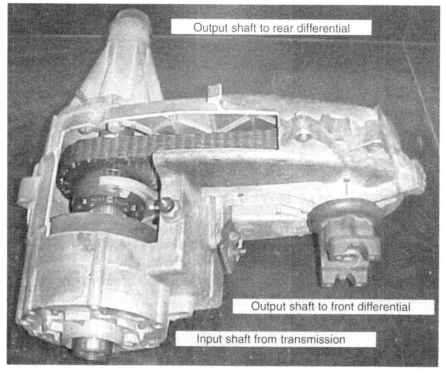

Figure 6.63. *A part-time four-wheel-drive transfer case.*

The transfer case also incorporates two selectable gear ratios—low and high. In four-wheel drive low, the vehicle has a limited top speed. However, because of the large gear ratio in low, the vehicle has a large amount of torque available at the drive wheels to enable the driver to extricate the vehicle from difficult situations.

The other feature of this system is that the front hubs usually are locked manually. In two-wheel drive, the front wheels spin freely around the spindles. When the driver desires four-wheel drive, the hubs must be locked manually onto the drive spindles for torque to be applied through the front wheels. The Isuzu Rodeo, Ford Bronco, and Dodge Ram all have part-time four-wheel drive.

6.9.2 On-Demand Four-Wheel Drive (4WD)

This is the next option in terms of increasing convenience and cost. An open differential is incorporated between the front and rear axles. The open differential absorbs shaft speed variations between the front and rear output shafts. However, being an open differential, it sends torque to the axle with least resistance. This system allows driving in four-wheel drive on dry pavement, but this will decrease the fuel economy of the vehicle. For this reason, it is referred to as on-demand. The driver may use two-wheel drive when there is no need for four-wheel drive. This system also will have automatic locking hubs that are either vacuum operated or electrically operated, saving the driver from a trip out of the cab in inclement weather to lock the hubs. Often "on-demand" (a configuration) is confused with "shift-on-the-fly" (an engagement method). The Chevrolet Blazer is an example of a vehicle with on-demand four-wheel drive.

6.9.3 Full-Time Four-Wheel Drive (4WD)

This is the highest cost option. This system has differentials everywhere, at both front and rear axles and in the transfer case. This allows the vehicle to be in four-wheel drive on dry pavement. The system allows for slip, but something had to be done about situations of very low traction—that is, the open differentials would send torque to the wheel with the least traction. Some vehicles, most notably the AM General Hummer, can lock all of the differentials. Other vehicles, such as the 1995 Jeep Grand Cherokee, have a viscous coupling that transmits power from the wheels that slip to the wheels that grip. In this category, the distinction between four-wheel drive and all-wheel drive begins to blur. For the purposes of this work, full-time four-wheel drive is applied to vehicles that still require the driver to select the four-wheel-drive option.

6.9.4 All-Wheel Drive (AWD)

For the purposes of this work, an all-wheel drive vehicle does not have a selectable transfer case. Generally, these vehicles are not intended for off-road use, but use four-wheel drive for inherent stability. Usually, they use viscous couplings to send power from the spinning wheels to the gripping wheels. The system operates automatically and requires no driver intervention.

This system also is used on high-performance cars to eliminate wheel spin caused by the enormous torque generated at the rear wheels.

6.10 Case Study: The Chrysler 42LE Automatic Transaxle

An example of a modern automatic transaxle is the Chrysler 42LE transmission. This transmission is an electronically controlled four-speed transaxle used in the Chrysler LH cars. Details of this transaxle are provided in a paper by Martin and Redinger (1993), and this case study will summarize the major details of this unit.

The 42LE transmission was designed to be mated to an engine producing 214 hp and 300 N-m of torque. Vehicle performance specifications were as follows:

a. Acceleration from a standing start   > 5.2 m/sec^2
b. 0–99.6 km/h (0–61.9 mph)   < 10 sec
c. Distance in 5 sec   > 45.7 m (149.9 ft)
d. Time to pass 64.4–96.6 km/h (40–60 mph)   < 5 sec
e. Grade at 88.5 km/h (55 mph), top gear   > 6%
f. Top vehicle speed   > 209 km/h (130 mph)

The design team was to achieve these parameters with zero defects within 185 weeks from concept to launch.

6.10.1 Configuration

The transaxle was to be mated to a vehicle with a longitudinal engine configuration, but the vehicle was to be front-wheel drive. This presented a packaging challenge to the designers, and their solution is shown in the cutaway view of the 42LE in Fig. 6.64.

The unit contains a hydraulic torque converter with an electronically controlled lockup clutch, a compact dual planetary gear set with clutch packs, a transfer chain drive, and a limited slip differential. Figure 6.65 shows a schematic of the transaxle.

Of note in this unit is the use of the electronically modulated converter clutch (EMCC). The clutch is engaged in third and fourth gears at turbine speeds of 1100–1500 rpm. Also, because of the engine layout coupled with the requirement for front-wheel drive, the designers made use of a transfer chain to send the torque forward from the transmission output.

6.10.2 Planetary Gear Set

Figure 6.66 shows the highly compact planetary set as assembled and in an exploded view. The key feature of this set is that the front ring and rear planetary carrier is a single piece, as well as the front ring and rear planetary carrier. Figure 6.67 shows a schematic of the clutches and gear sets. The unit provides four forward speeds (including an overdrive) and reverse.

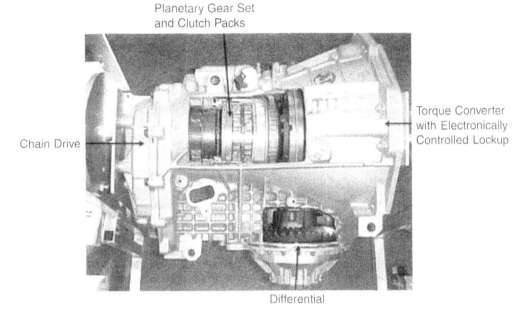

Figure 6.64. *Cutaway view of the Chrysler 42LE transaxle.*

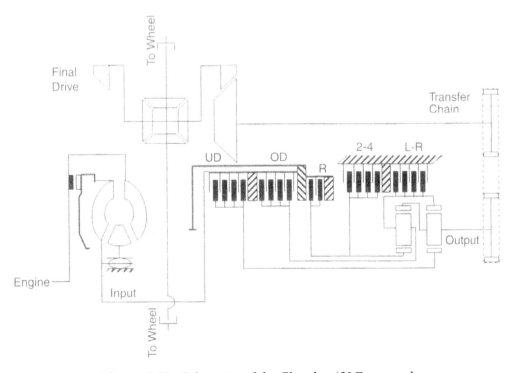

Figure 6.65. *Schematic of the Chrysler 42LE transaxle.*

Figure 6.66. *The Chrysler 42LE planetary gear set—assembled view (top) and exploded view (bottom).*

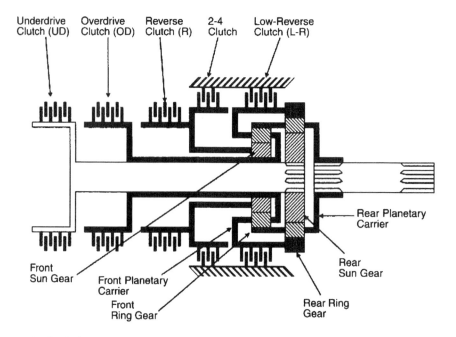

Figure 6.67. *Schematic of the Chrysler 42LE planetary gear sets and clutch packs.*

The transmission is electronically controlled and hydraulically actuated. The UD, OD, and R clutches are engaged as required and provide input to the gear set. The 2–4 and L–R clutches are applied as needed to fix a planetary element. Problem 6.6 at the end of this chapter summarizes how these clutches are used to obtain the desired gear ratios.

6.10.3 Chain Transfer Drive

The loads and speeds required for the transfer chain precluded the use of standard roller chains. Instead, the designers selected a Borg-Warner Hy-Vo chain with a rocker joint design. This chain exhibited a slightly higher efficiency than that of a roller chain, and it showed a reduction in wear due to the articulated joint motion. However, chain wear and the stretch produced by it had to be considered in sizing the case, to prevent chain slap noise.

6.10.4 Control System

The electronic control system is fully integrated within the vehicle and includes the transmission controller, engine control module, throttle position sensor, crank sensor, electronic gear selector position indicator, manual valve lever position sensor, body computer, ignition switch, starter relay, turbine speed sensor, and output speed sensor. The transmission controller contains adaptive logic to allow the transmission to "learn" driver shift preferences, as well as to alter shift patterns on the basis of manufacturing tolerances of the turbine and pump.

The 42LE met its targets for reliability and vehicle performance, and the design was completed ahead of schedule. Again, the interested student can find more details on this unit in the paper by Martin and Redinger (1993).

6.11 Problems

1. An automatic transmission receives as an input 422 ft-lb of torque from the torque converter. A multi-disc clutch is to be designed to transmit this torque to the planetary gear sets. The clutch rings have outer and inner diameters of 6.0 in. and 5.75 in., respectively. The friction material is to be sintered metal, operated wet, giving the surface a coefficient of friction of 0.08 with a maximum allowable applied pressure of 500 psi. How many two-sided discs are required for this application? Assume uniform wear. (Answer: Four.)

2. A centrifugal clutch is designed using three 0.5-oz. weights. The distance from the hinge pin to the cover plate joint (distance a) is 1 in., and the distance from the hinge pin to the weight (distance b) is 0.5 in. The weights are placed 6 in. from the axis of rotation. How much force is generated by the masses when the clutch is rotating at 5000 rpm? (Answer: 200 lb.)

3. An axial clutch is designed based on a fixed outer diameter, D. To maximize torque capacity, what should the inner diameter be, in terms of the outer diameter? Assume uniform wear. (Answer: d = 0.577D.)

4. A driveshaft is driven at its critical speed, and the radial deflection is found to be 0.25 in. at the midpoint of the shaft. When the shaft is driven at 80% of its critical speed, the deflection reduces to 0.08 in. Calculate the eccentricity of the shaft and its damping ratio. (Answer: e = 0.05", ζ = 0.1.)

5. A designer makes a 4-ft driveshaft from steel (E = 29 Msi, ρ = 0.00883 sl/in.^{3}). The shaft has a 3-in. outer diameter and a wall thickness of 1/8 in. Calculate the critical speed of the shaft. Assume the shaft is simply supported. (Answer: 5800 rpm.)

The designer decides to switch to aluminum for the shaft (E = 10 Msi, ρ = 0.00305 sl/in.^{3}). The size of the shaft remains the same. Calculate the new critical speed. (Answer: 5795 rpm.)

6. The Chrysler 42LE transmission has the following tooth counts for the planetary gear sets:

Front Set		Rear Set	
Ring	62	Ring	70
Planets	17	Planets	17
Sun	28	Sun	38

The following table lists the clutch activation for each gear, with the corresponding input, output, and fixed members. Calculate the gear ratio in each gear. (Hint: Recall that $\omega_{fpc} = \omega_{rr}$, and $\omega_{rpc} = \omega_{fr}$.) (Answer: First gear = 2.842:1; second gear = 1.573:1; third gear = 1.00:1; fourth gear = 0.689:1; reverse = −2.214:1.)

Gear	Elements Engaged	Input	Held	Output	Notes
1	UD, L–R	Rear sun	Front carrier	Rear carrier	$\omega_{rr} = \omega_{fpc} = 0$
2	UD, 2–4	Rear sun	Front sun	Front ring	
3	UD, OD				Unit locked
4	OD, 2–4	Front carrier	Front sun	Front ring	
Reverse	R, L–r	Front sun	Front carrier	Front ring	

7. The Ford Expedition has a 65.5-in. track and a 20-ft turning radius. The P265/70R17 tires have a nominal radius of 15 in. If the vehicle was designed with a solid rear axle (i.e., no differential), calculate the amount of wind-up in the rear axle when the vehicle performs a 180° turn at minimum radius. (Answer: 2.18 revolutions.)

8. A Ford Expedition is traveling at 75 mph in fourth gear (gear ratio = 0.67:1). The tachometer indicates an engine speed of 2100 rpm, and the rear wheels have a nominal radius of 15 in. What is the final drive ratio? (Answer: 3.73:1.)

9. The 1996 Dodge Viper GTS has the torque curve and gear ratios shown in the following table. The rear tires have a nominal radius of 13 in. Generate a plot of force at the road versus vehicle speed for each gear. Your result should appear similar to Fig. 6.1.

rpm	Torque [ft-lb]	Gear	Gear Ratio
1000	390	First	2.66
2000	440	Second	1.76
3000	445	Third	1.30
4000	470	Fourth	1.00
5000	410	Fifth	0.74
6000	330	Sixth	0.50
6500	290	Final Drive	3.07

Chapter 7

Steering Systems and Steering Dynamics

7.1 Introduction

Any mode of transportation used by people must have some means of control. For the automobile, two primary control systems are at the driver's disposal: (1) the steering system, and (2) the braking system. This chapter will focus on the steering system.

Most vehicles in service today have front-wheel steering, although a few vehicles have been marketed with four-wheel steering. Thus, the bulk of this chapter will discuss the front-wheel steering systems to include the following: a discussion of the steering mechanisms available, including power-assisted steering; the factors affecting wheel alignment; a simplified analysis of vehicle cornering dynamics; and the influences of front-wheel drive on steering response. The chapter concludes with a brief discussion of four-wheel steering and vehicle rollover behavior.

7.2 Steering Mechanisms

The fundamental problem in steering is to enable the vehicle to traverse an arc such that all four wheels travel about the identical center point. In the days of horse-drawn carriages, this was accomplished with the fifth-wheel system depicted in Fig. 7.1.

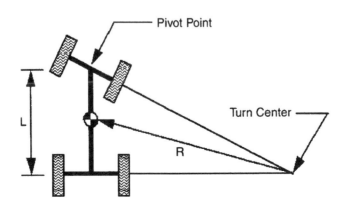

Figure 7.1. *Fifth-wheel steering.*

Although this system worked well for carriages, it soon proved unsuitable for automobiles. In addition to the high forces required of the driver to rotate the entire front axle, the system proved unstable, especially as vehicle speeds increased. The solution to this problem was developed by a German engineer named Lankensperger in 1817. Lankensperger had an inherent distrust of the German government, so he hired an agent in England to patent his idea. His chosen agent was a lawyer named Rudolph Ackerman. The lawyer secured the patent, but the system became known as the Ackerman system.

Figure 7.2 depicts the key features of this system. The end of each axle has a spindle that pivots around a kingpin. The linkages connecting the spindles form a trapezoid, with the base of the trapezoid formed by the rack and tie rods. The distance between the tie rod ends is less than the distance between the kingpins. The wheels are parallel to each other when they are in the straight-ahead position. However, when the wheels are turned, the inner wheel turns through a greater angle than the outer wheel. Figure 7.2 also shows that the layout is governed by the ratio of track (distance between the wheels) to wheelbase (distance between front and rear wheels). The Ackerman layout is accurate only in three positions: straight ahead, and at one position in each direction. The slight errors present in other positions are compensated for by the deflection of the pneumatic tires.

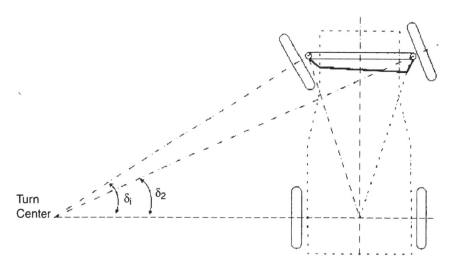

Figure 7.2. *Ackerman steering.*

For the purposes of this book, "steering mechanism" refers to those components required to realize the Ackerman system. Of course, all vehicles today use a steering wheel as the interface between the system and driver. (This has not always been the case. Early automobiles used a tiller.) The steering wheel rotates a column, and this column is the input to the steering mechanism. These mechanisms can be broadly grouped into two categories: (1) worm-type mechanisms, and (2) rack and pinion mechanisms.

7.2.1 Worm Systems

Figure 7.3 shows the steering linkages required by worm gear steering systems. The Pitman arm converts the rotational motion of the steering box output into side-to-side motion of the center link. The center link is tied to the steering arms by the tie rods, and the side-to-side motion causes the spindles to pivot around their respective steering axes (kingpins). To achieve Ackerman steering, the four-bar linkages must form a trapezoid instead of a parallelogram (refer to Fig. 7.2). Although all worm-type steering systems use linkages similar to these, the specifics of the steering boxes differ and are explained next.

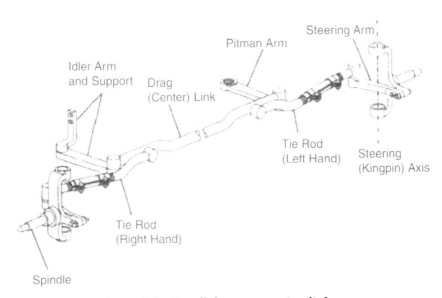

Figure 7.3. *Parallelogram steering linkages. Adapted from TM 9-8000 (1985).*

7.2.2 Worm and Sector

Figure 7.4 depicts a worm and sector system. The shaft to the Pitman arm is connected to a gear that meshes with a worm gear on the steering column. Because the Pitman shaft gear needs to rotate through only approximately 70°, only a sector of the gear is actually used. The worm gear is assembled on tapered roller bearings to absorb some thrust load, and an adjusting nut is provided to regulate the amount of end-play in the worm.

7.2.3 Worm and Roller

The worm and roller system (Fig. 7.5) is very similar to the worm and sector system. In this case, a roller is supported by ball bearings within the sector on the Pitman shaft. The bearings reduce sliding friction between the worm and sector. The worm also can be shaped similarly

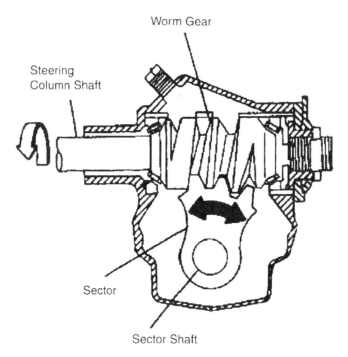

Figure 7.4. *Worm and sector steering gear. Adapted from TM 9-8000 (1985).*

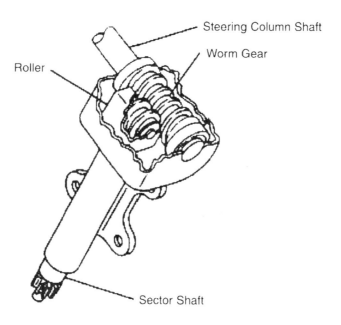

Figure 7.5. *Worm and roller steering gear. Adapted from TM 9-8000 (1985).*

to an hourglass, that is, tapered from each end to the center. This provides better contact between the worm and the roller, as well as a variable steering ratio. When the wheels are at the center (straight-ahead) position, the steering reduction ratio is high to provide better control. As the wheels are turned farther off-center, the ratio lowers. This gives better maneuverability during low-speed maneuvers such as parking.

7.2.4 Recirculating Ball

Figure 7.6 shows the recirculating ball system, another form of worm and nut system. In this system, a nut is meshed onto the worm gear by means of a continuous row of ball bearings. As the worm turns, the nut moves up and down the worm threads. The ball bearings not only reduce the friction between the worm and nut, but they greatly reduce the wear because the balls continually recirculate through the system, thereby preventing any one area from bearing the brunt of the wear.

The primary advantage of all worm-type steering systems is reduced steering effort on the part of the driver. However, due to the worm gear, the driver receives no feedback from the

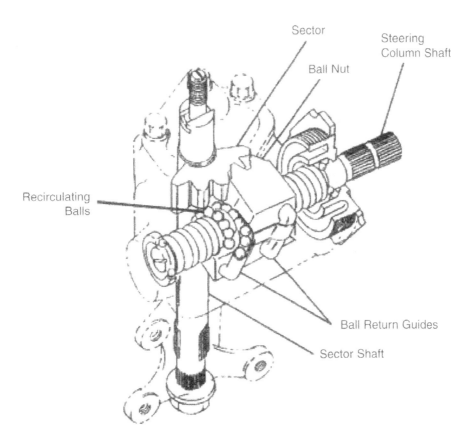

Figure 7.6. *Recirculating ball steering gear. Adapted from TM 9-8000 (1985).*

wheels. For these reasons, worm-type steering systems are found primarily on large vehicles such as luxury cars, sport utility vehicles, pickup trucks, and commercial vehicles.

7.2.5 Rack and Pinion Steering

The rack and pinion steering system is simpler, lighter, and generally cheaper than worm-type systems (Fig. 7.7). The steering column rotates a pinion gear that is meshed to a rack. The rack converts the rotary motion directly to side-to-side motion and is connected to the tie rods. The tie rods cause the wheels to pivot about the kingpins, thus turning the front wheels.

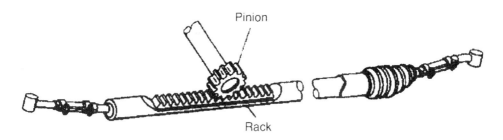

Figure 7.7. *Rack and pinion steering gear. Adapted from TM 9-8000 (1985).*

Rack and pinion systems have the advantage of providing feedback to the driver. Furthermore, rack and pinion systems tend to be more responsive to driver input, and for this reason, rack and pinion steering is found on most small and sports cars.

7.2.6 Power Steering

Many vehicles incorporate a power steering system, the purpose of which is to reduce the driver's effort to turn the steering wheel. The system usually is hydraulically operated, with hydraulic pressure provided by a pump driven by a belt from the crankshaft. Figure 7.8 shows a drive system that is an older, V-belt type. These systems used multiple V-belts to drive the various accessories on the front of the engine. Most new vehicles use a single, ribbed belt (serpentine belt) to drive all of the accessories.

The power steering pump contains an integral fluid reservoir, as well as the control and pressure regulating valves. The pump may be of the vane, tooth, or rotor type. Figure 7.9 shows one example of a vane-type pump.

The pump receives fluid from the reservoir, and because it is belt driven by the crank, the pump operates whenever the engine is running. Figure 7.10 shows a typical control valve. When the wheels are in the straight-ahead position, the spool valve is centered. This allows fluid under pressure to bypass the system and return to the reservoir. When the driver turns

Steering Systems and Steering Dynamics

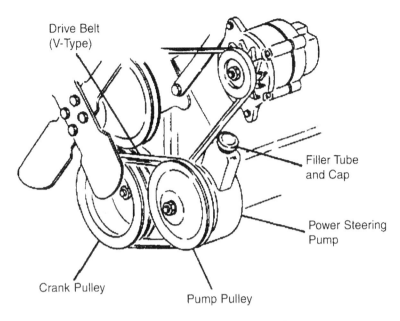

Figure 7.8. *Power steering pump drive. Adapted from TM 9-8000 (1985).*

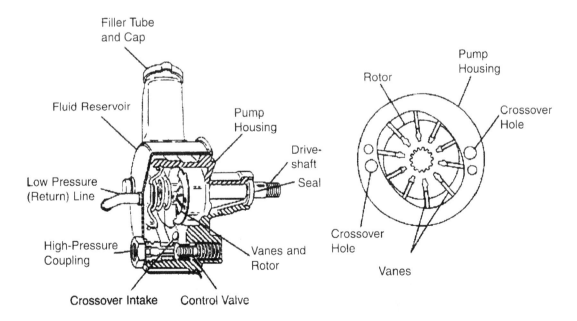

Figure 7.9. *Vane-type power steering pump. Adapted from TM 9-8000 (1985).*

the steering wheel, the spool valve is mechanically moved off-center. This allows fluid to be ported to the appropriate side of the cylinder unit and supplies the additional force to turn the wheels. The cylinder unit can be mounted on the steering column, in line with the rack, or integral to the recirculating ball gearbox. Figure 7.11 shows examples of each.

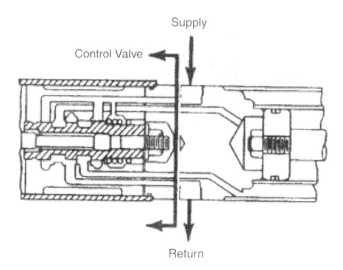

Figure 7.10. *Power steering control valve (TM 9-8000, 1985).*

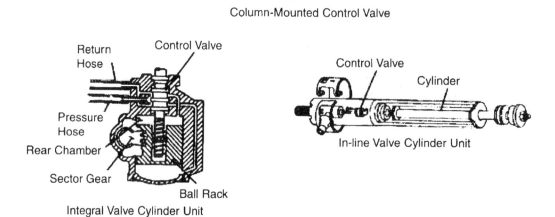

Figure 7.11. *Control valve configurations (TM 9-8000, 1985).*

7.3 Steering Dynamics

"Handling" is loosely used to describe the response of a car to the driver's inputs and the ease of control. As such, it is a subjective measure. A 1972 Cadillac may "handle like a pig," but it is a great car for taking Grandma to the store. A 1992 Corvette "handles great." A Porsche 911 may "be a handful," and in the hands of a novice driver may result in spin-outs and total loss of control. One of the most important factors influencing handling is the cornering behavior of a vehicle. The first step in understanding cornering performance is to analyze the low-speed turn behavior of a car.

7.3.1 Low-Speed Turning

A vehicle rolling through a low-speed turn requires that a perpendicular line through the front wheels pass through the same point. If this is not done, the front wheels "fight" each other, resulting in scuffing and tire wear. As shown in Fig. 7.2, this is accomplished by the Ackerman steering system. For the proper turn geometry under low-speed conditions (assuming small angles), the steer angles, δ, are given by

$$\delta_o = \frac{L}{R + \frac{t}{2}} \quad (7.1)$$

$$\delta_i = \frac{L}{R - \frac{t}{2}} \quad (7.2)$$

where

    δ = steer angle
    L = wheelbase
    R = turn radius
    t = track width

The average steer angle (small angle approximation) is called the Ackerman angle and is given by

$$\delta = L/R \, (\text{rad}) \quad (7.3)$$

$$\delta = 5.73 \, L/R \, (\text{deg}) \quad (7.4)$$

Thus, the Ackerman angle relates δ, the amount the front wheels must be turned, to R, the turn radius.

7.3.2 High-Speed Turning

At high speeds, lateral acceleration becomes a factor. To counteract this, the tires must generate lateral forces. To generate these forces, the tires experience a slip angle as shown in Fig. 7.12.

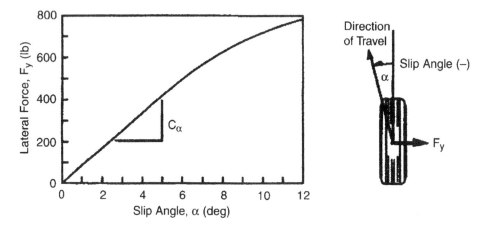

Figure 7.12. *Cornering force versus slip angle (Gillespie, 1994).*

For slip angles less than 5°, the cornering force varies linearly with the slip angle

$$F_y = C_\alpha \alpha \tag{7.5}$$

where C_α is defined as the cornering stiffness and is dependent on tire size and type (i.e., bias or radial), cord angle, wheel width, tread, load, inflation pressure, and so forth. For a high-speed turn, R is sufficiently large that it may assumed that $\delta_i \approx \delta_o$. Hence, the following analysis will use the bicycle model as shown in Fig. 7.13.

It is assumed that the vehicle is established in the turn (steady state) with a constant forward speed, V. From basic dynamics, the acceleration of the CG is

$$a_{CG} = \frac{mV^2}{R} \tag{7.6}$$

Thus, summing of the forces in the lateral direction yields

$$\sum F_y = F_r + F_f \cos(\delta_f - \alpha_f) = \frac{mV^2}{R} \tag{7.7}$$

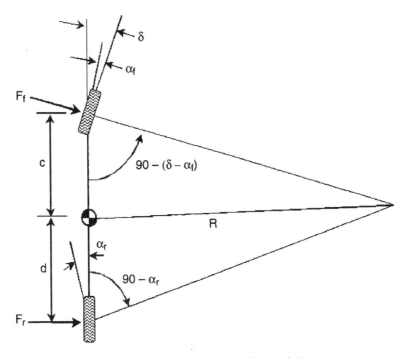

Figure 7.13. *The bicycle model.*

For small angles, $\cos\alpha \approx 1$, and Eq. 7.7 may be rewritten as

$$\sum F_y = F_r + F_f = \frac{mV^2}{R} \tag{7.8}$$

Because the vehicle is established in the turn (i.e., it is in equilibrium), the sum of the moments about the center of gravity must be equal to zero (again assuming small angles)

$$\sum M_{cg} = 0 = F_r d - F_f c \tag{7.9}$$

or

$$F_f = \frac{F_r d}{c} \tag{7.10}$$

Substituting Eq. 7.10 into Eq. 7.8, and noting that $(c + d) = L$ gives

$$F_r\left(1 + \frac{d}{c}\right) = F_r\left(\frac{L}{c}\right) = \frac{mV^2}{R} \tag{7.11}$$

or

$$F_r = \left(\frac{mc}{L}\right)\frac{v^2}{R} \tag{7.12}$$

Next, the car must be analyzed in the lateral plane, as shown in Fig. 7.14.

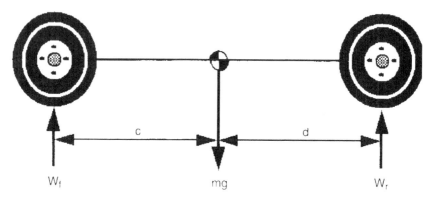

Figure 7.14. Side-view model.

Again, the sum of the moments about the front and rear wheel must equal zero

$$\sum M_f = 0 = W_r(c+d) - mgc \tag{7.13}$$

$$\sum M_r = 0 = mgd - W_f(c+d) \tag{7.14}$$

Thus,

$$W_f = \frac{mgd}{(c+d)} = \text{weight on the front axle} \tag{7.15}$$

$$W_r = \frac{mgc}{(c+d)} = \text{weight on the rear axle} \tag{7.16}$$

However, because $(c + d) = L$, Eqs. 7.15 and 7.16 can be rearranged, so that

$$\frac{md}{L} = \frac{W_f}{g}$$

$$\frac{mc}{L} = \frac{W_r}{g} \tag{7.17}$$

Substituting Eq. 7.17 (lower) into Eq. 7.12 gives

$$F_r = \frac{W_r V^2}{gR} \tag{7.18}$$

This also can be done to determine the force at the front

$$F_f = \frac{W_f V^2}{gR} \tag{7.19}$$

Finally, Eqs. 7.18 and 7.19 can be substituted into Eq. 7.5 to find the slip angles at the front and rear tires

$$\alpha_f = \frac{F_f}{C_{\alpha_f}} = \frac{W_f}{C_{\alpha_f}} \frac{V^2}{gR}$$

$$\alpha_r = \frac{F_r}{C_{\alpha_r}} = \frac{W_r}{C_{\alpha_r}} \frac{V^2}{gR} \tag{7.20}$$

Referring to Fig. 7.14, the sum of the interior angles of the triangle must equal 180°. Recall also that the vertex angle is nothing more than the Ackerman angle previously discussed in Section 7.3.1. Thus,

$$\sum \text{angles} = 180° = 57.3\left(\frac{L}{R}\right) + 90 - (\delta - \alpha_f) + 90 - \alpha_r \tag{7.21}$$

or

$$\delta = 57.3\frac{(L)}{R} + \alpha_f - \alpha_r \tag{7.22}$$

Substitution of Eq. 7.20 into Eq. 7.22 yields

$$\delta = 57.3\frac{L}{R} + \left[\frac{W_f}{C_{\alpha_f}} - \frac{W_r}{C_{\alpha_r}}\right]\frac{V^2}{gR} \tag{7.23}$$

Thus, Eq. 7.23 quantifies the angle that the front wheels must be turned to negotiate a turn of radius R at speed V for a given vehicle. Equation 7.23 often is written as

$$\delta = 57.3\frac{L}{R} + Ka_y \tag{7.24}$$

where

K = understeer gradient (deg/g)
a_y = lateral acceleration (g)

This equation is critical to the turning response of the car. It describes how the steer angle, δ, relates to the turn radius, R, or the lateral acceleration a_y. The key parameter in this equation is K. Each term in K represents the ratio of axle load (front or rear) to the cornering stiffness (front or rear). Three possibilities exist:

1. Neutral steer: $\quad \dfrac{W_f}{C_{\alpha_f}} = \dfrac{W_r}{C_{\alpha_r}} \Rightarrow K = 0$

On a turn of constant radius, the steer angle required to negotiate the turn is equal to the Ackerman angle, 57.3 L/R degrees, regardless of variations in the vehicle speed. Physically, as the vehicle negotiates the turn, the front and rear wheels generate slip angles at the same rate.

2. Understeer: $\quad \dfrac{W_f}{C_{\alpha_f}} > \dfrac{W_r}{C_{\alpha_r}} \Rightarrow K > 0$

On a turn of constant radius, the steer angle is proportional to K times the lateral acceleration in g's. Thus, to develop the side force required to maintain the turn, the front wheels must be turned to a greater angle as the speed increases. In a car with understeer, the lateral acceleration causes the front wheels to generate slip angles at a rate greater than the rear wheels. At the limit, the front wheels will break free first, causing the vehicle to depart the turn straight ahead. In racing, a car with understeer is said to "push."

3. Oversteer: $\quad \dfrac{W_f}{C_{\alpha_f}} < \dfrac{W_r}{C_{\alpha_r}} \Rightarrow K < 0$

On a turn of constant radius, the steer angle must decrease as speed increases. In this case, the lateral acceleration causes the rear wheels to develop slip angles at a rate greater than the front. As a result, the rear end slides outward, which causes the front wheels to turn inward. The front wheels then must be turned out of the turn to maintain a constant radius. A car with oversteer is said to be "loose."

Figure 7.15 graphically shows the relationship between steer angle and speed.

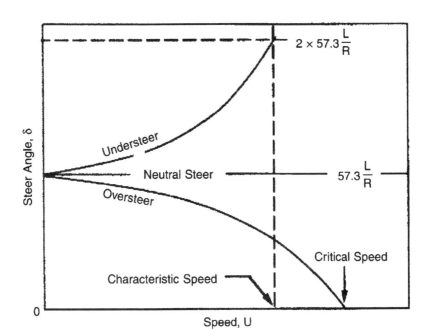

Figure 7.15. *Change of steer angle with speed (Gillespie, 1994).*

This graph illustrates the general trends of oversteer and understeer, although this is not tremendously useful in comparing cars. Thus, two parameters have been defined to help quantify the steering tendencies of a car. For an understeer vehicle, the understeer level is quantified by the characteristic speed. Characteristic speed is simply the speed at which the steer angle required to negotiate a turn is equal to twice the Ackerman angle. For an oversteer vehicle, a critical speed exists, above which the vehicle will become unstable. This speed is shown on Fig. 7.15 where the steer angle required to negotiate a turn goes to zero. Physically, this means that no input from the driver is required to turn the car, an undesirable feature when traveling down the highway. Hence, most vehicles are designed to exhibit understeer because this is safer than an oversteer car. Looking at Eq. 7.23, we can deduce several important factors in determining the steering qualities of a vehicle.

Because K is a function of axle loading, the manner and amount of the load placed in a vehicle is critical. For example, take a standard 1/2-ton pickup truck with understeer. If 1 ton of wood is loaded in the back, the second term in K, $\left(\dfrac{W_r}{C_{\alpha_r}}\right)$, has just increased greatly. Thus, the formerly docile pickup truck may (and probably will) become unstable at a very low speed. Such cases have been documented, with a corresponding loss of life. Mixing tires (bias and radials) affects the cornering stiffness terms. Because bias-ply tires generally have a lower cornering stiffness than radials, putting radials on the front and bias-ply tires on the rear also can lead to oversteer. Likewise, improperly inflated tires change the cornering stiffness and hence alter the steering characteristics of the car.

7.3.3 Effects of Tractive Forces

So far, the analysis has considered a car rolling through a turn. Obviously, a car is influenced by drag, rolling resistance, and so forth, and it needs some power from the engine to maintain its speed. These forces can be accounted for in the model. For the sake of generality, tractive forces will be applied to both the front and rear wheels, thus enabling the model to handle rear-wheel-drive, front-wheel-drive, and four-wheel-drive vehicles. The bicycle model now resembles Fig. 7.16.

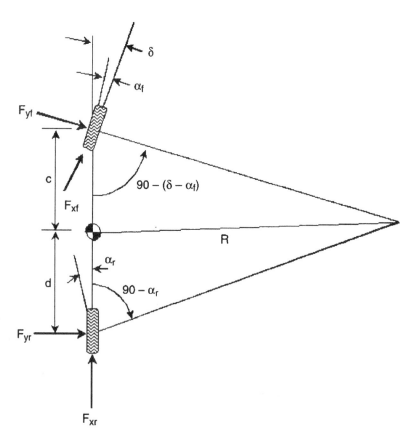

Figure 7.16. *The bicycle model with tractive forces.*

The development of the equation for steer angle follows the same course as previously described, and details of the development are contained in Gillespie (1994). The resulting equation is

$$\delta = \frac{57.3\frac{L}{R}}{1+\frac{F_{xf}}{C_{\alpha_f}}} + \left[\left(\frac{W_f}{C_{\alpha_f}} - \frac{W_r}{C_{\alpha_r}}\right) - \left(\frac{W_f}{C_{\alpha_f}}\frac{F_{xf}}{C_{\alpha_f}} - \frac{W_r}{C_{\alpha_r}}\frac{F_{xr}}{C_{\alpha_r}}\right)\right]\frac{V^2}{gR} \quad (7.25)$$

<---1--> <--------2--------> <--------------3-------------->

The three terms on the right side of Eq. 7.25 are as follows:

1. This is the Ackerman steer angle modified by the tractive force on the front wheel. For a rear-wheel-drive (RWD) vehicle, the term is unchanged from Eq. 7.23. For a front-wheel-drive (FWD) car, a positive tractive force acts to decrease the Ackerman angle (oversteer). If the front wheels spin (e.g., on ice or snow), a tractive force is still generated, but the cornering force goes to zero. In this case, the denominator becomes infinite, suggesting that turns of zero radius can be made with no steer angle. Although this equation would imply that front-wheel-drive cars exhibit oversteer, the effects are modified by factors that are not included in this model. The primary effects are driveline torque and tractive force influence on tire cornering stiffness and aligning torque. These effects will be discussed in Section 7.6. Furthermore, many other factors influence the understeer characteristics of a car, including camber change, roll steer, compliance steer, aligning torque, and so forth. Because these factors are intimately related to suspension geometry, they will be introduced in Chapter 8, although a full treatment of them is beyond the scope of this book. However, they are fully derived in the book by Gillespie (1994).

2. This is the understeer gradient (K) and is unchanged from the previous discussion.

3. This term represents the effect of tractive forces on the understeer characteristics of the vehicle. If F_{xf} is positive (FWD), the tractive forces tend to pull the car into the turn, resulting in slight oversteer. If F_{xf} is positive (RWD), the tractive force tends to cause understeer, by the same logic. This term also indicates why most cars are designed with understeer. If a driver takes a turn too fast, the natural reaction is to lift the throttle. This causes the rear tractive force to become negative (due to rolling resistance). This results in a reduction of the understeer gradient—a phenomenon known as throttle-off oversteer. Thus, the designer can use the natural reaction of the driver to help keep the car stabilized in the turn.

7.4 Wheel Alignment

In addition to allowing the vehicle to be turned, the steering system must be set up to allow the vehicle to track straight ahead without steering input from the driver. Thus, an important design factor for the vehicle is the wheel alignment. Four parameters are set by the designer, and these must be checked regularly to ensure they are within the original vehicle specifications. The four parameters discussed here are as follows:

1. Camber
2. Steering axis inclination (SAI)
3. Toe
4. Caster

7.4.1 Camber

Camber is the angle of the tire/wheel with respect to the vertical as viewed from the front of the vehicle, as shown in Fig. 7.17. Camber angles usually are very small, on the order of 1°; the camber angles shown in Fig. 7.17 are exaggerated. Positive camber is defined as the top of the wheel being tilted away from the vehicle, whereas negative camber tilts the top of the wheel toward the vehicle. Most vehicles use a small amount of positive camber, for reasons that will be discussed in the next section. However, some off-road vehicles and race cars have zero or slightly negative camber.

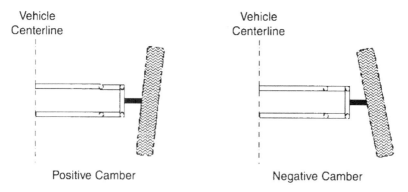

Figure 7.17. *Positive and negative camber (view from the front of the vehicle).*

7.4.2 Steering Axis Inclination (SAI)

Steering axis inclination (SAI) is the angle from the vertical defined by the centerline passing through the upper and lower ball joints. Usually, the upper ball joint is closer to the vehicle centerline than the lower, as shown in Fig. 7.18.

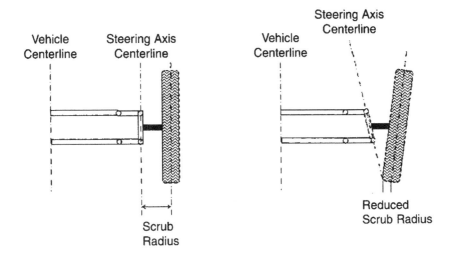

Figure 7.18. *Interaction of positive camber with steering axis inclination (SAI).*

Figure 7.18 also shows the advantage of combining positive camber with an inclined steering axis. If a vertical steering axis is combined with zero camber (left side of Fig. 7.18), any steering input requires the wheel to scrub in an arc around the steering axis. In addition to increasing driver effort, it causes increased tire wear. The combination of SAI and positive camber reduces the scrub radius (right side of Fig. 7.18). This reduces driver effort under low-speed turning conditions and minimizes tire wear. An additional benefit of this system is that the wheel arc is no longer parallel to the ground. Any turning of the wheel away from straight ahead causes it to arc toward the ground. Because the ground is not movable, this causes the front of the vehicle to be raised. This is not the minimum potential energy position for the vehicle; thus, the weight of the vehicle tends to turn the wheel back to the straight-ahead position. This phenomenon is very evident on most vehicles—merely turning the steering wheel to full lock while the vehicle is standing still will make the front end of the vehicle rise visibly. Although the stationary the weight of the vehicle may not be sufficient to rotate the wheels back to the straight-ahead position, as soon as the vehicle begins to move, the wheels will return to the straight-ahead position without driver input. Caster angle also contributes to this self-aligning torque and will be discussed in Section 7.4.4. Note that the diagrams in the preceding figures have been simplified to facilitate discussion. In practice, the wheel is dished so that the scrub radius is further reduced, as illustrated in Fig. 7.19.

7.4.3 Toe

Toe is defined as the difference of the distance between the leading edge of the wheels and the distance between the trailing edge of the wheels when viewed from above. Toe-in means the front of the wheels are closer than the rear; toe-out implies the opposite. Figure 7.20 shows both cases.

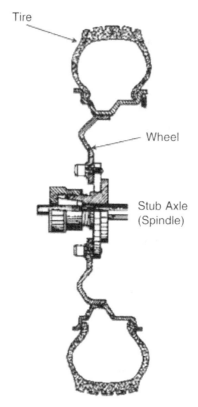

Figure 7.19. *Cross-sectional view of a wheel and tire assembly. Adapted from* Automotive Engineering *(1982).*

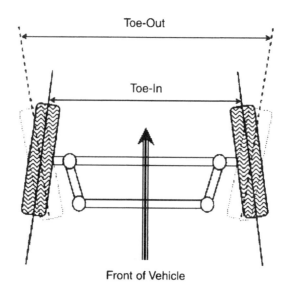

Figure 7.20. *Toe-in versus toe-out.*

For a rear-wheel-drive vehicle, the front wheels normally have a slight amount of toe-in. Figure 7.18 shows why this is true. When the vehicle begins to roll, rolling resistance produces a force through the tire contact patch perpendicular to the rolling axis. Due to the existence of the scrub radius, this force produces a torque around the steering axis that tends to cause the wheels to toe-out. The slight toe-in allows for this, and when rolling, the wheels align along the axis of the vehicle. Conversely, front-wheel-drive vehicles require slight toe-out. In this case, the tractive force of the front wheels produces a moment about the steering axis that tends to toe the wheels inward. In this case, proper toe-out absorbs this motion and allows the wheels to parallel the direction of motion of the vehicle.

7.4.4 Caster

Caster is the angle of the steering axis from the vertical as viewed from the side and is shown in Fig. 7.21. Positive caster is defined as the steering axis inclined toward the rear of the vehicle.

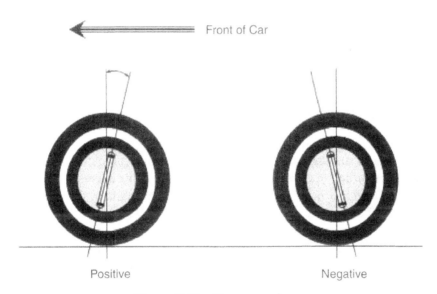

Figure 7.21. *Caster angle.*

With positive caster, the tire contact patch is aft of the intersection of the steering axis and the ground. This is a desirable feature for stability, as illustrated by Fig. 7.22.

When the wheel is turned, the cornering force acts perpendicular to the wheel axis and through the contact patch. This creates a torque about the steering axis that acts to center the wheel. Obviously, negative caster results in the opposite effect, and the wheel would tend to continue turning about the steering axis. The most common example of positive caster is a shopping cart. The wheels are free to turn around the steering axis, and when the cart is pushed straight ahead, the wheels self-align to the straight-ahead position.

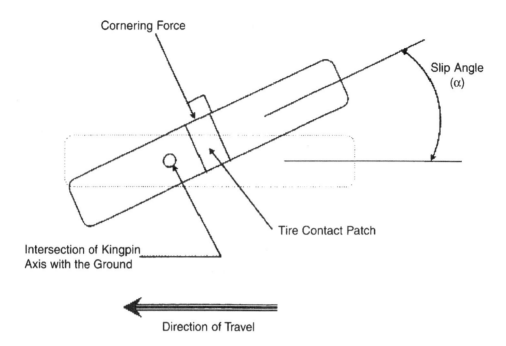

Figure 7.22. *Self-aligning torque generated by positive caster.*

7.4.5 Wheel Alignment

Over time, the alignment parameters previously discussed can begin to drift out of tolerance, as a result of wear of suspension components, impact with a curb, and so forth. The need for a wheel alignment usually is first indicated by uneven and/or excessive tire wear. Another signal is that the vehicle no longer tracks straight ahead without driver input. Table 7.1 summarizes the effects on the vehicle if these parameters are out of tolerance.

If any of these problems are noted on a vehicle, the only solution is to take the vehicle to a qualified alignment facility. Proper wheel alignment requires specialized measuring equipment because of the small angles involved. Adjustment of these parameters varies, depending on the vehicle. Most vehicles use shims to adjust camber and caster, whereas toe normally is adjusted via an adjustable tie rod. Regardless of the adjustment method, a wheel alignment is not within the capabilities of most do-it-yourself mechanics.

7.5 Steering Geometry Errors

As implied previously in this chapter, the action of the suspension system has a large effect on the handling qualities of a vehicle. The magnitude of these interactions goes beyond the scope of this book, and the interested reader will find them explained in sufficient detail in many good suspension or vehicle dynamics books (Gillespie, 1994; Bastow and Howard, 1993). However, two effects arising from the linkage relationship between the steering linkage and the control arms bear mention at this point, and these will be discussed here.

TABLE 7.1
EFFECT OF IMPROPER ALIGNMENT ON VEHICLE

Error in Alignment Parameter	Effect on Vehicle
Incorrect camber	• Tire wear • Ball joint or wheel bearing wear • Pulls to side with least negative camber
Caster too positive	• Hard steering • Wheel shimmy
Caster too negative	• High-speed instability/wandering
Unequal caster	• Pulls to side with least positive camber
Incorrect SAI	• Hard steering • Poor wheel return (to straight ahead) • Pulls to side with least inclination
Incorrect toe setting	• Tire wear • Vehicle pulls

It is a highly desirable feature for any car that as the front suspension deflects, there is no resultant input to the steering system. Figure 7.23 shows that the designer must put some thought into this.

In the datum condition, the instantaneous center for rotation of the axle/wheel assembly is found by projecting the lines of the upper and lower wishbones. Their intersection is the instantaneous center, I. For infinitesimal displacements of the wheel, the ball joint at the end of the rack can be anywhere along the line CI because this ensures that C will rotate about I.

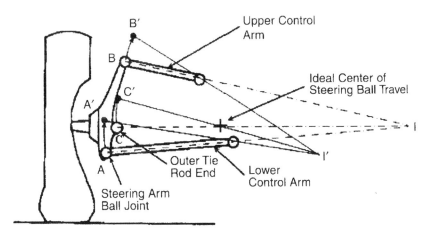

Figure 7.23. *Steering and suspension geometry for independent front suspension. Adapted from Gillespie (1994).*

If the wheel is displaced upward (jounce), point A moves to A', point B to B', and point C to C'. The instantaneous centers for the upper and lower links are found as indicated previously, by projecting back through the inner ball joints, and the new instantaneous center is I'. Likewise, the ball joint at the end of the rack can be anywhere along the line C'I'. If the end of the rack is located at the intersection of IC and I'C' (marked +), then there will be no bump-steer effect when the suspension is displaced from its datum position to the new position. A similar plot can be derived for the case of downward displacement (rebound) but has been omitted from Fig. 7.23 for clarity. In general, the location of the ball joint end will be correct only for two positions of suspension displacement when the vehicle is traveling in a straight line. The errors under other circumstances usually are small, compared to the compliance elsewhere in the steering/suspension system. Many computer-aided-design (CAD) programs have the capability of defining the ideal point for the tie rod end. Should this point be identified incorrectly, two effects can become apparent—toe change and roll steer.

One error is that the inboard joint could be too far inboard or outboard of the ideal point, as shown in Fig. 7.24.

Assuming Fig. 7.24 represents a rear view of the left wheel, the tie rod end is too far outboard and is not located on line C'I'. Suspension deflection upward will cause the wheel to turn to the left if the linkages are located aft of the kingpins. Likewise, this situation would cause the right wheel to turn to the right whenever the suspension deflects. Thus, the wheels will toe-out anytime the vehicle varies from its designed ride height. This makes it impossible to maintain the correct toe. If the tie rod ends are too far inboard, the opposite happens, and any suspension deflection causes the wheels to toe-in. The toe effects also will be apparent when the vehicle body rolls in a turn.

A second error arises when the tie rod end is above or below the ideal point, as shown in Fig. 7.25.

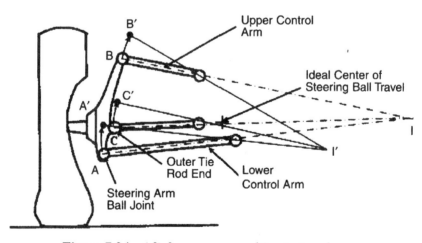

Figure 7.24. *A linkage error resulting in toe change. Adapted from Gillespie (1994).*

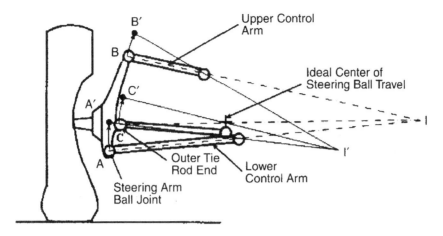

Figure 7.25. *A linkage error resulting in roll steer (Gillespie, 1994).*

Again, Fig. 7.25 depicts a rear view of the left wheel, and it is assumed that the steering linkages are aft of the kingpins. In this case, when the suspension goes into jounce, the wheel turns left, whereas rebound causes it to turn to the right. The opposite happens to the right wheel. That is, jounce induces a right turn, and rebound induces a left turn. Thus, when the vehicle body rolls in a turn, it causes both front wheels to turn in the same direction. For example, if the vehicle enters a right turn, the body rolls to the left and induces jounce on the left side and rebound on the right. This causes both wheels to turn left, out of the turn, and requires more steering input by the driver to maintain the turn. Thus, the effect adds understeer to the handling of the vehicle. If the tie rod end is above the ideal center, it adds oversteer to the response of the vehicle. Because this situation affects the vehicle handling response, it often is done intentionally to produce more or less understeer in a given vehicle. Thus, although it is classified as a geometry "error" here, it may be an intentional part of the vehicle design.

7.6 Front-Wheel-Drive Influences

Equation 7.25 indicates that the tractive forces of a front-wheel-drive vehicle would tend to induce oversteer. Anyone who has driven a front-wheel-drive car realizes that such vehicles generally do not exhibit such behavior and, in fact, usually exhibit a great deal of understeer. Although several mechanisms give rise to this, three will be examined here: torque about the driveline, loss of cornering stiffness due to tractive forces, and increase in aligning torque due to tractive forces.

7.6.1 Driveline Torque

The torque from the driveline through the CV joints on a front-wheel-drive car produces a torque about the steering axes, even when the wheel is going straight ahead. This torque can be calculated by reference to Fig. 7.26.

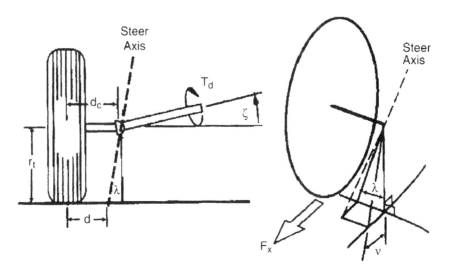

Figure 7.26. *Forces and moments on a front-wheel-drive wheel (Gillespie, 1994).*

If the moments caused by rolling resistance and the normal force on the tire are neglected, the moment about the steering axis is found to be (Gillespie, 1994)

$$M_{SA} = F_x d \cos v \cos \lambda + T_d \sin(\lambda + \zeta) \tag{7.26}$$

where

F_x = tractive force on the wheel
r = tire radius
d = distance from the center of the tire to the intersection of the steering axis and road
T_d = drive torque
λ = steering axis inclination (SAI) angle
v = caster angle
ζ = driveshaft angle

Because SAI and caster usually are small, it can be assumed that $\cos \approx 1$. Furthermore, the drive torque is equal to the tractive force times the tire radius. Thus, Eq. 7.26 can be written as

$$M_{SA} = F_x \left[d + r \sin(\lambda + \zeta) \right] \tag{7.27}$$

When the half shaft is horizontal (the normal case in straight-ahead driving), ζ is zero. Thus, the moment arms on each wheel are equal. The steering moments are equal and opposite, and no steering moment induces. However, when the vehicle enters a turn, the body will roll in the opposite direction. ζ on the outer half shaft will then decrease (become negative), while ζ on the inside will increase. This causes the steering moment on the inside wheel to become larger. This induces a moment in the steering system that opposes the turn and causes the wheels to turn out of the turn, thus adding understeer. The full effect of this influence varies, depending on the specific vehicle. However, the change in understeer from full-throttle to throttle-off generally is 1°/g (Gillespie, 1994). The effect can be reduced by minimizing body roll and stiffening the steering system.

The foregoing discussion assumes the half shafts are equal in length (equal ζ during straight-ahead driving). With many early front-wheel-drive cars with transversely mounted engines and offset transmissions, the half shafts were not equal in length. This gives rise to a phenomenon known as torque steer. *Autocar Magazine* performed road tests of several front-wheel-drive cars that had been modified with "go faster" kits, thus greatly increasing their horsepower. The tests found that most of the modified cars had a tendency to pull violently to the right when full engine power was applied in a low gear, and pulled equally violently to the left when the power was cut (Bastow and Howard, 1993). This was due entirely to the unequal-length half shafts, which produced different moment arms about the steering axes due to the variance in ζ. The solution to the problem is to either arrange for the transaxle to incorporate a layshaft with a bearing to provide equal-length half shafts, or to design the transmission so that it coincides with the wheel center height during acceleration. The latter option is difficult to achieve; therefore, the former solution is the one most often taken (Bastow and Howard, 1993). Because most front-wheel-drive cars were designed with fuel economy as the primary objective, torque steer was not a great problem, given the low power output of the engines. However, with the advent of high-powered front-wheel-drive vehicles such as the Dodge Intrepid, the designers must plan strategies for coping with torque steer.

7.6.2 Loss of Cornering Stiffness Due to Tractive Forces

It has long been known that tractive force on a tire reduces its cornering stiffness (Olley, 1946). This has been confirmed by recent testing, as shown in Fig. 7.27. This effect is more pronounced in bias-ply tires but is evident in radials. An increase in tractive force (due to throttle opening) reduces the cornering force of the tires. Thus, the wheels must be turned farther to negotiate the turn, which is understeer. The magnitude of this effect is estimated to be in the range 0–2°/g with a throttle change that produces an acceleration of 0.2g to a deceleration of 0.05g (Gillespie, 1994).

7.6.3 Increase in Aligning Torque Due to Tractive Forces

Aligning torque arises from the fact that the cornering forces produced by a tire are developed at a point aft of the center of the contact patch. This distance is known as pneumatic trail (Gillespie, 1994). Referring to Fig. 7.27, throttle application induces an increase in aligning torque. This tends to turn the wheels out of the turn, which requires the driver to increase the steering angle (understeer). This effect adds approximately 0.5–1°/g (Gillespie, 1994).

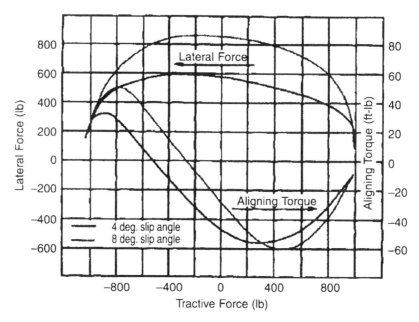

Figure 7.27. *Tractive force effects on lateral force and aligning torque (Gillespie, 1994).*

The combined effect of these three influences overrides the tendency toward oversteer as predicted by Eq. 7.25, and results in the understeer tendency of a front-wheel-drive car.

7.7 Four-Wheel Steering

In the early days of the automobile, someone occasionally would roll out a car with rear-wheel steering. Although a novel concept, it has long been known that such an arrangement is inherently unstable (Olley, 1946). If a car uses only the rear wheels to enter a turn, the initial motion must be away from the desired direction of the turn. This will establish the nose into the desired turn, but when established, the rear wheels must then be turned into the turn to generate the proper slip angles to maintain it. If the rear wheels reverse direction, the steering wheel must reverse also, which presents challenges to the driver. Furthermore, if the driver enters the turn too fast and is on the verge of a skid, nothing can be done to prevent it. To increase the turn radius, the rear wheels must be straightened. However, this increases the rear slip angle and induces the skid that the driver is attempting to avoid (Olley, 1946).

Despite the fact that rear-wheel steering has negative consequences for vehicle stability, the idea of four-wheel steering has recently come to the forefront as a means of enhancing maneuverability and stability. The analysis of a car with four-wheel steering must be undertaken in two separate regimes, namely, low-speed turns and high-speed turns. To maintain simplicity, the ensuing analysis will proceed with the bicycle model used previously (which assumes an average value for the inner and outer wheel steering angles).

7.7.1 Low-Speed Turns

The analysis of a low-speed turn can be done with reference to Fig. 7.28. In the low-speed case, the rear wheels are turned in the opposite direction as the front wheels. Normal practice is to turn the rear wheels in some proportion to the front wheels, or in other words,

$$\delta_r = \xi \delta_f \tag{7.28}$$

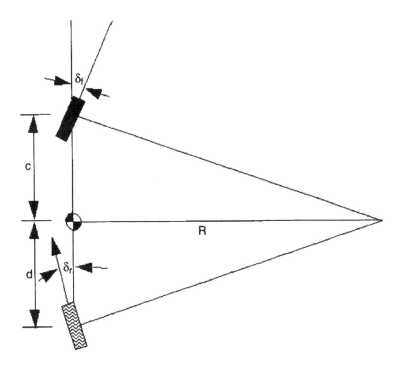

Figure 7.28. *Low-speed turning with four-wheel steering.*

Assuming small angles $(\tan \delta \approx \delta)$ and making use of Eq. 7.28 gives

$$\delta_f + \delta_r = \delta_f + \xi \delta_f = \delta_f (1+\xi) = \frac{L}{R} \tag{7.29}$$

Thus, the turn radius is given by (Gillespie, 1994)

$$R = \frac{L}{\delta_f (1+\xi)} \tag{7.30}$$

Recalling that for normal Ackerman front-wheel steering, the turn radius is given by $R = L/\delta_f$, Eq. 7.30 shows the reduction in turn radius provided by turning the rear wheels. For example, if 100% rear-wheel steering is used (i.e., the front and rear wheels are turned equally), the turn radius is reduced by half.

7.7.2 High-Speed Turns

The rear-wheel steering used in the low-speed case would be unmanageable in the high-speed case because the outward motion of the rear would produce excessive oversteer. In the high-speed case, the rear wheels are turned in the same direction as the front. This leads to the situation shown in Fig. 7.29.

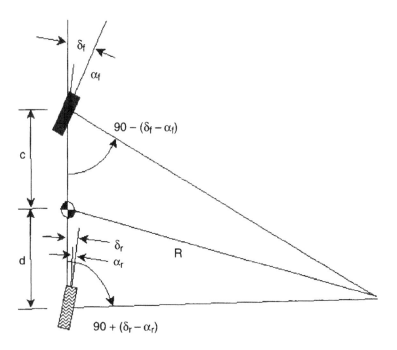

Figure 7.29. *Four-wheel steering during a high-speed turn.*

The analysis proceeds exactly as described in Section 7.3.2. Again assuming that the rear-wheel steering is proportional to the front, the resulting equation for the turn of the front wheels is

$$\delta_f = \frac{1}{1-\xi}\left\{57.3\frac{L}{R} + \left[\frac{W_f}{C_{\alpha_f}} - \frac{W_r}{C_{\alpha_r}}\right]\frac{V^2}{gR}\right\} \qquad (7.31)$$

Equation 7.31 reveals what is known intuitively, namely, that turning the rear wheels in phase with the front increases understeer. Also note that it is not desirable to have 100% rear-wheel steering in this case. If that were applied, the denominator of the leading term would go to zero, implying instability. What this implies physically is that when the front and rear wheels are turned equally, the car will tend to track tangentially off the turn. To hold the turn, the front wheels must be turned an infinite amount. With 100% steer, the rear wheels would follow suit, and in the limit, the car would translate sideways. This, of course, assumes that the wheels can generate the requisite slip angles, which is impossible.

7.7.3 Implementation of Four-Wheel Steering

Rear-wheel steering may be implemented passively or actively. The difference is the existence of some sort of control logic in determining how much (and in what direction) to turn the rear wheels. The simplest form of passive rear-wheel steering involves using linkages that are driven by the body motion to turn the rear wheels. Figure 7.30, which is not drawn to scale, schematically shows such a system.

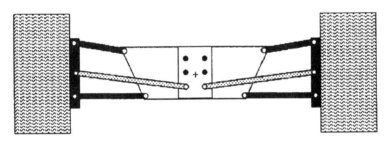

Figure 7.30. *Passive rear-wheel steering system (view from the rear of the car).*

In the system depicted in Fig. 7.30, the control rods (which are mounted behind the rear axle) may be connected to any of a series of holes in the body frame. The choice of connection point determines whether and in what direction the rear wheels are turned. In the configuration shown, body roll during a turn (about point "+") will cause lateral forces in the control rods. For example, a right turn, with corresponding body roll to the left, places the left control rod in tension and the right control rod in compression. This turns both rear wheels to the left, or out of phase with the front wheels. Conversely, if the control rods are attached at the top point, body roll causes the rear wheels to turn in phase with the front. The central connection point neutralizes the system (i.e., no rear-wheel steering). Such a system was implemented by the U.S. Air Force Academy on its 1999 Formula SAE car. The system worked as advertised, but the excessive oversteer resulting when the system was configured as in Fig. 7.30 made the car uncontrollable in the hands of most drivers, even at relatively slow speeds. In fact, the car generally spun out of control with a little as 0.3 lateral g's. Porsche developed a passive rear-wheel system in its Weissach axle (Bastow and Howard, 1993). The primary function of this system was to utilize the bushing compliance to induce understeer while braking in a turn.

However, passenger cars are increasingly moving toward active control of the rear-wheel steering. The principle advantage of active control is that the rear-wheel steering can be modulated for the given conditions. In other words, rear-wheel steering can be in or out of phase with the front, depending on the driver demand, speed and radius of turn, and so forth. One improvement over the Porsche system was developed by Nissan in its HICAS (High Capacity Active Control Suspension) system (Bastow and Howard, 1993), which is depicted in Fig. 7.31. The system incorporates steering linkages and knuckles, just as the front hubs. The system senses high cornering forces through steering wheel hydraulics. In the event that the front cornering power nears the limit, as determined by the reduced steering effort, the HICAS system reduces the rear-wheel slip angle through the actuation of hydraulic rams. The system can alter the rear-wheel slip angle by up to $\pm 0.5°$. Although this seems minute, the resulting change is sufficient to keep the car in balance through the turn (Bastow and Howard, 1993).

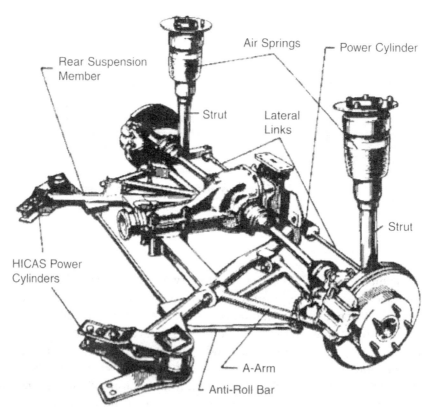

Figure 7.31. *The Nissan HICAS steering system (Bastow and Howard, 1993).*

In 1991, BMW introduced a completely active four-wheel steering system known as ARK (Active Rear Axle Kinematics), as shown in Fig. 7.32 (Bastow and Howard, 1993). The system uses a dedicated processor that receives inputs from the speedometer, steering wheel angle sensor, and front wheels (through the ABS sensors). Tests of the system during a violent

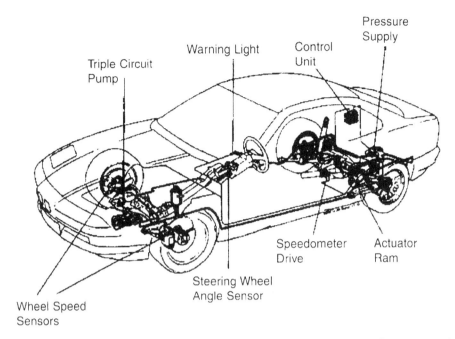

Figure 7.32. *The BMW ARK rear-wheel steering system (Bastow and Howard, 1993)*

lane-change maneuver showed that with ARK switched off, the car tended to oscillate two to three cycles before settling after the lane change. However, with ARK engaged, the vehicle remained flat throughout the maneuver and did not exhibit any heading oscillations after the maneuver (Bastow and Howard, 1993).

Given the results of the BMW and Nissan systems, the question to be asked is whether four-wheel steering is worth the added cost and complexity. A study by Lee (1995) attempted to answer this question. In the study, the performance of a vehicle during a lane-change maneuver was compared under three conditions: (1) with conventional two-wheel steering, (2) with four-wheel active steering control using an open-loop control algorithm, and (3) with four-wheel active steering control using a closed-loop control algorithm. The open-loop algorithm (4WSN in Fig. 7.33) is based on a vehicle model and computes a speed-dependent ratio between the front and rear wheels that attempts to achieve zero steady-state side velocity (Lee, 1995). The closed-loop algorithm (4WSY in Fig. 7.33) uses feed-through of the front-wheel steering command and feedback of the vehicle yaw rate (Lee, 1995). Figure 7.33 shows a comparison of vehicle yaw rate.

As Fig. 7.33 shows, the yaw rate with both four-wheel-steering algorithms is better damped than the two-wheel-steering case. However, the use of 4WS had little or no effect on the lateral track of the vehicle. Nevertheless, many road tests of four-wheel-steering vehicles result in drivers claiming that the car "feels better" (Gillespie, 1994). This is probably due to the increased yaw damping produced by the four-wheel-steering system. Although the effect on the maneuvering of the vehicle is minimal, the driver senses the lack of oscillation and interprets this as better handling. Thus, the question of cost and complexity comes down to the desire of the individual driver, as well as the depth of the driver's pocket.

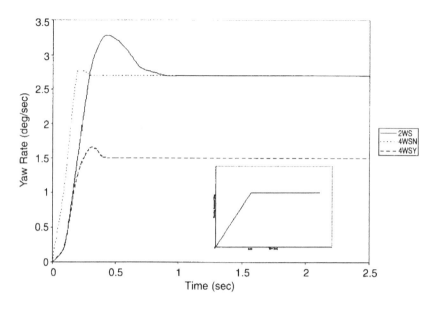

Figure 7.33. *Yaw rate versus time, 120-km/h (75-mph) lane change. Data taken from Lee (1995).*

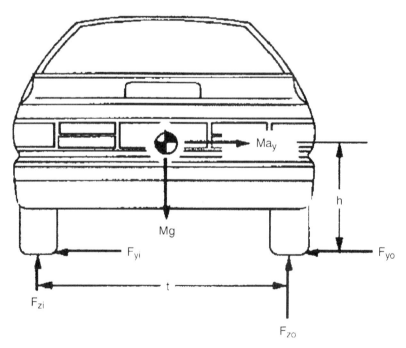

Figure 7.34. *Quasi-static rollover model (Gillespie, 1994).*

7.8 Vehicle Rollover

One aspect of cornering behavior that can be terrifying for a driver is vehicle rollover. Rollover is defined as the vehicle rotating 90° or more about its longitudinal axis (Gillespie, 1994) and can be caused by many factors. It can occur on a level surface if the tires can generate sufficient cornering force that the vehicle rolls before it slips. Any cross slope of the road also will excite (or inhibit) rollover. The most frequent cause of rollover is a skidding vehicle coming into contact with a surface irregularity such as a curb, dirt shoulder, or similar situation. The process is influenced by a large number of complex phenomena, and a detailed analysis goes beyond the scope of an introductory text. Nevertheless, some simple models exist that can aid one's understanding of vehicle rollover.

7.8.1 Quasi-Static Model

The simplest model for analyzing rollover assumes a rigid vehicle, as shown in Figure 7.34.

The analysis will assume the vehicle is on a level road, although cross slope may be accounted for in the model (Gillespie, 1994). At the point of incipient tip, the sum of the moments about the outer tire/road contact point must be zero

$$Ma_y h + F_{zi} t - Mg \frac{t}{2} = 0 \tag{7.32}$$

Furthermore, at incipient tip, the normal force on the inner wheel becomes zero. Thus, Eq. 7.32 may be solved for the lateral acceleration required to roll the vehicle, commonly known as the "rollover threshold"

$$\frac{a_y}{g} = \frac{t}{2h} \tag{7.33}$$

Although this rollover threshold is a simple first-order estimate, it generally requires lateral accelerations that are beyond the cornering capabilities of the tires. In other words, most vehicles will skid before reaching this rollover threshold.

7.8.2 Quasi-Static Rollover with Suspension

The static model neglects the compliance of the suspension and tires; hence, it overestimates the rollover threshold (Bernard et al., 1989). One addition to the model is to assume the body rolls about its "roll center," which is an imaginary point about which body roll occurs and through which suspension forces are transmitted to the sprung mass. Figure 7.35 shows the model. The analysis assumes the car is established in a steady turn on a level road, and it neglects any rolling of the axle.

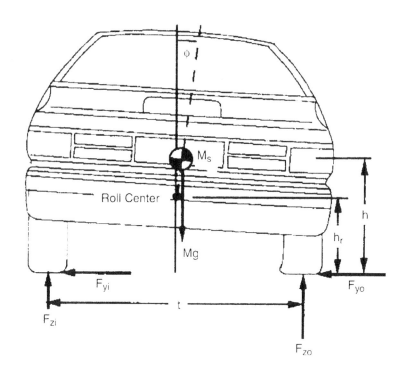

Figure 7.35. *Quasi-static model with suspension (Gillespie, 1994).*

Again, moments are summed about the outer wheel, and the normal force on the inner wheel goes to zero at incipient tip. The rollover threshold is given by

$$\frac{a_y}{g} = \frac{t}{2h} - \phi\left(1 - \frac{h_r}{h}\right) \tag{7.34}$$

Equation 7.34 is problematic because the body roll angle, ϕ, is a function of the lateral acceleration. Gillespie suggests replacing the body angle with the product of the roll rate, R_ϕ (rad/g), times the lateral acceleration. When this substitution is made, the roll threshold becomes (Gillespie, 1994)

$$\frac{a_y}{g} = \frac{t}{2h} \frac{1}{\left[1 + R_\phi\left(1 - \frac{h_r}{h}\right)\right]} \tag{7.35}$$

Typical values of roll rate and h_r/h for passenger cars are 0.1 rad/g and 0.5 (Gillespie, 1994). The second term then evaluates to 0.95, so the rollover threshold using this model is 5% below the "t over 2h" value predicted by the rigid car model.

7.8.3 Roll Model

The quasi-static models described here are valid only if the rate of change of lateral acceleration is much less that the roll response of the vehicle. To examine the effect of the vehicle roll response, the car can be modeled as a single-degree-of-freedom torsional system. To simplify the analysis, the model will use a composite stiffness and damping provided by the springs, shocks (also called dampers), and tires. The model is shown in Fig. 7.36, with the primary change being the addition of a mass moment of inertia about the longitudinal axis (I_{xx}).

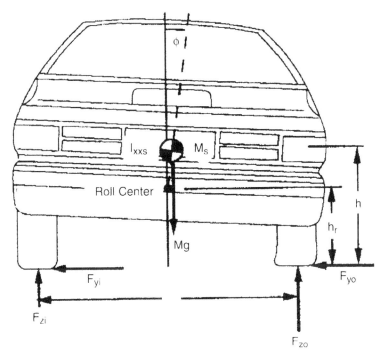

Figure 7.36. *Transient roll model (Gillespie, 1994).*

This model can be used to estimate the vehicle response to a sudden input of lateral acceleration. The differential equations of motion can be derived and solved using standard vibration techniques. The first case to be examined is the response to a step input, such as might occur when sliding off the road onto a dirt surface. The body roll angle as a function of time (t = 0 at the onset of the lateral acceleration) in this case is given by (Thompson, 1988)

$$\phi(t) = \frac{M_o}{K_t}\left[1 - \frac{e^{-\zeta\omega_n t}}{\sqrt{1-\zeta^2}}\cos(\omega_d t - \theta)\right] \quad (7.36)$$

where

M_o = static moment due to steady-state lateral acceleration (N-m)

K_t = composite torsional stiffness (N-m/rad)

ω_n = undamped natural frequency = $\sqrt{\dfrac{K_t}{I_{xx}}}$ (rad/sec)

ζ = damping ratio (% of critical)

ω_d = damped natural frequency = $\omega_n \sqrt{1-\zeta^2}$

θ = phase angle = $\tan^{-1}\left(\dfrac{\zeta}{\sqrt{1-\zeta^2}}\right)$

The vehicle response to the step input is characteristic of a damped oscillatory system and is shown in Fig. 7.37.

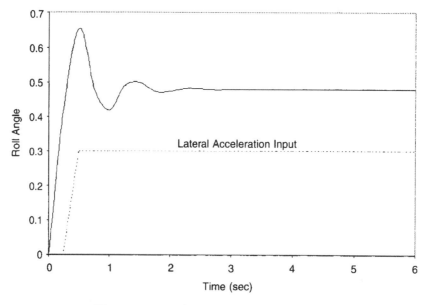

Figure 7.37. *The response to a step input.*

Assuming the vehicle is not critically (or over-) damped, the initial response "overshoots" the equilibrium position due to the roll inertia. Because of this overshoot, wheel liftoff can occur below the rollover threshold predicted by the quasi-static model. The rollover threshold is influenced heavily by the damping ratio, as shown in Fig. 7.38.

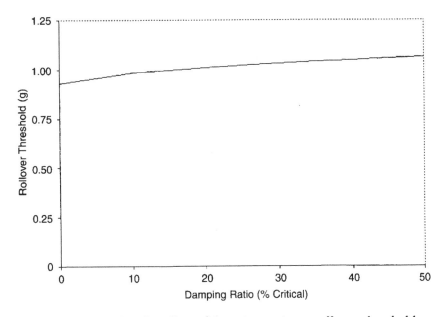

Figure 7.38. *The effect of damping ratio on rollover threshold.*

The vehicle roll response also may be forced by harmonic loading, which can occur during maneuvers such as lane changes. In fact, the infamous failure by the Mercedes-Benz A-Class in the Swedish "moose test" is a classic example of this type of rollover mode. Solving the equations of motion for forced harmonic vibration gives the rollover threshold as

$$\phi(t) = \frac{M_o}{I_{xx}\omega_n^2}\left[\frac{1}{\sqrt{\left[1-\left(\frac{\omega}{\omega_n}\right)^2\right]^2 + 4\zeta^2\left(\frac{\omega}{\omega_n}\right)^2}}\sin(\omega_d t - \theta)\right] \quad (7.37)$$

where the variables are as described for Eq. 7.36, except the phase angle, which is

$$\theta = \tan^{-1}\frac{2\zeta\left(\frac{\omega}{\omega_n}\right)}{1-\left(\frac{\omega}{\omega_n}\right)^2}$$

In this case, the vehicle roll response is most influenced by the input frequency, as shown in Fig. 7.39.

As shown in Fig. 7.39, the rollover threshold is lowest when the input frequency is at the resonant frequency of the vehicle. Experience shows that a lane change has a driving frequency of approximately 0.5 Hz (Gillespie, 1994). This is capable of exciting a resonant response in heavy trucks and leads to rollover. For cars, the resonant frequency is somewhat higher, although as the A-Class test shows, it can induce rollover.

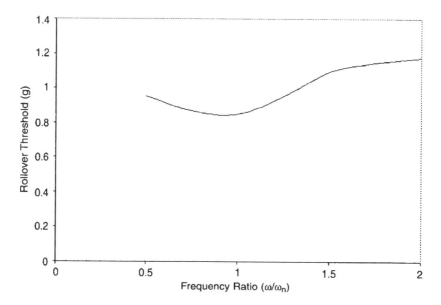

Figure 7.39. *The effect of frequency ratio on rollover threshold.*

Table 7.2 summarizes the rollover thresholds predicted by the models described here. The calculations were performed on a "typical" small car, with values either estimated or obtained from public sources. Although the actual thresholds thus are susceptible to error, the point of the table is to highlight the differences in predicted thresholds, depending on the model used. As the table shows, predicting the rollover threshold is not an easy task, and more complex

**TABLE 7.2
COMPARISON OF ROLLOVER THRESHOLD CALCULATIONS**

Model	Rollover Threshold (g)	% Reduction from Static
Quasi-Static	1.25	—
Quasi-Static Suspended	1.19	4.80
Step Input	0.92	26.4
Harmonic Forcing	0.80	36.0

models that incorporate yaw and roll modes exist (Gillespie, 1994). Also, the "tripping" rollover (occurring when a vehicle strikes a curb) is extremely difficult to model and analyze directly. Recent studies also suggest that the roll behavior of a vehicle is influenced strongly by its pitch response (Kawagoe et al., 1997).

With the increased popularity of sport utility vehicles (SUVs) and vans in the United States, rollover prevention has taken on an increased emphasis. Both of these types of vehicles have a higher center of gravity, which tends to increase the likelihood of rollover. A study by Wielenga and Chace (2000) examined an actual rollover accident involving a 1995 Chevrolet Astro van. The accident was caused by a four-steer maneuver that led to an excessive yaw rate, culminating in a rollover. The accident was reconstructed in a computer simulation, using the same steering inputs as the driver. The rollover prevention strategy adopted was anti-rollover braking (ARB). The ARB system engaged the front brakes when lateral acceleration exceeded 0.55g, with the brakes fully engaged by 0.6g. The braking force reduced the cornering power of the front tires, thus inducing understeer and reducing the yaw rate. The simulation indicated that the addition of the ARB control strategy would have prevented rollover in this case.

A different control strategy specifically targeted toward an off-road vehicle was examined in a study by Everett et al. (2000). The system involved replacing the stock anti-roll bars on a production SUV with a two-piece anti-roll bar that was hydraulically actuated. Hydraulic pressure was provided by an engine-driven pump, and the system induced torque in the bars that opposed body roll. The system provided good control in response to moderate steering inputs but could not eliminate body roll in response to rapid steering inputs. Control of body roll when the vehicle was fully loaded also was degraded. Nevertheless, the system gave promising results, and further work in refining the control algorithm was projected in the study.

7.9 Problems

1. The Ford Ranger Supercab has a 10-ft wheelbase and rides on P245/75R16 tires (same on the front and rear) that provide a cornering stiffness of 200 lb/°. The truck weighs 3650 lb and has a front/rear weight distribution of 55/45 (%).

   a. What is the understeer gradient for this truck? (Answer: 1.825 °/g.)

   b. What is the characteristic speed of the truck? (Answer: 68.5 mph.)

   c. If the owner puts bias-ply tires on the rear (cornering stiffness of 150 lb/°), what is the new understeer gradient? What has happened to the handling of this truck? Calculate the appropriate characteristic/critical speed. (Answer: –0.9125 °/g; the truck exhibits oversteer; critical speed = 96.9 mph.)

   d. Suppose the owner decides to put a half ton (1000 lb) of wood in the back of the truck. Assuming that all of the weight is applied to the rear axle and that the truck has the original radial tires on the rear, what is the critical speed for the truck? Neglect tractive forces in your analysis. Where should the center of gravity of the wood be located (distance forward of the rear axle) to ensure the truck does not exhibit oversteer? (Answer: 52.0 mph; 3.175 ft forward of the rear axle.)

2. Assume a driver has a car that normally exhibits understeer (tractive forces neglected). If this driver enters a turn too rapidly and applies the brakes, what happens to the handling characteristics of the car? Assume the brakes do not lock up, and consider that weight transfers to the front axle during braking. Use Eq. 7.25 to explain your answer.

3. You are the lead engineer for a new SUV that has failed the "moose test." Discuss three things you could change on the vehicle to enable it to pass the test. (Hint: What could you do to increase the natural frequency of the vehicle in roll?)

CHAPTER 8

Suspensions

8.1 Introduction

The suspension system comprises the interface between the vehicle body/frame and the road surface. (This statement assumes that the wheels and tires comprise part of the suspension system, which they indeed do.) Most people consider that the sole function of the suspension is to provide a comfortable ride. Although this is true, the system can be said to have three primary functions:

1. **Isolate passengers and cargo from vibration and shock.** It is desirable to make the passengers as comfortable as possible; thus, the suspension system must be able to absorb shocks and dampen vibration caused by irregularities in the road surface.

2. **Improve mobility.** The suspension provides clearance between the road and the bottom of the vehicle. It also provides lateral and longitudinal stability and resists chassis roll.

3. **Provide for vehicle control.** The suspension reacts to tire forces including acceleration, braking, and steering and forces. Furthermore, the suspension system is tasked to maintain the proper steer and camber angles relative to the road surface, as well as to keep all four tires in contact with the road while maneuvering.

The analysis of vehicle suspensions and their dynamic response is an extremely complicated task. Because this is an introductory text, many simplifying assumptions will be made in the analysis of suspension systems. Although the models introduced are simple, they nevertheless illustrate several important characteristics and design requirements for suspensions.

This chapter will begin with simplified vibrational analysis in both one and two degrees of freedom. Next, the primary components of the suspension will be discussed, with representative examples of current suspension systems. The chapter concludes with a discussion of the effect of suspension design on the dynamics of the vehicle.

8.2 Perception of Ride

Passenger opinion regarding what constitutes good ride quality obviously is extremely subjective. What one person considers the optimum ride may be completely unacceptable to another. The person who prefers sports cars will be appalled by the handling of a large luxury vehicle, whereas the owner of the luxury vehicle will be quite dissatisfied with the ride of a sports car.

Other factors come into play when people evaluate the ride quality of a vehicle. Certainly, the acoustic quality is a factor, and although not a direct result of the suspension, people object to noises, rattles, and squeaks in their vehicles. The "feel" of the seats is another important consideration and has an impact on the level of force or vibration transmitted to the occupant's body. The climate control system, while not at all influenced by the suspension design, influences perception of ride, too. If a person is uncomfortable because of the interior temperature, his or her subjective evaluation of ride quality will be affected. Thus, one of the challenges facing the suspension engineer is to take highly subjective evaluations and convert them into numerical standards.

Some debate exists as to what quality of motion people find objectionable. Displacement is not an issue. If it were, climbing steps does not in itself produce discomfort, although the effort required may. Likewise, velocity is not uncomfortable, as evidenced by pilots who operate aircraft at speeds greater than Mach 1, with no ill effects. Constant acceleration is felt as a constant force. While pulling a constant 8–9g in an aircraft is less than comfortable, the levels of acceleration produced by most passenger cars do not induce great discomfort.

However, the rate of change of acceleration, or the jerk, can produce discomfort. A parachutist feels discomfort at the moment of opening his or her parachute due to the shock, although it is usually mitigated by a profound sense of relief as the parachute inflates. But the jerk is not the only element that produces discomfort. The frequency of acceleration and its direction influence comfort. A car that pitches drastically when encountering a bump is seen as less comfortable than one that bounces in a more flat attitude, even if both motions continue for similar amounts of time.

A substantial body of literature is devoted to quantifying ride quality and human perception of ride. Studies and data have been collected by bodies such as the Society of Automotive Engineers (SAE) and the International Standards Organization (ISO), as well as by individual researchers. Gillespie (1994) provides a succinct overview of the literature. Although the sources are numerous, Gillespie concludes that there are no accepted standards for judging ride quality due to variables such as seat position, single versus multiple frequency inputs, multi-direction input, duration of exposure, and audible or ocular inputs.

The bottom line is that all of the research and comfort curves are a starting point for the suspension engineer. There is no substitute for the subjective evaluation provided by a road test. We could conclude that the suspension engineer should eliminate all vibration from the car, but this tends to be an infinite problem. As surely as one vibration is removed, the occupants become aware of another, more subtle vibration. As a result, suspension engineers appear to have solid job security for the foreseeable future.

8.3 Basic Vibrational Analysis

8.3.1 Single-Degree-of-Freedom Model (Quarter Car Model)

A vehicle consists of a multiple spring-mass-damper system that in reality has six degrees of freedom. Although the effective transverse and longitudinal stiffnesses of the suspension are much greater than the vertical stiffness, lateral and transverse compliance cannot generally be disregarded and may have a large impact on the vehicle dynamics. As a first simplifying assumption, the vehicle and suspension can be modeled in two dimensions, as shown in Fig. 8.1. The main mass consists of the vehicle itself and is comprised of all components that are supported by the springs. Hence, it is known as the sprung mass. Several components, such as axles, hubs, possibly the differential, and so forth, are not supported by the springs and are called the unsprung mass. The tires, being made of rubber, have inherent stiffness and damping and hence are modeled as a separate spring-damper system.

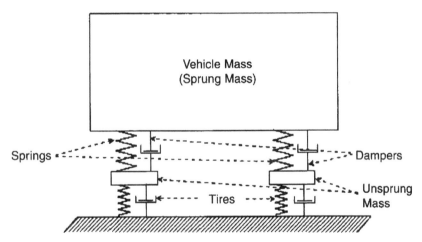

Figure 8.1. *Schematic representation of a vehicle as a spring-mass system.*

The primary motion of the vehicle mass is in the vertical direction. However, because of the separate springs and dampers at the front and rear, rotational motion usually results. At this point, it is instructive to examine a simple one-degree-of-freedom system to outline the basics of suspension analysis. In this case, it will be assumed that the tire stiffness is infinite, and the undamped motion of one spring will be examined. Figure 8.2 shows this model.

The equation of motion for the system can be obtained by applying Newton's Law and, in the case of unforced (free) vibration, is

$$\ddot{x} + \frac{k}{m}x = 0 \tag{8.1}$$

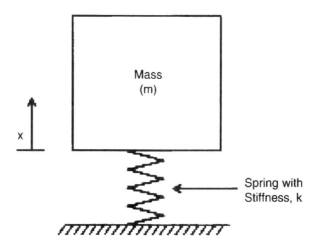

Figure 8.2. *Simple spring-mass system.*

The general solution for this linear differential equation is

$$x = A\cos(\omega_n t) + B\sin(\omega_n t) \tag{8.2}$$

where A and B are constants that depend on initial conditions, and ω_n is the natural frequency of the system and is defined as

$$\omega_n = \sqrt{\frac{k}{m}} \text{ (rad/sec)} \tag{8.3}$$

or

$$f_n = \frac{1}{2\pi}\sqrt{\frac{k}{m}} \text{ Hz} \tag{8.4}$$

Of course, this system, once disturbed from its datum position, would continue to oscillate at its natural frequency indefinitely. Although all real springs have some internal damping, a vehicle requires a more positive source of damping. Thus, the vehicle contains dampers. (In the United States, dampers are known as shock absorbers, although the name is misleading.) Figure 8.3 shows such a model.

Most automotive dampers can be modeled with sufficient accuracy by assuming they are viscous dampers. In other words, the damping force is proportional to the velocity of the displacements, or

$$F_d = c\dot{x} \tag{8.5}$$

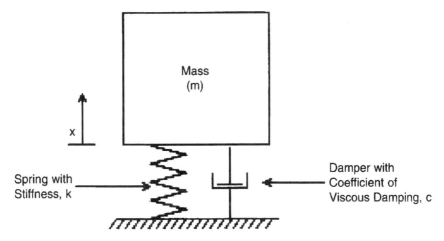

Figure 8.3. *Simple spring-mass-damper system.*

where c is the coefficient of viscous damping. In this case, the unforced (homogenous) equation of motion becomes

$$\ddot{x} + \frac{c}{m}\dot{x} + \frac{k}{m}x = 0 \tag{8.6}$$

and the general solution to Eq. 8.6 is

$$x = e^{-(c/2m)t}\left(Ae^{\sqrt{(c/2m)^2 - k/m}\,t} + Be^{-\sqrt{(c/2m)^2 - k/m}\,t}\right) \tag{8.7}$$

The first term in Eq. 8.7 is an exponentially decaying function of time. The terms in parentheses are dependent on whether the term under the radical is greater than, less than, or equal to zero. If the damping term $(c/2m)^2$ is greater than k/m, the terms in the radical are real numbers, and no oscillation is possible. If the damping term is less than k/m, the exponent becomes an imaginary number indicating oscillatory motion. The limiting case is when the damping term equals k/m. This case is known as critical damping, and the value of the critical damping coefficient is then

$$c_c = 2m\sqrt{\frac{k}{m}} = 2m\omega_n \tag{8.8}$$

Now, any damping condition can be expressed in relation to the critical damping. Thus, the damping ratio, ζ, is defined as

$$\zeta = \frac{c}{c_c} \tag{8.9}$$

A vehicle is normally underdamped ($\zeta < 1.0$); thus, Eq. 8.7 can be written as

$$x = Xe^{-\zeta \omega_n t} \sin\left(\sqrt{1-\zeta^2}\,\omega_n t + \phi\right) \tag{8.10}$$

where X and ϕ are arbitrary constants determined from the initial conditions. This equation indicates that the damped frequency of oscillation is modified in the presence of damping, and the damped natural frequency is given by

$$\omega_d = \sqrt{1-\zeta^2}\,\omega_n \tag{8.11}$$

The response to excitation is an exponentially decreasing sine wave, as depicted in Fig. 8.4.

In the case of forced vibration, the amplitude of the displacement is dependent on both the damping ratio and the excitation frequency. If harmonic forcing is assumed, Eq. 8.6 becomes

$$m\ddot{x} + c\dot{x} + kx = F_o \sin(\omega t) \tag{8.12}$$

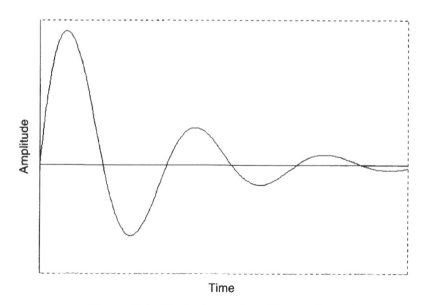

Figure 8.4. *Underdamped oscillation, $\zeta < 1.0$.*

where ω is the excitation frequency. The solution to Eq. 8.12 consists of a complementary function, which is the solution to the homogenous equation, and a particular solution (Thomson, 1988). The particular solution reduces to a steady-state oscillation at the excitation frequency, ω. The displacement can be nondimensionalized with respect to the static displacement (F_o/k), so that

$$\frac{Xk}{F_o} = \frac{1}{\sqrt{\left[1-\left(\frac{\omega}{\omega_n}\right)^2\right]^2 + \left[2\zeta\left(\frac{\omega}{\omega_n}\right)\right]^2}} \qquad (8.13)$$

Figure 8.5 shows a plot of Eq. 8.13 and illustrates the effect of damping on the nondimensional amplitude. As expected, displacements are largest when the damping is light and the system is excited near (or at) its natural frequency.

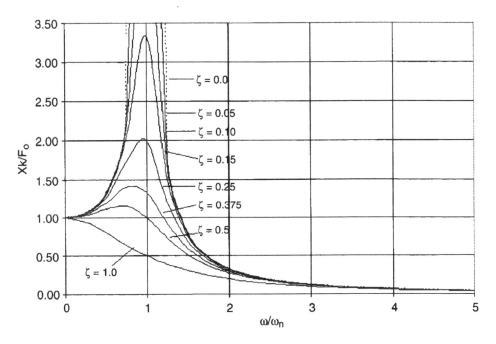

Figure 8.5. *Plot of Eq. 8.13.*

8.3.2 Two-Degrees-of-Freedom Model (Quarter Car Model)

In a vehicle, the excitation of the vehicle spring-mass system is provided by the motion of the tire/unsprung mass and can be analyzed by the same techniques used to analyze support motion. The details of such an analysis are contained in vibration texts (Thomson, 1988), and only the highlights will be discussed here. Referring to Fig. 8.1, and isolating one spring-mass-damper system, the displacement of the vehicle body will be defined by x, whereas the displacement of the unsprung mass will be designated as y, as shown in Fig. 8.6.

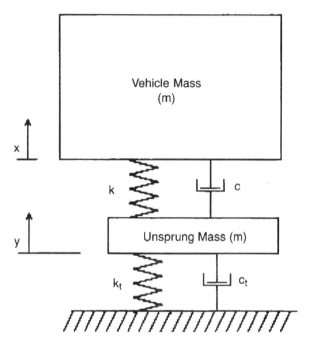

Figure 8.6. *Vehicle excited by the motion of the unsprung mass.*

The equation of motion for the vehicle mass is now

$$M\ddot{x} = -k(x-y) - c(\dot{x}-\dot{y}) \qquad (8.14)$$

Letting z = x − y, Eq. 8.14 can be written as

$$M\ddot{z} + c\dot{z} + kz = -m\ddot{y} \qquad (8.15)$$

At this point, it is most illustrative to assume that the motion of the unsprung mass is harmonic, which is not a bad assumption given that the tire and unsprung mass constitute a damped vibratory system. Before proceeding, the concept of transmissibility must be introduced. Transmissibility is defined as the ratio of the transmitted force to the ratio of the exciting force. Because in this case the exciting force is provided by the unsprung mass and tire, and as such is proportional to the displacement of the unsprung mass, the transmissibility is given by (Thomson, 1988)

$$TR = \left|\frac{F_t}{F_o}\right| = \left|\frac{X}{Y}\right| = \sqrt{\frac{1 + \left(2\zeta\frac{\omega}{\omega_n}\right)^2}{\left[1 - \left(\frac{\omega}{\omega_n}\right)^2\right]^2 + \left[2\zeta\frac{\omega}{\omega_n}\right]^2}} \qquad (8.16)$$

where

ω = natural frequency of the unsprung mass system
ω_n = natural frequency of the vehicle mass system

Figure 8.7 shows a plot of Eq. 8.16.

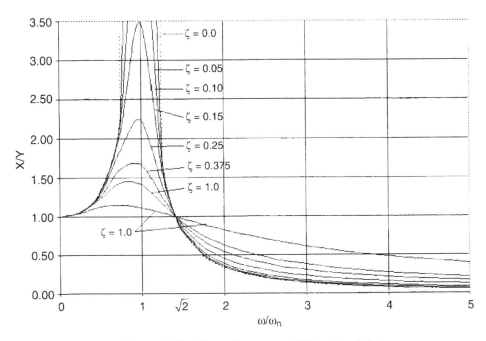

Figure 8.7. *Plot of transmissibility, Eq. 8.16.*

As mentioned in the introduction to this chapter, one of the functions of the suspension is to isolate passengers and cargo from vibration and shock. As shown in Fig. 8.7, as long as the frequency ratio $\left(\frac{\omega}{\omega_n}\right)$ is above $\sqrt{2}$, any displacement of the vehicle mass will be less than that of the unsprung mass. In fact, the natural frequency of the unsprung mass system should be much greater than that of the vehicle mass/suspension system. There are, of course, two ways to do this, recalling that the natural frequency of a spring-mass system is given by Eq. 8.3.

First, ensure that the stiffness of the tire is higher than that of the suspension springs. This usually is not an issue. The second way, and one of great importance to the vehicle designer, is to keep the unsprung mass as small as possible. As will be shown later, some suspension systems have a significant sprung mass, and such vehicles have a "bouncy" ride, the classic example being an unloaded dump truck.

8.3.3 Two-Degrees-of-Freedom Model (Half Car Model)

The quarter car model described in the previous sections is useful for illustrating basic concepts of suspension analysis. The next level of complexity in modeling the suspension is to develop a two-degrees-of-freedom model, as shown in Fig. 8.8. This model allows motion in bounce (x) and pitch (θ) directions, with both motions defined relative to the vehicle center of mass. The stiffness and damping are represented as an equivalent stiffness and damping. In other words, the stiffness and damping of the tire are considered part of the suspension stiffness and damping, and the unsprung mass is included in the vehicle mass. This model describes the rigid body bounce and pitch modes of the vehicle. Depending on the spring rates (front and rear) and the location of the center of gravity, the two modes will be coupled together to a varying degree.

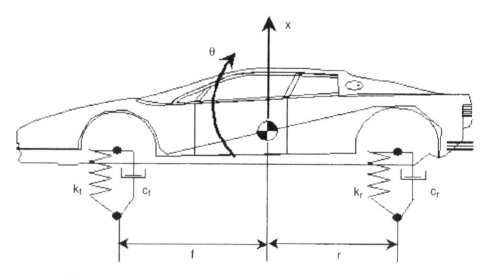

Figure 8.8. *Two-degrees-of-freedom rigid body vehicle model.*

Before proceeding, some terms require definition. The first is the dynamic index (DI) of the vehicle. The DI is defined as (Gillespie, 1994)

$$DI = \frac{k^2}{fr} \tag{8.17}$$

where f and r are the distances from the front and rear axles to the center of gravity, respectively, and k is the radius of gyration of the vehicle and is defined as

$$k = \sqrt{\frac{I}{M}} \tag{8.18}$$

Making use of Eq. 8.18, the equations of motion for the model shown in Fig. 8.8 are

$$\begin{bmatrix} m & 0 \\ 0 & mk^2 \end{bmatrix} \begin{Bmatrix} \ddot{x} \\ \ddot{\theta} \end{Bmatrix} + \begin{bmatrix} c_f + c_r & c_f f - c_r r \\ c_f f - c_r r & c_f f^2 + c_r r^2 \end{bmatrix} \begin{Bmatrix} \dot{x} \\ \dot{\theta} \end{Bmatrix}$$
$$+ \begin{bmatrix} k_f + k_r & k_f f - k_r r \\ k_f f - k_r r & k_f f^2 + k_r r^2 \end{bmatrix} \begin{Bmatrix} x \\ \theta \end{Bmatrix} \tag{8.19}$$
$$= \begin{Bmatrix} F(t) \\ M(t) \end{Bmatrix}$$

One method of obtaining the time response of Eq. 8.19 for various forcing functions is through modal expansion. The details of this method can be found in any vibrations text (Thomson, 1988) and will not be discussed here. The first step in the process is to solve for the system natural frequencies and modes, which are obtained from the homogeneous equation without damping. The equations of motion for this case then become

$$\begin{bmatrix} m & 0 \\ 0 & mk^2 \end{bmatrix} \begin{Bmatrix} \ddot{x} \\ \ddot{\theta} \end{Bmatrix} + \begin{bmatrix} k_f + k_r & k_f f - k_r r \\ k_f f - k_r r & k_f f^2 + k_r r^2 \end{bmatrix} \begin{Bmatrix} x \\ \theta \end{Bmatrix} = \begin{Bmatrix} 0 \\ 0 \end{Bmatrix} \tag{8.20}$$

In the absence of damping, the solution to the equations of motion will be sinusoidal. Thus, it is assumed that

$$x = X \sin(\omega t)$$
$$\theta = \Theta \sin(\omega t)$$

Differentiating these equations twice and substituting them into Eq. 8.20 gives

$$\begin{bmatrix} \dfrac{k_f + k_r}{m} - \omega^2 & \dfrac{k_f f - k_r r}{m} \\ \dfrac{k_f f - k_r r}{mk^2} & \dfrac{k_f f^2 + k_r r^2}{mk^2} - \omega^2 \end{bmatrix} \begin{Bmatrix} X \\ \Theta \end{Bmatrix} = \begin{Bmatrix} 0 \\ 0 \end{Bmatrix} \tag{8.21}$$

In the interest of convenience, the following notation will be defined:

$$A = \frac{k_f + k_r}{m}$$

$$B = \frac{k_f f - k_r r}{m}$$

$$C = \frac{k_f f^2 + k_r r^2}{mk^2}$$

Utilizing these parameters, Eq. 8.21 can be written as

$$\begin{bmatrix} A - \omega^2 & B \\ \dfrac{B}{k^2} & C - \omega^2 \end{bmatrix} \begin{Bmatrix} X \\ \Theta \end{Bmatrix} = \begin{Bmatrix} 0 \\ 0 \end{Bmatrix} \tag{8.22}$$

Because Eq. 8.22 must be true for all time, the determinant of the square matrix must be zero. (It is also true that X and Θ could be zero, but this is a trivial solution.) Expanding the determinant gives

$$\omega^4 - \omega^2(A+C) + \left(AC - \frac{B^2}{k^2}\right) = 0 \tag{8.23}$$

Thus,

$$\omega^2 = \frac{(A+C) \pm \sqrt{(A+C)^2 - 4\left(AC - \dfrac{B^2}{k^2}\right)}}{2} \tag{8.24}$$

These, then, are the two natural frequencies of the vehicle, and these two frequencies fall outside the uncoupled front and rear natural frequencies, which are given by

$$f_f = \frac{1}{2\pi}\sqrt{\frac{k_f}{m_f}}$$

$$f_r = \frac{1}{2\pi}\sqrt{\frac{k_r}{m_r}} \tag{8.25}$$

The natural frequencies given in Eq. 8.24 then can be substituted into either part of Eq. 8.23 to find the mode shapes. This results in two equations of the type

$$\left[[K]-[M]\omega_i^2\right]\{u_i\}=\{0\} \tag{8.26}$$

where $\{u_i\}$ is the mode vector and, in this case, is of the form

$$\{u_i\}=\begin{Bmatrix} X_i \\ \Theta_i \end{Bmatrix} \tag{8.27}$$

However, because Eq. 8.26 is homogeneous, only one element of the mode vector can be found in terms of the other element. Thus, if X is assumed to be unity, the modes shapes are given by

$$(u)_1 = \begin{Bmatrix} 1 \\ \dfrac{\omega_1^2-A}{B} \end{Bmatrix}$$

$$(u)_2 = \begin{Bmatrix} 1 \\ \dfrac{\omega_2^2-A}{B} \end{Bmatrix} \tag{8.28}$$

Another option is to solve for the mode shapes as amplitude ratios between X and Θ

$$\left(\dfrac{X}{\Theta}\right)_1 = \dfrac{B}{\omega_1^2-A}$$

$$\left(\dfrac{X}{\Theta}\right)_2 = \dfrac{B}{\omega_2^2-A} \tag{8.29}$$

When the modes are calculated as in Eq. 8.29, it will be seen that the two amplitude ratios always have opposite signs. If the amplitude ratio is positive, the node for that particular mode will be forward of the center of gravity (CG) by a distance of $\dfrac{X}{\Theta}$. If the ratio is negative, the node is aft of the CG by a distance of $\dfrac{X}{\Theta}$. If these distances fall outside the wheelbase, the motion is primarily bounce, whereas if the distance is inside the wheelbase, the motion is primarily pitch. Figure 8.9 illustrates the mode shapes. Also, note that for a dynamic

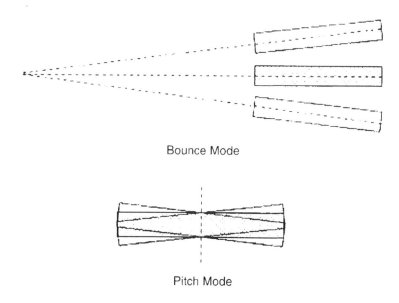

Figure 8.9. *Mode shapes.*

index of one, the nodes are centered over each axle, and an input at one suspension has no effect on the other.

Another required definition is the spring center. The spring center is that point along the longitudinal axis at which a vertical force produces only bounce and no pitch changes.

The technique of modal expansion is based on the fact that the natural modes of any system are orthogonal (Thompson, 1988). Thus, pre-multiplying the equation of motion by the transpose of either mode shape results in a set of uncoupled equations. However, as a rule, the damping matrix does not reduce to a diagonal matrix when multiplied by the mode shapes. Thus, in the interest of simplicity, it is assumed here that the damping matrix is proportional to the stiffness matrix, or

$$[C] = c_k [K] \tag{8.30}$$

Cars generally do not exhibit such proportional damping, but making this assumption allows the time response to be calculated and remains illustrative of basic suspension design principles. Furthermore, the time response also is a result of the type of forcing applied to the car (i.e., the type of road surface). Thus, this section will take the model depicted in Fig. 8.8 and assume a step input, as shown in Fig. 8.10. Such a situation is analogous to a car encountering a section of road being repaved. Although rare, it is useful because the non-periodic input will excite both natural modes.

A typical luxury car has been selected for modeling because such cars generally are designed with passenger isolation in mind. These types of vehicles usually exhibit natural frequencies on the order of 1 Hz and have damping ratios of 0.2 to 0.4 (Gillespie, 1994). Most modern

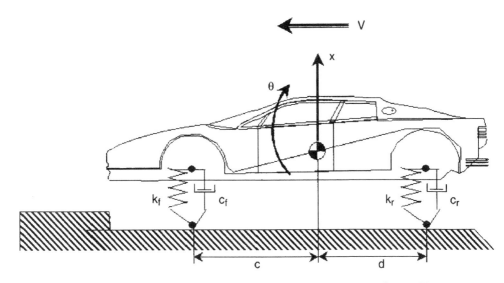

Figure 8.10. *Model encountering a step input at velocity V.*

cars also exhibit a dynamic index of unity and a weight distribution of nearly 50/50 front to rear (Gillespie, 1994). The vehicle chosen here has a mass of 1850 kg (4079 lb), a wheelbase of 2.93 m (9.61 ft), and a weight distribution of 55/45. For the first plot, the front and rear ride rates were selected to place the spring center at the center of gravity. The car is assumed to be traveling at 90 km/h (55 mph) and to hit a 5-cm (2-in.) step in the pavement. Figure 8.11

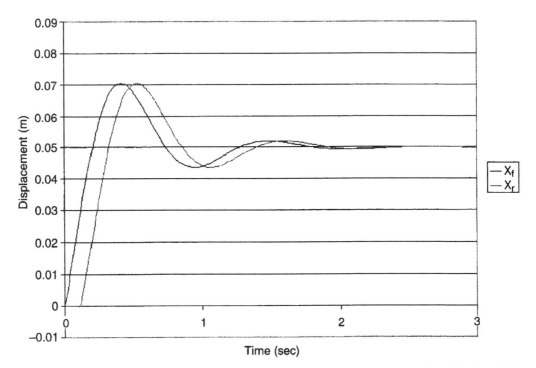

Figure 8.11. *Front and rear spring deflection, DI = 1.0, spring center at the CG, $\zeta = 0.35$.*

shows the deflection of the front and rear springs, whereas Fig. 8.12 shows the pitch and bounce response of the vehicle.

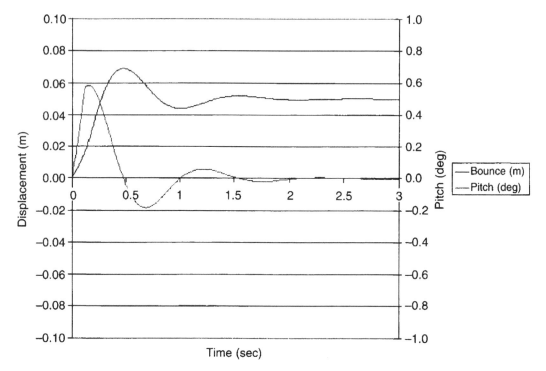

Figure 8.12. *Pitch and bounce response, DI = 1.0, spring center at the CG, $\zeta = 0.35$.*

As these figures show, the vehicle responds abruptly in pitch. Part of this abruptness is due to the current model, which assumes the wheels follow the step input exactly. In reality, the tire stiffness would "soften" the pitch response, but the trend remains. Because pitch oscillations are perceived as less desirable than bounce, this suspension setup would be unacceptable for such a car. As a first attempt at providing a flatter ride, the spring rates are changed to put the spring center 0.128 m (0.42 ft) ahead of the CG. To accomplish this, the rear ride rate is now less than that of the front. Figure 8.13 shows the resulting response.

This "solution" has exacerbated the pitch oscillations. Part of the problem is that the model assumes proportional damping; thus, softening the ride rate also lessens the damping ratio. Nonetheless, this case confirms Maurice Olley's contention that the rear ride rate should exceed that of the front by 30% (Gillespie, 1994). For the current model, if the same front and rear spring rates are selected, the ride rate of the rear exceeds that of the front by 10% due to the weight distribution. This places the spring center 0.147 m (0.482 ft) aft of the CG. Figure 8.14 shows the effect of this change on the response.

Here the car exhibits a much flatter response. Why this occurs is better illustrated by plotting the front and rear spring deflection, as shown in Fig. 8.15.

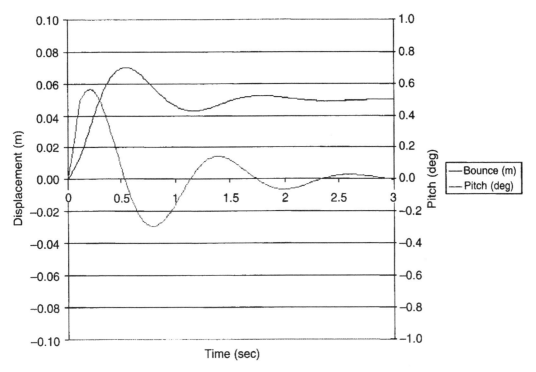

Figure 8.13. *Pitch and bounce response, DI = 1.0, spring center 0.128 m (0.42 ft) forward of the CG, $\zeta = 0.3$.*

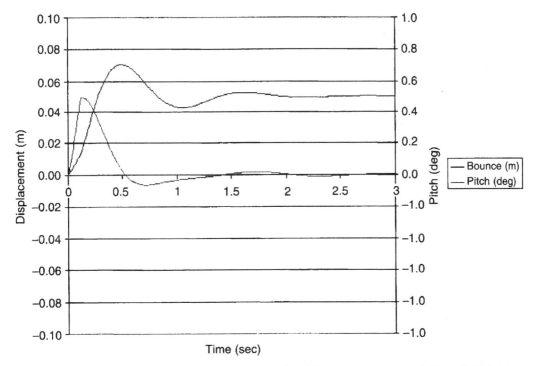

Figure 8.14. *Pitch and bounce response, DI = 1.0, spring center 0.147 m (0.482 ft) aft of the CG, $\zeta_{bounce} = 0.3$, $\zeta_{pitch} = 0.35$.*

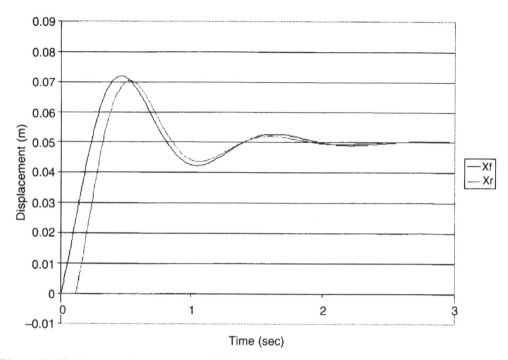

Figure 8.15. *Front and rear spring deflection, DI = 1.0, spring center 0.147 m (0.482 ft) aft of the C.G., $\zeta_{bounce} = 0.3$, $\zeta_{pitch} = 0.35$.*

As Fig. 8.15 shows, after approximately one cycle, the front and rear are moving in phase. This contributes to the flat ride and again confirms Olley's ride criteria. Finally, it is insightful to examine the effect of damping on the pitch response. For the following plot, the spring rates are unchanged; only the damping ratio is varied. Figure 8.16 shows the results.

On one hand, increasing the damping ratio leads to a flatter ride due to the more rapid decay in pitch oscillations. However, the initial pitch change when the rear wheels hit the step becomes much harsher, even allowing for the instantaneous input within the model. Thus, better body control can be had at the price of a more harsh ride.

Although this exercise has highlighted a very few design considerations, it should be obvious that the factors affecting the vehicle response are too numerous to model here and could include the type of road input, the vehicle speed, the height of the input, and the dynamic index of the vehicle. The analysis is further complicated by including the tire stiffness and the unsprung mass motion, and then can be increased again by modeling motion about all axes and in all ordinate directions. Finally, the ride characteristics depend to a great degree on the purpose of the specific vehicle. A suspension for a luxury car is, of necessity, quite different than that of a sports car, which will be vastly different from that of an off-road vehicle. The point is that although analysis of individual suspension components is relatively straightforward and implies a certain degree of exactness, the suspension system as a whole is dependent on such subjective terms as "ride" and "handling." Thus, the perfect system will never be found, and suspension engineers will always seek to build a better system.

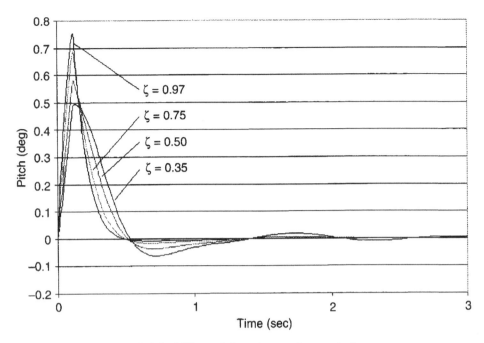

Figure 8.16. *Effect of damping ratio on pitch response.*

8.4 Suspension System Components

The primary components in the suspension system are the springs and the dampers (or shocks). Although there are only two primary components, there are several variations on the theme, and these will be discussed in the following sections.

8.4.1 Springs

The spring is the main component of the suspension system, and four types are primarily in use today: (1) leaf springs, (2) torsion bars, (3) coil springs, and (4) pneumatic (air) springs.

8.4.1.1 Leaf Springs

Figure 8.17 shows a typical leaf spring. Most early cars used this type of spring because leaf springs were used extensively on horse-drawn carriages, and early designers had some experience with them. The leaf spring shown in Fig. 8.17 is a multi-leaf type. This type of spring is made of a single elliptical spring with several smaller leaves attached to it with clamps. The leaves also are fixed rigidly by the center bolt, which prevents individual leaves from moving off-center during deflection. The additional leaves make the spring stiffer, allowing it to support greater loads. Furthermore, as the spring deflects, friction is generated between the leaves, resulting in some damping capability. Leaf springs also provide fore-and-aft location, as well as some lateral location, for the axle. Although leaf springs are simple and cheap, they tend to be heavy. Leaf springs also weaken with age and are susceptible to sag.

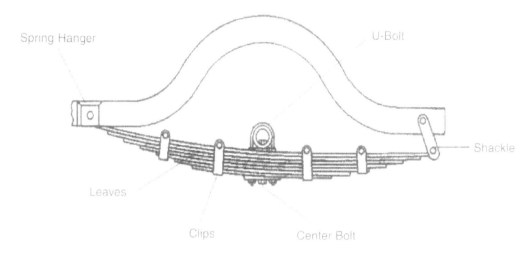

Figure 8.17. *A multi-leaf spring. Adapted from TM 9-8000 (1985).*

8.4.1.2 Torsion Bars

The torsion bar is a circular steel rod made of spring steel. One end of the rod is anchored to the frame, and loading is pure shear due to torsion. Figure 8.18 shows an example of a torsion bar. The torsion bar has very little inherent damping and therefore must be used in conjunction with dampers. As long as the bar remains in the elastic region, torque resistance will return the bar to its normal position upon unloading. The primary disadvantage of torsion bars is the axial space required for installation.

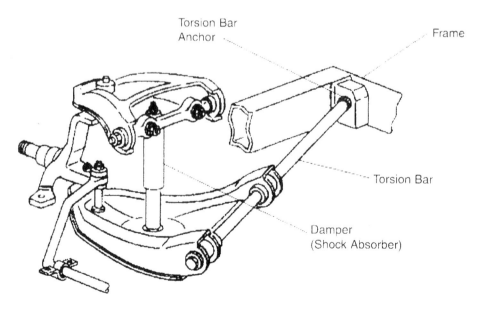

Figure 8.18. *A torsion bar suspension. Adapted from TM 9-8000 (1985).*

8.4.1.3 Coil Springs

Coil springs are basically torsion bars that have been wrapped into a coil. Figure 8.19 shows an example of a coil spring suspension. Similar to torsion bars, coil springs have little to no inherent damping and require the use of dampers. Coil springs are used widely in automotive applications due to their compact size. However, coil springs are not capable of providing any location of the axle; thus, they require control arms to limit longitudinal and lateral suspension motion.

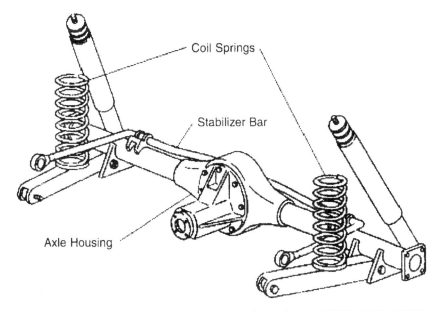

Figure 8.19. *A coil spring suspension. Adapted from TM 9-8000 (1985).*

Before analyzing coil springs, several terms must be defined. These terms are (reference Fig. 8.20):

1. Mean coil diameter, D: The center-to-center distance of the wire across the coil

2. Wire diameter, d

3. Pitch, p: The distance between successive coils on an uncompressed (free) spring

4. Spring index, C: $C = D/d$ and normally is greater than 3

5. Spring rate, k: $k = F/\delta$, where F is applied load and δ is deflection

6. Active coils: The number of coils not touching the support

7. Ends: Treatment of the last coil, which can be plain, squared, or ground, as shown in Fig. 8.21

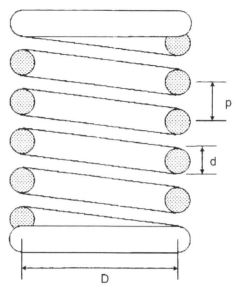

Figure 8.20. *Coil spring dimensions.*

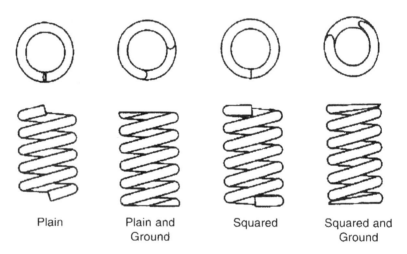

Figure 8.21. *Helical compression spring end treatments. Adapted from Krutz et al. (1994).*

The relationship among the total number of coils, the number of active coils, the free length, the solid length, and the pitch for a given spring can be determined from Table 8.1 and depends on the end treatment.

The spring analysis begins with reference to the spring loaded by a force, F, shown in Fig. 8.22. Taking a cut through one of the coils, the spring is seen to be acted upon by a direct shear and a torsional shear. The shear stress is a maximum on the inside of the coil, and the total shear stress is

TABLE 8.1
FORMULAS FOR SPRING DIMENSIONS
(N_t = TOTAL NUMBER OF COILS)
(ADAPTED FROM SHIGLEY AND MISCHKE, 2001)

	Plain	Plain and Ground	Squared	Squared and Ground
End coils, N_e =	0	1	2	2
Active coils, N_a =	N_t	$N_t - 1$	$N_t - 2$	$N_t - 2$
Free length, L_o =	$pN_a + d$	$p(N_a + 1)$	$pN_a + 3d$	$pN_a + 2d$
Solid length, L_s =	$d(N_t + 1)$	dN_t	$d(N_t + 1)$	dN_t
Pitch, p =	$(L_o - d)/N_a$	$L_o/(N_a+1)$	$(L_o - 3d)/N_a$	$(L_o - 2d)/N_a$

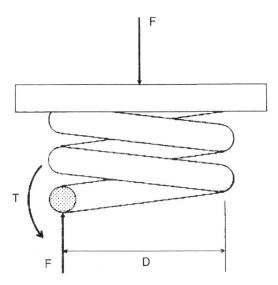

Figure 8.22. *Free-body diagram of a spring.*

$$\tau_{max} = \frac{Tr}{J} + \frac{F}{A} \tag{8.31}$$

Noting that $T = \dfrac{FD}{2}$, $r = \dfrac{d}{2}$, $J = \dfrac{\pi d^4}{32}$, and $A = \dfrac{\pi d^2}{4}$ and substituting these terms into Eq. 8.31 gives

$$\tau_{max} = \frac{8FD}{\pi d^3} = \frac{8FC}{\pi d^2} \tag{8.32}$$

However, the actual stress is larger than that predicted by the static analysis due to the curvature of the spring. A correction factor based on the work by Wahl (1963) is applied. Thus, the shear stress in the spring is

$$\tau_{max} = K_w \frac{8FD}{\pi d^3} = K_w \frac{8FC}{\pi d^2} \qquad (8.33)$$

where K_w is the Wahl factor and is given by

$$K_w = \frac{4C-1}{4C-4} + \frac{0.615}{C} \qquad (8.34)$$

The spring rate is given by

$$k = \frac{d^4 G}{8D^3 N_a} \qquad (8.35)$$

Other design considerations for coil springs include buckling and surge. These considerations are well treated in machine design texts.

8.4.1.4 Pneumatic (Air) Springs

Pneumatic suspension systems have been used in the United States on buses, trailers, and recently on passenger cars and sport utility vehicles (SUVs). The complete system consists of an air compressor, reservoir, control system, and gas springs, such as the type shown in Fig. 8.23. The unique factor of pneumatic systems is that the control system can modulate the spring pressure to provide a constant static deflection; in other words, the vehicle is self-leveling. Such a feature is particularly useful in vehicles for which their gross weight varies greatly, depending on the cargo load or a trailer being towed. The pneumatic, or gas, spring is a nonlinear spring, with a deflection curve as illustrated in Fig. 8.24.

Unlike a linear spring, the spring rate is not a constant and can be defined only as

$$k = \frac{dW}{dx} \qquad (8.36)$$

where W is the weight (or load) on the spring, and x is the deflection. Analysis of the gas spring can proceed with the following assumptions:

1. The air in the spring behaves as an ideal gas.
2. The spring is a closed system.
3. Due to the rubber enclosure, spring operation is reversible and adiabatic.

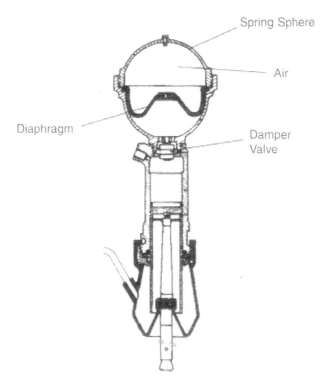

Figure 8.23. *A gas spring with an integral damper (hydragas suspension). Adapted from Bastow and Howard (1993).*

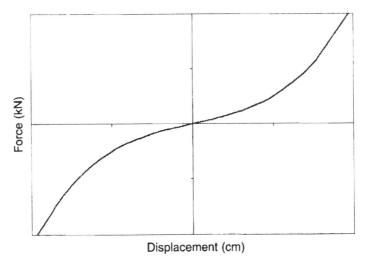

Figure 8.24. *A pneumatic spring force-deflection curve.*

Under these assumptions, the spring is modeled as a closed, piston-cylinder device. Thus, the load (weight) on the spring is balanced by the internal pressure acting over the area, or

$$W = PA \tag{8.37}$$

Because the area is assumed to be constant, it follows that

$$dW = A\,dP \tag{8.38}$$

For a reversible, adiabatic process

$$PV^k = C \tag{8.39}$$

where C is a constant, and k is the ratio of specific heats $\left(\dfrac{c_p}{c_v}\right)$. In reality, there will be some heat transfer from the gas. Thus, the actual polytropic exponent will lie somewhere between 1.0 (which is an isothermal process) and k, which for air or nitrogen is 1.4. Keeping the adiabatic assumption,

$$\begin{aligned}P &= CV^{-k} \\ dP &= -kCV^{-k-1}\end{aligned} \tag{8.40}$$

Combining Eqs. 8.37, 8.38, and 8.40 gives

$$\begin{aligned}W &= ACV^{-k} \\ dW &= -AkCV^{-k-1}dV\end{aligned} \tag{8.41}$$

Now, for any spring-mass system, the natural frequency (in hertz) is given by

$$f_n = \frac{1}{2\pi}\sqrt{\frac{k}{m}} = \frac{1}{2\pi}\sqrt{\frac{kg}{W}} \tag{8.42}$$

Using the spring rate for the gas spring (Eq. 8.36), Eq. 8.42 becomes

$$f_n = \frac{1}{2\pi}\sqrt{\frac{g}{W}\frac{dW}{dx}} \Rightarrow \frac{4\pi^2 f_n^2 dx}{g} = \frac{dW}{W} \tag{8.43}$$

Substituting the expressions in Eq. 8.41 into Eq. 8.43 yields

$$\frac{4\pi^2 f_n^2 dx}{g} = \frac{-AkCV^{-k-1}dV}{ACV^{-k}} = -k\frac{dV}{V} \tag{8.44}$$

If the spring deflects from some datum position, x_0, to some final position, x_2, the volume of the gas changes from V_0 to V_2. Thus, Eq. 8.44 may be integrated as

$$\int_{x_0}^{x_2} \frac{4\pi^2 f_n^2 dx}{g} = -k \int_{V_0}^{V_2} \frac{dV}{V} \tag{8.45}$$

which finally produces

$$f_n = \frac{1}{2\pi}\sqrt{\frac{-kg}{(x_2-x_0)}\ell n\left(\frac{V_2}{V_0}\right)} \tag{8.46}$$

Equation 8.46 allows the natural frequency of the system to be calculated for a given suspension travel and gas volume. For design, it is more likely that the suspension travel is defined, and a natural frequency is selected on the basis of the desired ride characteristics for the vehicle. Then Eq. 8.46 could be used to calculate the gas volumes (or piston areas) to provide the desired ride. The advantage of the air suspension is that as the load increases, the pressure also increases. Because this rise in pressure increases the stiffness of the spring, the system maintains a constant natural frequency as load increases.

8.4.2 Dampers (Shock Absorbers)

Most modern dampers are of the oil-filled telescoping type. They produce damping force by the action of fluid, usually oil, being forced through an orifice or valve. The dampers may be a single tube or a double tube, and Fig. 8.25 shows examples of each.

The twin tube damper is used on most passenger cars in the United States. Although twin tube dampers are heavier and tend to operate hotter than the mono tube types, they are easier to manufacture. The twin tube shock has an outer tube around the inner tube, and the space between them forms an oil reservoir. As the piston moves up and down, a valve in the bottom of the inner tube allows oil to flow into the reservoir.

In the mono tube damper, the only action is that of the fluid flowing through the valve in the piston. Most mono tube dampers have a volume of compressed gas below a floating piston. The gas moves the floating piston as the fluid volume changes. The purpose of this mechanism is to prevent foaming of the working fluid. Any air in the working fluid is compressible and passes through the valve easily. This greatly reduces the damping action of the shock.

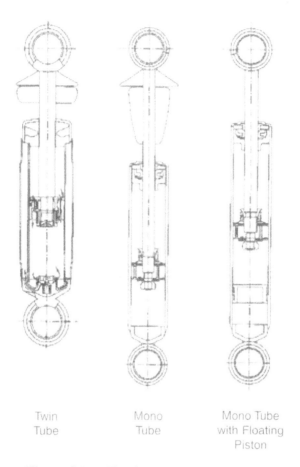

Figure 8.25. *Shock absorber construction. Adapted from Milliken and Milliken (1995).*

Regardless of the specific design, the dampers produce force proportional to the velocity of the piston. With multiple valves, the shocks can provide different levels of damping during compression or rebound. Bastow and Howard (1993) provide a complete chapter on damper characteristics, and they provide examples of the effect of damping ratio on vibration amplitude in the appendix. Milliken and Milliken (1995) also have excellent model results for various levels of damping.

8.5 Suspension Types

Classification of suspension types can be done by position (front or rear) or type (solid axle versus independent). The solid axle front suspension has practically disappeared from the passenger car. Thus, this work will group suspensions by type, with the understanding that the solid axle types generally are found only at the rear of the vehicle.

8.5.1 Solid Axle Suspensions

A solid axle has wheels mounted to each end of a rigid beam. Such systems generally are used when high load-carrying capability is required because they are very robust assemblies. They have the further advantage that as the suspension deflects, there is no camber change on the wheel due to the rigid connection. The downside to the arrangement is that the rigid connection results in a transmission of motion from one wheel to the other when the suspension deflects.

8.5.1.1 Hotchkiss Suspensions

The Hotchkiss drive was used extensively on passenger cars through the 1960s and is shown in Fig. 8.26. The system consists of a longitudinal driveshaft connected to a center differential by U-joints. The solid axle is mounted to the frame by longitudinally mounted leaf springs. Although the Hotchkiss suspension is simple, reliable, and rugged, it has been superseded by other designs for several reasons. First, as designers sought better ride qualities, the spring rates on the leaf springs dropped. This led to lateral stability difficulties because softening leaf springs requires that they be longer. Second, the longer leaf springs were susceptible to wind-up, especially as braking power and engine power began to rise. Finally, as front-wheel-drive cars became more prevalent, rear-wheel-drive cars were forced to adopt independent rear suspensions to attain similar ride and handling qualities. Nevertheless, the Hotchkiss drive is still used on many four-wheel-drive trucks and SUVs at both ends of the vehicle. One disadvantage of this suspension is that the stocky axles and differential contribute to a relatively large unsprung mass.

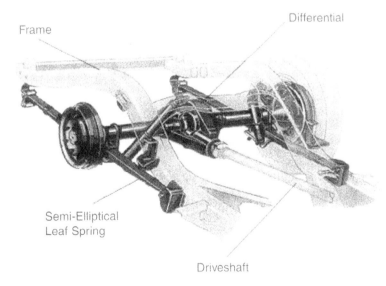

Figure 8.26. *A Hotchkiss drive. Adapted from Gillespie (1994).*

8.5.1.2 Four-Link Suspensions

The four-link rear suspension was conceived as a means of overcoming some of the limitations of the Hotchkiss drive and is shown in Fig. 8.27. The upper arms absorb braking and drive torques, while the lower arms provide location for the axle. The main advantage of the system is the use of coil or air springs, which provide a better ride than the leaf springs used on the Hotchkiss suspension.

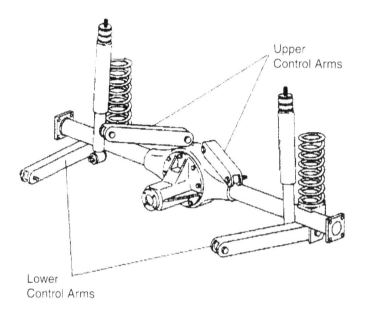

Figure 8.27. *A four-link rear suspension. Adapted from TM 9-8000 (1985).*

8.5.1.3 de Dion Suspensions

The de Dion axle, so named for its inventor, is an intermediate step between solid axles and independent suspension. The de Dion suspension has the differential mounted to the chassis, thus reducing the unsprung mass. The two wheels are connected by a hollow, sliding tube, which further reduces the unsprung mass. The disadvantage to the design is that if one wheel hits a bump, the system induces a rear-wheel steering effect. As one end of the axle is lifted, it induces a sideways motion of both tire contact points. This is resisted by the inertia of the rear end and the self aligning torque of the wheels. Although the effect is seen in independent suspensions, it affects only the wheel that hits the bump. Figure 8.28 shows a de Dion suspension.

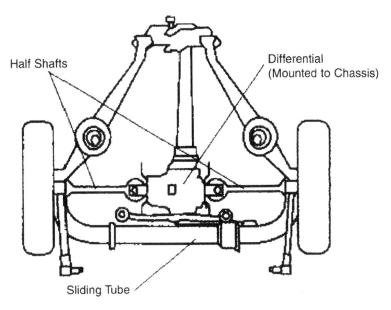

Figure 8.28. *A de Dion suspension. Adapted from Gillespie (1994).*

8.5.2 Independent Suspensions

Independent suspensions are used almost universally on the front due to the requirement for steering. The exceptions are four-wheel-drive vehicles; even then, many use independent suspensions in front.

8.5.2.1 Short-Long Arm Suspensions (SLA)

The SLA suspension, also called the A-arm or double wishbone suspension, has been prevalent on U.S. cars since World War II. Due to packaging requirements, the system lends itself particularly well to front-engined, rear-wheel-drive cars. Figure 8.29 shows such a system. These systems originally had equal-length upper and lower control arms, because such an arrangement precludes camber change when the suspension deflects. However, under cornering conditions, when suspension deflection is due to body roll, such a system promotes camber changes. Thus, most modern A-arm suspensions use a shorter control arm at the top. If the system is designed carefully, the resultant camber changes can be minimized while providing good camber qualities when cornering. The double wishbone suspension may be used on the front and rear of a vehicle.

8.5.2.2 MacPherson Struts

The rise in popularity of front-wheel-drive cars has led to greater use of the MacPherson strut suspension as shown in Fig. 8.30. The system was devised by Earle S. MacPherson, a Ford suspension engineer, in the 1940s (Bastow and Howard, 1993). The system consists of a strut

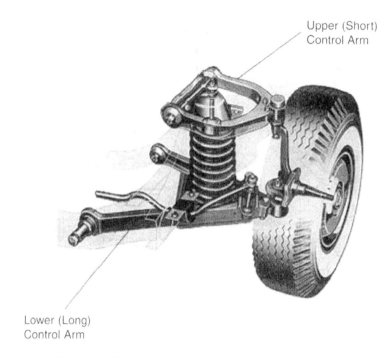

Figure 8.29. *A short-long arm (wishbone) suspension. Adapted from Gillespie (1994).*

(damper) connected to a lower control arm, with the upper end of the strut connected to the body or chassis. The system may have the coil spring concentric with the damper, or it may have a separate mounting location for the spring. The system also requires some means of longitudinal location, and such location may be achieved with a radius rod, wishbone-type lower control arm, or anti-roll linkages. Due to the small amount of space required for this system, it is ideal for front-wheel-drive cars that use transversely mounted engines, especially those with unibody construction. However, the system requires a larger amount of vertical space for installation and thus limits the designer's freedom to lower the hood height. MacPherson struts are used almost exclusively on the front.

8.5.2.3 Trailing Arm Suspensions

Trailing arm suspensions are used on the rear of vehicles. Pure trailing arm suspensions are used on high-performance cars, such as the Corvette shown in Fig. 8.31. The pivot axis of the control arms is perpendicular to the longitudinal axis of the vehicle; thus, the arms control squat and dive, and also absorb acceleration and braking forces. The differential usually is mounted to the chassis, reducing unsprung weight.

BMW and Mercedes-Benz have used semi-trailing arms, as shown in Fig. 8.32. The pivot axis is at an angle to the longitudinal axis of the vehicle. The angle varies from 18° (as in the system shown in Fig. 8.32) to as much as 25°. The semi-trailing arm system produces some camber change upon deflection, and this contributes a steering effect to the vehicle.

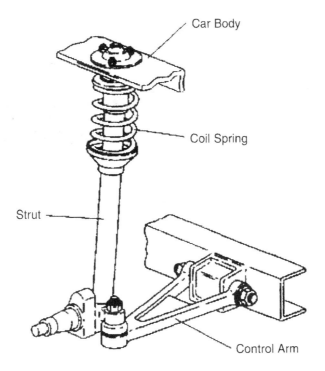

Figure 8.30. *A MacPherson strut suspension. Adapted from TM 9-8000 (1985).*

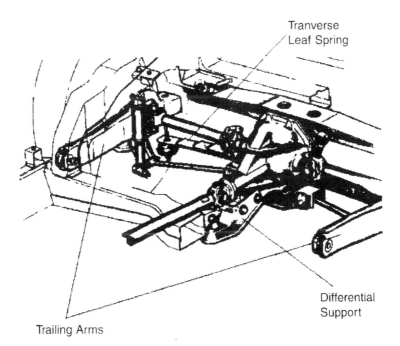

Figure 8.31. *A Corvette trailing arm suspension. Adapted from Gillespie (1994).*

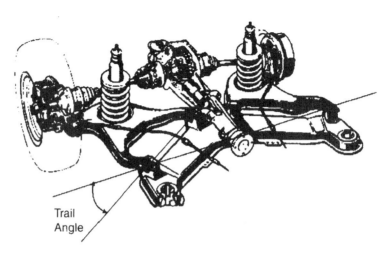

Figure 8.32. *A semi-trailing arm suspension. Adapted from Bastow and Howard (1993).*

8.5.2.4 Multi-Link Suspensions

Multi-link suspensions have ball joint connections at the ends of the control arms to eliminate bending loads. Most systems use four links, although Mercedes-Benz uses a five-link system. This over-constrains the motion but provides advantages in control of toe angles (Gillespie, 1994). Figure 8.33 shows an example of a multi-link suspension from the Jaguar XJ-40. This system evolved from Jaguar's double link suspension system and was designed to further

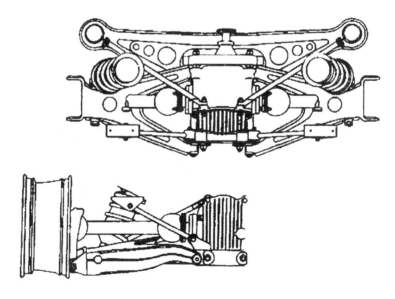

Figure 8.33. *The multi-link suspension of the Jaguar XJ-40. Adapted from Bastow and Howard (1993).*

reduce road noise (Bastow and Howard, 1993). The system uses the half shaft as an upper control link, and the entire system is mounted in a separate sub-frame. Variations of this system also were used on the XKE and XJS, although the XJS replaced the lower wishbone with a single linkage that further required a longitudinal radius rod for axle location.

8.5.2.5 Swing Arm Suspensions

The swing arm suspension is the easiest way to obtain an independent rear suspension. Figure 8.34 shows an example. The swing arm suspension was used most prominently on the Volkswagen Beetle. The major drawback to this system is the large camber change that results from suspension deflection. This results in unpredictable cornering performance.

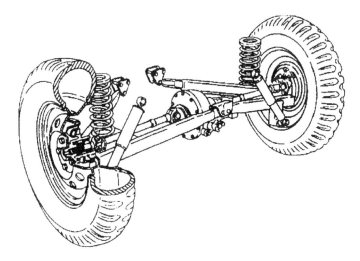

Figure 8.34. *A swing arm suspension. Adapted from TM 9-8000 (1985).*

8.6 Roll Center Analysis

An important concept relating to suspension analysis is that of the roll center. The suspension roll center may be thought of as the point through which lateral forces are transmitted to the sprung mass. Furthermore, lateral forces applied at the roll center cause no suspension roll (Gillespie, 1994). Each suspension (front and rear) will have its own roll center. A longitudinal line connecting the two roll centers defines the vehicle roll axis, and the sprung mass rotates around this axis with respect to the unsprung mass. Figure 8.35 illustrates the concepts.

Calculating the suspension roll center may be done with kinematics because all suspensions are merely linkages. However, a graphical solution often is easier and more intuitive, and this will be discussed here. Note that roll centers are instantaneous centers. As soon as the car rolls, the suspension kinematics drive the roll center to a different location. Nevertheless,

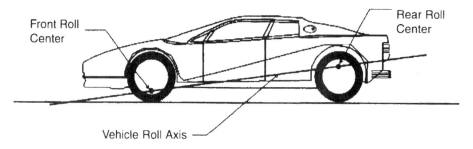

Figure 8.35. *Suspension roll centers and vehicle roll axis.*

calculation of the roll centers, and hence the vehicle roll axis, remains a worthwhile exercise because these centers define the relative motion between the sprung and unsprung masses, and hence have a great effect on the handling qualities of the vehicle.

The procedure described here will be demonstrated in subsequent sections for independent, wishbone-type suspensions, as well as MacPherson strut suspensions. Other suspension types can be analyzed by similar means, and the reader is directed to books such as Gillespie (1994) that offer a more complete treatment of the subject.

In general, the roll center may be found graphically, as illustrated in Fig. 8.36, by applying the following steps. In the procedure described here, the point of view taken is along the longitudinal axis of the vehicle.

1. Draw a straight line through the joints that connect each of the suspension linkages to the sprung mass.

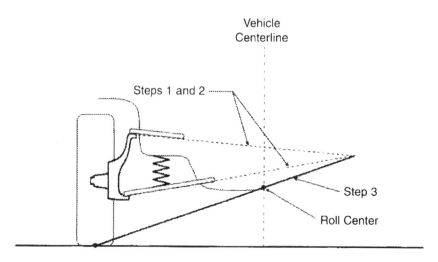

Figure 8.36. *Roll center diagram for positive swing arm suspension.*

2. Extend these lines to their point of intersection. This point defines the virtual reaction point of the linkages.

3. Draw a line from the center of the tire/road contact patch to the intersection described in step 2.

4. The point at which this line crosses the vehicle centerline is the roll center.

This procedure is illustrated in the following sections.

8.6.1 Wishbone Suspension Roll Center Calculation

Figure 8.36 shows the roll center calculation for an independent suspension with positive swing arm geometry. (The steps in the previously outlined procedure are shown in this figure.) The term "positive swing arm" is due to the fact that the roll center is above the ground. As the vehicle rolls in a turn, the suspension on the outer wheel is compressed (jounce), while the inner suspension extends (rebound). Because the suspension geometry is no longer symmetrical, the roll centers from each side no longer coincide. The roll center for the outer suspension moves downward, while that for the inner suspension moves upward. Because the outer wheel generates a higher lateral force, the overall effect is a lowering of the roll center.

Figure 8.37 shows the diagram for a suspension with negative swing arm geometry. In this case, the roll center is located below ground level. Both of the suspensions shown result in lateral tire scrub during suspension deflection.

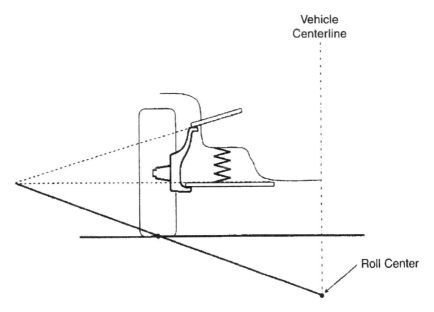

Figure 8.37. *Roll center diagram for a negative swing arm suspension.*

Another option is to have parallel links that are horizontal, as shown in Fig. 8.38. The procedure for determining the roll center is unchanged, the difference being that the virtual reaction point for the linkages is at infinity. Thus, the line from the tire-road contact patch also extends horizontally to infinity, and the roll center thus is located on the ground.

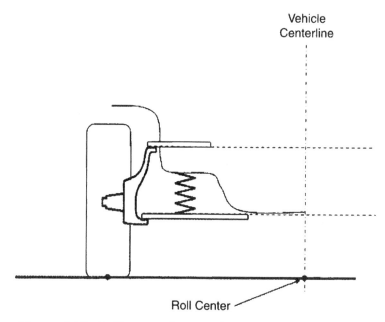

Figure 8.38. *Roll center diagram for a parallel link suspension.*

8.6.2 MacPherson Strut Suspension Roll Center Calculation

The MacPherson strut suspension has a lower control arm, but the upper link is provided by the strut. The same procedure applies, except that the line from the upper connection is drawn perpendicular to the strut axis, as shown in Fig. 8.39.

8.6.3 Hotchkiss Suspension Roll Center Calculation

The procedure for determining the roll center for a Hotchkiss suspension is a bit different than those outlined previously. However, the same basic principles apply. In the Hotchkiss suspension, the leaf spring connection points provide the lateral force reaction points. Thus, the roll center is found along a line connecting the two leaf spring-to-frame connection points. In this case, the intersection of this line with the centerline of the wheel defines the roll center, as shown in Fig. 8.40.

8.6.4 Vehicle Motion About the Roll Axis

Returning now to Fig. 8.35, few vehicles have identical suspensions on both the front and rear of the vehicle. As a result, the roll axis is inclined with respect to the longitudinal axis of the

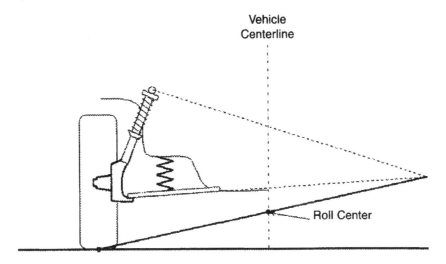

Figure 8.39. *Roll center diagram for a MacPherson strut suspension.*

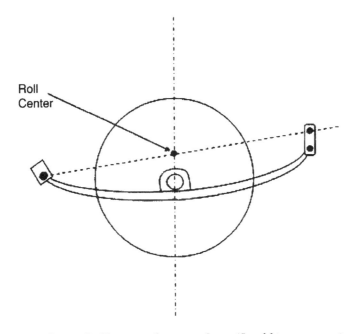

Figure 8.40. *Roll center diagram for a Hotchkiss suspension.*

vehicle. With most vehicles, the roll axis is inclined as shown in Fig. 8.35, with the front roll center lower than the rear. As the vehicle depicted negotiates a turn, the inertia of the vehicle mass will induce a roll of the sprung mass about the roll axis. Although the sprung mass is oriented parallel to the road surface, it is constrained by the suspension to roll about the roll axis. As Fig. 8.35 illustrates, the forward portion of the vehicle lies farther above the roll axis than the rear. Thus, angular displacement about the roll axis will cause a greater linear displacement of the front suspension than the rear.

As the vehicle turns, two moments are generated that tend to rotate the vehicle body toward the outside of the turn. Figure 8.41 illustrates both of these moments. When the vehicle is traveling straight ahead, the CG is at position 1 a distance $h_1 - h_r$ above the roll center. When the vehicle rolls in a turn, the CG moves to position 2. The first moment is created by the inertia of the vehicle, $\frac{mV^2}{r}$, where r is the turn radius, acting at a distance $h_2 - h_1$ above the roll center. As the CG is established at position 2, the weight of the vehicle adds to this overturning moment because the CG is no longer co-linear with the roll center. Thus, the second overturning moment is mgx. The magnitude of these moments will be different for the front and rear because the front and rear roll centers are at different heights.

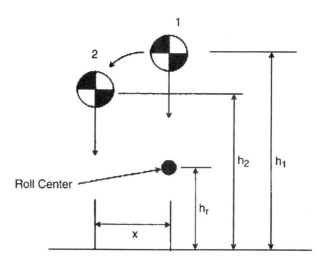

Figure 8.41. *Vehicle CG motion during cornering.*

Once established in a steady-state turn, these overturning moments ultimately are resisted by the tire contact patches at the road. However, the moments are transferred to the contact patches through the suspension linkages and the springs. (During the establishment of the roll when the sprung mass has some angular velocity about the roll axis, the dampers also provide a resistive moment.) Modeling this behavior for the whole vehicle is beyond the scope of this text.

Nevertheless, a half-car (front or rear) model is still useful for highlighting the effect of CG and roll center height on the vehicle body dynamics. The following model is developed under the assumption that the vehicle is established in a steady-state turn. The primary implication of this assumption is that the model gives no insight as to roll center motion while the turn is being established. The model also neglects the interaction between front and rear suspensions, and the tire stiffness also will be ignored. Although the model does account for the lateral shift of the CG, the vertical change due to body roll ($h_1 - h_2$ in Fig. 8.41) is assumed to be negligible. The objective of the model is to calculate the weight transfer and the body roll angle for the vehicle.

The development begins with reference to Fig. 8.42, which shows the entire vehicle traveling at a velocity V and established in a turn of radius R. The cornering forces (F_y) act at ground level, and the tires resist the overturning moments through the vertical reaction forces (F_z). The distance between the tire contact patches is t_w, and the sprung mass has a weight of W_s. The CG is at some height h above the ground, while the roll center height is h_r.

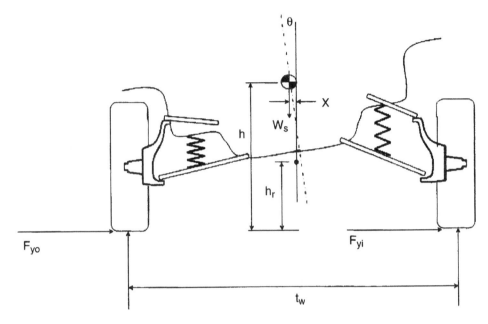

Figure 8.42. *Free-body diagram of a vehicle in a turn.*

If the body roll angles are assumed to be small, the lateral offset of the CG is

$$x = (h - h_r)\tan\theta \approx (h - h_r)\theta \tag{8.47}$$

Making use of Eq. 8.47, summing moments about the roll center gives

$$\sum M_{rc} = I\bar{\alpha} + \bar{r} \times m\bar{a}_{CG}$$
$$= -(F_{yo} + F_{yi})h_r - W_s\theta(h - h_r) + \frac{t_w}{2}(F_{zo} - F_{zi}) \tag{8.48}$$

where

$\bar{\alpha}$ = angular acceleration about the CG
$\bar{r}$ = position vector from the point of interest to the CG
$\bar{a}$ = acceleration of the vehicle CG

As the vehicle is in a steady-state turn, there is no angular acceleration. Furthermore, it can be shown that the total lateral cornering force (F_y) is

$$F_y = F_{yo} + F_{yi} = W_{tot} \frac{V^2}{Rg} \tag{8.49}$$

where W_{tot} is the total weight of the vehicle (sprung and unsprung mass). The reaction forces on the outside and inside of the turn are simply the static reaction force plus and minus the weight transferred due to the turn, or

$$F_{zo} - F_{zi} = (F_{static} + W_t) - (F_{static} - W_t) = 2W_t \tag{8.50}$$

assuming the car is balanced laterally under static conditions. Substituting Eqs. 8.49 and 8.50 into Eq. 8.48 yields

$$W_t = W_{tot} \frac{V^2}{Rg} \frac{h_r}{t_w} + W_s (h - h_r) \left[\frac{V^2}{Rg} + \theta \right] \tag{8.51}$$

Equation 8.51 calculates the weight transfer but is a function of the body roll angle. Thus, it is necessary to calculate the body roll angle, and this may be done with reference to Fig. 8.43, which is a free-body diagram of the sprung mass.

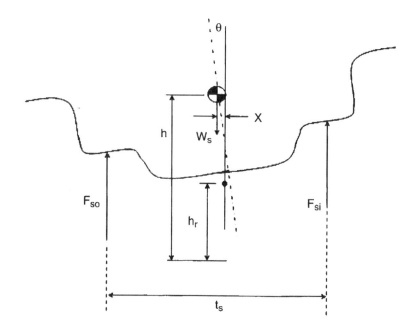

Figure 8.43. *Free-body diagram of sprung mass.*

In this case, the overturning moment is resisted by the spring forces, which are separated by a distance of t_s. Again summing moments about the roll center gives

$$\sum M_{rc} = I\bar{\alpha} + \bar{r} \times m\bar{a}_{CG} = -W_s\theta(h-h_r) + \frac{t_s}{2}(F_{so} - F_{si}) \qquad (8.52)$$

As in Eq. 8.50, the difference between the outer and inner spring forces is due to the weight transfer. Furthermore, the weight transfer can be related to the spring deflection by

$$W_t = K(\delta_o + \delta_i) = 2K\delta \qquad (8.53)$$

Using the small angle assumption again, it can be shown that

$$\delta = \frac{t_s\theta}{2} \qquad (8.54)$$

Substituting these relations into Eq. 8.52 gives

$$Kt_s^2\theta = W_s \frac{V^2}{Rg}(h-h_r) + W_s\theta(h-h_r) \qquad (8.55)$$

Finally, solving for the body roll angle

$$\theta = \frac{W_s \dfrac{V^2}{Rg}(h-h_r)}{Kt_s^2 - W_s(h-h_r)} \qquad (8.56)$$

Thus, the body roll angle is a function of the lateral acceleration $\left(\dfrac{V^2}{Rg}\right)$, the spring stiffness (K), and the difference in height between the CG and the roll center. Equation 8.56 now can be substituted into Eq. 8.51 to find the weight transfer

$$W_t = W_{tot}\frac{V^2}{Rg}\frac{h_r}{t_w} + \frac{Kt_s^2 W_s \dfrac{V^2}{Rg}(h-h_r)}{t_w\left[Kt_s^2 - W_s(h-h_r)\right]} \qquad (8.57)$$

Equation 8.57 indicates that there are two components to the weight transfer. The first term in the equation is the weight transferred through the suspension linkages, often referred to as the direct weight transfer. The second term is the weight transferred through the springs. Note, too, that the stiffness term, Kt_s^2, is the roll stiffness of the vehicle and has units of N-m/rad.

To examine the effects of the variables on weight transfer and body roll, a representative vehicle must be selected. For this study, the luxury car used in Section 8.3.3 will be used, and the rear suspension will be modeled. The car has a 61-in. (1.55-m) track width, and because it has an independent rear suspension, it will be assumed that the springs are effectively located at the outboard track location (Milliken and Milliken, 1995). The rear spring rate used to generate Fig. 8.14 will be used (35000 N/m, or 200 lb/in.). The total car mass is 1850 kg (4079 lb), and given the 55/45 weight distribution, the total rear mass in this analysis will be 832.5 kg (1835 lb). It also is assumed that the unsprung mass is 10% of the total; thus, the sprung mass is 750 kg (1653 lb). The car will be traveling around a turn with a lateral acceleration of 0.85g.

As a first analysis, the CG height was allowed to vary from ground level (h = 0, a physical impossibility) up to 1.5 times the stock CG height ($h/h_o = 1.5$). The calculation was repeated for four roll center heights, each of which was normalized to the stock CG height (h_r/h_o). The weight transfer is plotted as a percentage of the static weight on the wheel, for reasons that will be explained shortly. Figure 8.44 shows the results for the percent weight transfer, and Fig. 8.45 shows those for the body roll.

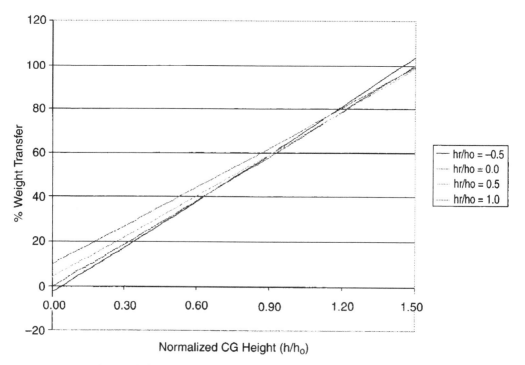

Figure 8.44. *Variation of weight transfer with CG height.*

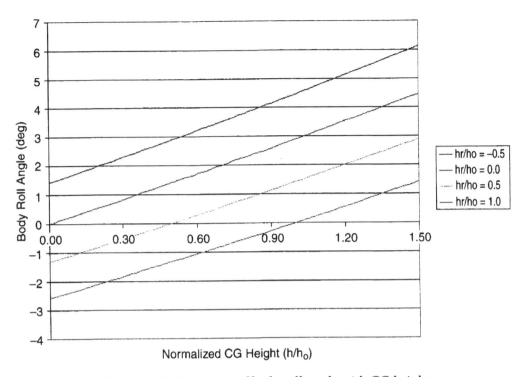

Figure 8.45. *Variation of body roll angle with CG height.*

Both figures indicate that the weight transfer and the body roll increase almost linearly with CG height. Thus, a low CG is desirable. One might expect that with a CG at ground level, no weight transfer would occur. However, recall that there is still direct weight transfer through the roll center. Thus, the only way to eliminate weight transfer is to have both the CG *and* the roll center at ground level. If the roll center is below ground level, the direct weight transfer is negative; in other words, weight transfers to the inside of the turn. Note that in this case, with a CG at ground level, the total weight transfer is negative. However, as the CG rises, the weight transfer through the springs is positive and quickly becomes larger than the direct weight transfer. Thus, even with a roll center below ground level, the total weight transfer is still to the outside of the turn.

One other useful result of the model is that it gives an indication of impending rollover. When the weight transfer is equal to the static weight on the wheel, the normal force on the inner wheel becomes zero, and the vehicle is at the point of incipient rollover. The point of incipient rollover is indicated on Fig. 8.44 when the percent weight transfer equals 100%. However, the model does not indicate whether the tires will break free prior to rollover, although this could be calculated knowing the coefficient of friction between the tire and the road.

Examining Fig. 8.45, the body roll is zero when the roll center coincides with the CG. When the roll center is above the CG, body roll is into the turn, similar to a motorcycle. Neither of these conditions is desirable from a human factors standpoint. Body roll toward the outside of the turn is a primary feedback mechanism to the driver. If the car remained level while negotiating a turn, the driver would have no indication of impending tire breakaway. A car that rolled into a turn also would require some adaptation by drivers because it is a completely unnatural phenomenon to most drivers.

Because the height of the CG is mostly fixed by the manufacturer, it is more instructive to examine variations in roll center height for a fixed CG. This has been done, and the results are shown in Fig. 8.46 for the stock CG height. As in the previous plots, the weight transfer is plotted as a percentage. The roll center height in Fig. 8.46 has been normalized by the CG height.

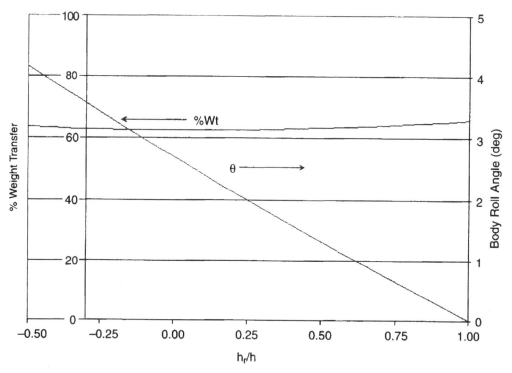

Figure 8.46. *Weight transfer and body roll versus roll center height for a fixed CG.*

Note that although the body roll decreases linearly as the roll center is raised, weight transfer has a minimum point. For this particular example, that minimum weight transfer occurs with the roll center at a point just above the ground. The tradeoff at work here is that between the direct weight transfer and the weight transferred through the springs. When the roll center is moved farther below ground level, the direct transfer is toward the inside of the turn. However, the effective moment arm ($h - h_r$) becomes larger, and the positive (toward the outside of the turn) weight transfer increases at a faster rate than does the negative direct weight transfer. Furthermore, by examining Fig. 8.44, it is seen that the CG height has a much larger effect on weight transfer than does the roll center height. Thus, lowering the CG remains the best way to minimize weight transfer, body roll, and the risk of rollover.

However, the fact that an optimal height for the roll center exists can be used to tune a suspension. For example, consider the luxury car used in this example. One way to minimize body roll is to use stiffer springs, but this results in a harsher ride. By optimizing the roll center height, the body roll can be minimized with softer springs, providing the luxury car owner with the compliant ride he or she desires.

Many other factors come into play when determining the "best" roll center height for a given vehicle. The biggest factor involved is the intended use of the vehicle. An off-road or four-wheel-drive vehicle will have a much different roll center requirement than will a Formula car. Again, the reader is reminded of the limitations of the foregoing model. By neglecting the interaction of the front and rear roll behavior, the model gives no insight into the effect of roll center height (front and rear) on the understeer characteristics of the vehicle, the tire angles (especially camber changes), or tire scrub during suspension deflection. Note that a high roll center on a car with independent suspension can lead to a phenomenon known as jacking. All of these effects are examined, modeled, and analyzed in more specialized books on suspensions and vehicle dynamics, such as Gillespie (1994) and Milliken and Milliken (1995).

8.7 Active Suspensions

Until recently, vehicle suspensions were designed with passive components and represented a compromise between passenger isolation and tire contact with the road. In the quest to improve suspension performance, much work has been done recently in designing and analyzing suspension systems that contain active components. Such systems span a range of complexity—from simple self-leveling suspensions to fully active systems. The benefits of an active suspension system were examined by Redfield (1987), and his work will be discussed here as a good introduction to the subject.

Redfield's work examined the performance potential of using variable rate suspension components, and it examined the relative advantage of using only active damping versus using a full-state control system. The datum for comparison was a standard family sedan with a passive suspension system. The work progressed from a simple one-degree-of-freedom model to a full pitch-heave model with sprung and unsprung masses. Although Redfield achieved good results with the complex models, such models exceed the scope of this introductory text. Instead, this work will focus on the results of his two-degrees-of-freedom (quarter car) model that includes the sprung mass and the tire stiffness. Figure 8.47 shows Redfield's model and the accompanying Bond graph. As shown in the figure, the model consists of sprung and unsprung masses (M and m), suspension spring and tire stiffnesses (K and k), the standard passive damper (B_p), and the control force (F_a). Inputs to the model were the road vertical velocity, V_o, and a disturbing force, F_d. In general, the control force was proportional to the sprung mass velocity; hence, the model utilizes only passive damping.

This model was used to examine the effects of suspension changes on sprung mass acceleration, suspension stroke, and tire force. However, in keeping with the focus of this chapter, only Redfield's results for sprung mass acceleration and tire force will be presented here. Both the acceleration and the tire force shown in Figs. 8.48 to 8.50 were normalized by the

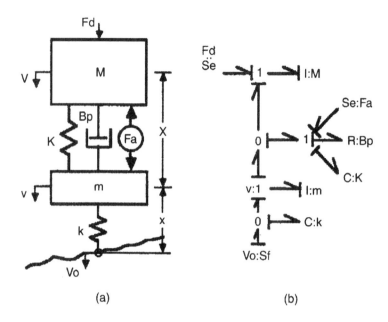

Figure 8.47. *(a) The two-degrees-of-freedom heave model, with (b) a Bond graph (Redfield, 1987).*

input road velocity (V_o). In each graph, either spring stiffness or damping ratio was varied from one-quarter of the value of the stock suspension to four times the value. The heavy line in all of the plots represents the frequency response with stock suspension components.

Figure 8.48 shows the frequency response of the sprung mass acceleration to passively changing suspension parameters. In each case, two resonance peaks appear. The first is near 1.5 Hz and corresponds to the sprung mass/suspension spring resonance; the second is near 11 Hz and corresponds to the unsprung mass/tire stiffness resonance. As the passive damping coefficient (B_{PN}) is increased (plot (a)), the magnitude of body acceleration decreases near the first resonant peak. However, if the damping is increased too much, higher accelerations appear at all frequencies above the sprung mass resonance. The final curve, in which the damping coefficient is four times the standard one, depicts a single resonance peak between the original peaks. This single peak indicates that the damper is so stiff that it behaves in the same way as a rigid connection between the sprung and unsprung masses. Thus, the peak is the resonance caused by the total vehicle weight acting on the tire stiffness.

Plot (b) in Fig. 8.48 shows the effect of increasing spring stiffness. As would be expected, the natural frequency of the sprung mass increases, and the stiffer springs also result in higher peak acceleration magnitudes. Thus, the ride becomes harsher. At frequencies at or above the tire resonance, stiffer springs have little to no effect on body acceleration.

Figure 8.49 shows the effects of the passive suspension changes on the tire force. Plot (a) again shows the result of increasing the damping coefficient, and the increases correspond to those shown in Fig. 8.47(a). In this case, increased damping reduces the magnitude of the tire

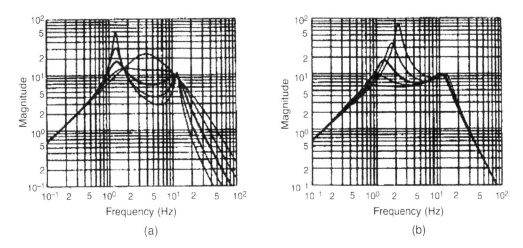

Figure 8.48. *Frequency response of sprung mass acceleration to changes in (a) passive damping and (b) spring stiffness (Redfield, 1987).*

force at both resonant peaks. However, this is not necessarily a good thing. Reducing the tire force at the tire resonance implies that the tire stands a higher probability of leaving the road surface. Thus, with the increased damping, wheel hop becomes a distinct possibility. Increasing the spring stiffness (plot (b)) increases tire force near the body resonance. Although this is desirable, the previous figure illustrates the effect of stiff springs on harshness, which may be undesirable depending on the type of vehicle. In short, this portion of Redfield's work highlighted the standard tradeoff of the suspension designer: a compliant ride comes at the cost of reduced contact between the road and tire.

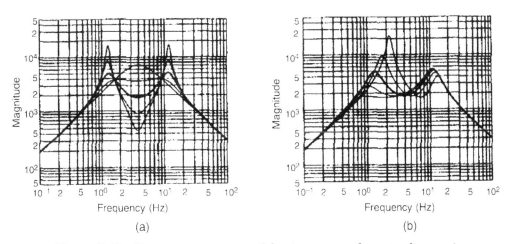

Figure 8.49. *Frequency response of the tire contact force to changes in (a) passive damping and (b) spring stiffness (Redfield, 1987).*

Figure 8.50 shows the results for the same suspension changes with the addition of active damping. The results indicate that the active damping system is the solution to the aforementioned compromise because it provides decreased amplitudes of the sprung mass near its resonance (a), while maintaining normal tire contact forces near the tire resonance (b).

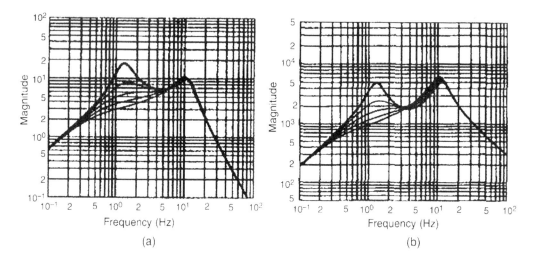

Figure 8.50. *Frequency response of (a) sprung mass acceleration and (b) tire contact force, with active damping (Redfield, 1987).*

Given the promising results of active damping, the logical step would be to implement a fully active suspension system with full-state control. Redfield ultimately did so with a more complex pitch-heave model that included the tire stiffness and unsprung mass. Again, this particular model is beyond the scope of this text.

However, Redfield reached a very important conclusion that is illustrated in Fig. 8.51. This figure shows a locus of optimized points, the optimization being performed between sprung mass acceleration and suspension stroke (a) and between tire contact force and suspension stroke (b). Plot (a) highlights another design tradeoff. If low body acceleration is desired, the suspension must have a longer stroke. The addition of active damping allows the designer to lower the body acceleration for a given stroke. The converse is even stronger—for a given acceleration, the stroke can be reduced significantly, a great advantage in packaging. Of course, for a "sportier" suspension, the necessary short stroke results in high accelerations, and the active damping has little effect at the upper left of the curve.

What is most interesting is that the implementation of a full-state control system adds little to the performance of the active damping system. The result is not entirely unexpected because the active damping was based on the velocity of the sprung mass. Furthermore, regarding the sprung mass, the problem is to dissipate energy. By actively changing the damping coefficient, the energy can be dissipated more effectively. The full-state control system adds the

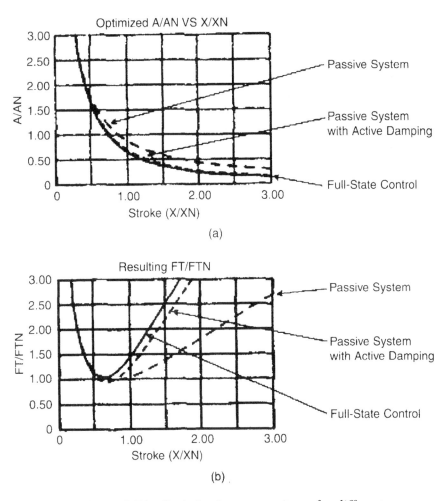

Figure 8.51. *Optimization comparisons for different control strategies (Redfield, 1987).*

capability for active actuation. However, because active actuation tends to add energy to a system, the full-state control in effect does nothing but actively damp the acceleration of the sprung mass.

Plot (b) in Fig. 8.51 shows the corresponding optimization for the tire contact force. In this case, the addition of full-state control provides some advantage over active damping. The reason for this is that if the tire should come off the road, the full-state control system can actively return it to the road surface, whereas an active damper cannot. Despite such an advantage, the designer must consider the cost of implementing a full-state control system and weigh this against the benefits.

8.8 Conclusions

As indicated, the purpose of this chapter was to lay a foundation for the basics of suspension analysis. The models presented here are useful for developing basic concepts and highlighting certain important principles. We hope this chapter has given the reader an appreciation for the complexity of suspension design and analysis. Obviously, anyone seriously considering a career as a suspension engineer must greatly expand his or her knowledge of this subject. The references cited here are highly recommended as good starting points for furthering that knowledge.

CHAPTER 9

Brakes and Tires

9.1 Introduction

When a car has been set in motion, the driver naturally is concerned that he or she is able to bring the vehicle safely to rest. This is the function of the braking system. Early cars, similar to many of their components, adopted braking systems from carriages. These systems used mechanical "scuff" brakes, which usually were blocks of wood wrapped in a friction material such as leather that rubbed on the wheels (Rinek and Cowan, 1997). As car technology advanced, most manufacturers began to use transmission braking, in which a band clamped around some rotating drum attached to the driveline (Rinek and Cowan, 1997). Some vehicles used both systems. For example, in 1996, the Jaguar-Daimler Heritage Trust rolled out its 1896 Daimler in celebration of the one-hundredth anniversary of the British auto industry. The braking system on this vehicle had a foot-pedal-operated system that applied braking to the wheels, and a long-handled lever that operated a band clamping a drum on the transmission (Cropley, 1996). The system stopped the car quite effectively, although the top speed of the vehicle was 24 mph. Systems employing a brake band have the advantage of self-energization (see Section 9.5), thus reducing driver effort.

The next leap in brake system design was the invention of the long shoe, internal expanding drum brake, usually accredited to Louis Renault. This type of brake was used in the United States on the 1910 Sears Model P car (Rinek and Cowan, 1997). Initially, these systems remained confined to the rear wheels, but by the 1920s, manufacturers had figured out how to mount brake drums to the front (steered) wheels (Rinek and Cowan, 1997). As vehicle speed and weight began to increase, some means of power assistance became necessary. Initial systems employed a vacuum booster to aid in the application of cables, which applied the brakes. Both the 1932 Cadillac and the 1932 Lincoln used vacuum-assisted, cable operated, four-wheel drum brake systems (Rinek and Cowan, 1997). However, hydraulically operated brakes eventually became the system of choice. Hydraulic brakes were first seen on the 1920 Duesenberg Eight (Rinek and Cowan, 1997), and Chrysler incorporated them into its B-70 model in 1924 (Rinek and Cowan, 1997). By the mid-1930s, most manufacturers had migrated to hydraulic brakes, and Chrysler began to experiment with vacuum-boosted hydraulic brakes in 1932 (Rinek and Cowan, 1997). However, it wasn't until the 1950s that volume production of vacuum-boosted hydraulic systems became the norm for passenger cars (Rinek and Cowan, 1997).

Although the drum brake was effective, heat buildup was a problem because the heavy, cast iron drum made an almost ideal heat sink. The drum also exhibited distortion under braking due to thermal stresses, as illustrated in Fig. 9.1. Note that the magnitude of the distortion is

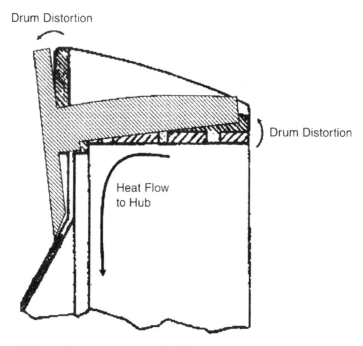

Figure 9.1. *Distortion of a brake drum during braking.*

exaggerated in the figure. Due to the heat flow from the rim of the drum to the hub, the drum would distort. Thus, the shoes would lose contact with the drum.

The solution was found in the disc brake. Disc brakes were available in the late 1800s but did not find their way onto the automobile until much later. English and French manufacturers began offering disc brakes in the 1950s, but another decade passed before American manufacturers followed suit. However, the disc brake has not completely eclipsed the drum brake, and drums continue to be found on late-model pickups and SUVs.

The next big development in braking was the development of antilock brake systems (ABS). Bosch obtained a patent for an electro-hydraulic ABS system in 1936 (Rinek and Cowan, 1997), and the systems initially were used on aircraft in the 1940s and 1950s. Although U.S. manufacturers experimented with ABS, the first production car with an ABS system was the 1971 Chrysler Imperial (Rinek and Cowan, 1997). Although the concept was not popular initially, the use of ABS has exploded in recent years. In 1991, only 10% of the vehicles sold in the United States were equipped with ABS. This number had grown to 50% by 1995 and is higher today (Rinek and Cowan, 1997). The trend now is toward using the ABS beyond only braking—for example, in traction control, stability control, and anti-rollover braking.

This brief history also provides a synopsis of the topics in this chapter. The chapter will begin with a discussion of vehicle dynamics under braking. An understanding of these dynamics will greatly aid in understanding the purpose of each component in a modern, hydraulic brake system—the topic of the next section. Next, the chapter develops relations for the torque capacity of both drum and disc brakes, and the brake sections conclude with a discussion of antilock braking. The chapter concludes with a discussion of tires—designations and types, plus a brief introduction to the dynamics of traction production on a tire.

9.2 Braking Dynamics

Until approximately 1965, the U.S. standard for braking was "30 feet from 20 miles per hour." Deciphering what this means requires a bit of calculus and basic dynamics. Recall that the definitions of velocity and acceleration are

$$V = \frac{ds}{dt} \quad \text{and} \quad a = \frac{dV}{dt}$$

These equations can be "solved" for dt and set equal to each other to give

$$ads = VdV \tag{9.1}$$

Integrating Eq. 9.1 yields (Note: 20 mph = 29.33 fps)

$$\int_0^{30\,ft} ads = \int_{29.33\,fps}^{0} VdV \Rightarrow a = -14.34 \frac{ft}{\sec^2} \tag{9.2}$$

Thus, the specification actually defines a required deceleration. Modern braking requirements and testing procedures are spelled out in the *SAE Handbook*, Vol. 2, Section 25. This section of the handbook contains requirements and testing procedures for individual braking components (J1652, J2430) as well as in-service and road tests for the braking system (J843, J201). The procedures define many specific tests, and they measure parameters including sustained deceleration, stopping distance, average and maximum torque, average and maximum pressure, and final rotor/drum temperature, among others. Without going into the details of every test, the maximum sustained deceleration required by the specifications is 0.65g. This test corresponds to a "panic" stop from 60 mph. "Normal" stopping tests specify a deceleration of 0.31g. One might ask why the required deceleration is not much higher. The answer is simply that it is unwise to require decelerations that would cause people or objects to be thrown into the front windshield.

For example, consider a Dodge Viper (1705 kg [3759 lb]) traveling at 80 mph (35.76 m/s). The car is to be stopped with the maximum sustained deceleration. The force required to bring the car to a stop is

$$F_b = 1705 \times 0.65 \times 9.81 = 10.87 \text{ kN} \tag{9.3}$$

The average power absorbed by the brakes is given by

$$P_{avg} = F_b \frac{V_o}{2} = \frac{10.87 \text{ kN} \times 35.76 \text{ m/s}}{2} = 194 \text{ kW} \tag{9.4}$$

The power at the onset of braking is twice this value, or 388 kW! Because the brakes basically convert kinetic energy to heat, this rate of energy dissipation naturally should raise concerns about heat dissipation from the system.

Furthermore, some interesting effects begin to appear when the car is braking. Recall from Section 7.3.2 that the static weight distribution of the car is given by

$$W_f = \frac{mgd}{(c+d)} = \text{weight on the front axle} \tag{9.5}$$

$$W_r = \frac{mgc}{(c+d)} = \text{weight on the rear axle} \tag{9.6}$$

The vehicle dynamics while braking can be analyzed by reference to the dynamic free-body diagram shown in Fig. 9.2.

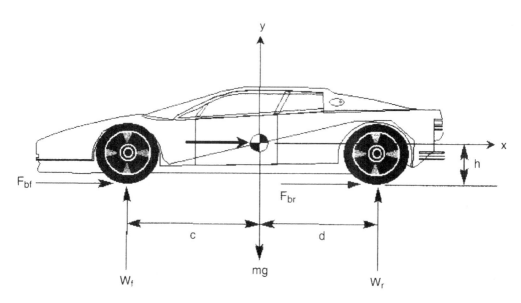

Figure 9.2. *Free-body diagram of a car under braking.*

Assuming that the car is established under steady-state braking, application of Newton's law provides the following equations:

$$\sum F_x = ma_x = F_{bf} + F_{br} \tag{9.7}$$

$$\sum F_y = 0 = W'_f + W'_r - mg \Rightarrow W'_r = mg - W'_f \tag{9.8}$$

$$\sum M_{CG} = 0 = h(F_{bf} + F_{br}) + W_r(d) - W_f(c) = ma_x h + W_r(d) - W_f(c) \tag{9.9}$$

These equations can be combined to determine the dynamic weight on the front and rear wheels during braking.

$$\begin{aligned} W'_f &= \frac{mgd}{c+d} + \frac{ma_x h}{c+d} \\ W'_r &= \frac{mgc}{c+d} - \frac{ma_x h}{c+d} \end{aligned} \tag{9.10}$$

Note that the first terms on the right sides of Eq. 9.10 are the static weights. Defining the second terms as the dynamic weight transfer, Eq. 9.10 can be rewritten as

$$\begin{aligned} W'_f &= W_f + W_d \\ W'_r &= W_r - W_d \end{aligned} \tag{9.11}$$

Thus, it is seen that under braking, weight is transferred to the front wheels. Again, consider the Viper undergoing a maximum deceleration braking from 80 mph. Statically, the car has a front/rear weight distribution of 49%/51%, a wheelbase of 2.45 m (8 ft), and the height of the center of gravity of 0.51 m (1.67 ft). Thus, under the preceding braking (0.65g), the weight distribution on the front and rear becomes

$$\begin{aligned} W'_f &= \frac{1705 \times 9.81 \times 1.2}{2.45} + \frac{1705 \times 0.65 \times 9.81 \times 0.51}{2.45} = 10.46 \text{ kN} \\ W'_r &= \frac{1705 \times 9.81 \times 1.25}{2.45} - \frac{1705 \times 0.65 \times 9.81 \times 0.51}{2.45} = 6.27 \text{ kN} \end{aligned} \tag{9.12}$$

Under braking, the car has a front/rear weight distribution of 62.5%/37.5%.

This has several implications. First, because the maximum braking force possible for a wheel is equal to the coefficient of friction times the normal force, the front wheels will have an increased capacity to provide braking force. In the preceding example, the front brakes can perform 62.5% of the braking during the stop. Second, given that the front wheels provide most of the braking, some system must be devised to apportion the application force between the front and rear. Finally, if the front brakes lock up, the vehicle loses steering, whereas if the rear locks up, the car will tend to swap ends due to the loss of cornering stiffness on the locked wheels. Thus, a system to prevent locking a wheel is desirable.

In summary, the problem facing the designer is to devise a system that can generate adequate braking forces on each axle while proportioning the braking force between the front and rear axles. The components in a modern braking system thus are designed to accomplish these functions.

9.3 Hydraulic Principles

As will be shown in later sections, the application forces required at the brake to enable the system to stop a vehicle within federal guidelines are substantial. Thus, some form of mechanical advantage must be provided to the driver. Large commercial vehicles use compressed air to provide this advantage, but all passenger cars use hydraulics to generate the requisite mechanical advantage. Hydraulics are based on Pascal's law, which basically states that for an incompressible fluid in a closed system, the pressure due to an applied force is uniform throughout the fluid. Figure 9.3 illustrates this law.

Pascal's law can be applied usefully if one of the pressure indicators is replaced by another piston with a greater surface area than the application piston. Figure 9.4 shows the results of this.

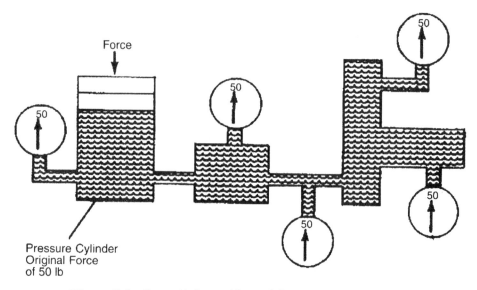

Figure 9.3. *Pascal's law. Adapted from TM 9-8000 (1985).*

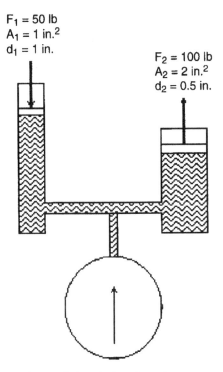

Figure 9.4. *Hydraulic advantage.*

The hydraulic system shown in Fig. 9.4 functions in the same way as a lever. With an application of 50 lb over a surface area of 1 in.^{2}, the pressure in the fluid rises to 50 psi. Because this pressure is constant throughout the fluid, the pressure in the 2 in.^{2} cylinder also is 50 psi. However, the force generated by the 2 in.^{2} piston is the pressure times the area, or 100 lb. On the other hand, the application piston moves twice as far as the power piston. The analogy, of course, is to a lever in which the long side moves twice as far as the short end, while the short end results in twice the lifting force. The primary advantage of a hydraulic system is that it occupies much less space for a given mechanical advantage than would a system of levers.

9.4 Brake System Components

This section will discuss the main system components from the brake pedal to the wheels. The components contained in the specific brake mechanisms (drums or discs) will be discussed in Sections 9.5 and 9.6.

9.4.1 Master Cylinder

Of course, the first component in the brake system is the brake pedal in the driver's compartment. Because the pedal is nothing more than a lever and provides some mechanical advantage, it will not be discussed here.

The output from the brake pedal is connected to the master cylinder. The function of the master cylinder is to provide a reservoir for the brake fluid, and it contains the driving piston in the hydraulic circuit. Initially, master cylinders contained a single piston that operated the brakes at all four wheels. The disadvantage to this system is that if a leak develops and the fluid drains away completely, the car is left with no braking whatsoever.

Modern vehicles utilize a dual master cylinder, as shown in Fig. 9.5. Each piston of the master cylinder is fed by its own reservoir and drives the brakes on two wheels. In this way, if one system develops a leak, braking will still be provided by the other two wheels. Most dual systems use an "X" configuration. In other words, one piston drives the left front and right rear wheels, while the other piston drives the right front and left rear wheels. This configuration is a compromise, in that regardless of which systems fails, braking is still provided by one front wheel.

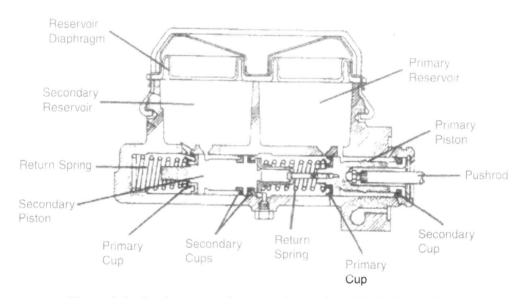

Figure 9.5. *Dual master cylinder. Adapted from TM 9-8000 (1985).*

Other options are to drive both front wheels from one of the master pistons and the rear brakes from the other. Obviously, if the front system fails in this system, braking is greatly reduced. Systems that operate one side or the other generally are not used because braking with one system lost would result in directional control problems.

9.4.2 Power Assistance

Many vehicles today, especially those with disc brakes, utilize some form of power assistance. The power booster is connected to the master cylinder, and it normally is driven by engine vacuum in a spark ignition engine, or a separate vacuum pump with a diesel engine. Figure 9.6 shows a typical power booster.

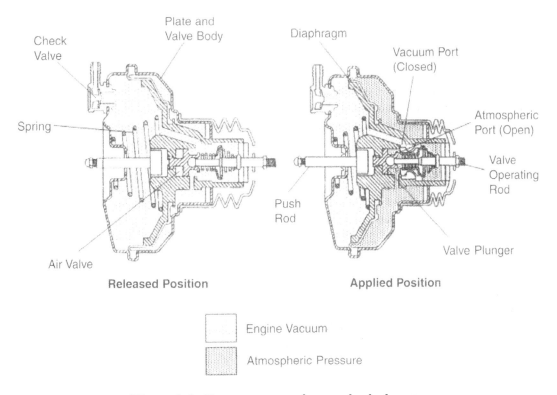

Figure 9.6. *Vacuum-operated power brake booster. Adapted from TM 9-8000 (1985).*

With the brake in the released position, the operating rod and spring push the valve plunger to the right. This seals the atmospheric port, allowing engine vacuum to be applied to both sides of the diaphragm. Thus, the spring is able to return the plunger to its initial position. When the brakes are applied, the pedal travel moves the plunger to the left. This seals the vacuum port and opens the atmospheric port. With vacuum on one side and atmospheric pressure on the other, the diaphragm applies a force to the left on the push rod. The push rod is connected to the master cylinder and thus operates the brakes. The vacuum check valve is a one-way valve that maintains engine vacuum to the left of the diaphragm. If engine vacuum is lost, the check valve prevents atmospheric pressure from developing on the left side of the diaphragm. This allows for power assist for a few brake applications, thus allowing the driver to stop the vehicle in the event of engine failure.

9.4.3 Combination Valve

The combination valve is so named because it incorporates two or three separate valves. All systems incorporate the proportioning valve and pressure differential switch, whereas vehicles with front discs and rear drums also incorporate a metering valve.

9.4.3.1 Proportioning Valve

As discussed in Section 9.2, the weight on the front wheels increases while the car is braking, enabling the front wheels to generate higher braking forces. The proportioning valve distributes the braking pressure so that brake pressure applied to the front brakes is higher than that applied to the rear. The amount of proportioning varies, depending on the vehicle in use; however, it is not uncommon for the front brakes to generate 70% of the braking force.

9.4.3.2 Pressure Differential Switch

Because most brake systems are dual master cylinder systems, a means must be provided to alert the driver to a pressure loss in one of the hydraulic circuits. The pressure differential switch is simply a plunger that is exposed to brake pressure from one system on one end, and the other system on the opposite end. If both systems are functional, brake application applies equal pressures to both ends of the plunger, and the plunger remains stationary. However, if one system has lost fluid (or is low on fluid), brake application results in different pressures being applied to opposite ends of the plunger, and the plunger is moved off-center. This motion engages a contact, which sends current to the warning light in the dash. This alerts the driver to the unsafe status of the braking system of that vehicle.

9.4.3.3 Metering Valve

As will be shown in Section 9.5, drum brakes are self-energizing. However, the self-energizing effect takes a finite amount of time to develop. As a result, drum brakes take slightly longer to activate than do disc brakes. It is desirable to have the rear brakes engage first because this provides better vehicle stability when braking. On vehicles with drum brakes on the rear and discs on the front, the discs up front would engage before the drums. Thus, the metering valve delays brake pressure to the front, which gives the rear drums time to activate.

9.5 Drum Brakes

Drum brakes are technically long shoe, internally expanding brakes. However, the name derives from the fact that the reaction surface in the system resembles a drum. Figure 9.7 shows a typical brake drum.

Normally, the brake drum is fixed to the wheel/hub and rotates with it. The shoes on the inside of the drum are expanded to make contact with the inner lining, thus generating the friction force that in turn generates a braking torque about the wheel. As has been mentioned, drum brakes are self-energizing. Before examining the shoes and their various arrangements, the phenomenon of self-energization must be understood. The simplest illustration of what this means is given by the common doorstop shown in Fig. 9.8.

As the door moves as indicated in Fig. 9.8, some activation force, F, must be applied to the stop to generate the friction required to stop the door motion. Note that the normal force opposite to the activation force is expressed as a pressure times the area of the doorstop. In

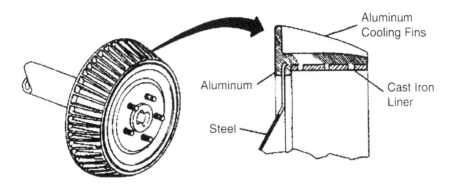

Figure 9.7. *A typical brake drum. Adapted from TM 9-8000 (1985).*

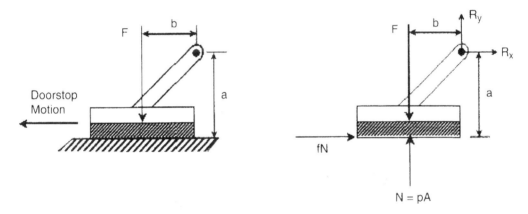

Figure 9.8. *A doorstop and corresponding free-body diagram.*

the case of the doorstop, or for short shoe brakes in general, it is safe to assume that the pressure is constant over the face of the friction surface. Furthermore, the force of friction is simply this normal force, N, times the coefficient of friction, f. Summing moments about the pin enables the required activation force to be calculated as

$$\sum M_o = 0 = Fb - Nb + fNa$$

which yields

$$F = N\left(1 - f\frac{a}{b}\right) = pA\left(1 - f\frac{a}{b}\right) \tag{9.13}$$

Equation 9.12 illustrates the self-energizing effect of the doorstop. Specifically, the motion of the door (in the direction indicated) results in a smaller activation force being required. If the hinge pin is constructed such that $f = \dfrac{b}{a}$, zero activation force is required to stop the door motion, and the stop is said to be self-locking. In the case where the door motion is reversed, Eq. 9.12 becomes

$$F = N\left(1 + f\frac{a}{b}\right) = pA\left(1 + f\frac{a}{b}\right) \qquad (9.14)$$

which indicates that to stop the door motion, an activation force is always required. Moving to the case of drum brake shoes, Fig. 9.9 shows a simple example of a system consisting of a single actuator and two individually anchored shoes.

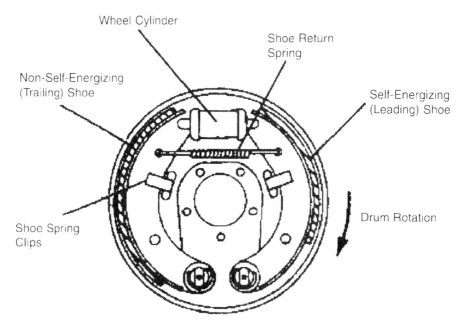

Figure 9.9. *A double-anchor, single-cylinder shoe arrangement. Adapted from TM 9-8000 (1985).*

In this case, the leading (right) shoe is self-energizing when the vehicle is traveling forward. Of course, the rear shoe is self-energizing when the car brakes in reverse, and this is one advantage of this type of arrangement. However, as a rule, cars do not travel very fast in reverse; hence, a system was desired that would allow both shoes to be self-energizing in the forward direction. Figure 9.10 shows two solutions to the problem.

The double-anchor, double-cylinder arrangement does indeed allow both shoes to be self-energizing. Each shoe has its own anchor pin that is secured to the backing plate. When the

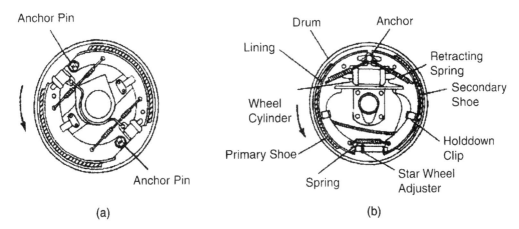

Figure 9.10. *Two brake drum configurations to give self-energization: (a) double-anchor double-cylinder, and (b) duo-servo brake. Adapted from TM 9-8000 (1985).*

car is moving forward, this arrangement allows both shoes to be self-energizing. The arrangement suffers from two disadvantages. First, when the vehicle is reversing, neither shoe is self-energizing; thus, stopping requires more pedal force from the driver. Second, the arrangement has two of everything (e.g., wheel cylinders, anchor pins), which adds to its cost and weight.

The duo-servo brake is the more popular solution. With this system, the self-energizing force is transmitted to both shoes, regardless of vehicle direction. Both shoes are actuated by a double-acting wheel cylinder, which is mounted to the backing plate at the top. When the brakes are applied, the wheel cylinder applies pressure to both shoes. When the shoes contact the drum, they both begin to move in the direction of rotation. One shoe is stopped by the anchor pin (the trailing shoe), while the other shoe is stopped by the star wheel adjuster link. Thus, the whole assembly "floats" and allows both shoes to become self-energizing.

This system also is self-adjusting. As the shoes wear, application of the brakes requires more motion of the shoe. The star adjuster is nothing more than a ratcheting device. When the wear becomes too large, the ratchet clicks to the next tang. Thus, the springs are not able to retract the shoes as far, so the motion required on the next engagement again is the same amount as for new shoes.

9.5.1 Analysis of Drum Brakes

The analysis of drum brakes is made with reference to Fig. 9.11 and proceeds with the following assumptions:

- The pivot is fixed

- The shoes are rigid

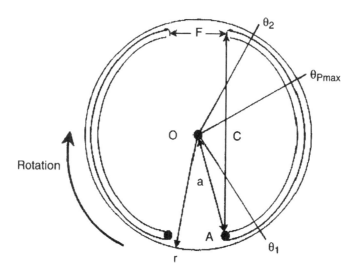

Figure 9.11. *Schematic of drum brake shoes.*

- Constant f

- b = Face width

- C = Distance from A to F

- a = Distance from A to O

- θ_1 and θ_2 are measured from the hinge pin and define the surface covered with friction material (may or may not cover the whole shoe)

It should be obvious that the pressure is not constant over the face of the shoe. The derivation of the pressure distribution is covered well in several machine design texts (Shigley and Mischke, 2001; Hamrock et al., 1999), but that is beyond the scope of this text. Referring to Fig. 9.11, the maximum pressure always occurs 90° from θ_1 (Shigley and Mischke, 2001; Hamrock et al., 1999). If θ_2 is less than 90°, then θ_2 is the point of maximum pressure. The pressure distribution thus is given by (Shigley and Mischke, 2001)

$$p = \frac{p_{max}}{\sin \theta_{P \, max}} \sin \theta \tag{9.15}$$

Figure 9.12 shows a free-body diagram of the self-energizing shoe. Both the normal force, dN, and the frictional force, μdN, have horizontal and vertical components as shown in the figure. The frictional forces produce a moment about pin A and have a moment arm of $r - a\cos\theta$. Assuming that the sum of the moments about the hinge pin is zero, the moment due to the frictional forces is found by integrating over the surface of the friction material. Making use of Eq. 9.15, the integration is

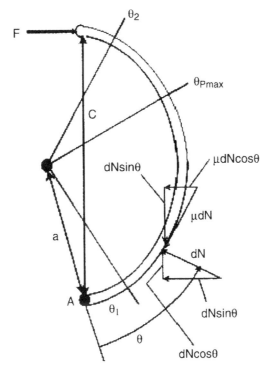

Figure 9.12. *Free-body diagram of a self-energizing shoe.*

$$M_f = \int \mu dN (r - a\cos\theta) = \frac{\mu p_{max} br}{\sin\theta_{P\,max}} \int_{\theta_1}^{\theta_2} \sin\theta (r - a\cos\theta) d\theta \qquad (9.16)$$

Expanding the integral gives

$$M_f = \frac{\mu p_{max} br}{\sin\theta_{max}} \left[-r\cos\theta \Big|_{\theta_1}^{\theta_2} - \frac{a}{2} \sin^2\theta \Big|_{\theta_1}^{\theta_2} \right] \qquad (9.17)$$

The same procedure is used to determine the moment produced by the normal force

$$M_N = \int dN(a\sin\theta) = \frac{p_{max} bra}{\sin\theta_{P\,max}} \int_{\theta_1}^{\theta_2} \sin^2\theta d\theta \qquad (9.18)$$

or

$$M_N = \frac{p_{max} bra}{\sin\theta_{max}} \left(\frac{\theta}{2} - \frac{1}{4}\sin 2\theta \right) \Big|_{\theta_1}^{\theta_2} \qquad (9.19)$$

Note that θ in Eq. 9.19 *must* be expressed in radians. Because the shoe is in equilibrium, and the moments caused by the friction forces and normal force are opposite in sign, the application force must balance these moments. Thus, the activation force is given by

$$F = \frac{M_N - M_f}{C} \tag{9.20}$$

Examination of Eq. 9.20 again shows the self-energizing effect on the leading shoe. If the analysis is repeated for the trailing (non-self-energizing) shoe, the activation force required is

$$F = \frac{M_N + M_f}{C} \tag{9.21}$$

The torque applied to the drum by the shoe is the sum of the frictional forces, fdN, times the drum radius, r,

$$T = \int \mu r dN = \frac{\mu p_{max} b r^2}{\sin \theta_{p\,max}} \int_{\theta_1}^{\theta_2} \sin\theta d\theta \tag{9.22}$$

or

$$T = \frac{\mu p_{max} b r^2}{\sin \theta_{max}} (\cos\theta_1 - \cos\theta_2) \tag{9.23}$$

9.5.2 Example

Given: A drum brake as shown in Fig. 9.13, with the following dimensions and friction characteristics.

$a = 90$ mm (3.54 in.)
$b = 30$ mm (1.18 in.)
$D = 280$ mm (11.02 in.)
$F = 1000$ N
$\mu = 0.3$

Find: p_{max} for each shoe and the total torque produced by the brake.

Solution: From the given information,

$\theta_1 = 0°$
$\theta_2 = 120°$
$r = \dfrac{D}{2} = 140$ mm (5.5 in.)

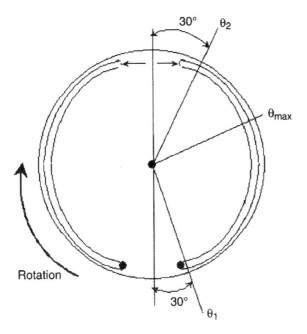

Figure 9.13. *Example problem.*

and recalling that

$$q_{Pmax} = 90°$$

First, calculate the moments as a function of maximum pressure, and calculate C:

$$M_f = \frac{0.3 p_{max}(0.03)(0.14)}{\sin 90}\left[-0.14\cos 120 + 0.14\cos 0 - \frac{0.09}{2}\sin^2 120 + \frac{0.09}{2}\sin^2 0\right]$$

$$M_f = 2.22 \times 10^{-4} p_{max}$$

$$M_N = \frac{p_{max}(0.03)(0.14)(0.09)}{\sin 90}\left(\frac{120}{2}\left(\frac{\pi}{180}\right) - \frac{1}{4}\sin 240\right)$$

$$M_N = 4.78 \times 10^{-4} p_{max}$$

$$C = 2(0.09\cos 30) = 0.156 \text{ m}$$

For the self energizing shoe,

$$F = \frac{M_N - M_f}{C} \Rightarrow 1000 = \frac{2.56 \times 10^{-4} p_{max}}{0.156} \Rightarrow p_{max} = 609 \text{ kPa}$$

$$T = \frac{0.3(609 \times 10^3)(0.03)(0.14)^2}{\sin 90}(\cos 0 - \cos 120) = 161 \text{ N-m}$$

For the non-self-energizing shoe,

$$F = \frac{M_N + M_f}{C} \Rightarrow 1000 = \frac{7.00 \times 10^{-4} p_{max}}{0.156} \Rightarrow p_{max} = 223 \text{ kPa}$$

$$T = \frac{0.3(223 \times 10^3)(0.03)(0.14)^2}{\sin 90}(\cos 0 - \cos 120) = 59 \text{ N-m}$$

Thus, for the brake, the total torque is 220 N-m.

9.6 Disc Brakes

Although drum brakes have the advantages of self-energization and ease of parking brake incorporation, they suffer from several disadvantages. Their heat dissipation is problematic, and drum brakes are prone to brake fade as the drum becomes hot due to extended or frequent heavy braking. Also, drum brakes are very sensitive to moisture or contamination inside the drum. Any water in the drum rapidly vaporizes under braking, causing the coefficient of friction of the shoe to become nearly zero.

On the other hand, disc brakes do not suffer these handicaps. The rotors can be vented to aid heat dissipation, and any water or contamination of the rotor is quickly removed by the scraping action of the pads. Figure 9.14 shows a typical disc brake system.

9.6.1 Disc Brake Components

The components of the disc brake are discussed in the next subsections.

9.6.1.1 Brake Disc

The brake disc, also called the rotor, is connected to the wheel hub. Figure 9.15 shows a typical example. The rotor provides the friction surface for the pads, thus generating the

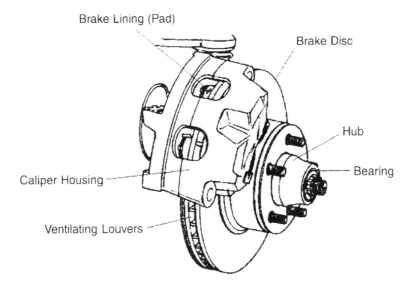

Figure 9.14. *A typical disc brake. Adapted from TM 9-8000 (1985).*

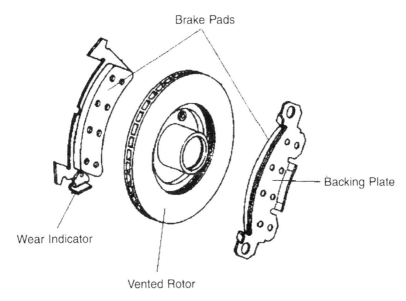

Figure 9.15. *A vented rotor and brake pads. Adapted from TM 9-8000 (1985).*

braking torque. Rotors usually are vented as shown in Fig. 9.15 to aid in the dissipation of heat. Some rotors also are cross drilled to save weight. High-performance brakes now are using carbon fiber as a rotor material. Carbon fiber provides good, fade-free performance when the material has been heated. Many Formula 1 teams use carbon fiber brakes, and the driver must ride the brake during the warm-up laps to bring them up to operating temperature.

The performance of these brakes under such demanding conditions is attested to by the fact that the rotors often glow red hot after the brakes have been applied during a race. The wheels are connected to the rotor by the lugs.

9.6.1.2 Brake Pads

The brake pads, also shown in Fig. 9.15, consist of a stamped steel backing plate to which the friction material is attached. The material, also called the lining, may be bonded to the plate with adhesive, or it may be riveted. Most disc brakes also contain a wear indicator. This indicator is a small tab of spring steel, the edge of which is set to a predetermined height below the surface of the new pad. When the pad wears to the point where it should be replaced, the spring steel begins to rub on the rotor when the brakes are applied. This produces an irritating squeal that is intended to motivate the driver to have the brake pads replaced. Should the driver ignore the warning, the brakes will continue to function to the point where no lining material remains. The author has had the experience of a fairly new disc brake pad disintegrating during a stop. During the subsequent trip home, the rivet heads remaining on the backing plate provided more than adequate stopping power, although at great damage to the rotor.

9.6.1.3 Caliper

The brake caliper houses the pistons, and these pistons apply the activation force to the brake pads. The caliper may house as few as one or as many as six pistons, depending on the specific vehicle in question. Calipers fall into two categories: (1) fixed, or (2) floating. Figure 9.16 shows a fixed caliper.

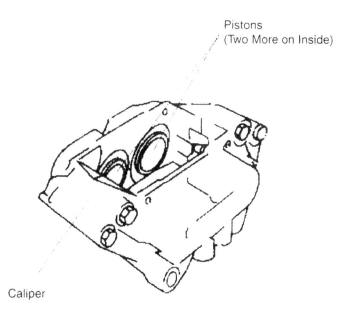

Figure 9.16. *A fixed brake caliper. Adapted from TM 9-8000 (1985).*

As its name implies, the fixed caliper is rigidly connected to its mounting surface. The fixed caliper thus requires a minimum of two pistons, one on each side. When the brakes are applied, each piston drives its corresponding brake pad into contact with the rotor. Figure 9.17 shows a floating caliper.

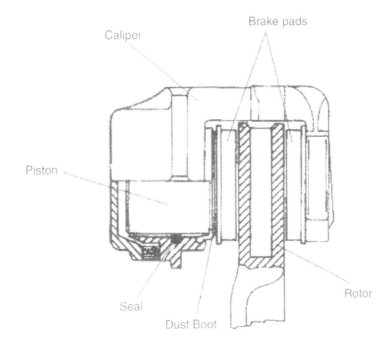

Figure 9.17. *A floating brake caliper. Adapted from TM 9-8000 (1985).*

The floating caliper can slide side-to-side on its mounting surface. Thus, pistons are required on only one side. When the brakes are applied, the piston drives its pad into contact with the rotor. This results in a reaction force that causes the caliper to slide away from the rotor (to the left in Fig. 9.17). This sliding motion brings the opposite pad into contact with the rotor, and the brakes then are fully applied. The "floating" design works well as long as there is no corrosion on the caliper pins. Corrosion may cause the caliper to bind on the pin, resulting in only the inboard pad being applied. Most passenger cars use sliding calipers because fewer components are involved than with a fixed caliper. On the other hand, most high-performance cars use the more expensive fixed caliper design with multiple pistons on either side of the caliper to generate the higher application forces that the performance of the vehicle requires.

9.6.2 Disc Brake Analysis

The analysis of the torque capacity of a disc brake proceeds straightforwardly when one recognizes that a disc brake is merely an axial clutch that is somewhat less than a complete circle (Fig. 9.18).

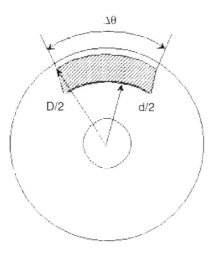

Figure 9.18. *Disc brake analysis.*

The analysis thus follows the logic contained in Chapter 6 for the axial clutch, except that the limits of integration are defined by the arc length of the pad, $\Delta\theta$, instead of going all the way around the disc (2π). Assuming constant wear, the application force is

$$F = \frac{\Delta\theta}{4} p_a d(D-d) \tag{9.24}$$

where $\Delta\theta$ must be in radians. The braking torque as a function of pressure and force are

$$T = \frac{\Delta\theta}{16} \mu p_a d(D^2 - d^2)$$

$$T = \frac{F\mu}{4}(D+d) \tag{9.25}$$

Note that the expression for torque capacity as a function of application force (the second equation in Eq. 9.25) is identical to that for an axial clutch (Eq. 6.7), and this leads many to ask why a clutch contains an entire ring of friction material. Would it not save weight and cost to use a disc brake to transmit engine torque, or would it not provide better braking to use an annulus of friction material on the rotor?

The answer is "No" for several reasons. First, although it is true that the torque capacity of the clutch and brake are equal for a given application force, that same force translates into a significantly higher pressure on the disc brake pad due to its smaller area. Thus, if a disc brake pad were to be used to transmit engine torque in the clutch, the activation force required to prevent slippage would be high enough to cause the lining material to disintegrate. For the converse, a full-circle clutch would be a poor choice for a brake precisely because brakes are designed to slip and thus generate heat. The full-circle brake would greatly inhibit cooling and would quickly result in fading brakes.

9.6.3 Heat Dissipation from Disc Brakes

The fact that brakes convert kinetic energy into heat has been mentioned, but it is instructive to consider exactly how much energy is involved during a single stop. Again using the Dodge Viper as an example, if uniform deceleration is assumed in its stop from 80 mph, the time to stop is given by

$$t = \frac{V_o}{a} = \frac{35.76 \text{ m/s}}{0.65 \times 9.81 \text{ m/s}^2} = 5.61 \text{ sec} \tag{9.26}$$

Because it was already determined that the average power absorbed by the brakes was 194 kW, the energy absorbed by the brakes during the stop amounts to

$$E = P \times t = 194 \text{ kW} \times 5.61 \text{ sec} = 1088 \text{ kJ} \tag{9.26}$$

Another way to determine the energy absorbed is to calculate the change in kinetic energy of the vehicle. Because the final velocity is zero, the change in energy is

$$\Delta KE = \frac{1}{2} m V_o^2 + \frac{1}{2} I \omega_o^2 \tag{9.27}$$

This method requires some knowledge of the components of the car, or at least a reasonable guess. For now, it will be assumed that only the rotation of the wheels/rotors is important; any other rotating mass will be neglected. If the Viper rides on P275/40ZR17 tires and the deformation of the tire is neglected, the rolling radius of the tire is 0.326 m (1.07 in.). Thus, at 80 mph, the angular velocity of the tires is 109.7 rad/sec. (See the section on tire designations for hints on how to calculate tire radius from the preceding information.) Also, if the mass moment of inertia for the wheel/tire is assumed to be 0.124 kg-m^2, then the change in kinetic energy for the Viper is

$$\Delta KE = \frac{1}{2} 1705 (35.76)^2 + \frac{1}{2} 0.124 (109.7)^2 = 1091 \text{ kJ} \tag{9.28}$$

which is sufficiently close to the previous estimate.

Because all of this energy is converted into heat, it is of great importance to the designer to understand where the heat will go, as well as to devise a means of dissipating it as rapidly as possible. Thus, it is instructive to estimate the temperature change of the rotor during this stop.

As an upper bound, it will be assumed that all of this energy is absorbed by the rotor. Furthermore, the energy will be absorbed into that portion of the rotor that is swept by the disc brake pad. The Viper has vented discs, making it difficult to exactly calculate the mass of the annulus. However, by making use of rotors from other vehicles, it is estimated that the annulus swept by the pads has a mass of 3.0 kg (6.6 lb). The specific heat at constant volume for steel is 0.475 KJ/kg-K (Lindeburg, 1998). Now, as calculated, 62.5% of the braking is done by the front brakes, and it is assumed that each front rotor contributes equally. Thus, the energy absorbed by a single front rotor is

$$E_{rotor} = 1088 \text{ kJ} \times \frac{0.625}{2} = 340 \text{ kJ} \tag{9.29}$$

Using the equation

$$E = mc_v \Delta T \tag{9.30}$$

the temperature change of the rotor for the example Viper stop is

$$\Delta T = \frac{E}{mc_v} = \frac{340 \text{ kJ}}{3 \text{ kg}(0.475 \text{ kJ/kg}-K)} = 239 \text{ K} \tag{9.31}$$

Although this overestimates the temperature rise in the rotor, such a temperature rise should not be a problem for the rotor. However, recall that this is the temperature rise for a single stop. Using the same assumptions, what would be the temperature rise for a Viper descending from the top of Pike's Peak into Manitou Springs if the driver were to "ride" the brakes all the way down? In this case, it is the potential energy of the vehicle that must be dissipated. Because Pike's Peak stands 1723 m (5653 ft) above Manitou Springs, the potential energy that must be absorbed is

$$E = mg\Delta h = 1705 \text{ kg} \times 9.81 \text{ m/s}^2 \times 1524 \text{ m} = 25.5 \text{ mJ} \tag{9.32}$$

In this case, the driver would not produce a deceleration of 0.65g, so the static weight distribution will be used to calculate braking percentage. Thus, the energy absorbed by one front rotor is

$$E_{rotor} = 25.5 \text{ mJ} \times \frac{0.41}{2} = 5.23 \text{ mJ} \tag{9.33}$$

and the temperature change would be

$$\Delta T = \frac{5.23 \text{ mJ}}{3 \text{ kg}(0.475 \text{ kJ/kg}-K)} = 3670 \text{ K} \tag{9.34}$$

Obviously, this is nonsense because the rotors would melt completely when the vehicle was only halfway down the mountain. However, it is useful in pointing out that heat transfer mechanisms are hard at work during braking. The single stop example is close to reality simply because the time scale involved is short enough that the heat does not have much time to transfer away from the rotor. For the case of multiple stops or the Pike's Peak example, all three modes of heat transfer come into play. Heat conducts into the brake pads, pistons, caliper, hubs, and so forth. Radiation also plays a role in dissipating the heat. The major factor is convection. Most disc rotors are vented, and some are cross drilled, all in an effort to aid natural convection of the heat caused by braking. Although this topic goes beyond the scope of this work, a more thorough treatment may be found in the book by Limpert (1999).

9.7 Antilock Brake Systems (ABS)

When a driver applies the brakes, the shoes/pads cause the rotating wheel to slow down relative to the ground. This generates slipping between the road and the tire, and this slip generates the braking forces on the vehicle. As the driver increases brake pressure, the slip increases and generates higher braking forces. This process is limited by the static coefficient of friction between the road and the tire. Beyond that point, the slip increases uncontrollably, and at 100% slip, the tire-road is operating at its dynamic coefficient of friction, and the wheel is locked. (This effect is shown later in Fig. 9.25 in the tire section of this chapter.)

Recall from Chapter 7 that the lateral force developed by a tire is a function of slip angle. When the wheel is locked, the lateral force generated decreases markedly (as shown in Fig. 9.26 in the next section). Thus, locking the wheels while braking has two effects. First, because the dynamic coefficient of friction is lower, the braking force generated is slightly lower, thus slightly increasing braking distance. More importantly, the locked wheel does not produce much lateral force. Thus, no steering is available on a locked front wheel, whereas a locked rear wheel is unstable because the rear wheels cannot resist the rapid increase in yaw velocity induced by steering inputs.

The function of antilock brake systems (ABS) is to provide controlled braking under all conditions. The system senses wheel lockup and momentarily reduces braking pressure to the affected wheel. Under most conditions, ABS results in a decreased stopping distance by keeping the tire-road at the maximum coefficient of friction. However, its primary benefit is that vehicle control is maintained throughout the stop by inhibiting lockup of any wheel. Figure 9.19 shows a plot of brake pressure, wheel velocity, and slip for a wheel with ABS applied.

As shown in Fig. 9.19, wheel slip increases linearly with brake pressure until the static coefficient of friction is exceeded. The wheel slip then begins to increase rapidly. The ABS senses this increase, reduces brake pressure, and thus increases wheel velocity (i.e., the wheel is no longer locked up). When the system senses that the wheel is rolling again, brake pressure is reapplied, and the process repeats itself.

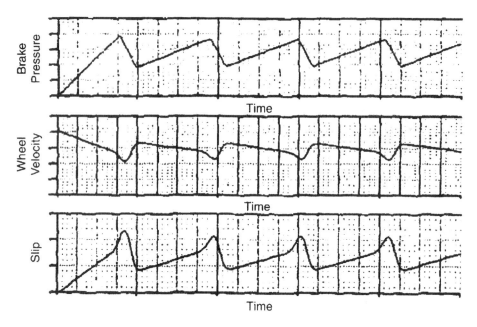

Figure 9.19. *Antilock brake system (ABS) response for a single wheel (SAE J2246, 1992).*

An antilock brake system can be classified by the number of channels used by the system, a channel being one input to the controller. Hence, a two-channel ABS has two inputs to the controller. Early systems, and some used on SUVs, have ABS on only the rear brakes and tend to be two-channel systems. A three-channel system receives inputs from each front wheel and from a rear axle sensor, as shown in Fig. 9.20.

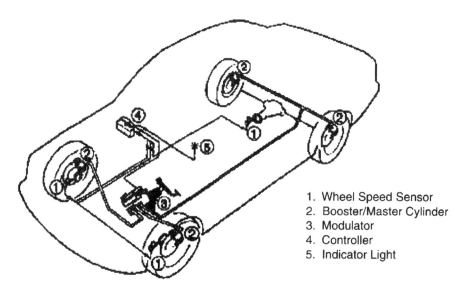

1. Wheel Speed Sensor
2. Booster/Master Cylinder
3. Modulator
4. Controller
5. Indicator Light

Figure 9.20. *A generic three-channel antilock brake system (ABS) (SAE J2246, 1992).*

The disadvantage of a three-channel system is that one rear wheel could lock without being detected by the system. Thus, most modern ABS use four-channel systems, wherein each wheel has its own speed sensor providing an input to the controller. The specific components and control strategies used in ABS vary by manufacturer.

The basic concept of ABS is simple to understand, but several complications confront the designer when implementing such a system. One complication is what is known as "split coefficient" braking. In this case, one side of the vehicle is traveling on a low-friction surface while the wheels on the other side are in contact with a much higher coefficient of friction. In the absence of any input to the controller other than wheel speed, the ABS will result in imbalanced forces side to side, which in turn causes the vehicle to turn toward the side with the higher coefficient of friction. Thus, many systems also incorporate "yaw control." The yaw sensor detects the buildup of yaw velocity under a split coefficient braking situation and modulates the brake pressure on the high-friction side to keep the vehicle tracking straight ahead. A further evolution of this system is full vehicle yaw control, sometimes called stability control or anti-rollover braking. Such systems not only detect vehicle yaw, but also measure the driver input through the steering wheel. This input tells the controller where the driver wants to go. The yaw sensor calculates where the car is actually going, and the controller modulates the brake pressure to ensure that the car does go where the driver intends it to go. Figure 9.21 shows a control schematic for such a system.

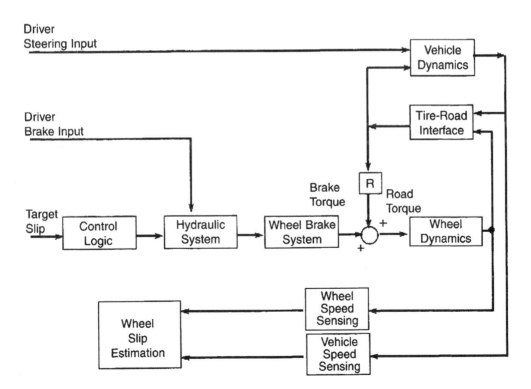

Figure 9.21. *A yaw control system block diagram. Adapted from SAE J2246 (1992).*

In a similar manner, ABS is now incorporated into stability control systems that prevent wheel spin during acceleration. Such systems currently are seen on high-end cars and enhance vehicle control when starting to move on slippery surfaces. Such systems also are used on high-performance cars, especially those with enough horsepower to generate significant wheel spin on dry pavement.

9.8 Tires

To this point in this chapter, the tire has been treated in much the same way that the average driver treats his or her tires, namely, it is assumed to be there, doing its job, and not much thought has gone into how it does that job. The truth is that the tire has a substantial job to perform, and a great deal of engineering has gone into the development of the modern car tire. The tire represents the sole point of contact between the vehicle and the road. Hence, all acceleration, braking, and steering forces must pass through those four small patches of rubber. In addition, the tire forms a component of the suspension system, and in and of itself provides stiffness and damping, thus impacting the ride and handling characteristics of the vehicle.

The credit for inventing the pneumatic tire goes to Robert Thomson, and he received a patent for his invention in England in 1845 (Woehrle, 1995a). In the United States, John Dunlop independently invented a pneumatic tire for his son's tricycle, and he received his patent in 1888 (Woehrle, 1995a). These early tires were difficult to repair. Thomson's tire had 70 bolts that had to be removed to repair the tire. This led to the development of the "clincher" tire by William Bartlett (Woehrle, 1995a). His design became the standard for tires and incorporated a set of wires at the end of the sidewall to provide a stiff surface. This surface was mounted to a lip in the rim.

Initially, reinforcement material for the tire consisted of fabric with a square weave. This quickly proved unreliable because the square weave caused a sawing action as the tire deformed (Woehrle, 1995a). This led to the bias-ply (or cross-ply) tire, in which the reinforcing cords ran diagonally from bead to bead. The radial tire was patented by Christian Hamilton Gray and Thomas Sloper in 1913; however, it was not put into use until Michelin produced the Michelin X tire in 1948 (Woehrle, 1995a). In radial tires, the reinforcing cords run perpendicularly across from bead to bead.

Early tires also were white because that is the natural color of rubber. Carbon black was added to rubber by Sidney Mote of the India Rubber, Gutta Percha, and Telegraph works in 1904, although the motivation was to improve the strength and hardness of the rubber as opposed to cosmetics (Woehrle, 1995a).

Early tires were quite naturally rather slender due to previous experience with horse-drawn carriages. As the speed and power of cars increased, the need for better-performing tires was obvious. Although the early tires had aspect ratios of 100%, this gradually was reduced over time. (See Section 9.8.2 for the definition of aspect ratio.) In 1923, Michelin introduced the "balloon" tire, with an aspect ratio of 98% and an inflation pressure of 2 bar (28 psi)

(Woehrle, 1995b). By the 1950s, aspect ratios had been reduced to 80%, with further reductions to 70% or 60% by the mid-1970s. This trend has continued until today, and tires now are available with aspect ratios as low as 35%. The lower aspect ratios improve high-speed and handling performance, and also allow for larger-diameter discs and brakes.

The next hurdle for tire development is the elimination of the spare tire. "Run-flat" tires are already coming into the marketplace, albeit on only high-end cars to date. As the technology and manufacturing processes advance, the cost of such tires inevitably will drop, allowing their use on more mainstream vehicles. Thus, the days of the spare tire seem numbered.

9.8.1 Tire Construction

As mentioned, tires basically fall into two categories of construction: (1) bias, and (2) radial. Figure 9.22 shows an example of a bias-ply tire.

As shown in Fig. 9.22, the cords of the plies in a bias-ply tire run diagonally from bead to bead. This results in a tire with good sidewall strength, a smooth ride, and adequate handling. Bias-ply tires also are cheaper to manufacture. However, bias-ply tires suffer from tread squirm, and they run hotter than other types of tire. This results in increased wear and a higher potential for failure.

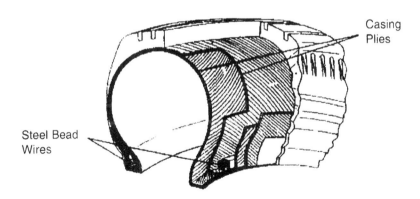

Figure 9.22. *Bias-ply (cross-ply) tire construction. Adapted from TM 9-8000 (1985).*

Initially, the cord materials were natural materials, such as cotton or linen. The first manmade material to be used was rayon, and this was superceded by nylon (Woehrle, 1995a). Nylon eventually died out due to its tendency for "flat spotting" (Woehrle, 1995a). When a car with nylon-reinforced tires remained stationary for even a brief time, the tire would deform. The deformity would remain for only a short distance when the car was driven, but until the tire regained its round shape, it produced an annoying thump. In a competitive market, this resulted in a poor first impression and hurt the sales of cars so equipped.

A follow-on to the bias-ply tire was the belted bias tire. This tire contained the usual bias plies, but they were reinforced with circumferential belts, initially made of Fiberglas® (Woehrle, 1995a). These tires ran cooler than regular bias-ply tires and provided better tread life and stopping power. However, they also produced a stiffer ride and were more expensive than bias-ply tires.

The other category of tire construction is the radial tire, shown in Fig. 9.23. The plies in this tire ran directly across the tire from bead to bead. Radial tires provide the longest tread life because they run cooler, and they also provide excellent grip. They are more expensive than bias-ply tires, and the softer sidewall is more susceptible to punctures. Furthermore, radial tires exhibit lower rolling resistance, which translates into increased fuel economy for the vehicle.

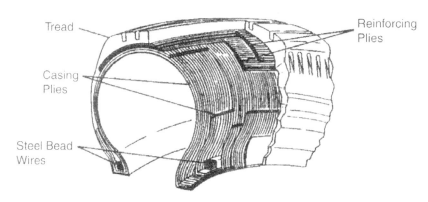

Figure 9.23. *Radial tire construction. Adapted from TM 9-8000 (1985).*

Radial tires require some type of circumferential belt for reinforcement. Fiberglas® has been used, but the most popular choice has been steel belts.

9.8.2 Tire Designations

A wealth of information can be found on the sidewall of every tire, when one understands the meaning of the designations. Prior to 1967, tires were designated by a series of numbers, such as 8.5-15. The first number was the cross-sectional width of an inflated tire, in inches; the second number was the rim diameter, again in inches. After 1967, a half-alphanumeric, half-metric designation system was devised. This gave way in 1977 to the P-metric designation used today. Figure 9.24 shows an example of the P-metric designation.

In the P-metric system, the first letter designates the type of tire: passenger car, light truck, or temporary (lightweight spare). Next, the section width is given in millimeters and is the width of the inflated tire. Then, the aspect ratio is the ratio of the section height (tread to

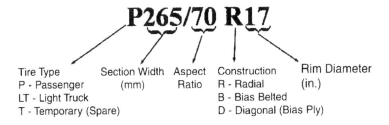

Figure 9.24. *P-metric tire designation.*

bead) to the section width. Thus, a tire that has a low aspect ratio also has a low profile (shorter sidewall). The tire construction then is designated by an R, B, or D, for radial, belted-bias, or bias-ply (diagonal) construction, respectively. The final number is the rim diameter in inches. The designation may be followed by M, S, or M + S, designating a tire designed for use in mud, snow, or both. Also, this designation may include a speed rating with the tire construction, such as P295/35ZR18. In this case, the Z is the speed rating. Table 9.1 shows other speed ratings.

TABLE 9.1
TIRE SPEED RATINGS

Symbol	Maximum Speed, mph
F	50
G	56
J	62
K	68
L	75
M	81
N	87
P	93
Q	100
R	106
S	112
T	118
U	124
H	130
V	149
Z	+149

Many cars are incapable of achieving the speed ratings of their tires. Furthermore, in the United States, such speeds would be illegal. Regardless, many cars are equipped with speed-rated tires. Although the high-speed-rated tires provide some improvement in handling, many drivers want the designation despite the inability of the car to achieve the speed. Thus, it becomes a matter of marketing. Nevertheless, it is extremely unwise to operate a tire above its speed rating. Many light-truck tires have a speed rating of 75 mph, with the caveat that the inflation pressure must be increased if those speeds are to be sustained. Thus, an SUV traveling at 85 mph on the highway runs a greater risk of catastrophic tire failure. As the tire rolls, the sidewall deforms when it arrives at the bottom of the tire due to the vehicle weight. When it rotates past this point, it springs back to its normal shape. Because the tire has its own stiffness and damping, at a certain speed the deformation and release will correspond to the resonant frequency of the tire. Such a resonant failure is rare, but the primary reason to avoid overspeeding your tires is that rubber exhibits hysteresis. As the sidewalls deform and release, the rubber generates heat. If the rate of deformation is too great, the rate of heat buildup likewise is too great. This weakens the tire and can lead to premature failure, particularly when operating in hot climates.

In the United States, the U.S. Department of Transportation (DOT) requires the Uniform Tire Quality Grading System to be molded into the sidewall of each tire. These quality ratings are intended to provide consumers with information regarding the relative performance of the tire in three areas: (1) treadwear, (2) traction, and (3) resistance to high temperature. The treadwear designation is a number that compares the wear rate of the tires subjected to a standard government test. Thus, a tire with a treadwear rating of 200 would wear twice as well as a tire with a rating of 100 on the standard government test. Of course, nobody drives his or her car in accordance with the test procedures; therefore, the treadwear rating is only a qualitative comparison, and actual tread life will depend heavily on the driving technique of the operator.

The traction rating is given as A, B, or C, and is an indication of the ability of the tire to stop on wet pavement. Again, the test procedure is rigorously spelled out by the DOT; thus, the traction rating also is dependent on driver technique. Tires with a C traction rating tend to perform very poorly on wet surfaces.

The temperature rating also is given as an A, B, or C rating. In this case, a C rating indicates that the tire meets the DOT standards for temperature resistance. Ratings of A and B indicate performance that exceeds the standard.

This leads to the second important source of tire information—the tire placard. This placard usually is located on the door jamb of the driver's side of the vehicle. The placard lists the gross vehicle weight rating (GVWR), as well as the gross axle weight rating (GAWR) for both axles. It is important for the driver to ensure that the vehicle is within these limits, and this is more often an issue for pickup trucks or SUVs than for passenger cars.

The placard also gives the recommended cold inflation pressure for the front and rear tires. It is important to measure tire pressure when the tires are cold. As mentioned, when the tire has traveled even a few miles, the sidewall flex causes heat to build up. The increased temperature of the tire results in increased pressure. Thus, taking a measurement of tire pressure when the tires are hot will lead to an erroneous high pressure reading. Drivers should check tire pressure on a monthly basis at minimum. Tires will tend to lose 1 psi of pressure per month. Driving on under-inflated tires causes increased sidewall flex and a corresponding increase in heat.

9.8.3 Tire Force Generation

The forces generated by a tire do not act through a point at the tire-road interface. Instead, the forces are distributed over the contact patch of the tire. (The contact patch is also called the footprint, and its size is a function of vertical load and tire pressure.) Furthermore, the forces are not uniform across the patch in either the lateral or longitudinal directions.

Two mechanisms are at work in generating tire forces. The first is adhesion. This is the friction developed by a rolling tire that is caused by the bonding between the tire tread and the aggregate in the road surface (Gillespie, 1994). In other words, on a molecular level, the rubber tends to flow into and over the "peaks and valleys" on the road surface. The friction produced by this mechanism is greatly reduced by water on the road.

The second mechanism is hysteresis and is the energy lost by the tire as it is deformed by the aggregate in the road (Gillespie, 1994). This mechanism is not affected to the same degree by water on the road; thus, tires with high hysteresis rubber in the tread tend to perform better in the rain. Both of these mechanisms require wheel slip in order to be generated. Wheel slip usually is defined as a nondimensional percentage and is given by

$$\text{Slip (\%)} = \frac{V - r\omega}{V} \quad (9.24)$$

where

V = vehicle velocity
r = wheel radius
ω = angular velocity of wheel (rad/sec)

As mentioned in the previous section, as the driver increases brake pressure, the slip angle increases, and the braking force also increases up to the point of the maximum (static) coefficient of friction. When that point is reached, the slip increases uncontrollably, and the tire rapidly approaches a locked condition and operates at its dynamic coefficient of friction. This is shown graphically in Fig. 9.25.

Figure 9.25 shows this relationship between slip and coefficient of friction in the longitudinal direction—that is, under braking conditions, for dry pavement. The curve maintains this shape on wet or icy roads, although the magnitude of the coefficient is proportionately lower. When a tire is turned as while steering, the lateral force generation is not quite so straightforward. The contact patch deforms due to the shear developed by the slip angle, α, which is the

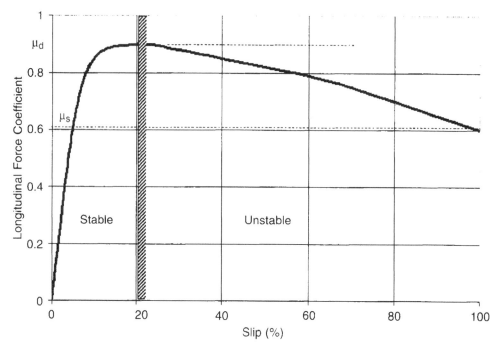

Figure 9.25. *Longitudinal coefficient of friction versus slip. Data taken from SAE J2246 (1992).*

same slip angle defined in Chapter 7 in the development of the steering dynamics equations. Figure 9.26 shows the coefficient curve for the lateral case.

In this case, the initial slope of the curve is defined to be the cornering stiffness of the tire, C_α, and for reasonable slip angles, the tire generates significant lateral forces. The problem for the tire occurs when the wheel locks up. Figure 9.25 shows that the longitudinal force, which is the braking force, drops off when the wheel locks. Of greater importance is the degree to which the lateral force declines under locked conditions, and this is shown in Fig. 9.27.

Figure 9.27 gives graphical proof of the value of ABS. Even for large slip angles (α), when the wheel locks, the lateral force coefficient drops to well below 20% of the normal force on the tire. Anyone who has slammed on the brakes (without ABS) while negotiating a turn will attest to the principal as the car plows straight ahead off the curve.

The lateral cornering stiffness is further affected by several variables of the tire, including tire construction, inflation pressure, rim size, tire aspect ratio, and load on the tire. Figures 9.28 and 9.29 show these effects for several tires.

These figures show how much better the average radial tire produces lateral forces than a similar bias-ply tire. This also helps to explain why putting radial tires on the front of the car and bias-ply tires on the rear is not good for the handling characteristics of the vehicle (reference Problem 7.1).

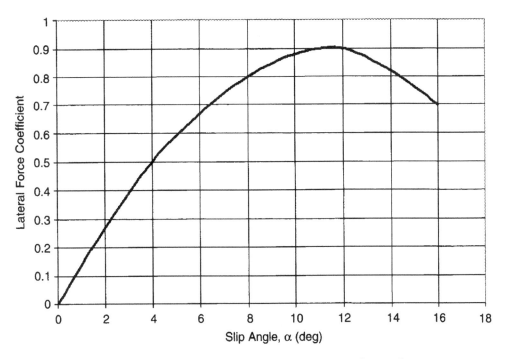

Figure 9.26. *Lateral force coefficient versus slip angle. Data taken from SAE J2246 (1992).*

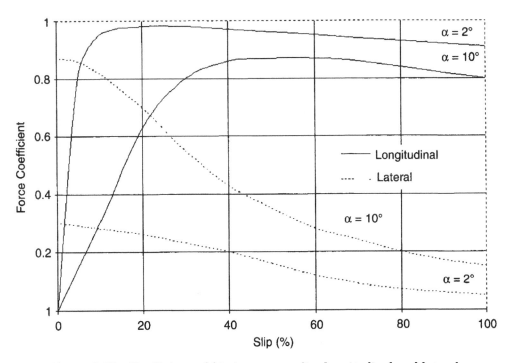

Figure 9.27. *Coefficient of friction versus slip, longitudinal and lateral. Data taken from SAE J2246 (1992).*

432 | *Automotive Engineering Fundamentals*

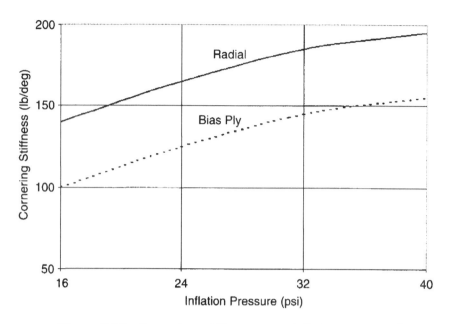

Figure 9.28. *Cornering stiffness versus inflation pressure for radial and bias-ply tires. Data taken from Gillespie (1994).*

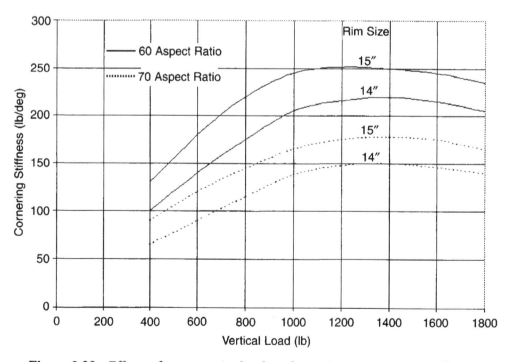

Figure 9.29. *Effects of aspect ratio, load, and rim size on cornering stiffness. Data taken from Gillespie (1994).*

Furthermore, Fig. 9.29 helps to explain some of the mystery of race car setup and adjustment. Adjustments to the aero wings on a race car alter the downforce on the front or rear of the car. Thus, it also affects the cornering stiffness of the tires, which in turn alters the understeer gradient of the vehicle. The disadvantage to dialing in more downforce is that it produces drag in addition to downforce, thus slowing the car. As a result, it is now common practice on many race circuits for the crew chief to vary the tire pressure. This, too, alters the cornering stiffness of the tire. It also affects the rolling resistance of the tire but may decrease it. Even if the rolling resistance increases, the magnitude of the resultant resistive force is much less than the additional drag produced by additional downforce.

9.9 Summary

This chapter has given a brief overview of brake systems, the analysis of brake mechanisms, and tire construction and performance. Because these topics are complex, the engineer engaged in brake or tire development is directed to the sources cited in this chapter for more in-depth treatment of these topics. Nonetheless, the techniques discussed here provide a solid foundation from which to begin study of these critical automotive components.

9.10 Problems

1. A truck with a mass of 6800 kg (14,991 lb) has a brake system capable of exerting an instantaneous braking effort of 670 kW at 40 mph (17.88 m/s). While traveling at this speed, the driver sees in his path an obstacle that is 45 m (148 ft) away. Assuming the driver's reaction time is three-quarters of a second, and assuming constant deceleration, will the truck stop before hitting the obstacle? (Answer: Yes; total stopping distance is 42.4 m [139 ft])

2. A car weighing 3220 lb goes from 60 mph to a stop in 180 ft. If the CG is 24 in. off the ground, the wheelbase is 120 in., and the car has a static weight distribution of 50/50, calculate the vertical forces on the front and rear axles during the stop. (Answer: 2040 lb front, 1180 lb rear)

3. For the drum brake in Section 9.5.2, determine the maximum coefficient of friction between the shoe and drum before the shoe becomes self-locking. (Answer: 1.64)

4. Derive the equations for the application force and torque capacity of a disc brake as functions of application force and pressure for the case of uniform pressure. (Hint: Refer to Chapter 6 for the torque capacity of a clutch with uniform pressure.)

Answer:

$$F = \frac{\Delta\theta}{8} p_a d\left(D^2 - d^2\right)$$

$$T = \frac{\Delta\theta}{24} \mu p_a \left(D^3 - d^3\right)$$

$$T = \frac{F\mu}{3} \frac{\left(D^3 - d^3\right)}{\left(D^2 - d^2\right)}$$

5. The owner of a 1997 Corvette with P245/45ZR18 tires on the rear wants to replace them with 275-width tires but does not want to change the rim diameter or worry about speedometer errors. What aspect ratio tire should the owner buy? (Answer: 40 series tires—P275/40ZR18)

Chapter 10

Vehicle Aerodynamics

10.1 Introduction

Motor vehicle aerodynamics is a complex subject because of the interaction between the air flow and the ground, and the complicated geometrical shapes that are involved. Aerodynamics is important because it affects both vehicle stability and fuel consumption. Fuel economy obviously is improved by reducing the aerodynamic drag, and the benefits can be calculated easily for constant-speed operation. However, the actual benefit for normal use will be less, because drag reduction does not significantly reduce the energy required for acceleration at normal speeds.

Road vehicle aerodynamics has been treated by Barnard (1996), who gives a very readable account, and also by Hucho (1998a), who gives a particularly comprehensive treatment. The main developments with vehicle aerodynamics probably occurred during the early 1980s, and the use of low-drag vehicles has now become common. A notable exception is off-road vehicles (e.g., sport utility vehicles) that have a very boxy shape; however, even here, drag reductions can be achieved by techniques that include corner rounding. The development of low-drag vehicle shapes is now more rapid because of greater past experience and better computational techniques.

Drag reduction is not the only aerodynamic consideration. The air flow also will affect the aerodynamic lift forces and the position of the center of pressure, both of which can have a profound effect on vehicle handling and stability. Although the presence of the ground has only a slight effect on the drag forces, it has a profound effect on the lift forces.

The aerodynamic designer also should consider the way in which the air flow controls the water and dirt deposition patterns on the glass and lamp surfaces. In addition, it is important to minimize any wind noise and to design for the ventilation flows. The air flow for engine cooling is the most significant, and the air flows for the passenger compartment, brakes, and transmission cooling are all much less significant.

The comparison of drag data from different tests normally should be avoided, because the absolute values will depend on the details of the experiment; the reasons for this are explained in the next section. However, it is valid and appropriate to examine the changes in drag as a result of changes to the vehicle shape in a given sequence of tests.

In general, vehicles are still designed by body stylists, and aerodynamicists then develop refinements to the shape to give reductions in drag and other aerodynamic improvements.

However, because the stylists are becoming more conscious of the desirability of low-drag vehicles, the basic vehicle shapes also are becoming more streamlined.

Because of the highly complex, three-dimensional, time-variant nature of the flow around a vehicle, it is impossible to computer-model the complete flow fully. Numerical techniques can be used to predict the main features of a flow or to examine some small aspect of the flow in a key area. This means that the experimental techniques applied to models in wind tunnels are very important, and these are discussed in Section 10.2.2, after a treatment of the fundamentals of vehicle aerodynamics. Because experimental testing is time-consuming and expensive, as much refinement as possible is achieved by computer modeling. This is increasing in importance with the ever-greater capabilities of computers and their programs.

Because passenger and commercial vehicles have such radically different shapes, they are the subjects of separate sections. Commercial vehicles are much less streamlined than passenger vehicles, and, in general, aerodynamic drag is less significant because the speeds are lower and the rolling resistance is more significant due to the greater weights. An additional complication that is common with trucks is the tractor trailer combination. The aerodynamic behavior depends on the spacing between the two bodies. The two extremes are zero separation, where the behavior is that of a single body, and infinite separation, where there is no "slip-streaming" effect and the drag will be that of the two bodies in isolation.

10.2 Essential Aerodynamics

10.2.1 Introduction, Definitions, and Sources of Drag

Consider a vehicle moving in a straight line on horizontal ground; the air flow is dependent on the vehicle speed and the ambient wind, as shown in Fig. 10.1. The wind has a non-uniform velocity profile because of the local topography and the earth boundary layer, and in general, the velocity will fluctuate in both magnitude and direction. The aerodynamic forces and moments act at the center of pressure. For clarity, the aerodynamic moments have been omitted from Fig. 10.1. Unlike the center of gravity, the center of pressure is not fixed but depends on the air flow; the center of pressure tends to move forward at high velocities. The aerodynamic forces are resolved in the manner shown by Fig. 10.1, because the component in the direction of the vehicle motion must be overcome by the tractive effort, not the component of force in the direction of the air motion.

Also shown in Fig. 10.1 is the lateral force coefficient center—the center of action for the lateral force coefficients from the front and rear tires. For stable operation at all speeds, the lateral force coefficient center must be behind the center of gravity (Ellis, 1969). As with the center of pressure, the center of the lateral force coefficient is not fixed, but will depend on the load transfer characteristics of both axles and the effects of traction at the driven axle. The vehicle will be stable if the center of pressure is behind the lateral force coefficient center. If the center of pressure is in front of the center of gravity, then a dynamic instability can arise: any divergence from the desired course introduces a turning moment about the center of gravity, which tends to increase the divergence further. This can be alleviated by changing the slip

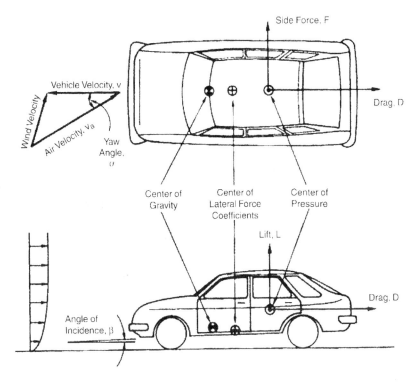

Figure 10.1. *Aerodynamic forces on a vehicle in a real environment, showing the relationships among the vehicle velocity, the (absolute) wind velocity, and the air or relative velocity (Stone, 1989).*

angles of the tires and thus the position of the lateral force coefficient center. The effect of aerodynamics on vehicle stability is described by Ward (1985) and discussed in great detail by Buchheim et al. (1985).

Because the vehicle and air velocity are not co-linear, there is a yaw angle, α, and a resultant side force. The lift force is a result of the asymmetrical flow above and below the vehicle, an effect that evidently will be influenced very strongly by the presence of the ground and the angle of incidence (defined in Fig. 10.1).

The drag and lift characteristics of a body are described by the dimensionless drag and lift coefficients, C_d and C_l. These are defined by the following equations:

$$\text{Drag, } D = \frac{1}{2}\rho v^2 A C_d$$

$$\text{Lift, } L = \frac{1}{2}\rho v^2 A C_l \tag{10.1}$$

where

ρ = air density
v = relative velocity or air velocity
A = vehicle frontal area

The term $\frac{1}{2}\rho v^2$ sometimes is called the dynamic pressure because it is the pressure rise that would occur if an incompressible flow of velocity v were brought to rest without friction.

When drag coefficients are compared, care is needed to check that the velocity and area are defined consistently. The velocity may be the vehicle speed, and the area may or may not include the area bounded by the wheels, the ground, and the underside of the vehicle. Thus, often it is better to quote the product of the frontal area and the drag coefficient, AC_d. The drag coefficient also depends slightly on the Reynolds number (effectively velocity), and this is shown in Fig. 10.2. However, Barnard (1996) shows that at Reynolds numbers above 2×10^6, the drag coefficient remains constant.

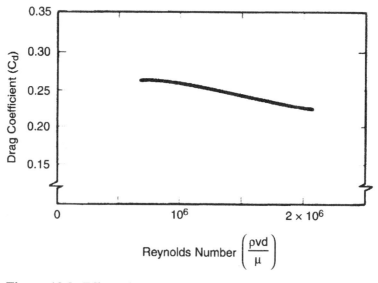

Figure 10.2 *Effect of Reynolds number on the drag coefficient. After Hucho (1978).*

The drag coefficient also is influenced by the cooling flows, the vehicle ventilation (especially if the windows are open), the ground effect, and any additions such as roof racks. An empty roof rack can increase the drag by 10%, and a bicycle on the roof can increase the drag by approximately 40% (Hucho, 1998b).

Although these effects are important, the influence of yaw angle on the drag coefficient is much more significant. Figure 10.3 shows the effect of the yaw angle on the drag coefficient for a typical car (Sovran, 1978). The ratio of the drag coefficient to that at zero yaw angle has been used here, to eliminate the problems of definition associated with absolute values of drag coefficient. Furthermore, the increase in drag of 55% is typical for a range of commercial and private vehicles. The drag increases with a non-zero yaw angle because of the way the flow will separate from the side of the vehicle. Because there generally will be a wind that is not in the direction of the vehicle motion, the sensitivity of the drag coefficient to yaw is very important. Indeed, a reduction in zero yaw drag at the expense of the peak drag occurring at a lower yaw angle is undesirable. Conversely, an increase in zero yaw drag may be beneficial for real-world fuel economy if it reduces the maximum drag coefficient when the flow is yawed.

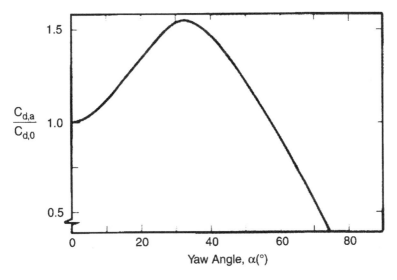

Figure 10.3. *The effect of yaw angle on drag coefficient (Sovran, 1978). Published with permission from the Plenum Publishing Corporation.*

Any tests that are designed to reveal the true drag and lift forces must take into account the ground effect, and the only way that this can be modeled properly is by having a moving ground plane. Bearman (1978) describes a series of experiments on an idealized vehicle model in which the ground clearance (between both a stationary and a moving ground) was varied. Figure 10.4 shows the results, which demonstrate (as has already been stated) that the effect of the ground is more pronounced on the lift forces than on the drag forces. The lift coefficient is very sensitive to the angle of incidence (β), especially at the low ground clearances that are found in automotive applications. In comparison, with large ground clearances, the angle of incidence and the ground motion all have a comparatively small effect on the drag. For both lift and drag, the effect of the moving ground is to decrease the effective angle of incidence, thus increasing the downforce (negative lift) and increasing the drag.

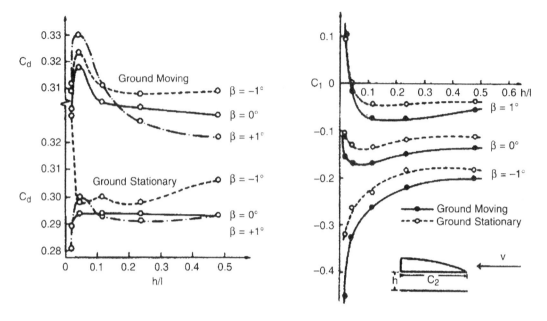

Figure 10.4. *The effect of ground clearance, angle of incidence, and ground motion on the drag and lift coefficients (Bearman, 1978). Published with permission from the Plenum Publishing Corporation.*

The higher velocity of the flow over the vehicle roof results in a lower pressure than under the vehicle body, where the flow velocity is low. According to potential flow theory (which can be used to describe the flow outside the boundary layer), the pressure difference between the topside and underside of the vehicle leads to circulation and a lift force. Furthermore, the presence of circulation implies vorticity, and because vorticity has to be preserved, there will be two trailing vortices as shown in Fig. 10.5. The interrelation between lift and drag is highly

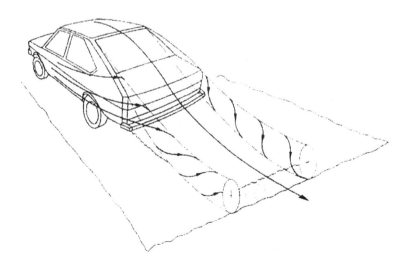

Figure 10.5. *Flow field around a car, showing the trailing vortices (Hucho, 1998b).*

complex; suffice to say that there are examples of modifications decreasing both lift and drag, increasing both lift and drag, and increasing drag while decreasing lift. Theoretically, the minimum drag would be expected to occur with zero lift.

Decreasing the lift force is of great importance in racing cars, where negative lift (or downforce) is produced at the expense of increased drag. Clearly, the maximum speed is reduced, but the high downforce enables much greater cornering speeds. Dominy and Dominy (1984) show how downforce is produced on a racing car (Fig. 10.6), and they state that the downforce can be three times the vehicle weight at a speed of 270 km/h (170 mph). The downforce arises from the inverted aerofoil (which obviously increases the drag), and the low-pressure region produced by the diffuser-like geometry under the body on either side of the driver. To minimize the inflow of air from the sides, it is possible to use flexible skirts, or if these are banned, then it is necessary to have very close control of the ground clearance. As with more conventional vehicles, a downforce is produced by the low-pressure region behind the vehicle "feeding" under the body. Because this low-pressure region is one of the sources of drag, it is clear that the downforce can be increased only at the expense of greater drag.

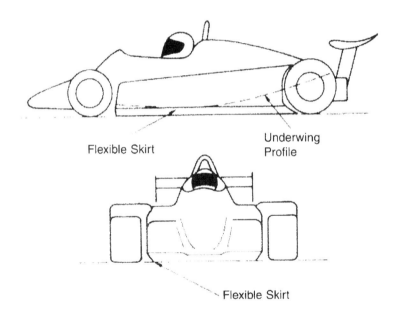

Figure 10.6. *Diagram of a Formula 1 racing car, illustrating the flexible skirts and the underside wing profile (Dominy and Dominy, 1984). Produced from the Proceedings of the Institution of Mechanical Engineers by permission of the Council of the Institution of Mechanical Engineers.*

To understand the methods used to reduce drag, first it will be necessary to discuss the mechanisms that produce drag and how these contribute to the total drag force. Only the theory will be outlined here, because this will be sufficient to provide clear definitions of the terms that are used. Much fuller treatments can be found in many engineering fluid mechanics books, such as Massey (1983).

When a fluid flows over a surface at constant speed, a drag force will be produced that consists of two parts: (1) skin friction drag (D_f) caused by viscous effects at the surface, and (2) pressure drag (D_p) as a result of the pressure distribution from the main flow (including the wake) acting on the body surface. Figure 10.7 shows the flow over part of a surface, with the resultant pressure distribution. Consider area dA at point P; the component of drag due to the pressure distribution is P sinϕ dA, and the component of drag due to the skin friction is t_w cosϕ dA. Thus, for a complete body, the total drag is

$$D = D_f + D_p = \int_A \tau_w \cos\phi \, dA + \int_A p \sin\phi \, dA \qquad (10.2)$$

where the wall shear stress is $\tau_w = \mu (d\mu/dy)_w$.

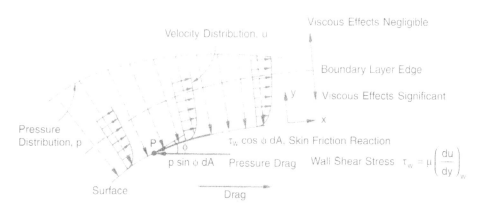

Figure 10.7. *The influence of pressure and velocity distributions on drag (Stone, 1989).*

In theory, the pressure distribution can be found by assuming inviscid flow outside the boundary layer and solving the potential flow equations. The shape of the boundary layer and the velocity distribution can be described by empirical correlations. In practice, separation occurs (i.e., the flow does not adhere to the surface), and because the position of separation and the nature of the subsequent flow are difficult to predict, complete numerical solutions are not always possible. The point of separation can vary with the Reynolds number (flow velocity), and this accounts for the slight variation in drag coefficient already seen in Fig. 10.2.

Separation occurs where there is a rapid change in the surface direction, or where the pressure is increasing in the direction of the flow (a positive pressure gradient); this is illustrated in Fig. 10.8. The positive pressure gradient tends to reverse the direction of the flow, and this is most significant at the base of the boundary layer where the fluid momentum is smallest. Reverse flow will occur where the velocity gradient away from the wall is zero. Separation

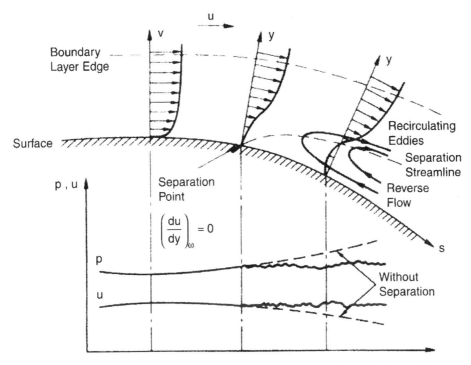

Figure 10.8. *The effect of flow separation on the velocity and pressure distribution (Stone, 1989).*

prevents a further rise in pressure, as can be seen from Fig. 10.8, and this will have an adverse effect on the pressure drag. The reversed flow next to the surface will reduce the surface drag only slightly. Whether or not the separated flow will reattach to the body will depend on the subsequent surface geometry. The reversed flow forms large irregular eddies that dissipate energy from the mainstream by viscous action.

The drag coefficient for a streamlined body with no separation will be approximately 0.05, and this is due almost entirely to surface drag. For a realistically shaped vehicle body, there will be separation, and the lowest drag coefficient feasible is likely to be approximately 0.110. Because any separation profoundly increases the drag, separation should be reduced even at the expense of increased skin friction drag. Turbulence in the boundary layer increases the skin friction, but because the momentum in the fluid close to the surface is greater, this delays the onset of separation and can lead to an overall reduction in drag. Other means of boundary layer control are possible. Some examples are boundary layer suction or flow injection, as shown in Fig. 10.9. Obviously, any gains in drag reduction must be balanced against the energy cost associated with providing the drag reduction. However, there is scope for using ventilation or cooling flows in this way.

It was stated at the beginning of this section that the center of pressure tends to move forward as the vehicle speed increases. The center of pressure will depend on a summation of the dynamic pressure terms, which are a function of velocity squared, and on the nature of the separated flow. The points of separation in the external flow are essentially fixed, but the

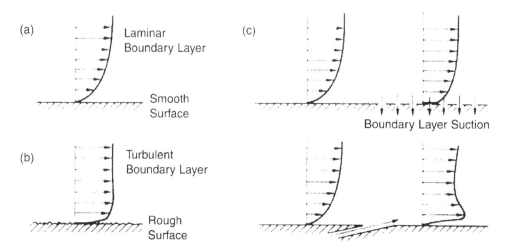

Figure 10.9. *Methods of boundary layer control: (a) boundary layer injection, (b) a turbulent boundary layer, and (c) boundary layer suction (Stone, 1989).*

positions of flow reattachment will tend to move back along the vehicle as the speed increases. The pressure recovery pattern downstream of separation thus is variable.

For an actual vehicle, it is difficult to apportion the source of drag between the skin friction drag and the pressure drag; this is especially true for the flow under the vehicle. Are the wheels, transmission, and suspension elements to be treated as rough surfaces that contribute to skin friction? Or are they to be treated as bodies that contribute to the pressure drag? Table 10.1 gives an approximate breakdown of drag for cars.

TABLE 10.1
BREAKDOWN OF THE CONTRIBUTIONS TO CAR DRAG COEFFICIENTS

	Rectangular Three Box Sedan of the 1970s	Streamlined Hatchback, Typical of the 1980s
Idealized vehicle shape	0.25	0.15
Vehicle with wheels, transmission, and suspension	0.35	0.25
With air flow to the radiator	0.40	0.27
With surface irregularities caused by body trim, doors, and glass	0.45	0.30

The trailing vortices shown in Fig. 10.5 are a result of the circulation around the car, which also causes the lift forces. The vortices obviously contain kinetic energy, and this is obtained from the mainstream as part of the work in overcoming the drag force. This component is termed "induced drag" and forms part of the pressure drag. In aerofoil theory, reducing the lift force to zero should minimize the induced drag.

Another part of the pressure drag arises from what is called "internal drag." Internal drag arises from the loss of momentum in the flows that are used for cooling and ventilation. The way that these flows exit from the vehicle also must be considered, because they can have either an adverse or beneficial effect on both the skin friction and the pressure drag. The flow through the radiator is an order of magnitude larger than any other flow; therefore, it is the only one discussed here. Consider a flow rate of air Q_r to the radiator area A_r. If the vehicle velocity is v, and the momentum from the flow is entirely dissipated, then the drag force, D_r, is given by

$$D_r = Q_r \rho v \tag{10.3}$$

and the radiator drag coefficient, C_{dr}, is

$$C_{dr} = \frac{D_r}{\frac{1}{2}\rho v^2 A} = \frac{Q_r \rho_r v}{\frac{1}{2}\rho v^2 A}$$

If $Q_r = A_r v_r$, where v_r is the flow velocity into the radiator, then

$$C_{dr} = \frac{2 v_r A_r}{vA} \tag{10.4}$$

For typical values of v_r/v and A_r/A, Hucho (1978) reports that

$$0.01 < C_{dr} < 0.06$$

10.2.2 Experimental Techniques

The drag and lift coefficients, and any other information about the flow or pressure distribution around a vehicle, must be determined experimentally. The most controlled conditions will occur in wind tunnels, but these also must be representative of atmospheric conditions.

Because the cost of wind tunnels increases with size, much experimental work is conducted with models. As all the features from a full-size vehicle can neither be copied onto a model nor adequately scaled down (e.g., surface roughness), these will be one source of the discrepancies between the model and full-size tests. The presence of an object in the wind tunnel also will modify the flow in the tunnel. Where the cross-sectional area of the model is only a few percent of the working section of the tunnel, the effect of the blockage can be neglected. Also in a wind tunnel, there will be boundary layers on the tunnel walls, which will influence the mainstream flow. The effects of boundary layers and working section blockage can be allowed for, and the methods are described in books on wind tunnel testing techniques, such as Pankhurst and Holder (1952).

Already there are two conflicting requirements: (1) that the model should be large to give more representative results, and (2) that the model should be small to minimize the blockage effects. In addition, the flow should have dynamic similarity. To give the same ratio of inertia to viscous friction forces, the Reynolds numbers (Re) must be the same, as given by

$$\mathrm{Re} = \frac{\rho v d}{\mu} \qquad (10.5)$$

where

ρ = fluid density
v = flow velocity
d = characteristic body dimension
μ = dynamic viscosity

If the wind tunnel is operating with air at atmospheric conditions, then a quarter-scale model will require flows at four times the full-scale speed. Although the drag coefficient is fairly insensitive to the Reynolds number (Fig. 10.2), the position of any flow separation is likely to vary, and this will change the position of the center of pressure.

If a vehicle is designed to travel at 135 km/h (85 mph), then the Mach number (Ma) is 0.1, and the effects of air compressibility are negligible. (Mach number is the ratio of air velocity to the velocity of sound, $\frac{v}{\sqrt{[\gamma RT]}}$.)

For a quarter-scale model at the same Reynolds number, the Mach number would be 0.4, and the effects of compressibility could be significant. This can necessitate testing at the scale of twelve inches to the foot (that is, full size), or using pressurized wind tunnels to increase the air density so that the model and full-scale velocities can be equal. Another (expensive) alternative is to use a cryogenic wind tunnel to reduce the viscosity of the air. This is the subject of Problem 10.3.

Figure 10.10 shows a typical wind tunnel. The aim is to produce a uniform flow with a low level of turbulence (local, small-scale velocity fluctuations). Return circuits are common on all but the smallest wind tunnels because the kinetic energy of the air is preserved and the power input thus is minimized. There are two reasons for accelerating and then decelerating the flow. First, by placing the fan in the slow-speed section of an open-circuit wind tunnel,

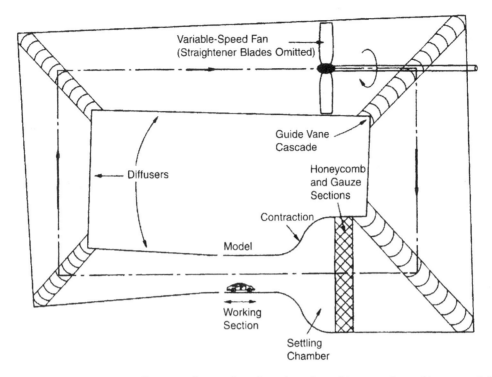

Figure 10.10. *A return flow wind tunnel with a closed working section. (An essential element that is not shown here is a heat exchanger after the fan to stop the air temperature from rising.) (Stone, 1989)*

the power input is minimized. Second, the contraction accelerates the main stream without changing the scale of the turbulence; thus, the significance of the turbulence introduced by the fan is reduced. Downstream of the working or test section of the tunnel, it may be open to the atmosphere if ambient conditions are to prevail in the test section.

The model should be mounted in such a way that permits the forces and moments to be measured by a balance; the balance usually is mounted outside the tunnel. If necessary, corrections must be applied to allow for the loading on the model support. To investigate the ground effect, a moving belt is needed. To prevent the belt surface from lifting upward, it may be necessary to apply suction to the underside of the belt.

If the turbulence level is made too low, it is then possible to add turbulence by means of grids and screens, so that the correct turbulence levels are obtained. The turbulence level is important, because a turbulent flow will cause an earlier transition from a laminar to turbulent boundary layer; this increases the skin friction drag. However, the boundary layer on a vehicle will be mostly turbulent, and, in any case, the skin friction drag is a small component of the total drag. More significantly, though, the increased turbulence will tend to delay separation of the flow (Figs. 10.8 and 10.9), which can have a notable effect on reducing the pressure drag. To investigate the performance of the vehicle in a wind, it also may be necessary to modify the boundary layer of the wind tunnel. By placing objects on the floor upstream of the working section, a more realistic representation of the earth boundary layer can be obtained. The vehicle needs to be placed on a turntable so that the effects of yaw can be studied.

Barnard (1996) presents worldwide data for 13 automotive wind tunnels, with working section areas of 12–84 m^2, and maximum speeds of the order of 200 km/h (124 mph). The majority are of the return flow type (Fig. 10.10), and even so, the power requirements can exceed 1MW (see Problem 10.5). Hucho (1998b) discusses the difference in drag coefficient reported when the same vehicle is tested in different wind tunnels. The standard deviation typically is 2% of the mean value, with closed-section wind tunnels (as in Fig. 10.10) giving consistently higher drag coefficients than tunnels with a break in the working section to ensure ambient static pressure in the test section.

The pressure distribution on the model can be found from surface pressure tappings. The diameter of these should be as small as possible, so that the pressure measurement refers to a point. The tappings often are made by hypodermic tubing, which then is connected to an appropriate manometer or pressure transducer. Such tappings must be flush with the surface; otherwise, turbulence will be introduced, and invalid readings might be obtained.

Flow visualization is a useful technique with vehicle aerodynamics; a practical description of many techniques is given by Merzkirch (1974). Smoke can be used to identify both streamlines and the regions of flow separation where there is recirculation. In practice, these techniques would be used in separate tests, but the combined effect is shown in Fig. 10.11. The smoke often is vaporized oil or kerosene, and its density and velocity should be compatible with the flow. The "rake" that delivers this flow must be streamlined; otherwise, flow disturbances will be introduced. To show regions of recirculation, the smoke can be admitted by a tubular "wand." This also can be used to identify single streamlines.

Tufts of wool or other fibers can be glued to the model, and these show the local surface flows. This is useful when trying to predict water and dirt deposition patterns. Tufts also can be mounted on a grid downstream of the model to show the flow structure in the wake. Surface flow effects of a longer time-scale can be investigated by covering the model with oil or

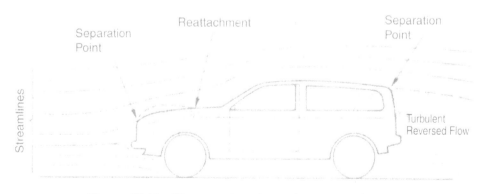

Figure 10.11. *The use of smoke to show streamlines and regions of separation (Stone, 1989).*

applying the oil in discrete dots. Sometimes pigments are added to the oil, such as lamp black or titanium dioxide. Photography is an important technique with all forms of flow visualization. Different types of information can be recorded by using short exposures, long exposures, or video. Both experience and experiment usually are necessary. Occasionally, experiments are conducted with models in water flows, which can facilitate some aspects of flow visualization.

An alternative approach for finding drag coefficients is to conduct coast-down tests on full-size vehicles. This approach also provides information on the rolling resistance of the vehicles, but the experimental difficulties can be significant. The tests should be conducted on a straight road of known inclination (preferably constant or zero), under windless conditions. Even wind speeds that are low compared with vehicle speeds produce a flow with yaw, and it has already been shown that the drag coefficient is very sensitive to the yaw angle. Furthermore, even if there is a wind with constant velocity, the yaw angle will change as the vehicle decelerates. The forces on the vehicle control the deceleration as

$$M_{eff}\frac{dv}{dt} + R + \frac{1}{2}\rho v^2 A C_d + Mg\sin\theta + U = 0 \qquad (10.6)$$

where

$$M_{eff} = M + \frac{I}{r^2}$$

and

- M = vehicle mass
- I = inertia of the roadwheels and drivetrain referred to the wheel axis
- r = wheel radius
- R = rolling resistance (assumed to be independent of speed)
- θ = inclination of the road-upward in the direction of travel taken as positive
- U = unsteady flow term, which is negligible in coast-down tests

The experiment is likely to record position or speed as a function of time, rather than deceleration as a function of velocity. Thus, the data must be differentiated, or Eq. 10.6 must be integrated. Both approaches are discussed by Evans and Zemroch (1984), with a method of fitting the data to Eq. 10.6, to determine the drag coefficient and rolling resistance by linear regression. The tests must be planned carefully, with coast-down from a range of vehicle speeds in both directions. A coast-down test is the subject of Problem 10.4.

10.3 Automobile Aerodynamics

10.3.1 *The Significance of Aerodynamic Drag*

The general aerodynamic considerations have already been discussed in the previous section. Therefore, the two key aspects in this section are how reductions in automobile drag coefficients are obtained and their effects on fuel economy.

First, it will be useful to compare the rolling resistance and the aerodynamic resistance. Figure 10.12 shows the results of this comparison. A constant rolling resistance of 225 N is assumed, and the aerodynamic resistance (drag) has been plotted for two cases: $C_d = 0.33$, and $C_d = 0.45$. In each case, the frontal area is assumed to be 2.25 m². These values are typical of a small car of the mid-1990s and mid-1970s, respectively. Because aerodynamic resistance is proportional to speed squared, the resistance is insignificant at low speeds but increases rapidly and becomes significant at high speeds. The aerodynamic resistance (drag, D) equals the rolling resistance at 80 km/h (50 mph) and 70 km/h (43 mph) for the cases where $C_d = 0.33$ and 0.45, respectively. Thus, at higher speeds, the reductions in drag will have the greatest effect on automobile performance; this is most evident for the maximum speed. The power (W) required to propel a vehicle is the product of the tractive force (N) and speed (m/s). On Fig. 10.12, the constant power lines thus are rectangular hyperbolas. For a given power

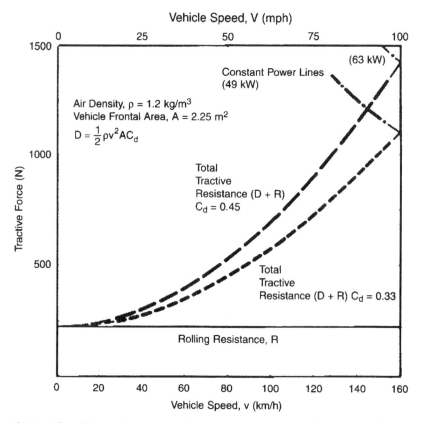

Figure 10.12. *The effect of aerodynamic drag on vehicle performance (Stone, 1989).*

(49 kW), the reduction in drag from $C_d = 0.45$ to $C_d = 0.33$ will allow an increase in the maximum speed from 145 to 160 km/h (90 to 99 mph). Alternatively, if the vehicle with $C_d = 0.45$ is required to travel at 160 km/h (99 mph), then 63 kW will be required. It is important to remember that the power used in overcoming aerodynamic drag is proportional to the speed cubed.

The effect of aerodynamic drag reductions on vehicle acceleration is small, apart from speeds approaching the maximum vehicle speed. The difference between the total tractive resistance and the tractive force available from the powertrain is used to accelerate the vehicle. This difference will reduce as the speed increases, because the tractive resistance increases and the available tractive force (F) reduces (assuming constant power). Consider the acceleration from 0 to 100 km/h (0 to 62 mph). The maximum possible tractive force available at 100 km/h (62 mph) with 49 kW maximum power is 1765 N, whereas the tractive resistance is 694 N. This leaves a balance of 1071 N for acceleration. Reducing the aerodynamic drag coefficient to $C_d = 0.33$ increases the force available for acceleration at 100 km/h (62 mph) by 11.7%. This represents an upper bound on the reduction in acceleration time, because at lower speeds, the reduction in drag will be even less significant. This case is considered further in Example 10.1 of Section 10.7, where the equations of motion are solved. This exact analysis shows a 4.2% reduction in the acceleration time.

The effect of the reduced aerodynamic drag on fuel economy evidently will be most significant at the highest vehicle speeds. A simple argument would suggest that a 10% reduction in total tractive resistance would give a 10% reduction in fuel economy. The supposition here is that the powertrain efficiency remains the same, as a result of changing the transmission ratios and/or reducing the size of the engine. Nor is any account taken here of the second-order effects. For example, reducing the size of the engine gives a weight savings throughout all the powertrain components, and a weight reduction will reduce the rolling resistance. On this simple basis, the potential fuel savings are shown in Table 10.2.

TABLE 10.2
EFFECT OF REDUCING AERODYNAMIC DRAG
FROM $C_d = 0.45$ TO $C_d = 0.33$
FOR CONSTANT SPEED FUEL CONSUMPTION
(A = 2.25 m², ρ = 1.2 kg/m³, R = 225 N)

Speed (km/h)	50	80	120	160
Reduction in fuel consumption	6.7%	15.2%	20.0%	22.5%

These fuel savings are not achieved in practice because vehicles are not driven at constant speeds. The requirements to model the fuel consumption of a vehicle are discussed further in Section 11.2. In particular, the following steps are necessary:

a. The driving pattern must be defined (speed as a function of time).
b. The powertrain efficiency must be defined.
c. The aerodynamic and rolling resistance must be defined.

Hucho (1978) demonstrated that correlations exist among frontal areas, mass, and power for European cars, and this enables a widely applicable fuel economy model to be developed. Hucho assumes a given engine efficiency map and a fixed transmission efficiency of 90%. The fuel consumption is a weighted average from the highway and urban driving cycles. Figure 10.13 shows the results from this model for three different sizes of vehicles. The figure suggests that a reduction in drag coefficient from 0.45 to 0.33 would lead to an overall gain in fuel economy of 9%.

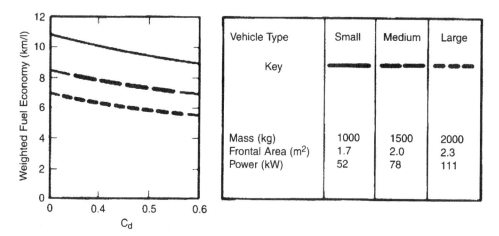

Figure 10.13. *The effect of the drag coefficient on vehicle fuel economy. Adapted from Hucho (1978).*

10.3.2 Factors Influencing Aerodynamic Drag

The aerodynamic drag of a vehicle will depend on both the overall shape of the vehicle (e.g., whether it is a notchback or a hatchback), and body details such as the gutters at the edge of the windshield or the wheel trim. Despite the apparent dissimilarities among vehicles, if cars are grouped by size into small, medium, and large sizes, and by body type as either notchback or hatchback, then in each group, the centerline cross sections and wheelbase sizes are remarkably similar (Hucho, 1978). Nonetheless, in each category, there is a significant variation in drag coefficient, which must be attributed to differences in detail design.

In hatchback cars, the angle of inclination of the rear window determines whether separation occurs at the top or bottom of the rear window. Naturally, this has a strong influence on the drag coefficient, as shown by Fig. 10.14. Evidently, when separation occurs below the rear window, dirt deposition on the rear window will be a less serious problem. The height of cars is almost independent of size, and the height of the bottom edge of the rear window is dictated by visibility requirements. Thus, longer cars can accommodate smaller angles of inclination for the rear window, which leads to lower drag coefficients.

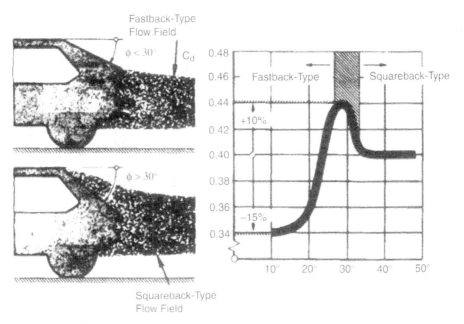

Figure 10.14. *The influence of rear-window inclination on the drag coefficient, C_d, and the region of separation (Hucho et al., 1976).*

The nose geometry also can be important through its influence on the position of flow separation on the hood (bonnet), as shown by Fig. 10.15. This is apart from the arrangement of the air flow into the engine compartment, which has been discussed in Section 10.2.1. Hucho et al. (1976) also give results for drag reductions as a result of many minor design changes. For example:

a. Rounding the transition from the roof to the rear windows can give a 9% drag reduction.

b. Reducing the width of the car to the rear can give a 13% drag reduction.

Furthermore, by ensuring that separation does not occur over the bonnet and windshield, not only is the drag coefficient minimized, but the pressure rise at the base of the windshield is increased. This pressure rise is important because it is the source of ventilation for the passenger compartment.

Attention to detail is vital if the drag coefficient of a streamlined basic body shape is not to be increased inordinately. Typical of the approach that leads to a low drag coefficient is the use of flush-mounted glass, recessed windshield wipers, optimized rearview mirrors, low-turbulence wheel trims, and effective door seals.

One significant detail is the design of the A-pillar, the pillar that separates the windshield from the side window. Not only does the design affect the aerodynamic performance, but it

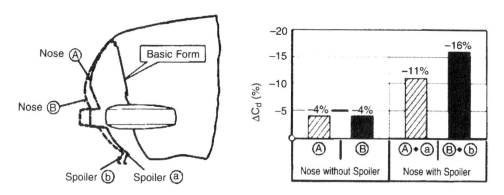

Figure 10.15. *The influence of nose geometry on the drag coefficient (Hucho et al., 1976).*

also affects the flow of water from the windshield to the side window, and the turbulence around the side window. This turbulence inside the region of flow separation produces "wind noise."

Figure 10.16 shows five different designs of A-pillars, with the drag coefficient and features associated with each design. There is a 10% variation in drag coefficient. Furthermore, the A-pillar also must be designed to prevent water from dripping into the car when the front doors are opened.

The drag coefficients of future vehicles will be governed by the interior accommodation requirements, public taste, and the cost-effectiveness of any aerodynamic refinements. Future possibilities often are illustrated by concept cars that are developed by motor manufacturers, and a significant example of this type is the Ford Probe IV vehicle shown in Fig. 10.17.

The design specification was for a fully functional vehicle with seating for four passengers and a drag coefficient below 0.2; the vehicle is described by Santer and Gleason (1983) and Peterson and Holka (1983). Many radical design features have been adopted in the Ford Probe IV, including the following:

a. Totally enclosed rear wheels, accounting for a 9% reduction in drag

b. Front wheels totally enclosed by a flexible membrane, to give a 5% drag reduction

c. Strakes to smooth the flows to and from the wheels

d. A completely smooth underbody

e. A rear-mounted cooling system

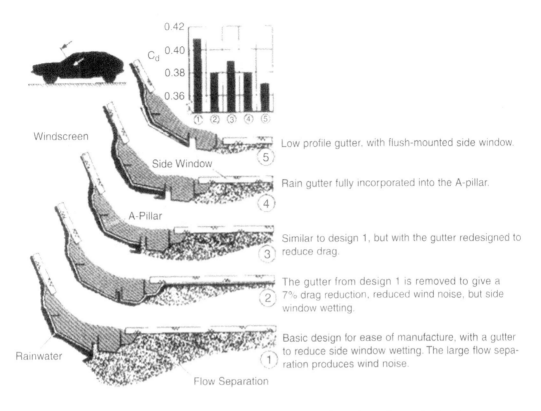

Figure 10.16. *Design of the A-pillar to obtain low drag, low wind noise, and proper flow of rainwater (Hucho et al., 1976).*

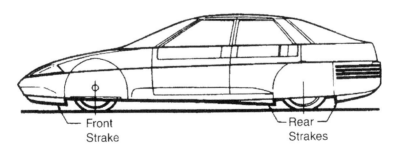

Figure 10.17. *The Ford Probe IV concept car, $C_d = 0.15$ (Santer and Gleason, 1983).*

f. Control of the vehicle ride height (equal lowering of the front and rear by 30 mm [1.2 in.] reduces the drag by 5%)

g. A transversely mounted engine inclined at 70° to the vertical, to give a low bonnet line

The result of these measures is a car with a drag coefficient of 0.15.

10.4 Truck and Bus Aerodynamics

10.4.1 *The Significance of Aerodynamic Drag*

At the start of the previous section, it was shown that the aerodynamic resistance and rolling resistance are comparable for a car traveling at approximately 75 km/h (47 mph). For a truck or bus, the drag coefficient is approximately twice that of a car, and the frontal area also is four to five times greater. Thus, the aerodynamic resistance at a given speed can be ten times greater for a truck or bus than for a car.

The rolling resistance of a truck or bus likewise is much greater than that of a car. Furthermore, the rolling resistance is highly dependent on vehicle weight, and the payload of a truck or bus varies more widely than for a car. Consequently, the rolling resistance cannot be assumed as constant. Table 10.3 shows some typical values of rolling resistance. Trucks operate with higher tire pressures, which lowers the rolling resistance coefficient.

TABLE 10.3
ROLLING RESISTANCE AS A FUNCTION OF MASS FOR VARIOUS VEHICLES

	Car	Truck	
		Unladen	Laden
Mass of vehicle (kg)	1,000	11,000	33,000
Rolling resistance of vehicle (N)	225	1,050	2,250

If a value of $AC_d = 5.7$ m² is assumed for the truck shown in Table 10.3, then the variation of aerodynamic resistance with speed will be as shown in Fig. 10.18. For this particular case, the aerodynamic resistance becomes equal to the rolling resistance at 63 km/h (39 mph) for an unladen vehicle, and at 92 km/h (57 mph) for a laden vehicle. Thus, aerodynamic resistance is significant for any vehicle cruising at motorway speeds, and for unladen or lightly loaded vehicles at significantly lower speeds. Evidently, the aerodynamic design will be particularly significant for high-speed coaches because the payload is small (say, 4000 kg [8818 lb] for 50 people). Also shown on Fig. 10.18 is the 100-kW power line. If this power is available at the rear wheels through appropriate gear ratios, then the laden vehicle can travel at 85 km/h (53 mph), and the unladen vehicle can travel at 98 km/h (61 mph).

10.4.2 *Factors Influencing Aerodynamic Drag*

The drag coefficient for a rectangular box that is typical of commercial vehicle shapes is approximately 0.9. Of course, this increases for flows of non-zero yaw angle, and the drag coefficient can become greater than unity. This simple rectangular shape is closest to that of a coach. Trucks with integral cabs and bodies and tractor trailer combinations have radically different shapes; therefore, these will be discussed afterward.

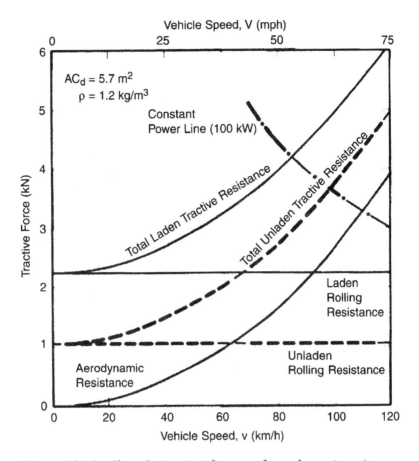

Figure 10.18. *The relative significance of aerodynamic resistance and rolling resistance for a truck (Stone, 1989).*

A bus does not have a purely rectangular shape, and these departures from a simple rectangular shape have a significant effect on drag reduction, as follows:

a. The vehicle front is convex in plan view.
b. The windshield is inclined to the vertical.
c. All corners between the front face and the sides are well rounded.

A well-known example of this is the work on the Volkswagen Microbus conducted by Moeller (1951) [also reported by Schlichting (1960)]. By making the front of this vehicle convex, with an inclined windshield, the drag coefficient is reduced to 0.76. However, there is still separation occurring with the flow onto the top and sides of the vehicle. This separation is eliminated by corner rounding, and the drag coefficient then is reduced to 0.42.

Corner rounding is an important technique in drag reduction, and the radius of curvature is quite small in many cases. Typically, $\frac{r}{b}$ is approximately 0.1 (where r is the radius of curvature, and b is the breadth of the object in the plane of measurement). Because the drag reduction occurs through reducing or preventing flow separation, when the radius of curvature is sufficient to prevent flow separation, any further increase in the radius of curvature will not lead to any further drag reduction. However, the onset of separation occurs earlier with higher Reynolds numbers (effectively flow velocity), and the minimum radius of curvature to prevent flow separation increases with increasing Reynolds numbers. Several examples of this can be found in Hucho et al. (1976).

There are essentially two types of truck: (1) those built on a single chassis, and (2) tractor trailer combinations. Tractor trailer combinations are articulated to improve maneuverability, and this necessitates an air gap between the units. The behavior of truck trailer combinations has many similarities with tractor trailer combinations; thus, these will not be discussed separately. Mason and Beebe (1978) report a series of tests in which the spacing between the tractor and the trailer was varied. The trailer used in the tests had a simple rectangular shape, with no corner rounding; the tractor had the same body width and well rounded corners ($\frac{r}{b} \approx 0.1$). Figure 10.19 shows some significant results. The drag coefficient of the isolated trailer is 0.92, and the flow is characterized by a separation bubble above the leading edge of the trailer. When the tractor is placed directly in front of the trailer, the drag coefficient is minimized ($C_d = 0.72$), and this limiting case corresponds to a single-chassis truck. As the gap between the tractor and the trailer is increased, the drag coefficient increases. When a gap exists, the flow above the tractor roof divides, and there is some downflow (with associated turbulence) between the tractor and trailer. Under these conditions, the drag coefficient will be the sum of the two drag coefficients measured in isolation.

When the tractor is placed directly ahead of the trailer, the drag reduction is caused by the tractor smoothing the air flow onto the trailer. The drag increase as the separation increases is attributed to a downflow, and this increase in drag can be virtually eliminated by using a horizontal plate to inhibit the downflow. Figure 10.20 illustrates this. A vertical plate along the centerline between the tractor and trailer also can be beneficial, especially for reducing the cross flow when there is yaw. Barnard (1996) reports typical reductions in the drag coefficient of approximately 0.05 with no yaw, and a reduction approaching 0.2 with 10° yaw.

To minimize the drag of a tractor trailer combination, the flow from the tractor must be matched to the trailer. In general, tractors are both narrower and lower than trailers. With reference to Fig. 10.21, where high- and low-drag combinations are illustrated, it can be seen that corner rounding does not always lead to drag reduction. The guiding principle is to minimize the effect of flow separation and its associated drag.

The height of the tractor also must be matched to the trailer. The simplest means of achieving this is a vertical plate, as shown in Fig. 10.22. In practice, this would not be satisfactory because for non-zero yaw angles, the drag reduction benefits would be lost. In general, a yaw angle of 15° increases the drag of a tractor trailer combination by 35–45%.

An unmatched tractor trailer combination will have a drag coefficient that corresponds almost exactly to that of an isolated trailer; this also holds for non-zero yaw angle flows. Corner rounding on the trailer appears to reduce the drag of the tractor trailer combination only if the

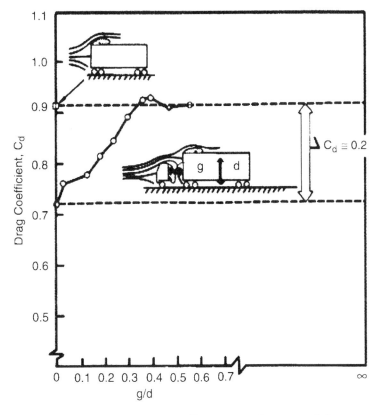

Figure 10.19. *The variation in drag coefficient as a function of the tractor trailer separation (Mason and Beebe, 1978). Published with permission from the Plenum Publishing Corporation.*

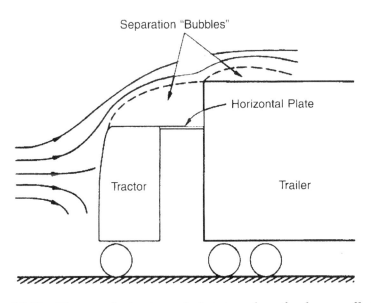

Figure 10.20. *The use of a horizontal plate to reduce the drag coefficient by eliminating the downflow between the tractor and trailer (Stone, 1989).*

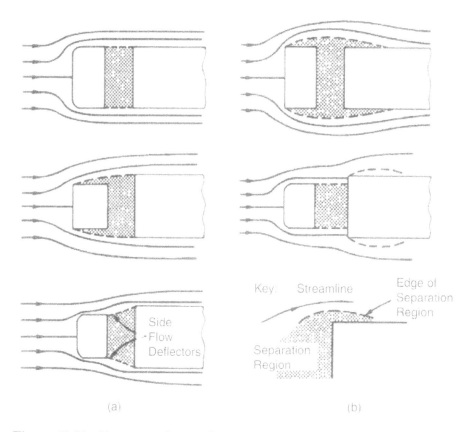

Figure 10.21. *Tractor trailer combinations with (a) low drag and (b) high drag.*

Figure 10.22. *The use of a vertical plate to match the air flow from the tractor to the trailer to reduce drag.*

combination is otherwise unmatched. The use of a fairing to match the flow can give a reduction in drag coefficient of approximately 0.2. However, this benefit is lost entirely by the time the flow has a yaw angle of 20°. If a gap seal is used to prevent lateral flow, then the reduction in drag associated with the fairing is maintained.

A significant example of a low-drag truck is the General Motors Aero Astro, which is described by Gregg (1983). This truck uses a roof-mounted deflector and side gap fillers to produce a drag coefficient in the range 0.4–0.5. The other area for improvement is in a reduction of the pressure drag associated with the separation flow behind the vehicle. The most effective device is a vertical plate, and Fig. 10.23 illustrates the way this modifies the flow. Mason and Beebe (1978) also report the increase in pressure recovery on the rear face that is caused by the vertical plate. The pressure recovery attributable to the vertical plate leads to a reduction in drag coefficient of 0.04 for a trailer and 0.09 for a bus. Such a vertical plate also might help to reduce the spray from wet road surfaces.

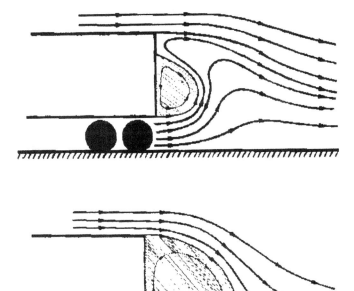

Figure 10.23. *The use of a vertical plate to increase the pressure recovery behind a vehicle, thereby reducing drag (Mason and Beebe, 1978). Published with permission from the Plenum Publishing Corporation.*

10.5 Aerodynamics of Open Vehicles

In the 1920s and 1930s when cars had separate chassis, closed sedans (saloons) were more expensive than tourers, which had fabric hoods. When unitary designs were introduced with no separate chassis, then open-top cars became more expensive, and a weight penalty usually was associated with reinforcing the body panels to provide enough stiffness, especially torsional stiffness. A significant number of soft-top cars now are available, despite such cars tending to be noisier, less secure, more expensive, and lower in performance than their closed-body counterparts.

Barnard (1996) reviewed drag data for open and closed cars, pointing out that for the old Volkswagen Beetle, lowering the hood increased the drag coefficient from 0.50 to 0.68. However, a more typical increase is 20%, but even this can be reduced by appropriate attention to detail.

Figure 10.24 shows the flows in a vertical plane around an open car as broken lines, but there also is a significant air flow into the cabin from around the side of the windshield and over the door (solid lines), which will contribute significantly to drag. The flow separating from the top of the windshield causes a large separation bubble that tends to be unstable. The resulting variation in the reattachment point causes unpleasant low-frequency pressure fluctuations in the cabin, which are known as buffeting. Furthermore, there is a substantial downdraft for the rear-seat passengers. The reattachment point can be stabilized by using a deflector at the top of the windshield and raising the hood cover. A more radical option is to have a deflector screen (inclined forward) behind the rear seats. Raising the side screens reduces the side flows, and this will give a drag reduction of the order of 0.03.

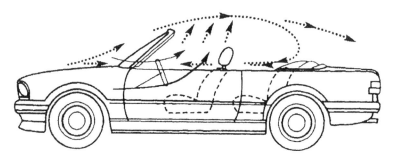

Figure 10.24. *Air flow in an open-top car (Callister and George, 1998).*

When the hood is up, the drag coefficient tends to be higher than in the parent saloon car because there is less control over the form of a fabric hood. In the case of the Rover 200, the soft-top vehicle with the hood up had a drag coefficient of 0.37, compared to 0.34–0.36 for the parent vehicle. With the soft-top lowered (but the side screens up), the drag coefficient was 0.39. This demonstrates that it is possible to develop open-top cars with only a slight drag penalty.

Motorcycle drag data tends to be more scattered because of the influence of the driver and clothing. For road-going motorcycles, the drag coefficients tend to be in the range 0.55–0.65 when there is fairing, but as high as 0.77 when there is no fairing (Bayer, 1998). There is less variation in the frontal area, with values usually in the range 0.7–0.8 m^2. Bayer also discusses the issues affecting stability, lift, and drag, and the influences of clothing, helmets, driver position, and passengers.

10.6 Numerical Prediction of Aerodynamic Performance

It should be self-evident that the aerodynamic testing and refinement of vehicle shapes is a time-consuming and expensive activity. The traditional experimental approach can be optimized by drawing on past experience—for instance, the degree of corner rounding, the bonnet and rear window slopes, the extent of windshield wrap-around, and the details of glass fixing. However, this approach still requires experiments to determine the aerodynamic characteristics of a basic vehicle shape.

An alternative approach is to solve the fluid flow equations numerically, by using the techniques developed for predicting the flow over aerofoils and through turbomachinery. Unfortunately, vehicle geometries are much more complex because they are three-dimensional non-streamlined bodies adjacent to a surface; consequently, the application of numerical methods to vehicles is more difficult and limited.

The approach in computational fluid dynamics (CFD) is to separate the flow into two parts: (1) the main stream, where viscous effects are negligible, and (2) the boundary layer, where viscous effects are significant. The flow in the main stream is assumed to be subject to only the effects of pressure and momentum (i.e., a potential flow), and the fluid flow equations thus simplify to a single linear partial differential equation (Laplace's equation). The boundary layer is very small compared with any body dimension; therefore, the effect on the body shape is negligible. However, when separation occurs, the flow is no longer attached to the body, and the flow can no longer be described by simple analytical expressions. Initially, numerical solutions were feasible only prior to the occurrence of flow separation. By implication, it is important to identify where separation will occur, to be able to define the extent of validity for any numerical solution. The point of separation can be determined by correlations that consider the down-wind rate of change of static pressure and the Reynolds number. However, it is still necessary to predict the point of reattachment.

Computational techniques have been reviewed thoroughly by Ahmed (1998). The first CFD method to be used successfully was the panel or boundary element method, a linear technique in which the surface of a vehicle is divided by a grid into a series of panels. The three-dimensional flow equations are linearized with matching boundary conditions between neighboring panels, and the flow equations are solved to give the velocity and pressure at the center of each panel.

The hierarchy of non-linear methods is as follows:

a. Inviscid Euler methods

b. Time-averaged viscous (Reynolds averaged Navier-Stokes [RANS]) methods

c. Unsteady viscous, using either large eddy simulation (LES) methods or direct numerical simulation (DNS) methods

Euler methods account for vorticity but neglect viscosity. The Navier-Stokes equations give the relationships among pressure, momentum, and viscous forces in three-dimensional space, and these must be solved alongside the continuity and energy conservation equations. The time-averaged methods (RANS) rely on sub-models to describe the effect of turbulence, and the results are critically dependent on the choice and use of the turbulence sub-models. The time averaging thus computes mean values of pressure velocity and force.

Large eddy simulation (LES) models calculate the large eddies as part of the flow, whereas small eddies are assumed to have a "universal" character that does not depend on the local flow. Of greatest complexity are direct numerical simulation (DNS) methods that make no assumptions about the turbulence and solve the full unsteady Navier-Stokes equations. Direct numerical simulation (DNS) is capable of giving excellent results (agreement with tests of the order of 1–2%, which is comparable with the accuracy of the experimental measurements) but at the cost of a grid with a million points and approximately 24 hours of computation on a Cray supercomputer.

10.7 Conclusions

Drag reduction can have a significant effect on the fuel economy of all types of vehicle, especially at high speeds. The improvements to the steady-state fuel consumption are greater than those attained in normal road use because negligible energy savings are obtained during low- and medium-speed acceleration. The effects of reducing drag are a significant increase in the maximum speed of a vehicle but a negligible reduction in most acceleration times.

When considering vehicle drag results, care should be taken in making comparisons among different tests. The drag coefficient will depend on how the area is defined, the nature of any turbulence, the Reynolds number, the velocity distribution of the incident flow, the accuracy of any body details (especially on a reduced-scale model), the internal flows, and the presence of the ground. Remember also that vehicles usually are subject to a cross wind and that drag coefficients rise markedly for non-zero yaw angles.

Vehicle aerodynamics is not only concerned with drag reduction. The stability and handling of a vehicle will depend on the position of the center of pressure and the sign and magnitude of any lift forces. Although the ground effect (the relative motion between the vehicle and the ground) has a small effect on drag, it has a profound effect on lift—even to the extent of creating a downforce. The aerodynamic design also should consider the water and dirt deposition patterns on the glass surfaces, wind noise, and the various cooling and ventilation flows.

The three-dimensional flow around a vehicle is highly complex, and complete numerical solutions have not been achieved yet. Consequently, the wind tunnel testing of models and full-size vehicles remains a vital part of their aerodynamic development. Because this is a lengthy and costly process, past experience with techniques such as corner rounding is very important. A key aspect in aerodynamic development is the control and nature of the flow separation. In this context, flow visualization is important because smoke traces enable streamlines and regions of separation to be identified.

It has been shown here that current drag levels can be reduced further, with figures as low as $C_d = 0.15$ for a car, and approximately three times this number for coaches and trucks. However, whether or not such vehicles become commercial propositions will depend on consumer attitudes toward the appearance of the vehicles, as well as the additional manufacturing costs.

10.8 Examples

Example 10.1

Calculate the 0–100 km/h (0–62 mph) acceleration time for a vehicle of effective mass (m) 1000 kg (2205 lb), rolling resistance (R) 225 N for two cases: (a) $C_d = 0.45$, and (b) $C_d = 0.33$. As in Fig. 10.12, $A = 2.25$ m², and $\rho = 1.2$ kg/m³.

Assume that the available tractive force (F) is constant at 3530 N between 0 and 50 km/h (0 and 31 mph), and 1765 N between 50 and 100 km/h (31 and 62 mph).

Solution:

Applying Newton's second law to the motion of the car gives

$$m\frac{dv}{dt} = F - \left(R + \frac{1}{2}\rho v^2 A C_d\right)$$

or

$$\frac{dv}{dt} = \frac{F-R}{m} - \frac{\rho v^2 A C_d}{2m} \tag{10.7}$$

Rearranging in a form suitable for integration gives

$$\frac{dv}{\left(\dfrac{F-R}{m} - \dfrac{\rho v^2 A C_d}{2m}\right)} = dt$$

Let

$$\alpha^2 = \frac{2(F-R)}{\rho A C_d}$$

and multiply both sides by $\dfrac{\rho A C_d}{2m}$

$$\frac{dv}{(\alpha^2 - v^2)} = \frac{\rho A C_d}{2m} dt$$

$$\left(\frac{1}{(\alpha - v)} \times \frac{1}{(\alpha + v)}\right) dv = \frac{\rho A C_d}{2m} dt$$

This equation now can be integrated to give

$$\frac{1}{2\alpha}\left[\ln\left(\frac{\alpha + v}{\alpha - v}\right)\right]_{v_2}^{v_1} = \frac{\rho A C_d}{2m}[t]_{t_2}^{t_1}$$

$$\frac{\rho A C_d}{2m} = \frac{1.2 \times 2.25 \times 0.45}{2 \times 1000} = 0.6075 \times 10^{-3}$$

Case (a): $C_d = 0.45$

Because F takes two constant values

$$\alpha = \sqrt{\frac{2(F - R)}{\rho A C_d}}$$

$$\alpha_1 = \sqrt{\frac{2(530 - 225)}{1.2 \times 2.25 \times 0.45}} = 73.76$$

$$\alpha_2 = \sqrt{\frac{2(1765 - 225)}{1.2 \times 2.25 \times 0.45}} = 50.35$$

The acceleration time must be evaluated in two stages:

(i) 0–13.89 m/s (50 km/h [31 mph]) in time t_a, and

(ii) 13.89–27.78 m/s (100 km/h [62 mph]) in time t_b

(i)

$$\frac{1}{2\alpha_1}\left[\ln\left(\frac{\alpha_1+v}{\alpha_1-v}\right)\right]_0^{13.89} = \frac{\rho A C_d}{2m}[t]_0^{t_a}$$

$$t_a = \frac{1}{2\times 73.36\times 0.6075\times 10^{-3}}\left[\ln\left(\frac{73.76+13.89}{73.76-13.89}\right)-0\right] = 4.25 \text{ s}$$

(ii)

$$\frac{1}{2\alpha_2}\left[\ln\left(\frac{\alpha_2+v}{\alpha_2-v}\right)\right]_{13.89}^{27.28} = \frac{\rho A C_d}{2m}[t]_0^{t_b}$$

$$t_b = \frac{1}{2\times 50.35\times 0.6075\times 10^{-3}}\left[\ln\left(\frac{50.35+27.28}{50.35-27.28}\right)-\ln\left(\frac{50.35+13.89}{50.35-13.89}\right)\right] = 11.04 \text{ s}$$

Total acceleration time is

$$t_a + t_b = 4.25 + 11.04 = 15.29 \text{ s}$$

Case (b): $C_d = 0.33$

Thus,

$$\alpha_1 = 86.13$$

and

$$\alpha_2 = 58.80$$

Again, the acceleration is evaluated in two stages:

$$\frac{\rho A C_d}{2m} = \frac{1.2\times 2.25\times 0.33}{2\times 1000} = 0.4455\times 10^{-3}$$

$$t_a = \frac{1}{2\times 86.136\times 0.4455\times 10^{-3}}\left[\ln\left(\frac{86.13+13.89}{86.13-13.89}\right)-0\right] = 4.24 \text{ s}$$

$$t_b = \frac{1}{2\times 58.80 \times 0.4455 \times 10^{-3}}\left[\ln\left(\frac{58.80+27.28}{58.80-27.28}\right)-\ln\left(\frac{58.80+13.89}{58.80-13.89}\right)\right]=10.41 \text{ s}$$

The total acceleration time is now 14.65 s, a reduction of 0.64 s, or 4.2%. As would be expected for the reasons given in the discussion of Fig. 10.12, the differences in acceleration times for 0–50 km/h (0–31 mph) are negligible.

In general, the available tractive force also will be a function of speed, and this usually will prevent the equation of motion being solved analytically. Instead, a numerical approach will be needed, using the equations derived here. The acceleration times for a change of velocity must be evaluated over a time increment for which the available tractive force (F) can be treated as a constant. The acceleration times for all the velocity increments then can be summed to give the total acceleration time.

Example 10.2

For the vehicle with parameters defined by Fig. 10.12, there is a residual tractive force (the difference between tractive force available from the engine and the total tractive resistance) of 550 N when traveling at 120 km/h (75 mph). What is the maximum headwind into which the speed of 120 km/h (75 mph) can be maintained?

Solution:

Drag is

$$D = \frac{1}{2}\rho A C_d v^2$$

Thus,

$$D_{120+w} - D_{120} = 550 \text{ N}$$

where D_{120} is the drag at 120 km/h (75 mph), and D_{120+w} is the drag at 120 km/h (75 mph) with a headwind of speed w.

Thus,

$$550 = \frac{1}{2}\rho A C_d \frac{(120+w)^2 - 120^2}{\left(60^2 \times 10^{-3}\right)^2}$$

Vehicle Aerodynamics

$$550 = \frac{\frac{1}{2} \times 1.2 \times 2.25 \times 0.33}{\left(60^2 \times 10^{-3}\right)^2} \times \left(240w + w^2\right)$$

or

$$w^2 + 240w - 16 \times 10^{-3} = 0$$

of which the positive solution is w = 54 km/h (34 mph).

Alternatively, at 120 km/h (75 mph):

Drag is

$$D_{120} = \frac{1}{2}\rho A C_d v^2 = \frac{1}{2} \times 1.2 \times 2.25 \times \left(60 \times 10^3 / 60^2\right)^2$$

$$= 495 \text{ N}$$

Thus, the total force available to overcome drag is

$$495 + 550 = 1045 \text{ N}$$

$$1045 = \frac{1}{2}\rho A C_d \frac{(120+w)^2}{\left(60^2 \times 10^{-3}\right)^2}$$

$$(120+w) = \frac{60^2}{10^3 \sqrt{\dfrac{1045}{\frac{1}{2} \times 1.2 \times 2.25 \times 0.33}}}$$

As done previously, taking the positive value gives w = 54 km/h (34 mph).

10.9 Discussion Points

10.1 Identify the types of aerodynamic drag, and state their relative significance.

10.2 Apart from drag reduction, what should be considered in the aerodynamic design of vehicles?

10.3 State the principal techniques for reducing aerodynamic drag of passenger cars.

10.4 The MIRA wind tunnel has an open circuit, with a working section of 4.42 by 7.94 m (14.5 by 26 ft) and a maximum speed of 133 km/h (83 mph). Four fans are located after a divergent section at the exit of the tunnel. Suppose that the annular exit area from each fan has an inner diameter of 2 m (6.6 ft) and an outer diameter of 5.5 m (18 ft), and that a constant air density of 1.2 kg/m³ can be assumed throughout the wind tunnel. What is the loss of power associated with the unrecovered kinetic energy of the flow leaving the fans?

Answer: 192 kW (The electrical input to the fans is actually 970 kW, because of frictional losses at the entry and in the ducts, and the finite fan efficiency.)

10.5 A proposal is being considered for a quarter-scale wind tunnel that will obtain matching of the Reynolds number and Mach number for an ambient temperature of 25°C (77°F), by means of operating at low temperatures at ambient pressure. The temperature dependence of the viscosity of air can be described by Sutherland's Law, with the following numerical values:

$$\mu = 8.14 \times \frac{413.15}{T+120} \times \left(\frac{T}{293.15}\right)^{3/2} \times 10^{-6} \text{ Nsm}$$

At what temperature should the tunnel be operated?

Answer: 111 K (The liquefaction of oxygen at –183°C [361°F] and nitrogen at –196°C [385°F] present practical limitations.)

10.6 A coast-down test has been performed on a sport utility vehicle to give the velocity history as shown in Fig. 10.25.

Using the following data, estimate the value of AC_d:

Vehicle mass	1850 kg (4079 lb)
Effective vehicle mass	1900 kg (4189 lb)
Rolling resistance coefficient (assumed to be independent of speed)	0.015
Inclination of the road-upward in the direction of travel taken as positive	0
Density of air	1.2 kg/m³

Answer: 3.5 m²

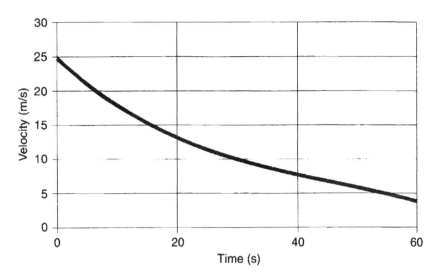

Figure 10.25. *Graph for Discussion Point 10.6.*

10.7 If the value of AC_d for the vehicle discussed in Fig. 10.18 is reduced from 5.7 m² to 4.1 m², calculate the potential improvements in the steady-state fuel consumption for the laden and unladen cases, at speeds of 60 and 100 km/h (37 and 62 mph), assuming the same powertrain efficiency. For the laden vehicle, what is the increase in maximum speed with 100-kW propulsive power?

Answers:

Reductions in Fuel Consumption (%)	Unladen	Laden
60 km/h (37 mph)	10.4	8.3
100 km/h (62 mph)	20.0	15.3

Maximum laden speed increases from 23.8 to 25.7 m/s (85.6 to 92.5 km/h [53.2 to 57.5 mph]).

10.8 What are the methods for reducing the drag of trucks with separate tractors and trailers?

10.9 A high-performance car has a drag coefficient of 0.37 and a frontal area of 2.44 m². The effective mass of the vehicle and driver is 2000 kg (4409 lb), and the ambient conditions are an air temperature of 20°C (68°F) and a pressure of 101.325 kN/m². In the speed range 100–200 km/h (62–124 mph), the rolling resistance is 340 N and the mean torque at the rear wheels is 1123 Nm; the effective diameter of the tires is 0.58 m (1.9 ft).

Show that if the drag coefficient were reduced to zero, there would be a 29% reduction in the 100–200 km/h (62–124 mph) acceleration time.

CHAPTER 11

Transmission Matching and Vehicle Performance

11.1 Introduction

To provide an appreciation of the issues involved in matching the transmission to a vehicle, use is made here of an extended worked example. This simple case study is restricted to steady-state operation and is concerned with sizing the engine for a given top-speed performance, determining the gear ratio required for a specified hill-starting performance, and calculating the fuel economy for constant-speed operation with different transmission options. This example demonstrates the need for computer modeling, which is introduced first in general terms. The chapter ends with an overview of the ADVISOR modeling software, a versatile package that is available free of charge.

11.2 Transmission Matching

Unfortunately, the power requirements of vehicles are characterized by part-load operation. Thus, it is necessary to consider not only the engine efficiency, but how it is matched to the vehicle through the transmission system. Because the principles in matching the gearbox and engine are essentially the same for any vehicle, it will be sufficient to discuss only one vehicle. The example used here is a vehicle with the specification as shown in Table 11.1.

TABLE 11.1
EXAMPLE VEHICLE SPECIFICATION

Rolling resistance, R	225 N
Drag coefficient, C_d	0.33
Frontal area, A	2.25 m²
Required top speed	160 km/h
Mass	925 kg

The tractive force (F) is a function of speed (v) for this vehicle

$$F = R + \frac{1}{2}\rho v^2 A C_d$$

The first step is to determine the power requirement (and thus the engine size) for maximum speed. The speed torque characteristics of the engine then allow the "top gear" gear ratio to be determined as shown here in Section 11.2.1. The gear ratio (expressed as a ratio of vehicle speed to engine speed) then enables the final drive ratio (the back axle ratio in a rear-wheel-drive vehicle) to be determined for a particular size of tire. Section 11.2.2 shows how an overdrive gear ratio allows better fuel economy to be obtained, and this theme is continued in Section 11.2.3 with the discussion of continuously variable transmissions (CVT). This section also introduces the concept of gearbox span (the ratio between the highest and lowest gear ratios), which is treated more fully in Section 11.2.4. Section 11.2.4 shows how the lowest gear ratio must be determined on the basis of starting the vehicle on an incline, and that the span of a CVT limits the extent of the available "overdrive ratio" and thus the fuel economy.

11.2.1 Selecting the Engine Size and Final Drive Ratio for Maximum Speed

At a speed of 160 km/h (99 mph), a power of 49 kW is required (brake power, $W_b = F \times v$). For the sake of this discussion, a manual gearbox is assumed, but the same general principles apply for automatic transmissions. The term "top gear" here refers to either the third gear in an automatic gearbox or the fourth gear in a manual gearbox. Similarly, the term "overdrive" here refers to either the fourth gear in an automatic gearbox or the fifth gear in a manual gearbox. To travel at 160 km/h (99 mph), a power of 49 kW is needed at the wheels (W_w). To determine the necessary engine power (W_b), divide by the product of all the transmission efficiencies ($\Pi \eta$).

$$W_b = \frac{W_w}{\Pi \eta} \tag{11.1}$$

Assuming efficiencies of top gear 95% and final drive 98%, then

$$W_b = \frac{49}{0.98 \times 0.95} = 52.6 \text{ kW}$$

Suppose a four-stroke spark ignition engine to be used has the engine map defined by Fig. 11.1. The contours show the engine efficiency expressed in terms of the brake specific fuel consumption (sfc, which is inversely proportional to the brake efficiency, η_b), because this facilitates estimation of the fuel consumption.

$$\eta_b = \frac{3600 (\text{s/h})}{\text{sfc}(\text{kg/kWh}) \times \text{CV}(\text{kJ/kg})} \times 100\%$$

Assume a calorific value (CV) of 44,000 kJ/kg for gasoline.

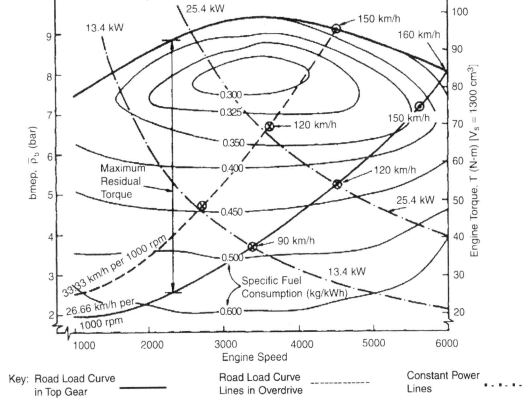

Figure 11.1. *Road load curves and constant power lines added to an engine fuel consumption map for a spark ignition engine (Stone, 1989).*

If the top speed of the vehicle (160 km/h [99 mph]) is to coincide with the maximum engine speed (6000 rpm), then the overall gearing ratio must give 26.7 km/h (16.6 mph) per 1000 rpm. At 6000 rpm, the brake mean effective pressure (bmep, p_b) is 8.1 bar. If the brake power (W_b) required is 52.6 kW, the swept volume necessary can be found from

$$V_s = \frac{W_b}{p_b \times N'} = \frac{52.6 \cdot 10^3}{8.1 \times 10^5 \times 6000/120} = 1300 \, \text{cm}^3$$

where N' is the number of cycles per second.

Now that the swept volume has been determined, the bmep axis on Fig. 11.1 can be recalibrated as a torque, T,

Power $\qquad W_b = p_b \times V_s \times N' = T \times \omega$

For a four-stroke engine,

$$\omega = 4\pi \times N'$$

$$T = \frac{p_b \times V_s}{4\pi} \quad \text{(N-m)}$$

Because the gearing ratios and efficiencies have been defined such that the maximum power of the engine corresponds to 160 km/h (99 mph), the total tractive resistance curve (i.e., the propulsive force as a function of speed) can be scaled to give the road load curve (i.e., the engine torque required for propulsion as a function of engine speed), as shown in Fig. 11.1. This scaling automatically incorporates the transmission efficiencies because they were used in defining the maximum power requirement of the engine.

Also identified on Fig. 11.1 are the points on the road load curve that correspond to speeds of 90 and 120 km/h (56 and 75 mph). The difference in height between the road load curve and the maximum torque of the engine represents the torque that is available for acceleration and overcoming headwinds or gradients. In the case of 120 km/h (99 mph), there is a balance of 41.8 N-m. The torque (T) can be converted into a tractive effort because the overall efficiency and gearing ratios are known.

Gearing Ratio (gr)   26.67 km/h (16.57 mph) per 1000 rpm

$$= \frac{26.67}{60} = 0.444 \text{ m/rev}$$

$$\frac{0.444}{2\pi} = 0.07074 \text{ m/radian}$$

The residual tractive force available is

$$\frac{T}{gr} \times \eta_{gearbox} \times \eta_{final\,drive} = \frac{41.8}{0.07074} \times 0.95 \times 0.98 = 550 \text{ N}$$

Because the vehicle mass is 925 kg ((2039 lb), its weight is 9074 N. Thus, 120 km/h (99 mph) can be maintained up a gradient of $\frac{550}{9074} = 6\%$. If this gradient is exceeded, the vehicle will slow down until sufficient torque is available to maintain a constant speed. As the speed reduces, the torque required for steady level running is given by the road load curve, and the torque available is determined from the engine torque curve. The rate at which this difference increases as speed reduces is referred to as the torque backup. A high torque backup gives a vehicle good driveability because both the speed reduction when gradients are met and the need for gear-changing are minimized. The maximum residual torque available in top gear for hill climbing occurs at 2200 rpm (which corresponds to 59 km/h [37 mph]). If the speed reduces beyond this point, the torque difference decreases. Assuming the gradient remains unchanged, the engine soon would stall. In practice, a gear change would be made long before this point is met because a driver normally would attempt to maintain speed by operating the engine close to the maximum power point of the engine.

By interpolation on Fig. 11.1, the specific fuel consumption of the engine can be estimated as 0.43 kg/kWh at 120 km/h (75 mph) and 0.49 kg/kWh at 90 km/h (56 mph). The power requirement at each operating point can be found from the product of torque and speed. Because the specific fuel consumption also is known, it is possible to calculate the steady-state fuel economy at each speed. Table 11.2 summarizes these results.

TABLE 11.2
TOP GEAR PERFORMANCE FIGURES
FOR VEHICLE AS DEFINED BY TABLE 11.1 AND FIG. 11.1
(ALL VALUES IN TERMS OF ENGINE OUTPUT)

Vehicle Speed v (km/h)	Engine Speed (rpm) ω (rad/s)	Torque T (N-m)	Power W_b T × ω (kW)	sfc (kg/kWh)	Fuel Economy v/(bsfc × W_b) (km/kg)
90	3375	38	13.4	0.49	13.7
120	4500	54	25.4	0.43	11.0
150	5625	74.5	43.9	0.375	9.1

Figure 11.1 clearly shows that none of these operating points is close to the area of the highest engine efficiency. Because power is the product of torque and speed, lines of constant power appear as hyperbolas on Fig. 11.1. The operating point for minimum fuel consumption is where these constant power hyperbolas just touch the surface defined by the specific fuel consumption contours. This optimal economy operating line has been added in Fig. 11.2. However, if a vehicle with a conventional manual or automatic gearbox were designed to operate at these points, there would be so little torque backup that the vehicle could not be driven. The only way for the optimal economy operating line to be followed is by means of a continuously variable transmission system. A less radical option is to use an overdrive ratio because this moves the engine operating point closer to the regions of lower fuel consumption.

11.2.2 Use of Overdrive Ratios to Improve Fuel Economy

Suppose a gearbox with an overdrive ratio of 0.8:1 and an efficiency of 95% was to be fitted. From this information, it is possible to draw the additional road load line shown in Fig.11.1 as follows. The overall gearing ratio is now 33.33 km/h (20.71 mph) per 1000 rpm, and this

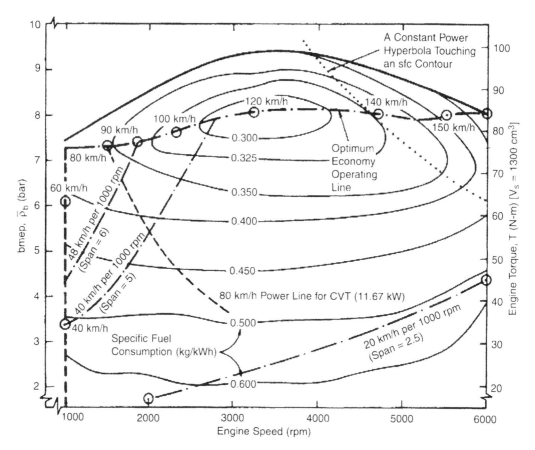

Figure 11.2. *Locus for operation of a continuously variable transmision (CVT) on the fuel consumption map of Fig. 11.1 (Stone, 1989).*

enables the x-axis to be scaled directly. When scaling the y-axis, due account must be taken of the reduced efficiency of the overdrive ratio (95%) compared with the direct drive (98%). From Eq. 11.1,

$$\text{Torque} = \text{Tractive force} \times \text{Gearing ratio} \times \frac{0.98}{0.95}$$

This reduction in efficiency also accounts for the slight increase in power requirements to 26.2 kW at 120 km/h (75 mph) and 13.8 kW at 90 km/h (56 mph).

Figure 11.1 shows that the overdrive road load curve intersects the maximum engine torque curve at 4500 rpm. Because the gearing ratio is 33.33 km/h (20.71 mph) per 1000 rpm, the top speed in overdrive will be 150 km/h (93 mph). This is quite a small reduction in top speed from when top gear is used.

At 120 km/h (75 mph), the residual engine torque available is 29.2 N-m (compared with 41.8 N-m in top gear). Taking into account the transmission efficiency, this implies an ability to overcome 4.2% gradients or 40-km/h (25-mph) headwinds. As previously mentioned, the specific fuel consumption can be found at the operating points by interpolation on Fig. 11.1, and the fuel consumption can be calculated. Due account must be taken of the increased engine power output requirement that is caused by the reduced transmission system efficiency in overdrive. Table 11.3 summarizes the results.

TABLE 11.3
OVERDRIVE PERFORMANCE FIGURES
FOR VEHICLE AS DEFINED BY FIG. 11.1
(ALL VALUES IN TERMS OF ENGINE OUTPUT)

Vehicle Speed v (km/h)	Power W_b (kW)	sfc (kg/kWh)	Fuel Economy $v/(bsfc \times W_b)$ (km/kg)	Improvement in Fuel Economy Compared with Top Gear (%)
90	13.8	0.445	14.7	7.3
120	26.2	0.335	13.7	24.5
150	45.2	0.420	7.9	−13.2

The use of an overdrive gear ratio can be seen to give a 24.5% improvement in fuel economy at 120 km/h (75 mph), and a 7.3% improvement in fuel economy at 90 km/h (56 mph) in this particular case. At 150 km/h (93 mph), the fuel consumption is worse in overdrive because the engine is operating at full throttle with a rich air-fuel mixture and a correspondingly high specific fuel consumption. Of course, additional benefits of using an overdrive ratio are the reduced engine noise and wear.

11.2.3 Use of Continuously Variable Transmissions (CVT) to Improve Performance

First, continuously variable transmissions (CVT) tend to have a lower mechanical efficiency than conventional gearboxes. Second, CVTs have only a finite span (the ratio of the minimum to maximum gear ratios). The lowest gear ratio is determined by the fully laden hill-start requirement; thus, the finite span may not be able to permit the engine to operate at high-torque/low-speed combinations. Furthermore, at the minimum engine operating speed of 1000 rpm, the minimum fuel consumption operating point corresponds to approximately 8 kW—sufficient power to propel the vehicle at approximately 70 km/h (43.5 mph). Thus, even a CVT system will not enable the lowest fuel consumption to be obtained at low vehicle speeds. The solution is a hybrid vehicle, in which at low powers (less than, say, 8 kW in this example) the engine is not used, but a battery/electric motor system is used. Hybrid vehicles are discussed further in Section 12.2.

It has already been stated that the optimal efficiency for a given power output is where the constant power line just touches a specific fuel consumption contour. The locus of these points defines the optimum operating line for a vehicle, as shown in Fig. 11.2. The only way this locus can be followed is by means of a CVT.

Assuming an efficiency of 90% for the CVT and 95% for the final drive, the power that is required to propel a vehicle can be calculated. The operating points for the CVT then can be found by interpolation on Fig. 11.2. However, because this engine cannot run below 1000 rpm, at speeds of approximately 70 km/h (43.5 mph) and lower, the engine must operate at part throttle.

At 120 km/h (75 mph), it may appear that the residual torque available is only 14.6 N-m, but this is not the case with a CVT. If more power is required at a given speed, the reduction ratio in the CVT increases so that the engine speed can rise, and the operating point moves to the right on the locus. At 120 km/h (75 mph), the overall gearing ratio gives 38.4 km/h (23.9 mph) per 1000 rpm. To obtain maximum torque at this road speed, maximum power is needed (6000 rpm), and the gearing ratio would reduce to 20 km/h (12.4 mph) per 1000 rpm. To calculate the residual torque at this speed, it is simplest to refer to the tractive effort at the road wheels.

Tractive effort required for 120 km/h (75 mph) 722 N

Maximum power available at the engine 52.6 kW

Maximum power at the wheels is found by including the transmission efficiencies $52.6 \times 0.95 \times 0.9 = 45.0$ kW

$$\text{Maximum tractive effort} = \frac{\text{Maximum power}}{\text{Speed}} \qquad (11.2)$$

$$= \frac{45 \times 10^3}{120 \times 10^3 / (60 \times 60)} = 1350 \text{ N}$$

Subtracting the tractive effort required for 120 km/h (75 mph) on level ground with no headwind gives $1350 - 722 = 628$. Thus, 628 N is available for overcoming headwinds or gradients when the gearing ratio is 20 km/h (12.4) per 1000 rpm. This implies an ability to climb 6.9% $\left(\frac{628}{(925 \times 9.81)}\right)$ gradients. When compared with the performance in top gear of a fixed ratio gearbox, this is only a slight improvement because the CVT efficiency is lower and the difference in the gearing ratios is only slight. Figure 11.2 also shows that the top speed is reduced by 8 km/h (5 mph) because of the inefficiency of the CVT.

The main attraction of a CVT is the potential for improvements to vehicle fuel economy. As already stated, the power requirement at each speed can be calculated, and the specific fuel consumption can be found from interpolation on Fig. 11.2. Table 11.4 summarizes the results.

TABLE 11.4
PERFORMANCE OF A VEHICLE WITH A CVT
AS DEFINED BY FIG. 11.2
(ALL VALUES IN TERMS OF ENGINE OUTPUT)

Vehicle Speed v (km/h)	Power W_b (kW)	bsfc (kg/kWh)	Fuel Economy v/(bsfc × W_b) (km/kg)
90	14.6	0.332	18.6
120	27.7	0.290	14.9
150	47.8	0.420	7.5

For convenience, the fuel economy figures, with an indication of the gradient climbing ability and maximum velocity, are summarized in Table 11.5 for the three different transmission systems that have been discussed.

TABLE 11.5
COMPARISON OF VEHICLE PERFORMANCE
FOR THREE DIFFERENT TRANSMISSION SYSTEMS

Vehicle Speed (km/h)	Fuel Economy (km/kg)		
	Top Gear	Overdrive	CVT
90	13.7	14.7	18.6
120	11.0	13.7	14.9
150	9.1	7.9	7.5
Gradient (%) that can be ascended at 120 km/h (75 mph)	6.0	4.2	6.9
Maximum speed (km/h)	160.0	150.0	154.0

In summary, using an overdrive or CVT improves the vehicle fuel economy at all but the highest speeds, with the CVT providing the largest gains. At the highest speeds, the lower efficiency of an overdrive ratio or a CVT leads to the engine operating in a less efficient regime, thereby leading to a reduction in vehicle fuel economy.

The results shown in Table 11.4 should be viewed with caution because the engine map that has been used is typical but arbitrary, and constant values of efficiency have been assumed for the transmission elements. In practice, the map from a specific engine would be combined with a transmission system that should have efficiency recorded as a function of speed, load, and reduction ratio. At this level, the calculations are most effectively performed by a computer, but the logic and approach are, of course, the same as adopted here.

11.2.4 Gearbox Span

The span of a gearbox is the ratio between the highest and lowest reduction ratios. In the introduction to this chapter, it was stated that the ability to start ascending a gradient (say, 33%) determined the lowest gear ratio. This requirement will be discussed for a car with a manual gearbox. In the case of an automatic gearbox, the torque converter will reduce the size of the reduction ratio needed, and it will suffice here to quote some typical results for comparison.

Figure 11.1 shows that the engine produces a maximum torque of 99 N-m at 3500 rpm. It will be assumed here that the clutch can transmit this torque with whatever slip is necessary because initially the gearbox and final drive will be stationary. From Table 11.1, we have the following:

> Vehicle mass   925 kg
> Vehicle weight $925 \times g = 9074$ N

Then, the tractive force necessary on a 33% gradient is

$$0.33 \times 9074 + \text{rolling resistance N}$$

$$= 2994 + 225 = 3219 \text{ N}$$

Allowing for the transmissions efficiencies, then the gearing ratio (gr) is

$$gr = \frac{T}{F} \times \eta_{\text{gearbox}} \times \eta_{\text{final drive}}$$

$$gr = \frac{99}{3219} 0.95 \times 0.98$$

$$gr = 0.0286 \text{ m/radian}$$

This compares with 0.07074 m/radian in top gear, implying a reduction ratio of 2.47:1 for the lowest gear because top gear is invariably a direct drive. In practice, the reduction ratio between first and fourth gears will be greater to reduce the peak demand on the clutch and provide some torque in reserve. If the starting requirement was based on an engine torque of 73 N-m (by using a reduction ratio of 3.36), then Fig. 11.1 shows that the clutch could be fully engaged at 1000 rpm as opposed to 3500 rpm. The ratio of 3.36 also provides a margin for starting the car on steeper slopes and when the car is more heavily loaded.

Table 11.6 gives some typical gearbox ratios with the step-up ratios. It may seem strange that the step-up ratios are not constant because this would have given a geometric progression of gear ratios. A geometric progression theoretically would allow optimum acceleration by using the same position of the torque curve in each gear. However, in each case here, the step-up ratio into top gear is smaller than the other step-up ratios. This occurs because the third gear of a manual gearbox (or second gear for an automatic gearbox) is used most frequently after top gear, for overcoming gradients or acceleration, and the step-up ratio associated with a geometric progression would be greater than optimum. An added advantage is the improved fuel economy in third gear when there is a small step-up ratio into top gear.

TABLE 11.6
MANUAL AND AUTOMATIC GEARBOX RATIOS

Five-Speed Manual Gearbox			Four-Speed Automatic Gearbox		
Gear	Reduction Ratio	Step-Up Ratio	Gear	Reduction Ratio	Step-Up Ratio
1	3.36	1.6	1	2.39	1.65
2	2.10	1.5	2	1.45	1.45
3	1.40	1.4	3	1.00	1.25
4	1.00	1.25	4	0.80	
5	0.80				

Because all the gear ratios have now been defined, it is possible to compute the tractive force available for each gear ratio as a function of vehicle speed. This is shown in Fig. 11.3, with the tractive force that would be available with a CVT. Other than for first gear, the CVT tractive force curve lies below the maxima of the individual gear ratios because the CVT efficiency is assumed to be 90%.

Another limitation with a CVT is the span (i.e., the range of ratios) that can be achieved (usually only approximately five or six in a single stage). Because the largest reduction ratio is determined by the hill-starting requirement, the finite span of a CVT will determine what the smallest reduction ratio will be. That is, the finite span limits the extent of the available overdrive gearing.

In Fig. 11.2, the starting requirement is for an overall gearing ratio of 8 km/h (5 mph) per 1000 rpm. Consider this vehicle being driven at 120 km/h (75 mph) on level ground with no headwind, with the driver wishing to reduce the speed gradually to 80 km/h (50 mph). Initially, the throttle would remain fully open, and as the vehicle slowed, the engine would slow faster as the overall gearing ratio reduced. However, if the span of this gearbox was 5, then at 112 km/h (70 mph), the gearing ratio would become its maximum of 40 km/h (25 mph) per 1000 rpm. Any further reduction in the vehicle speed then would be accomplished by a corresponding reduction in the engine speed, and the throttle closing as the operating point

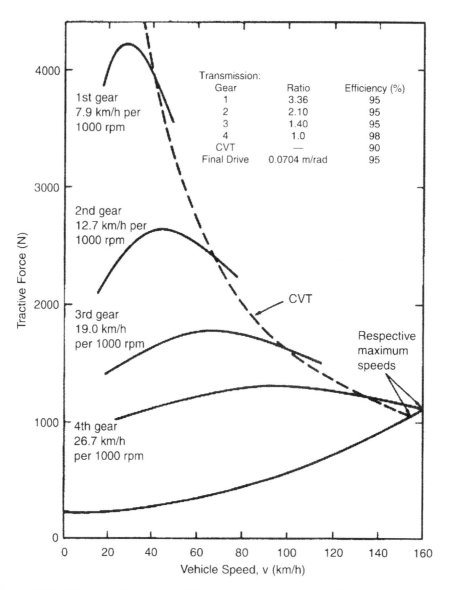

Figure 11.3. *Tractive force available as a function of speed for the vehicle defined in Table 11.1, engine defined by Fig. 11.1, and gearbox ratios of Table 11.6.*

followed the 40 km/h (25 mph) per 1000 rpm road load curve to the 80 km/h (50 mph) operating point (i.e., an engine speed of 2000 rpm).

Referring again to Fig. 11.2, it can be seen that the full benefits of a CVT are not attainable with a span of either 5 or 6. In this particular case, a span of 8 is necessary to follow the optimum locus. The limited span also will affect the fuel economy. For example, if the span is 6, then optimum fuel economy cannot be attained in the speed range 48–90 km/h (30–56 mph). The effect on fuel economy at 80 km/h (50 mph) is shown as a function of span in Table 11.7.

TABLE 11.7
EFFECT OF A CONTINUOUSLY VARIABLE TRANSMISSION (CVT) ON MEDIUM-SPEED FUEL ECONOMY AT 80 KM/H (50 MPH)

	CVT Span			Fixed Ratios	
	5	6	>6.6	Top (3.36)	Overdrive (4.2)
Power (W_b) (kW)	11.67	11.67	11.67	10.72	11.06
Specific Fuel Consumption (sfc) (kg/kWh)	0.42	0.375	0.36	0.515	0.475
Fuel economy (km/kg)	16.3	18.3	19.0	14.6	15.2

Table 11.7 shows that if a CVT is to offer significant fuel economy improvements, it must be able to operate over a wide span of gear ratios. This is partially to offset the inherently lower efficiency.

This discussion has not included automatic gearboxes because the torque converter characteristics are difficult to define. Furthermore, because the trend now is toward automatic gearboxes with "lockup" torque converters (as discussed in Section 6.5), the steady-state treatment is the same as for manual gearboxes.

Finally, when a comparison is made between the performance of spark ignition and compression ignition engines, it is important to remember that there are slight differences in the calorific values of the fuels, but significant differences in their densities and thus their volumetric calorific values. Table 11.8 shows some typical values that can be used for making comparisons between diesel and gasoline engines and vehicles.

TABLE 11.8
TYPICAL DENSITIES AND CALORIFIC VALUES FOR GASOLINE AND DIESEL FUELS

	Gasoline	Diesel
Calorific value (MJ/kg)	44	42
Density (kg/m^3)	750	900
Volumetric calorific value (MJ/L)	33	38

11.3 Computer Modeling

11.3.1 Introduction

Powertrain optimization evidently is not possible by analytical solution of closed form equations. Indeed, the simple examples developed in Section 4.2 are useful as illustrations of a method, but a method that would quickly become tedious. Spreadsheets can be used quite readily for estimating acceleration times by treating the system as quasi-steady. It also is possible to compute steady-speed fuel consumption data, but this requires interpolation within the specific fuel consumption data. In general, this is not easy to do with spreadsheets, but it can be done (e.g., the Vauxhall case study in Chapter 12). Likewise, it is possible to model drive cycles, but this can be a triumph of ingenuity over common sense.

Powertrain optimization lends itself to computer modeling, and the requirements for a model would include the following:

- **For steady-state operation**

    a. Access to engine maps showing the torque output, emissions, and specific fuel consumption, and a method of interpolation

    b. A method for deducing wind resistance as a function of area, drag coefficient, ambient conditions, and vehicle speed

    c. A method for deducing rolling resistance as a function of vehicle weight, speed, and tire type

    d. A way of specifying the gearing ratios and their efficiencies as a function of the operating condition

- **For transient operation**

    e. Moment of inertia of the engine and moment of inertia of the gearbox in different gears

    f. Time constants for gear changing and engine response (especially if turbocharged)

    g. Modifications to the engine maps (for emissions and fuel consumption) while the load/speed operating point is changing

- **For cycle simulation**

    h. A method of specifying the required velocity as a function of time

    i. Algorithms for deducing the gear changes for different driving characteristics (e.g., economy or sport)

j. Allowance for the fuel consumption and emissions on overrun

k. Allowance for changes to engine fuel consumption maps during warm-up

- **Emissions**

l. Steady-state engine emissions

m. Allowances for the effects of transient operation during acceleration and braking

n. Allowance for engine performance during warm-up

Such a computer model must have a means of specifying the basic vehicle to be "tested" and the changes to be made during the optimization. Examples of this approach can be found in many papers, and some early examples include Porter (1979), Thring (1981), and Lorenz and Peterreins (1984). This approach enables tradeoffs to be investigated without recourse to the testing of numerous vehicles. Obviously, it remains necessary to test the final vehicles to validate the model and to obtain emissions certification.

An example that might be investigated is the tradeoff between fuel consumption and performance for a given vehicle, with different capacity engines and final drive ratios. The results will depend on the definition of performance and fuel consumption. Performance may be specified as the time taken to travel 400 m (1312 ft) from rest, or to accelerate between specified speeds. Equally, fuel consumption may be at a specified speed or speeds, over a specified cycle, or some weighted combination.

Figure 11.4 shows the predicted tradeoff between fuel economy and performance for a General Motors "X" car with different transmission ratios and engine sizes. The fuel economy is a 55/45 weighted average of the U.S. Environmental Protection Agency (EPA) urban and highway driving cycles. For any particular fuel economy requirement, there is an optimum engine size and transmission ratio. With reference to points A and B on Fig 11.4, both represent different engine/transmission combinations with the same fuel economy. However, the vehicle with the larger engine and the lower final drive ratio offers better performance.

The optimization will not depend solely on the final drive ratio and engine size, but also on the intermediate gear ratios. For example, consider the steady-speed fuel economy and acceleration from 0 to 100 km/h (0 to 62 mph). The steady-state fuel economy will be good if there is a low reduction ratio final drive. This will not aid acceleration but can be compensated for by having high reduction ratios in the intermediate gears. In other words, a gearbox with a large span is needed. Figure 11.5 illustrates this more generally. However, as the ratio span increases, the number of discrete gear ratios needed increases, if progression through the gears is to remain smooth.

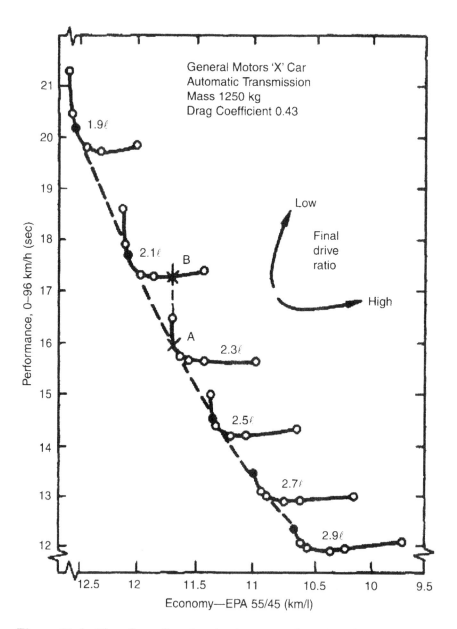

Figure 11.4. *The effect of engine displacement (liters) and final drive ratio on performance and fuel economy. Adapted from Porter (1979).*

11.3.2 ADVISOR (ADvanced VehIcle SimulatOR)

Many vehicle modeling packages have been developed, including GPSIM and DRIVESIM at General Motors Corporation, DYNMOD at Ford Motor Company, DYNASTY and ENTERPRISE at Caterpillar Incorporated, and SIMPLEV at the Idaho National Engineering and Environmental Laboratory. Most models have been developed for only in-house use, by manufacturers, but some are available to buy or are free in some cases. One such free package

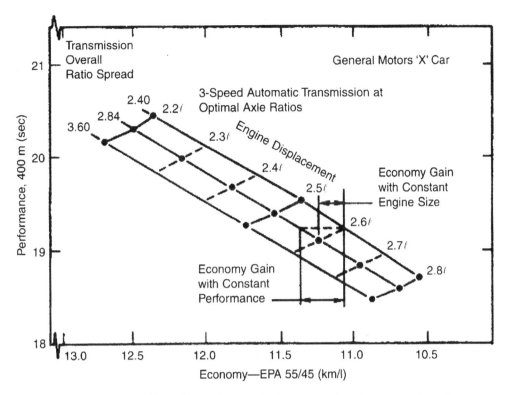

Figure 11.5. *The effect of transmission overall ratio spread (span) on performance and fuel economy (Porter, 1979).*

is the ADVISOR model (http://www.ctts.nrel.gov/analysis/), developed by the National Renewable Energy Laboratory (NREL) to allow system-level analysis and tradeoff studies of vehicles, including electric, hybrid, and fuel cell vehicles.

Combinations of different fuel converters, energy storage devices, transmissions, and electric machines must be assessed, and accurate estimates of fuel consumption and exhaust emissions obtained. Accurate real-time simulations also can be used for developing the first stage of the vehicle control system. This is especially true of hybrid vehicles that have extra freedom in defining the engine operating point.

ADVISOR (ADvanced VehIcle SimulatOR) was written by the NREL in November 1994. It served as a quick analysis tool, in support of the hybrid vehicle propulsion system subcontracts with the auto industry for the U.S. Department of Energy. In January 1998, ADVISOR 2.0 was released on the World Wide Web. It can be run over the Web or downloaded onto a computer that has MATLAB® installed. The current version is ADVISOR 3.2.

ADVISOR runs in the graphical, object-oriented programming language of MATLAB/ Simulink®. Simulink is a graphical environment that uses a library of simple building blocks to define a model. Blocks can be linear or nonlinear, and modeled in continuous or sampled time. User-defined blocks also can be created. MATLAB is the platform on which Simulink runs, and it provides analytical tools and plotting functions to help visualize results.

ADVISOR has a graphical user interface (GUI) to facilitate putting vehicles and test procedures together. The ADVISOR GUI also contains "auto-sizing" features, parametric testing, acceleration tests, and gradability tests, as well as graph plotting features to analyze a multitude of different variables. ADVISOR is essentially a backward-facing model, taking required speed as an input and determining the powertrain powers, speeds, and torques required to meet the vehicle speed. When the requested vehicle speed has been fed backwards all the way up the powertrain, however, the resulting component powers, torques, and speeds then are fed forward down the powertrain, and the achieved vehicle speed is obtained. All being well, the requested and achieved speeds will be the same.

The following configurations are modeled in ADVISOR:

Conventional: With choices of prime mover and gearbox type and ratios

Electric: With choices of electric motor types, gearboxes and ratios, and battery systems

Fuel Cell: With choices of fuel cell system (including allowances for the losses associated with different fuels), electric motor types, gearboxes and ratios, and battery system (if any)

Hybrid: Series and parallel, with choices of prime mover, gearbox type and ratios, electric motor/generator types, gearboxes and ratios, and battery (or other energy storage) system

The configuration is selected from the user interface, for which a typical screen is shown in Fig. 11.6. There is a choice of configurations and then a selection of different components to use within it. Components can be added or removed from the lists, and the data files (".m" files, which, for example, provide data on the engine efficiency and emissions) can be edited. The data files can be either experimental data or data obtained from simulations. Component size and efficiency can be adjusted as needed, and ADVISOR can attempt to "AutoSize" the modules according to certain criteria. Graphs of component efficiencies can be viewed and any variables edited.

ADVISOR contains a number of standard U.S. and European test cycles that can be run any number of times and smoothed with a filter if required. Other more specific test procedures (involving more than one cycle) are included, as are acceleration tests and road tests. A parametric study also can be conducted across a maximum of three variables. The selection is made through the simulation setup screen shown in Fig. 11.7. This screen also is where initial conditions, such as battery state of charge and exhaust temperatures, are defined.

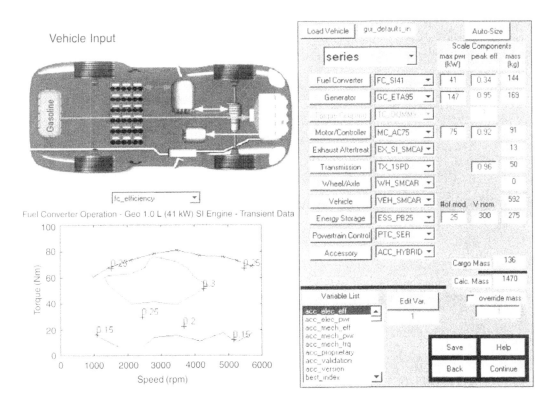

Figure 11.6. *An ADVISOR screen for defining the vehicle configuration.*

Test results are displayed and analyzed on the test results screen, as shown in Fig. 11.8. The standard plots are of speed requested against speed achieved, battery state of charge, emissions, and gear ratio, but any other variables used in the simulation (e.g., motor torque requested and motor torque achieved) also can be viewed. The standard MATLAB plotting controls are included in a separate window. Acceleration and grade tests (if selected) also are shown here, with a complete energy use analysis. The "Output Check Plots" button produces approximately ten more graphs of fuel converter and motor operating points, efficiencies, and so forth.

ADVISOR also is useful in the development of control strategies because the control strategy defined within Simulink can be downloaded into the hardware controller of a vehicle. This is particlularly useful for more complex vehicles, such as those with CVT or hybrid technology.

11.4 Conclusions

The simple matching of engine size and overall gearing ratio to obtain a specified maximum speed, and the consequential selection of a first gear ratio for hill starting, can be done

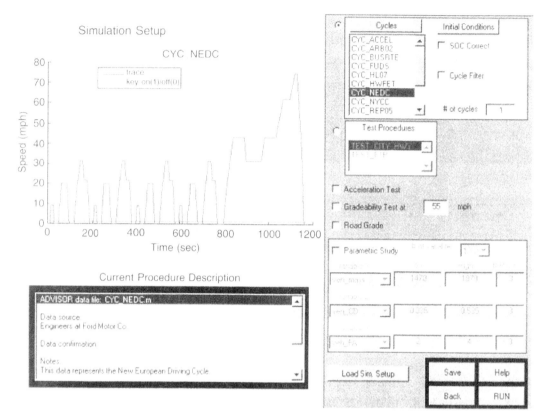

Figure 11.7. *An ADVISOR simulation setup screen.*

manually. The estimation of steady-state fuel consumption requires interpolation, and this quickly becomes tedious, especially when intermediate gear ratios are being considered. Thus, the use of computer modeling is essential, and sophisticated models are readily available. These enable parametric studies to be conducted for vehicle optimization and to make predictions of fuel consumption and emissions over specified drive cycles. The models must be able to simulate conventional vehicles, electric vehicles (with batteries, fuel cells, or a combination of the two), and hybrid vehicles.

Transmission Matching and Vehicle Performance | 493

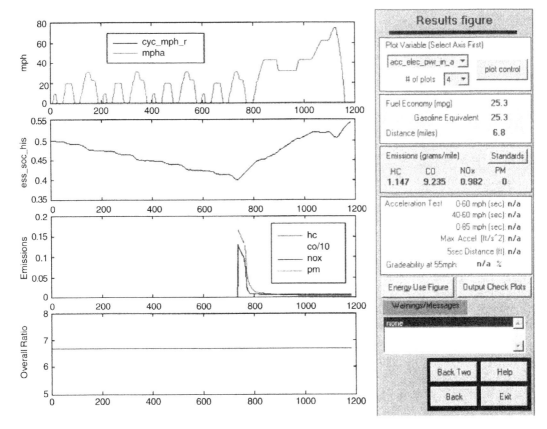

Figure 11.8. *An ADVISOR test results screen.*

CHAPTER 12

Alternative Vehicles and Case Studies

This chapter combines coverage of alternative vehicles and case studies, and starts with a discussion of electric vehicles. The advantages of electric vehicles are their low local emissions, but they suffer from a low energy storage capability and inferior performance. However, these disadvantages can be eliminated by hybrid vehicles, which are able to achieve the best attributes of both electric and conventional vehicles. This naturally leads to a discussion of the Toyota Prius hybrid vehicle as a case study because this was the first hybrid vehicle to enter production. This case study is preceded by an examination of the Vauxhall 14-40, a vehicle introduced in 1922. Despite its age, the 14-40 demonstrates conventional vehicle technology and emphasizes the improvements that have occurred over the last 80 years.

12.1 Electric Vehicles

12.1.1 Introduction

For almost a century, electric vehicles have been dependent on lead-acid batteries, with their poor specific energy storage—a ton of lead-acid batteries stores as much energy as approximately 3 liters (0.8 gal) of gasoline. Of course, this is not a fair comparison because the conversion efficiency of chemical energy to mechanical work is a factor of approximately four lower than the electrical conversion efficiency. Nonetheless, this illustrates the problems with energy storage, which limit a practical vehicle to a range of approximately 100 km (62 mi) and a maximum speed of 100 km/hr (62 mph).

In 1899, the Belgian driver Camille Jenatzy (1868–1913) set a world land speed record of 106 km/h (65.8 mph) in an electric car. The first electric cars were manufactured by Magnus Volk (England) in 1888, and by William Morrisson (United States) in 1890. Electric cars were popular until around 1915 because many journeys were short. Likewise, electric cars were easier to drive. By 1920, roads had been improved, and expectations of speed and endurance (coupled with the development of better engines and gearboxes) led to the demise of electric vehicles. In 1921, there were approximately 9 million vehicles in the United States, of which only 0.2% were electric (Georgano, 1997). Electric vehicles are widely used when the range and maximum speed are not limitations, including places such as airports, warehouses, golf courses, and urban deliveries in the United Kingdom.

Energy storage is not only a matter of how much energy can be stored per unit mass (specific energy, usually expressed as Wh/kg), but there is also the question of how rapidly the energy can be released—the specific power (W/kg). Figure 12.1 illustrates the specific power and specific energy capabilities of various energy storage systems. Both axes are on log scales, and this emphasizes the limitations of batteries.

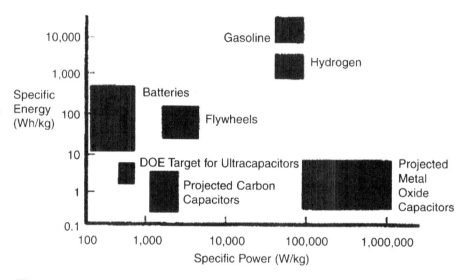

Figure 12.1. *Specific power and specific energy capabilities of various energy storage systems (http://www.ott.doe.gov/oaat/storage.html).*

The choice of an electric motor is more straightforward. Historically, brushed DC motors were used for ease of control, but the need for brush maintenance is a major disadvantage. With the development of solid-state controllers, both AC induction motors and brushless DC motors have become competitive in terms of cost, low maintenance, controllability, and efficiency. Figure 12.2 shows that the efficiency of a typical brushless DC motor falls between 85 and 95% for most of its operating envelope.

Section 12.1.2 discusses various battery technologies, and Section 12.1.3 reviews some typical electric vehicles.

12.1.2 Battery Types

Only an overview of battery types and their performance can be presented here. More details can be found in books such as those by Bernt (1997), Crompton (1995), or Rand et al. (1998).

Unfortunately, no current battery technology has demonstrated an economically acceptable combination of power, specific energy, efficiency, and life cycle. In general, batteries use toxic materials; thus, it is essential at the design stage to incorporate recyclability. Technology also is needed to accurately determine the battery state of charge. Additional battery attributes that are needed include a low self-discharge rate, high charge acceptance (to

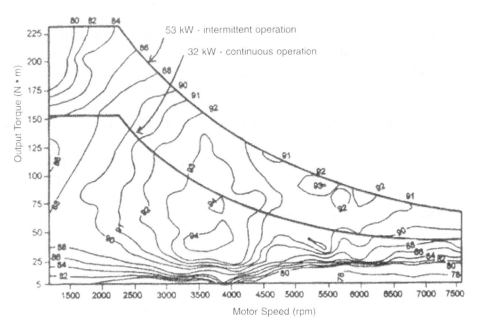

Figure 12.2. *Torque/speed characteristics and efficiency of the Unique Mobility Caliber EV 53 brushless DC motor.*

maximize regenerative braking utilization and short recharging time), no memory effects (i.e., partial discharging followed by recharging must not reduce the energy storage capacity), and a long cycle life. Table 12.1 summarizes the capabilities of various battery technologies.

TABLE 12.1
PERFORMANCE OF VARIOUS BATTERY TYPES, WITH PROJECTIONS FOR THE NEXT THREE TO FIVE YEARS ( ) (ASHTON, 1998), AND THEORETICAL LIMIT ON SPECIFIC ENERGY (RAND ET AL., 1998) [ ]

Battery Type	Specific Energy Storage, Wh/kg	Specific Power (for 30 s at 80% Capacity), W/kg	Specific Cost, $/kWh	Cycle Life (Charges and Discharges to 80% of Capacity)
Lead-acid	35 (55) [171]	200 (450)	125 (75)	450 (2000)
Nickel-cadmium (NiCd)	40 (57) [217]	175 (220)	600 (110)	1250 (1650)
Nickel-metal hydride (NiMH)	70 (120)	150 (220)	540 (115)	1500 (2200)
Lithium ion	120 (200)	300 (350)	600 (200+)	1200 (3500)

Of equal importance to the energy storage and power capabilities of a battery is its efficiency. Unfortunately, such data are difficult to establish because battery efficiency depends on many parameters, including its state of charge, temperature, age, and the rate of charge/discharge. The losses in a battery usually are dominated by the ohmic loss (i.e., the resistance to the flow of both electrons within the conductors and ions within the electrolyte); therefore, the voltage falls almost linearly with current. Because the power is the product of voltage and current, the efficiency will fall slightly faster than linear when plotted against power. For a nickel-metal hydride (NiMH) battery with a rating of 40 kW, the efficiency might be 70% at rated power and 87% at 20 kW. Similar arguments apply to the recharging, so a slow recharging is advantageous. Although this can be achieved with overnight recharging at home, it is not suitable for urban vehicles being used in a pool system or taxis that require rapid recharging. Practical batteries are considered now in detail.

12.1.2.1 Lead-Acid Batteries

Lead-acid batteries are currently used in commercially available electric vehicles (EVs). Despite continuous development since 1859, the possibility of further development still exists to increase the specific power and energy. Lead-acid batteries are selected for their low cost, high reliability, and an established recycling infrastructure. However, problems including low energy density, poor cold-temperature performance, and low cycle life limit their desirability.

The lead-acid cell consists of a metallic lead anode and a lead oxide (PbO_2) cathode held in a sulfuric acid (H_2SO_4) and water electrolyte. The discharge of the battery is through the chemical reaction

$$PbO_2 + Pb + 2H_2SO_4 \rightarrow 2PbSO_4 + 2H_2O$$

The electron transfer between the lead and the sulfuric acid is passed through an external electrical connection, thus creating a current. In recharging the cell, the reaction is reversed. Lead-acid batteries have been used as car batteries for many years and can be regarded as a mature technology. The lead-acid battery is suited to traction application because it is capable of a high power output. However, due to the relatively low energy density, lead-acid batteries become large and heavy to meet the energy storage requirements.

12.1.2.2 Nickel-Cadmium (NiCd) Batteries

Nickel-cadmium (NiCd) batteries are used routinely in communication and medical equipment and offer reasonable energy and power capabilities. They have a longer cycle life than lead-acid batteries and can be recharged quickly. The battery has been used successfully in developmental EVs. The main problems with nickel-cadmium batteries are high raw-material costs, recyclability, the toxicity of cadmium, and temperature limitations on recharging. Their performance does not appear to be significantly better than that of lead-acid batteries, and the energy storage can be compromised by partial discharges—referred to as memory effects.

12.1.2.3 Nickel-Metal Hydride (NiMH) Batteries

Nickel-metal hydride (NiMH) batteries currently are used in computers, medical equipment, and other applications. They have greater specific energy and specific power capabilities than lead-acid or nickel-cadmium batteries, but they are more expensive. The components are recyclable, so the main challenges with nickel-metal hydride batteries are their high cost, the high temperature they create during charging, the need to control hydrogen loss, their poor charge retention, and their low cell efficiency.

Metal hydrides have been developed for high hydrogen storage densities and can be incorporated directly as a negative electrode, with a nickel hydroxyoxide (NiOOH) positive electrode and a potassium/lithium hydroxide electrolyte. The electrolyte and positive electrode had been extensively developed for use in nickel-cadmium cells.

The electrochemical reaction is

$$MH_x + NiOOH + H_2O \rightarrow MH_{x-1} + Ni(OH)_2 + H_2O$$

During discharge, hydroxyl (OH^-) ions are generated at the nickel hydroxyoxide positive electrode and consumed at the metal hydride negative electrode. The converse is true for water molecules, which means that the overall concentration of the electrolyte does not vary during charging/discharging. There are local variations, and care must be taken to ensure that the flow of ions across the separator is high enough to prevent the electrolyte "drying out" locally.

The conductivity of the electrolyte remains constant through the charge/discharge cycle because the concentration remains constant. In addition, there is no loss of structural material from the electrodes; thus, they do not change their electrical characteristics. These two details give the cell very stable voltage operating characteristics over almost the full range of charge and discharge.

12.1.2.4 Lithium Ion (Li-Ion)/Lithium Polymer Batteries

The best prospects for future electric and hybrid electric vehicle battery technology probably come from lithium battery chemistries. Lithium is the lightest and most reactive of the metals, and its ionic structure means that it freely gives up one of its three electrons to produce an electric current. Several types of lithium chemistry batteries are being developed. The two most promising of these appear to be the lithium ion (Li-ion) type and a further enhancement of this, the lithium polymer type.

The Li-ion battery construction is similar to that of other batteries except for the lack of any rare earth metals that are a major environmental problem when disposal or recycling of the batteries becomes necessary. The battery discharges by the passage of electrons from the lithiated metal oxide to the carbonaceous anode by current flowing via the external electrical circuit. Li-ion represents a general principle, not a particular system. For example, lithium/aluminum/iron sulfide has been used for vehicle batteries. Li-ion batteries have a very linear discharge characteristic, and this facilitates monitoring the state of charge. The charge/discharge efficiency of Li-ion batteries is approximately 80%, which compares favorably with nickel-cadmium batteries (approximately 65%) but unfavorably with nickel-metal hydride batteries (approximately 90%). Although the materials used are non-toxic, a concern with the use of lithium is, of course, its flammability.

Lithium polymer batteries use a solid polymer electrolyte, and the battery can be constructed similar to a capacitor, by rolling up the anode, polymer electrode, composite cathode, current collector from the cathode, and insulator. This results in a large surface area for the electrodes (to give a high current density) and a low ohmic loss.

12.1.3 Types of Electric Vehicles

In 1996, General Motors became the first major automotive manufacturer in recent times to market an electric vehicle. Table 12.2 shows the General Motors specification for the EV1.

**TABLE 12.2
GENERAL MOTORS EV1
(VAUXHALL, 1998)**

Body Style	Two-seater
Mass	1350 kg
Motor Rating	102 kW
Battery Capacity	16.2 kWh
Mass	533 kg
Recharge:	
220V/6.6 kW	3 h
(15–95% Charge) 110V/1.2 kW	15 h
Range with 85% Discharge:	
Urban Cycle	112 km
Motorway	145 km
Acceleration (0–100 km/h [0–62 mph])	<9 s
Top Speed (Regulated)	129 km/h
Drag Coefficient	0.19
Frontal Area	1.89 m^3
AC_d	0.36 m^3

The EV1 uses a three-phase AC induction motor with an integral (fixed-ratio) reduction gearbox and differential. It has a peak rating of 103 kW, which probably is most significant for its regenerative braking capability that extends the range of the vehicle by up to 20%. The motor has a maximum speed of 13,000 rpm, and the system mass is 68 kg (150 lb), with a service interval of 160,000 km (99,424 mi).

The EV1 was introduced with lead-acid batteries, but NiMH batteries became available in 1998. With NiMH bateries, a range of almost 600 km (373 mi) was achieved, and a 0–60 mph acceleration time of 7.7 s was achieved with lead-acid batteries. In 1997, a prototype EV1 obtained the world land speed electric car record with 295 km/h (183.3 mph).

Figure 12.3 shows how the battery pack is accommodated within the chassis of the EV1. There is an onboard 110-V battery recharger, or a fixed 220-V 30-A recharger that transfers the power inductively—obviating the need for high-current electrical connections.

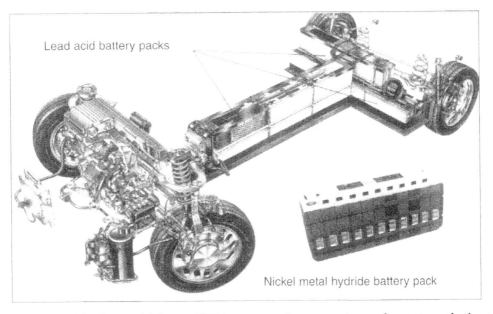

Figure 12.3. *The General Motors EV1 battery and powertrain configuration; the heat pump used for climate control can be seen in front of the lowest wheel (Vauxhall, 1998).*

The cost of the EV1 is high. General Motors has a leasing package (approximately 30% of vehicles in the United States are leased), but by March 1999, only approximately 600 EV1 vehicles had been leased (NEL, 1999). Note that General Motors has invested approximately $1 billion in the development of the EV1 and has a time-scale of approximately 10 years to determine its success. Other manufacturers have introduced electric vehicles, including the Chrysler EPIC, Ford Ranger EV, Chevrolet S10, Nissan Altra EV, Honda EV Plus, Toyota RAV4, PIVCO City Car, and Nissan Prairie Joy. Recently, Japanese manufacturers announced a number of small electric vehicles, as summarized in Table 12.3 (Yamaguchi, 2000).

Ford announced a prototype version of its Ka, the e-Ka, using Li-ion batteries that give a comparable performance to a gasoline-engine vehicle, albeit with a range of 200 km (124 mi)

TABLE 12.3
SMALL JAPANESE ELECTRIC VEHICLES

Manufacturer:	Nissan	Toyota	Mitsubishi
Model:	Hypermini	e-Com	MEEV-II
Seats	2	2	2
Mass (kg)	840	770	640
Motor (kW/Nm)	24/130	18.5/76	—
Motor Type	AC Synchronous	AC synchronous	AC synchronous
Battery Type	Li-ion	Ni-MH	Li-ion
Battery Specification	90 Wh/kg	100 km range	145 km range
Length (m)	2.65	2.79	2.60
Width (m)	1.475	1.475	1.48

at 80 km/h (50 mph). The battery packs weigh 280 kg (617 lb) and have a power density of 126 Wh/kg (Broge, 2000). DaimlerChrysler has been developing an electric version of its A-Class vehicle (using the same induction motor as in the fuel cell powered NECAR III). Interestingly, the company is developing its own battery technology, based on a sodium/nickel/chloride ion system. This ZEBRA battery has achieved a power density of 155 W/kg and an energy density of 81 Wh/kg, which compares favorably with NiMH technology (Anonymous, 1998). On test, the ZEBRA battery achieved the equivalent of a 200,000-km (124,280-mi) life, but the battery must operate in the temperature range 270–350°C (518–662°F).

12.1.4 Conclusions About Electric Vehicles

Despite improvements in battery technology, electric vehicles remain handicapped by battery range, initial cost, and durability. These shortcomings can be avoided by hybrid electric vehicles, the subject of the next section. Nonetheless, electric vehicles will become increasingly common as a result of fiscal incentives, legislation that allows only zero emission vehicles in sensitive areas, and falling costs as production increases. However, remember that most electricity is generated from fossil fuels; thus, electric vehicles merely change the location where emissions are produced. Furthermore, on a "well to wheel basis," the efficiency is below that of a diesel engine vehicle.

12.2 Hybrid Electric Vehicles

12.2.1 Introduction

Hybrid vehicles offer potential for significantly reduced fuel consumption and emissions during normal operation because of the scope for operating the prime mover at its optimum and the ability to meet sudden power demands from a combination of the prime mover and stored energy. Hybrid vehicles also can operate with zero emissions in sensitive urban environments. Thus, despite the extra cost associated with hybrid vehicle systems, there will be circumstances in which the increased capital cost is justified. Compared to fuel cells, these vehicles use existing technology and can be produced more cheaply. They also offer the highest "well to wheel" efficiency.

Parallel and series hybrid configurations are well established, but the Toyota Prius (the first commercially introduced hybrid vehicle) uses a dual hybrid system that combines features of series and parallel hybrid operation.

Figure 12.4 shows the two basic types of hybrid electric vehicles: (1) series, and (2) parallel. Series hybrids (Fig. 12.4a) have no mechanical connection between the engine and the road. The engine, or power unit (PU), instead drives a generator (G), producing electricity that then is used to propel the vehicle via an electric motor (M). Any power excess or shortage is routed to onboard batteries (B), which also allow the vehicle to run as an EV.

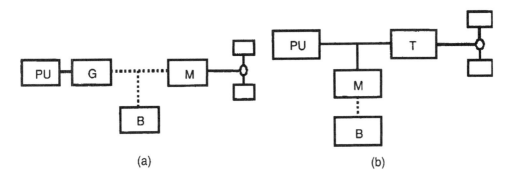

Figure 12.4. *The two basic types of hybrid electric vehicles: (a) series, and (b) parallel.*

The advantages of the series system are as follows:

a. The power unit (PU) is not mechanically coupled to the road and thus does not have to meet the instantaneous demands at the wheels. A wider range of fuel converters, such as fuel cells and gas turbines, can be used, allowing for greater potential efficiency.

b. The PU must meet only average power demands and thus can be run at an optimum operating point.

c. The configuration is simple to implement.

The disadvantages of the series system are as follows:

a. All of the power to the wheels must come from the electric motor alone.

b. All power produced by the PU must be converted to electric energy and then back to mechanical energy, incurring significant losses.

c. Both a generator and a motor are required, making the configuration heavy.

Thus, the series system is best suited to prime movers that have a very different operating speed to that of the axles, and an efficiency that is very sensitive to the operating point.

The series system is well suited to gas turbine applications; Longee (1998) describes such a system.

Figure 12.4(b) shows the parallel hybrid configuration. Here, the power unit (PU) is mechanically coupled to the wheels through a transmission, and an electric machine is used to supplement torque available.

The advantages of the parallel system are as follows:

a. Most of the power is delivered mechanically, thereby avoiding electrical losses.

b. Peak performance is met using both systems, so that the electric machine can be kept small.

c. Only one electric machine is required.

The disadvantages of the parallel system are as follows:

a. The engine cannot always run at its optimum operating point.

b. A mechanical transmission is required.

c. The configuration is more difficult to implement, with mechanical couplings and a more complicated control system.

The parallel system is appropriate when the PU is a reciprocating engine, because its efficiency is less sensitive to the operating point than a gas turbine. Also, there are efficiency gains in using mechanical power transmission compared to electrical power transmission (as in the series hybrid system). The Honda Integrated Motor Assist (IMA) hybrid vehicle (Insight, introduced in October 1999) uses a parallel configuration (Yonehara, 2000). The Insight is a two-seater with a drag coefficient of 0.25 and a mass of 820 kg (1808 lb). The 1-liter spark ignition engine has an output of 52 kW/92 Nm, whereas the brushless DC motor has an output of 10 kW/49 Nm. The nickel-metal hydride (Ni-MH) battery has a mass of only 20 kg (44 lb), which implies that extended electrical operation is not intended. The electric motor is only 60 mm (2.4 in.) thick and is installed between the engine and the gearbox. The vehicle is said to have a performance equivalent to that of a 1.5-liter-engine vehicle. This type of hybrid, referred to as a MYBRID (Mild or Minimum hYBRID), is favored by some of the major manufacturers. It is intended to assist only during transients and to operate ancillaries (NEL, 1999).

However, as mentioned, the first hybrid vehicle to enter commercial production was the Toyota Prius (Yamaguchi, 1997). The Prius uses a dual hybrid system that combines series and parallel hybrid operation. This is explained in the next section.

12.2.2 Dual Hybrid Systems

If the parallel system is modified by the addition of a second electrical machine (which is equivalent to adding a mechanical power transmission route to the series system), the result is a system that allows transmission of the prime mover power through two parallel routes: (1) electrically, and (2) mechanically. This is equivalent to the use of a mechanical shunt transmission with a continuously variable transmission (CVT) to give an infinitely variable transmission (IVT) (Ironside and Stubbs, 1981). The result is a transmission that enables the engine to operate at a high efficiency for a wider range of vehicle operating points. A well-documented example of this configuration is the "dual" hybrid system developed by Equos Research (Yamaguchi et al., 1996) and used in the Toyota Prius. Figure 12.5 shows this system.

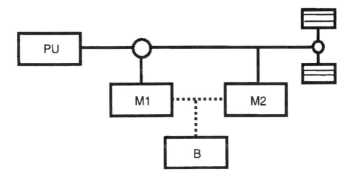

Figure 12.5. *A dual hybrid configuration using planetary gears (PG) to couple the power unit (PU) and one of the electric machines (M1).*

The planetary gears act as a "torque divider," sending a proportion of the engine power mechanically to the wheels and driving an electric machine (M1) with the remainder. Consequently, the configuration acts simultaneously as a parallel and a series hybrid. Engine speed is controlled using Machine 1, removing the need for a transmission, a clutch, or a starter motor. Machine 2 acts in the same way as the motor in a parallel system, supplementing or absorbing torque as required. The diagrams in Fig. 12.6 show the possible modes of operation, and each mode is explained next.

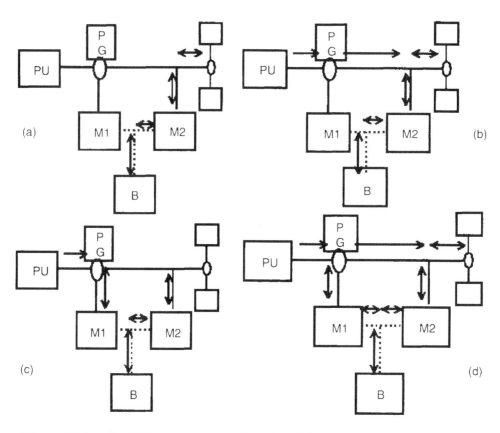

Figure 12.6. *The different operational modes of the Toyota Prius hybrid vehicle: (a) electric, (b) parallel, (c) charging, and (d) dual.*

Electric Mode. (Fig. 12.6a) The engine is switched off, and Machine 1 acts as a "virtual clutch," keeping the engine speed at zero. Torque and regenerative braking are provided by Machine 2.

Parallel Mode. (Fig. 12.6b) Machine 1 is stationary (perhaps with a brake applied), and the configuration is a simple parallel one, with a fixed engine-to-road gear ratio.

Charging Mode. (Fig. 12.6c) The vehicle is stationary, and all of the engine power is used to drive Machine 1 and charge the batteries. Torque is still transferred to the wheels, allowing the car to "creep."

Dual Mode. (Fig. 12.6d) Some power is used to drive the wheels directly, while the remainder powers Machine 1. The speed of Machine 1 determines the engine operating speed.

The charging and parallel modes are effectively subsets of the dual mode, and this continuity in control is the real strength of the configuration. The dual hybrid configuration combines the advantages of both series and parallel, as follows:

a. Optimal engine operating point at all times.

b. Much of the power (especially at cruising speeds) is delivered mechanically to the wheels, thereby increasing efficiency.

c. Charging is possible, even when the vehicle is stationary.

d. The combined torque of the engine and Machine 2 is available, improving performance.

Compared to a series hybrid (where the electrical machines must be rated for the prime mover and the vehicle power requirement), only a fraction of the prime mover power is transmitted electrically in the dual hybrid system. The main difficulty with the dual hybrid is in the design of a control system, which must resolve the two degrees of freedom (engine speed and engine torque) and the associated transients into an optimal and robust control strategy. System modeling is essential for optimizing this.

12.3 Case Studies

12.3.1 Introduction

Two vehicles vastly separated by time and technology have been selected as case studies here. The first is the Vauxhall 14-40, which originally was produced as the M-type in 1922. The second vehicle is the 1998 Toyota Prius. In their respective times, these vehicles were considered to be at the forefront of technology. For this reason, these two vehicles have been selected as case studies here. Likewise, the contrast between the two provides an elegant picture of how automotive technology has advanced in the 77 years separating these two cars.

12.3.2 The Vauxhall 14-40

12.3.2.1 Introduction

The Vauxhall Iron Works was founded in 1857 by Alexander Wilson and established a strong reputation for steam engines and pumps (Vauxhall, 1999). In 1897, the company developed its first gasoline engine to propel a small river launch named *Jaberwock*. The first production car was introduced in mid-1903 and had a 5-hp gasoline engine, two forward gears (no reverse), and a hand throttle with tiller steering. The success of this first model led to the development of larger and more powerful vehicles.

In the winter of 1907–1908, the Royal Automobile Club (RAC) published the rules for a 2,000-mile trial to be held over 13 days, and Vauxhall decided to enter. The A-type was designed by the 24-year-old Laurence Pomeroy (later to become chief engineer) for the 2,000-mile trial and was ready in June 1908. The 2,000-mile trial included timed hill-climbs in Scotland and the Lake District, and a 200-mile race at the newly opened Brooklands race track. The Vauxhall A-type took first place, with a convincing victory over Rolls-Royce.

At Brooklands in December 1909, an A-type set the record for the flying half-mile at 88.6 mph (subsequently increased to 100.08 mph) and averaged 10 laps at 81.33 mph. This success led to the design of a number of sporting tourers that had considerable commercial and racing success, perhaps culminating in the E-type (more widely known as the 30-98) that was designed by Pomeroy and built in 13 weeks in 1913. Its debut was at the Shelsley Walsh hill-climb, where it set a record that stood for 15 years.

The 30-98 continued in production until 1927, and its guaranteed 100-mph performance meant that it could outperform Bentleys. The 30-98 had four cylinders with a bore/stroke of 98/150 mm, to give a displacement of 4.5 liters. The "30" refers to the RAC horsepower rating, which was based on the piston area and was used for tax ratings. The "98" is the bore in millimeters and the approximate horsepower.

Vauxhall had established itself as a manufacturer of high-performance cars, but the company decided that it also should enter the middle-class market. For the 1921 motor show, it introduced the M-type, which with slight modifications became the LM-type in 1925. Both vehicles are commonly referred to as 14-40s. The M-type cost £595, which is equivalent to approximately $90,000 in 2002.

12.3.2.2 Specifications

The 14-40 used a four-cylinder in-line engine with a bore of 75 mm and a stroke of 130 mm to give a displacement of 2.3 liters. The engine used side valves and a 4.8:1 compression ratio, to give a maximum power of 43 bhp (32.1 kW) at 2400 rpm. The clutch and gearbox were in unitary construction with the engine, using a single plate clutch and three-speed gearbox (four speeds for the LM-type), connected via a torque tube to the spiral bevel rear axle (4.5:1). The wheelbase was 114 in. (117 in. for the LM) with a track of 50 in. The M-type had only rear-wheel brakes, but the LM type also had front-wheel brakes. The solid front axle was supported by half-elliptic leaf springs, with cantilevered leaf springs for the rear axle; there were no dampers. The maximum vehicle speed was 55–60 mph, depending on body type, there being four/five-seater tourer, saloon, and two-seater tourer options.

12.3.2.3 Engine Design and Performance

The engine design is commonly attributed to Harry Ricardo, although in his book (1924) that provides extensive details of the engine, he attributes the design to the Vauxhall chief engineer, C.E. King. However, Vauxhall was Ricardo's client, so this attribution might be a courtesy. The engine undoubtedly has the influence of Ricardo.

Figure 12.7 shows a cross section of the Vauxhall 14-hp engine (Ricardo, 1924). The side valves and Ricardo turbulent combustion chamber are clearly visible, with the light alloy pistons. The four cylinders are cast as a single unit in cast iron, with detachable cylinder head, crankcase, and sump all in cast aluminum. The connecting rods have a center distance of 260 mm (10 in.) (giving L/R = 4), and the big-ends are too large to fit through the cylinder bore. All the crankshaft journals have a diameter of 1.75 in. (44.45 mm).

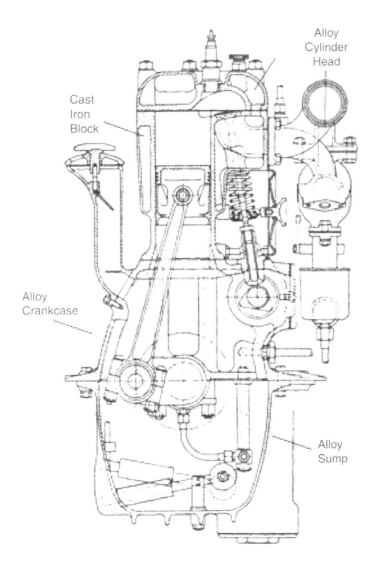

Figure 12.7. *Transverse cross section of the Vauxhall 14-hp engine. Adapted from Ricardo (1924).*

The inlet valves have a port diameter of 1.4 in. (35.6 mm) with a lift of 0.35 in. (8.9 mm). The exhaust valves have the same lift but a port diameter of 1.31 in. (33.3 mm). Also shown in Figure 12.7 is the hot-spot arrangement between the inlet and exhaust manifolds. This facilitated minimal use of the strangler (choke) with the contemporary low-volatility fuels.

The fixed venturi updraft carburetor reduced the likelihood of fuel leaks leading to engine compartment fires. Figure 12.8 shows a longitudinal cross section of part of the Vauxhall 14-hp engine (Ricardo, 1924). The drive to the camshaft is via an intermediate gear that is located on an eccentric journal, which enables some control of the mesh with the crankshaft and camshaft gears. The intermediate gear also drives the magneto and accessories pulley at

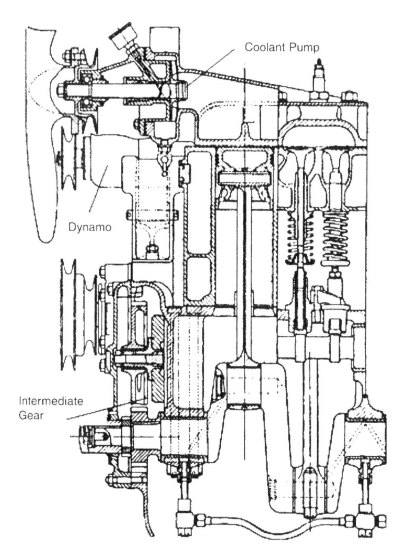

Figure 12.8. *Longitudinal cross section of part of the Vauxhall 14-hp engine. Adapted from Ricardo (1924).*

engine speed. The whole of this assembly has an outer bore that is eccentric to its shaft axis, so that rotation of the assembly controls the gear mesh. The magneto is a two-pole design with two cam lobes separated by 180° for the contact breaker. There is then a 2:1 reduction gear, so that the HT distributor is driven at half engine speed. A disadvantage of this system is that there is no external indication for a pair of cylinders as to which is on the gas exchange top center as opposed to being at the end of compression! Thus, it is necessary to check the camshaft position. Ignition advance is controlled manually by a lever on the steering wheel. A fixed ignition timing was used only for approximately the first decade of automobiles, whereas automatic speed advance was adopted in the late 1920s and vacuum advance in the mid-1930s.

Another lever on the steering wheel is a hand throttle. This provides idle speed control and is particularly important after starting because the strangler (choke) has no mechanism to increase the idle speed.

Neither of the V belts for the dynamo or coolant pump/fan have tensioner pulleys. The dynamo has a cylindrical casing with a single field winding, so the rotor is eccentric to the casing. The dynamo belt thus is tensioned by rotating the dynamo body. The centers for the cooling V belt are fixed, of course, but the pulley incorporates a large-diameter screw thread, and the belt is tensioned by varying the separation of the pulley sides. This is an idea that perhaps appears better on the drawing board than it is to use, because of friction between the belt and pulley sides. The use of a coolant pump is commendable because many vehicles until the 1940s relied solely on the thermo-syphon effect.

The crankshaft has three main bearings, and the forced lubrication system uses a plunger pump operated by an eccentric circle cam on the camshaft. The relief flow from the pressure relief valve is used to lubricate the timing gears. There are no lubricant filters—only some gauze strainers—which is one reason why the white-metalled bearings for the main bearings and connecting rods might need reworking after 10,000 miles. Another contribution toward high engine wear is the absence of an air filter, an innovation that was uncommon until around 1930 on quality cars.

The pistons in aluminum alloy use two compression rings and an oil control ring. These pistons (Fig. 12.7) are designed for low friction, because in the region around the piston pin, there is no contact with the cylinder walls. Instead, there is a contact surface around the base of the piston. The pistons have a slit (almost vertical) on the anti-thrust side of the piston skirt to allow for thermal expansion.

The compression ratio of 4.8:1 is, of course, very low compared to current practice, but it is a reflection of the fuel quality that was available. Even overhead valve engines would be limited to these low compression ratios. The octane rating scale was not proposed by Edgar until 1927 (Edgar, 1927). However, prior to this, Ricardo had devised a toluene number (Ricardo, 1992) around 1919, based on mixtures of n-heptane and toluene. Vincent (1992) suggests that fuels in the 1920s had an octane rating in the range 60–70. It was well known that different fuel components had different anti-knock ratings. However, until the end of the 1920s, gasoline was mostly the result of distillation, and the anti-knock quality of the fuel depended on the origins of the crude oil.

Work by Ricardo (for Shell) showed in 1917 that gasoline distilled from Borneo crude oil contained high levels of aromatics (notably benzene, toluene, and xylene). This fuel had a higher octane rating that permitted a unity increase in the compression ratio, with a 10% increase in power and slightly greater reduction in fuel consumption. This fuel enabled Alcock and Brown to cross the Atlantic in 1919. They used a modified Rolls-Royce Eagle engine, in which the compression ratio was increased from 5:1 to 6:1 (Ricardo, 1992).

In 1921, Midgely and Boyd (working at the General Motors Research Laboratories) discovered the anti-knock properties of tetra ethyl lead. However, concern about health hazards restricted its use in the United States until 1926. In the United Kingdom, lead additives were not introduced until 1928 (Owen and Coley, 1995).

With an octane rating of 60–70, the highest useful compression ratio would be limited to approximately 5, and the Ricardo turbulent head combustion system provides side-valve engines with comparable burn rates and valve areas to two-valve overhead valve arrangements. The side-valve configuration has the advantages of simplicity of cylinder head design, ease of camshaft installation, and simple but stiff valve actuation. The development of the turbulent head combustion system has been described in detail by Ricardo (1992).

Around 1919, Ricardo commenced a series of trials in which the squish area and compression ratio were varied, in what is now known as a Ricardo turbulent head (Fig. 12.7). The engine was fitted with a Hopkinson optical indicator, which showed that with the optimum squish area, the rate of pressure rise was approximately 2.5 bar/°ca. This rapid combustion led to an improvement of approximately 10% in both fuel economy and output. Ricardo also found that the compression ratio could be raised from 4:1 to 5:1 for knock-free operation, and this contributed an additional 10% in both fuel economy and output.

The other area in which gasoline has changed is its volatility. Vincent (1992) prepared a critical review concerning the use of modern gasolines in older engines, from reference to key technical sources over the last 70 years. In 1908 (Blount, 1908), an analysis of commercially available gasolines indicated that they had a much higher initial boiling point (60–65°C [140–149°F]) than current fuels (30–35°C [86–95°F]). BS4040 requires the percentage evaporated at 70°C (158°F) (E70) to be between 10 and 45%. A typical figure might be 25 to 35%, with appreciable amounts (say, 15 to 20%) boiling below 50°C (122°F). Test data presented in 1924 by Ricardo (1924) indicates similar levels of volatility.

In the 1930s, the demand was for an increasing proportion of gasoline to be derived from crude oil. This meant that the heavier fractions from the crude oil were used, both as direct additions and as "cracked" spirit, which consisted of lower-boiling-point hydrocarbons. When these were added to the distillate fuels, there was an increase in the volatility range, as reported by Garner (1936–1937) and Fossett (1943–1944). Figure 12.9 illustrates the changes in volatility. In 1908, the boiling point range of the fuels was very limited, and evaporation would not occur below approximately 55°C (131°F). By 1924, the range of volatilities had increased, but the fuel continued to vaporize over a fairly narrow temperature range.

The higher initial boiling point and increased temperature requirement for 10% vaporization (or low front-end volatility) exacerbated cold starting and increased the need for fuel enrichment during warm-up. To reduce the time needed for warm-up, the 14-hp Vauxhall engine incorporated a heat exchanger at the junction between the inlet and exhaust manifolds (Fig. 12.7). Note that in 1938, Vauxhall incorporated a bi-metallic spring-operated flap that controlled a hot spot, which helped to achieve very low fuel consumption and rapid warm-up.

The increased volatility of modern fuels now can cause vapor lock problems on older vehicles. Often the fuel tank was mounted on the bulkhead within the engine compartment, and the fuel could reach quite high temperatures. A common alternative was to use a vacuum system to

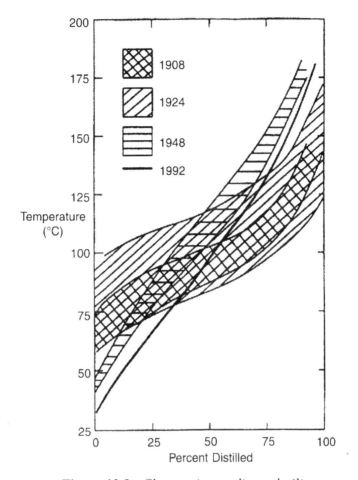

Figure 12.9. *Changes in gasoline volatility.*

draw fuel into a holding tank within the engine compartment. Indeed, the Vauxhall 14-40 uses an Autovac system (Fig. 12.10) that relies on inlet manifold suction to draw fuel into a chamber that empties into an integral holding tank.

12.3.2.4 Engine Performance

The engine data presented by Ricardo (1924) has been replotted here in SI units as Figs. 12.11 and 12.12. Above 2000 rpm, the bmep falls rapidly, and this initially is due to rising frictional losses. However, above 2500 rpm, the imep also falls rapidly, and this will be attributable to a fall in volumetric efficiency. The valve timings are quite conventional, with the exception of an early exhaust valve opening:

Exhaust valve opening	62° bbdc
Exhaust valve closing	8° atdc
Inlet valve opening	13° btdc
Inlet valve closing	48° abdc

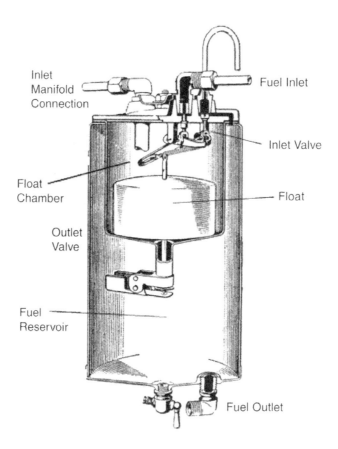

Figure 12.10. *The Autovac fuel pump system. Adapted from Judge (1924).*

At 2500 rpm, the mean piston speed is a comparatively modest 10.8 m/s. If a mean valve flow coefficient is conservatively estimated at 0.25, then this leads to a Mach index of 0.64 and a volumetric efficiency of greater than 80%. However, the volumetric efficiency then falls rapidly as the piston speed is increased (Stone, 1999). Also, the inlet manifold and induction pipe from the carburetor are of comparatively small cross-sectional area to aid mixture preparation. Ricardo (1924) indicates that at 2500 rpm, the mixture velocity in the induction pipe corresponds to approximately 75 m/s, which will have a dynamic pressure of 3.4 kPa. Thus, the fall in volumetric efficiency at high engine speeds is more likely to be attributable to losses in the induction system than to pressure drop across the inlet valve.

The data in Fig. 12.12 for the part-load fuel consumption also have been modeled by assuming that as the throttle is closed, all of the following occur:

a. The gross indicated specific fuel consumption remains constant at the full throttle value.

b. The charge trapped is proportional to the inlet manifold pressure.

c. The pumping work is increased by an amount equal to the manifold depression.

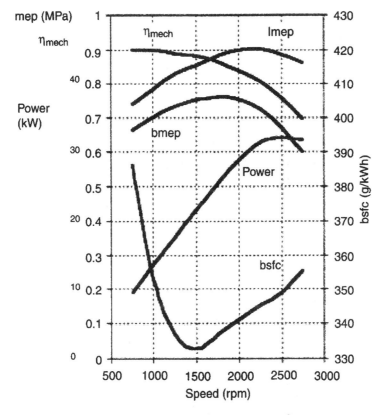

Figure 12.11. *Vauxhall 14-hp engine performance.*

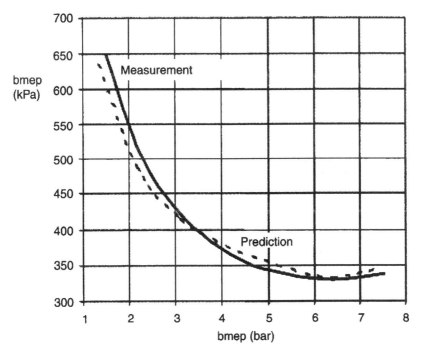

Figure 12.12 *Part-load fuel consumption data at 1600 rpm, measurements and predictions.*

Figure 12.12 shows that this simple approach gives good agreement except at low loads, where air-fuel ratio enrichment will have been used to compensate for the increased levels of residuals. This approach has been used to estimate the brake specific fuel consumption (bsfc) map for the engine (Fig. 12.13), based on the full-throttle bsfc data in Fig. 12.11.

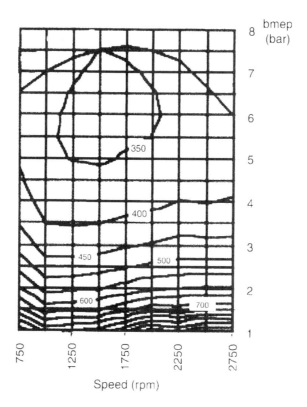

Figure 12.13. *Brake specific fuel consumption map (bsfc – g/kWh) for the 14-hp Vauxhall engine.*

A comparison with modern engine performance is instructive and will be done using several criteria. Comparisons will be made with the Rover MG K series family of engines that are reviewed by Stone (1999). First, the maximum bmep of approximately 7.5 bar can be compared with a modern SI engine, for which a bmep of 10 bar would be considered low, 11 bar more representative of an optimized engine, and 12 bar being achievable by utilizing variable geometry induction systems or variable valve timing. The 1100 and 1400 cm^3 displacement K series engines have a bore of 75 mm (as does the Vauxhall 14-40), but the strokes are 59 and 79 mm, compared with 130 mm. As would be expected, the maximum power speed of the engines is approximately double that of the Vauxhall. (Thus, the maximum mean piston speeds are comparable.) However, the specific outputs of the K series engines are approximately 55 kW/L, compared with 14 kW/L for the Vauxhall. This is due primarily to the limited maximum engine speed of the Vauxhall.

Long-stroke engines were developed because the annual road tax in the United Kingdom was based on piston area—the 14-40 had an RAC horsepower rating of 14 and attracted a tax of £14 pa. Figure 12.11 shows that the 14-40 engine had an fmep of approximately 1.5 bar at mid-engine speeds, which is approximately twice that of the K series engines at the same mean piston speed. It is well known that power is proportional to piston area, whereas torque is proportional to the swept volume. The long stroke and (relatively) large displacement of the 14-40 engine thus provides a high torque, minimizing the need for gear changing. The conservative valve timings and massive flywheel (c 6200 kgcm^2, 32 kg [70.5 lb]) permit an idle speed of approximately 200 rpm and an ability to drive from 5 mph in top gear. This is an advantage when the gearbox is without synchromesh. The lower compression ratio (5:1 as opposed to 10.5:1) gives approximately a 25% penalty on the indicated efficiency, and the higher frictional losses will have a particularly adverse effect on the part-load fuel economy. The minimum specific fuel consumption of the 14-40 (Fig. 12.13) is approximately 330 g/kWh compared with approximately 250 g/kWh for the K series engines. At part load (e.g., 2 bar bmep), the difference is more pronounced: 550 g/kWh for the 14-40, and 400 g/kWh for a K series engine.

Finally, remember that oil consumption will have fallen by at least an order of magnitude. Instead of measuring consumption in miles per gallon, miles per pint will be used, and often there will be no need now to top-up the oil between oil changes.

12.3.2.5 Vehicle Design and Performance

Figure 12.14 shows the unitary construction of the engine, clutch, and gearbox.

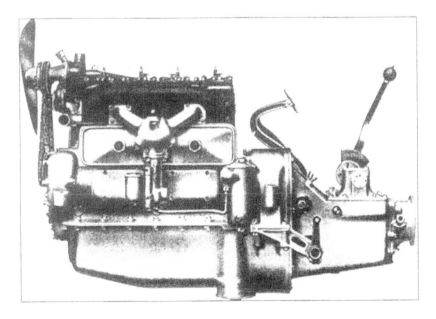

Figure 12.14. *Unitary construction of the engine, clutch, and gearbox.*

The clutch is a single plate with two friction faces, with ball-bearing races for the clutch release bearing and the gearbox input shaft. The gearbox (three forward gears and reverse) is of the sliding mesh type without any synchromesh. Synchromesh was first introduced by Cadillac in 1929; in the United Kingdom, Vauxhall (by then owned by General Motors) introduced synchromesh in 1932 (Lord Montagu, 1978). Synchromesh did not become common on all forward speeds until the 1950s. The gears are straight-cut, which leads to considerable whine in reverse, first, and second gears.

The propeller shaft is contained within a torque tube that is connected rigidly to the back axle casing. The torque tube is connected to the gearbox casing by a spherical bearing to permit back axle articulation. The torque tube adds to the unsprung mass but eliminates the need to have Panhard rods for maintaining the rear axle alignment. The propeller shaft is driven by a universal joint that has steel journals running in bronze bearings, with lubrication from the gearbox oil. The Edward Hardy and Clarence Spicer partnership for the manufacture of needle roller bearing universal joints (U-joints) was not established until 1925. The final drive uses spiral bevel gears with a 4.5:1 reduction ratio. Straight bevel gears (or for quieter but less efficient operation, a worm and wheel) were the norm until around 1920. Hypoid gears with their non-intersecting axes that allowed the body to sit lower in the frame (developed by the Gleason company) were not used until around 1940.

There are only rear-wheel brakes (34-cm [13.4-in.] diameter drums) with separate trailing and leading shoes for the foot brake and hand brake. The absence of front-wheel brakes mattered less when traffic densities were lower and did lead to a light front axle. Steering is by a worm and sector, acting via a drop arm and drag link to the off-side wheel. Both the drag link and track rod incorporate spring-loaded ball joints to reduce the effect of shock loads from the wheels.

The engine torque/speed curve can be converted to a tractive effort/vehicle speed plot. Table 12.4 shows the necessary data and assumptions for Fig. 12.15. Hucho (1998b) indicates that the drag coefficient for a carriage style saloon of the 1920s is approximately 0.8.

TABLE 12.4
VEHICLE DATA FOR THE VAUXHALL 14-40

Transmission Efficiency:	
First, Second Gears	0.90
Top Gear	0.95
Drag, AC_d	2.4 m^2
Test Mass	1200 kg
Air Density	1.2 kg/m^3
Rolling Resistance Coefficient	0.015

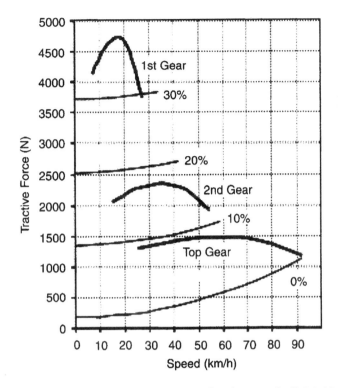

Figure 12.15. *Tractive force curves for the Vauxhall 14-40; see data in Table 12.4.*

Figure 12.15 shows the tractive force available in the three different gears, and the tractive effort required in still air for level ground and three gradients. Figure 12.15 indicates a maximum speed of 95 km/h (59 mph) on level ground, and a good ability to climb a 30% gradient. Somewhat less satisfactory is the wide spacing between the gears, a shortcoming that was solved by the introduction of a four-speed gearbox for the LM-type in 1925.

At 30 mph (48 km/h), Ricardo (1924) reports a fuel consumption of 29 mpg (23 mi/U.S. gal, or 9.8 L/100 km). At 30 mph in top gear, the engine is operating at 1429 rpm and a bmep of 2.22 bar (6.08 kW), and the fuel consumption map (Fig. 12.13) indicates a specific fuel consumption of 508 g/kWh. This would equate to 8.6 L/100 km (33 mpg or 26 mi/U.S. gal), which is reasonable agreement in view of the underestimate of the fuel consumption shown in Fig. 12.12 at low loads.

Figure 12.16 shows the vehicle acceleration performance. The standing quarter-mile takes 41 s, with 0–30 mph (48 km/h) taking 25 s, and 0–50 mph (80 km/h) taking 78 s. The 30–50 mph (48–80 km/h) performance in top gear is 54 s. Figure 12.17 shows the velocity/time history of the Vauxhall 14-40.

Comparisons now will be made with the Rover 800, a five-seater sedan that uses the 2.5-liter V-6 version of the K series engine. This vehicle is heavier (1465 kg [3330 lb]) but with

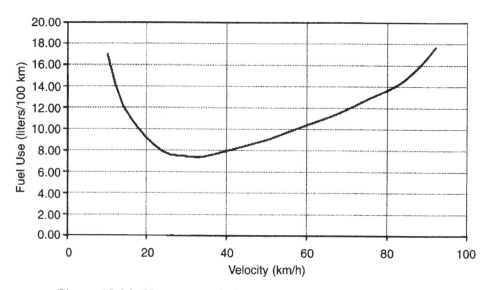

Figure 12.16. *Variation in fuel consumption with vehicle speed.*

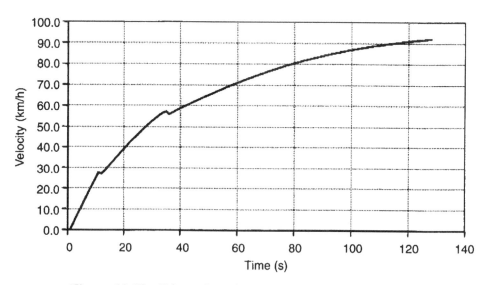

Figure 12.17. *Velocity/time history of the Vauxhall 14-40.*

significantly lower drag ($AC_d = 0.66$ m^2) and a significantly more powerful engine (129 kW compared with 33 kW). The Rover 800 has a substantially lower minimum steady-state fuel consumption of approximately 5 L/100km in the speed range 30–60 km/h (19–37 mph) (compared with 8 L/100 km in the speed range 25–40 km/h [15.5–25 mph] for the Vauxhall 14-40). The hill-climbing abilities are comparable, although the Vauxhall would be much slower, of course. The Rover 800 has a top speed of 220 km/h (137 mph) (compared with 95 km/h [59 mph] for the 14-40). The standing quarter-mile for the Rover 800 takes 17 s compared with 41 s for the 14-40, whereas for 0–50 mph (0–80 km/h), the Rover 800 takes 6.8 s, compared with 78 s for the Vauxhall. Finally, the 30–50 mph (48–80 km/h) performance for the Rover 800 is 6.6 s in fourth gear (but approximately 1.4 s in second gear), whereas the Vauxhall takes 54 s!

12.3.2.6 Conclusions

The Vauxhall 14-40 embodied many of the best contemporary design ideas, but vehicle design obviously has evolved considerably over the last 80 years. Front-wheel brakes became common by the mid-1920s, as did four-speed gearboxes. Dampers would continue to be an optional extra throughout the 1920s.

Vehicle performance was limited by the large aerodynamic drag, and this had an adverse effect on the high-speed fuel consumption. The engine bmep was quite reasonable, but the maximum engine speed limited the power output. The low compression ratio (4.8:1) was entirely appropriate for its time, which meant that engine performance was not limited by the choice of a side-valve design.

Perhaps the greatest changes have occurred in the areas of cost, safety, durability, and comfort (especially noise and ventilation), all of which are beyond the scope of this book.

12.3.3 The Toyota Prius

Figure 12.18 shows the dual hybrid configuration adopted in the Toyota Prius. The terminology of generator and motor for the electrical machines refers to their primary function because both must be able to act as either a motor or a generator. Figure 12.18 shows how the engine is connected to the planet carrier of the epicyclic gear box, with the generator (Machine 1) connected to the sun gear, and the motor (Machine 2) connected to the annulus. There is a fixed reduction gearbox (not shown) for power transmission to the road wheels. Table 12.5 gives the Toyota Prius specifications.

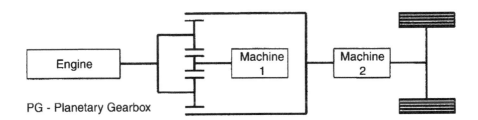

Figure 12.18. *The epicyclic gearbox configuration used in the Toyota Prius dual hybrid configuration.*

The performance of the hybrid vehicle is superior to the equivalent conventional vehicle (Toyota Corolla 1.5-liter automatic) in terms of acceleration and fuel economy (Hermance and Sasaki, 1998), as shown in Table 12.6.

TABLE 12.5
TOYOTA PRIUS SPECIFICATIONS
(MERCER, 1998, AND HERMANCE AND SASAKI, 1998)

Length: 4.275 m	Width: 1.695 m	Height: 1.49 m
Engine: 1.5-L four-cylinder	DOHC	Spark ignition
	13.5:1 compression ratio	42.6 kW @ 4000 rpm
Motor: Permanent magnet DC, 30 kW @ 940–6000 rpm		
Generator: Permanent magnet DC, 15 kW @ 5500 rpm		
Battery: Nickel-metal hydride—40 12-V units, 6.5 Ah rating		
Planetary gear ratio (ring/sun): 2.6:1		Final drive: 3.93:1
Maximum speed:	142 km/h (engine alone)	161 km/h (hybrid)

TABLE 12.6
PERFORMANCE COMPARISON FOR
HYBRID AND CONVENTIONAL VEHICLES

	Hybrid	Conventional
Acceleration (40–70 km/h)	5 s	>6 s
Fuel Economy (Japanese 10–15 Mode)	3.57 L/100 km	7.14 L/100 km

The spark ignition engine in the Prius has been optimized for high efficiency. By limiting the engine speed to 4000 rpm, low-mass and low-friction components can be used. Furthermore, a variable valve timing system is used to give a compression ratio of 9:1 but a much higher expansion ratio (up to 14:1). The peak thermal efficiency is in the region of 38% (Hermance and Sasaki, 1998). Figure 12.19 shows that the engine gearbox and two electrical machines have been made into a compact arrangement in the Prius.

12.3.4 *Modeling the Dual Configuration*

The dual configuration allows power to be transmitted through two parallel paths (mechanically or electrically) from the prime mover to the driving wheels. Because the electrical path has an infinitely variable transmission ratio, there is considerable extra flexibility in choosing the engine operating point. Of course, the electrical machines must operate within speed and torque/power constraints. To establish the steady-state operating strategy, a "pre-simulation" is undertaken to provide look-up tables (in terms of required power and wheel speed) that are used in the subsequent simulation.

Figure 12.20 shows an engine efficiency map (plotted against torque and speed) with a constant power line shown, for which the circle represents the operating point with a conventional transmission. If the electrical system had no limitations and no losses, then the optimum operating point would be where the power line is tangential to an efficiency contour. However,

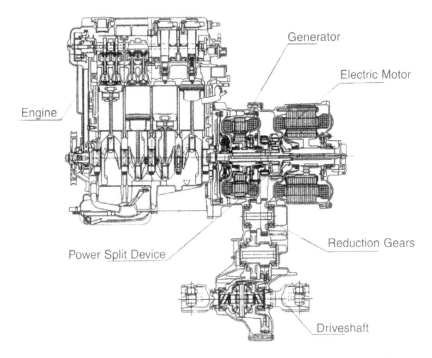

Figure 12.19. *The engine, gearbox, and two electrical machines used in the Toyota Prius.*

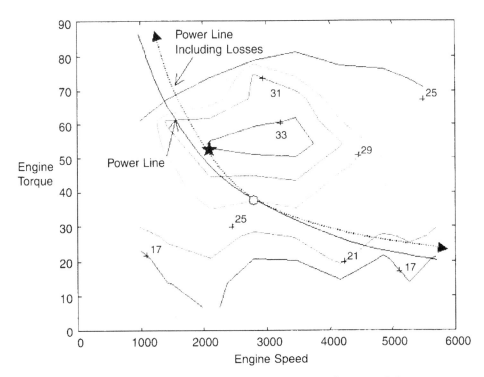

Figure 12.20. *Engine brake efficiency map showing the selection of the optimum engine operating point (the star) for a given power requirement.*

because of losses and practical limitations, it is necessary to compute a power line including losses, and to find where this touches the efficiency contours (shown by a star in Fig. 12.20).

Figure 12.21 shows how the dual hybrid system is able to modify the engine operating regime, so that there is a substantial increase in the region where there is a brake specific fuel consumption (bsfc) below 300 g/kWh. Because a dual hybrid vehicle also will have some form of energy storage (e.g., a battery), there is no need to operate the engine at low power outputs (say, 7.5 kW in this case), thereby yielding a further efficiency gain.

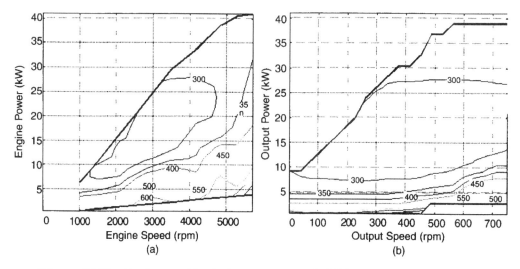

Figure 12.21. *Brake specific fuel consumption (g/kWh) maps for the (a) engine and (b) dual hybrid powertrain system.*

12.4 Conclusions

The reciprocating internal combustion engine presents a formidable challenge for fuel cells. The fuel economy, emissions, and specific output of reciprocating engines all continue to improve, and the large volumes for manufacture lead to competitive pricing. Electric vehicles remain handicapped by their battery technology, which leads to a high cost and limited range. (Their acceleration performance can be remarkable, but the maximum speeds tend to be limited.) In most cases, electric vehicles merely represent a relocation of the emissions source.

In contrast to fuel cells, hybrid vehicles use comparatively well-established technology and are now being marketed. The manufacturing cost premium (perhaps a factor of two for the Prius) would be reduced with larger-scale manufacture and less sophisticated systems, or systems with a less powerful electric mode (MYBRIDs). Hybrid vehicles have the highest "well to wheel" efficiency and present the greatest competition for fuel cell powered vehicles.

CHAPTER 13

References

Adler, U. (1988), "Starter Installations," in *Automotive Electric/Electronic Systems*, Adler, U. and Bauer, H., Eds., Robert Bosch, Stuttgart, 1998.

Adler, U. (1994), Ed., *Diesel Fuel Injection*, Robert Bosch, Stuttgart, 1994.

Ahmed, S.R. (1998), "Computational Fluid Dynamics," in *Aerodynamics of Road Vehicles*, Fourth Ed., Hucho, W.-H., Ed., Society of Automotive Engineers, Warrendale, PA, 1998.

Allard, A. (1982), *Turbocharging and Supercharging*, Patrick Stephens, Cambridge, 1982.

Ando, H. (1996), "Combustion Control Strategies for Gasoline Engines," Institution of Mechanical Engineers, Seminar Publication, *Lean Burn Combustion Engines*, Paper No. S433/001/96, pp. 3–17, Institution of Mechanical Engineers, London, 1996.

Annand, W.J.D. and Roe, G.E. (1974), *Gas Flow in the Internal Combustion Engine*, Foulis, Yeovil, 1974.

Anonymous (1998), "Attractive Alternatives," *Daimler Benz HighTech Report 1998*.

Appleby, A. and Foulkes, F. (1989), *Fuel Cell Handbook,*, Van Nostrand Reinhold, New York, 1989.

Arai, M. and Miyashita, S. (1990), "Particulate Regeneration Improvement on Actual Vehicle Under Various Conditions," Paper No. C394/012, Institution of Mechanical Engineers, Conf. Proc., *Automotive Power Systems—Environment and Conservation*, MEP, London, 1990.

Armstrong, A. (2000), "ICE Hybrids—Good Enough?" in *Fuel Cells 2000*, USA.

Ashley, S. (2001), "Fuel Cells Start to Look Real," in *Automotive Engineering*, Vol. 109, No. 3, 2001, pp. 64–80.

Ashton, R. (1998), "Energy Storage," in *Design of a Hybrid Electric Vehicle*, Project Report, University of Oxford, Department of Engineering Science, UK, 1998.

Automotive Engineering Vol. II: Power Train and Running Gear, Department of Engineering, U.S. Military Academy, West Point, NY, 1982.

Barnard, R.H. (1996), *Road Vehicle Aerodynamic Design*, Addison Wesley Longman, Harlow, 1996.

Bastow, D. and Howard, G.P. (1993), *Car Suspension and Handling*, Third Ed., Society of Automotive Engineers, Warrendale, PA, 1993.

Batt, R.J., McMillan, J.A., and Bradbury, I.P. (1996), "Lubricity Additives—Performance and No-Harm Effects in Low Sulphur Fuels," SAE Paper No. 961943, Society of Automotive Engineers, Warrendale, PA, 1996.

Bayer, B. (1998), "Motorcycles," in *Aerodynamics of Road Vehicles*, Fourth Ed., Hucho, W.-H., Ed., Society of Automotive Engineers, Warrendale, PA, 1998.

Bearman, P.W. (1978), "Turbulence and Ground Effects," in *Aerodynamic Drag Mechanisms*, Sovran. G., Morel, T., and Mason, W.T., Eds., Plenum Press, New York, 1978.

Bernard, J., Shannan, J., and Vanderploeg, M. (1989), "Vehicle Rollover on Smooth Surfaces," SAE Paper No. 891991, Society of Automotive Engineers, Warrendale, PA, 1989.

Bernt, D. (1997), *A Handbook of Battery Technology*, Second Ed., Research Studies Press, 1997.

Blackmore, D.R. and Thomas, A. (1977), *Fuel Economy of the Gasoline Engine*, Macmillan, London, 1977.

Blount, B. (1908), "Composition of Commercial Petrols," in *Proc. of the Inst. Automobile Engineers*, Vol. 3, 1908–1909, p. 301.

Boatto, P., Boccaletti, C., Cerri, G., and Malvicino, C. (2000), *Internal Combustion Engine Waste Heat Potential for an Automotive Absorption System of Air Conditioning, Parts 1 and 2*, Proc. Institution of Mechanical Engineers, Part D, Vol. 214, No. D8, 2000, pp. 979–990.

Booker, D.R. (2000), "Particulate Matter Sizing," keynote address at the Institution of Mechanical Engineers Conf., "Computational and Experimental Methods in Reciprocating Engines," London, November 2, 2000.

Bostock, L.G. and Cooper, L. (1992), "Turbocharging the Ford 2.5 HSDI Diesel Engine," Institution of Mechanical Engineers Seminar, "Diesel Fuel Injection Systems," April 14–15, 1992, MEP, London.

Brogan, M.S., Will, N.S., Twigg, M.V., Wilkins, A.J.J., Jordan, K., and Brisley, R.J. (1998), "Advances in DeNOX Catalyst Technology for Stage IV Emissions Levels," in *Future Engine Systems and Technologies,* Institution of Mechanical Engineers Seminar Publication, PEP, London, 1998.

Broge, J.L. (2000), "Ford's Prototype Electric Ka," in *Automotive Engineering,* Vol. 108, No. 7, pp. 28–32.

Buchheim, J. Maretzke and R. Piatek (1985), "The Control of Aerodynamic Parameters Influencing Vehicle Dynamics," SAE Paper No. 850279, Society of Automotive Engineers, Warrendale, PA, 1985.

Callister, J.R. and George, A.R. (1998), "Wind Noise," in *Aerodynamics of Road Vehicles*, Fourth Ed., Hucho, W.-H., Ed., Society of Automotive Engineers, Warrendale, PA, 1998.

Campbell, C. (1978), *The Sports Car*, Fourth Ed., Chapman and Hall, London (same location; name changed), 1978.

Clymer, F. (1950), *Early American Automobiles*, McGraw-Hill, New York, 1950.

Connolly, F.T. (1994), "Direct Estimation of Cyclic Combustion Pressure Variability Using Engine Speed Fluctuations in Internal Combustion Engines," SAE Paper No. 940143, Society of Automotive Engineers, Warrendale, PA, 1994.

Correa, S.M. (1992), "A Review of NOx Formulation Under Gas-Turbine Combustion Conditions," in *Combust. Sci. and Tech.*, Vol. 87, 1992, pp. 329–362.

Crandall, R.W., Gruenspecht, H.K., Keeler, T.E., and Lave, L.B., *Regulating the Automobile*, The Brookings Institution, Washington, DC, 1986.

Crompton, T.R. (1995), *Battery Reference Book*, Second Ed., Butterworth Heinmann, 1995.

Cropley, S. (1996), "HP Source," in *Autocar*, November 1996, pp. 106–111.

Dentis, L., Mannoni, A., and Parrino, M. (1999), "HC Refrigerants: An Ecological Solution for Automotive A/C Systems," Paper No. C543/006/99, *Vehicle Thermal Management Systems*, Institution of Mechanical Engineers, London, 1999.

Dominy, J.A. and Dominy, R.G. (1984), "Aerodynamic Influences on the Performance of the Grand Prix Racing Car," Proc. Institution of Mechanical Engineers, Vol. 198D, No. 12, 1984, pp. 1–7.

Eade, D., Hurley, R.G., Rutter, B., Inman, G., and Bakshi, R. (1996), "Exhaust Gas Ignition," in *Automotive Engineering*, Vol. 104, No. 4, 1996, pp. 70–73.

Eastham, D.R., Parker, D.D., and Summerton, K. (1995), "Developments in Tri-Metal Bearings," Paper 2, T&N Symposium, 1995.

Eastwood, P. (2000), *Critical Topics in Exhaust Gas Aftertreatment*, Research Studies Press, Baldock, 2000.

Edgar, G. (1927), "Measurement of Knock Characteristics of Gasoline in Terms of Standard Fuel," in *Ind. Eng. Chemistry*, Vol. 19, 1927, p. 109.

Ellis, J.R. (1969), *Vehicle Dynamics,* London Business Books, 1969.

Enga, B.E., Buchman, M.F., and Lichtenstein, I.E. (1982), "Catalytic Control of Diesel Particulates," SAE Paper No. 820184 (also in SAE P-107), Society of Automotive Engineers, Warrendale, PA, 1982.

Ermisch, N., Dorenkamp, R., Neyer, D., Hilbig, J., and Scheliga, W. (2000), "The Powertrain of the 3-L-Lupo," in *Dieselmotorentechnik 2000,* Bargende, M. and Essers, U., Eds., Expert Verlag, 2000.

Evans, E.M. and Zemroch, P.J. (1984), "Measurement of the Aerodynamic and Rolling Resistances of Road Tanker Vehicles from Coast-Down Tests," Proc. Institution of Mechanical Engineers, Vol. 198D, No. 11, 1984, pp. 211–18.

Everett, N.R., Brown, M.D., and Crolla, D.A. (2000), "Investigation of a Roll Control System for an Off-Road Vehicle," SAE Paper 2000-01-1646, Society of Automotive Engineers, Warrendale, PA, 2000.

Felger, G. (1988), "LH-Jetronic," in *Automotive Electric/Electronic Systems,* Adler, U. and Bauer, H., Eds., Robert Bosch, Stuttgart, 1988.

Ferguson, C.R. (2001), *Internal Combustion Engines,* 2nd Ed., John Wiley and Son, 1986.

Forlani, E. and Ferrati, E. (1987), "Microelectronics in Electronic Ignition—Status and Evolution," 16th ISATA Proc., Paper No. 87002.

Fossett, H. (1943–1944), "Petrol—Its Development, Past, Present and Future—With Some Notes on the Potentialities of High Octane Fuels for Road Vehicles," Proc. Inst. of Automobile Engineers, Vol. 38, 1943–1944, p. 89.

Frankl, G., Barker, B.G., and Timms, C.T. (1989), "Electronic Unit Injectors for Direct Injection Engines," Institution of Mechanical Engineers Seminar, "Diesel Fuel Injection Systems," MEP, London, 1989.

Gardiner, W.C. (2000), *Gas Phase Combustion Chemistry,* 2nd Ed., Springer-Verlag, New York, 2000.

Garner, F.H. (1936–1937), "Research in Relation to the Motor Vehicle. Section 11. Fuels and Lubricants," Proc. Inst. of Automobile Engineers, Vol. 31, 1936–1937, p. 710.

Garrett, K. (1990), "Fuel Quality, Diesel Emissions, and the City Filter," in *Automotive Engineering,* Vol. 15, No. 5, 1990, pp. 51 and 55.

Georgano, G.N. (1997), *Vintage Cars 1886 to 1930,* Tiger Books Intl., 1997.

Gillespie, T.D. (1994), *Fundamentals of Vehicle Dynamics*, Society of Automotive Engineers, Warrendale, PA, 1994.

Gregg, W.W. (1983), "GMC Aero Astra Body Panels," SAE Paper No. 831003 (also in SAE SP-545), Society of Automotive Engineers, Warrendale, PA, 1983.

Gruden, D. and Kuper, P.F. (1987), "Heat Balance of Modern Passenger Car SI Engines," in "XIXth Intl. Symposium, Intl. Centre for Heat and Mass Transfer, Heat and Mass Transfer in Gasoline and Diesel Engines," Dubrovnik, 1987.

Gunther, D. (1988), "Exhaust Emissions Engineering," in *Automotive Electric/Electronic Systems*, Adler, U. and Bauer, H., Eds., Robert Bosch, Stuttgart, 1988.

Hahn, H.W. (1986), "Improving the Overall Efficiency of Trucks and Buses," Proc.1., 1986.

Hahne, D. (1984), "A Continuously Variable Automatic Transmission for Small Front Wheel Drive Cars," Paper No. C2/84, *Driveline '84*, Institution of Mechanical Engineers Conf. Proc. 1984-1, MEP, London.

Hamrock, B.J., Jacobson, B.O., and Schmid, S.R. (1999), *Fundamentals of Machine Elements*, McGraw-Hill, New York, 1999.

Hawker, P.N. (1995), "Diesel Emission Control Technology," in *Platinum Metal Review*, Vol. 39, No. 1, 1995, pp. 2–8.

Heimrich, M.J., Albu, S., and Osborn, J. (1991), "Electrically Heated Catalyst System Conversions on Two Current-Technology Vehicles," SAE Paper No. 910612, Society of Automotive Engineers, Warrendale, PA, 1991.

Heisler, H. (1995), *Advanced Engine Technology*, Edward Arnold, London, 1995.

Hendricks, E., Chevalier, A., Jensen, M., Sorenson, S.C., Trumpy, D., and Asik, J. (1996), "Modelling of the Intake Manifold Filling Dynamics," SAE Paper No. 960037, Society of Automotive Engineers, Warrendale, PA, 1996.

Hendricks, E., Vesterholm, T., Kaidantzis, P., Rasmussen, P., and Jensen, M. (1993), "Non-Linear Transient Fuel Film Compensation," SAE Paper No. 930767, Society of Automotive Engineers, Warrendale, PA, 1993.

Hermance, D. and Sasaki, S. (1998), "Hybrid Electric Vehicles Take to the Streets," in *IEEE Spectrum*, November 1998, pp. 48–52.

Herzog, P. (1998), "HSDI Diesel Engine Developments Towards Euro IV," Institution of Mechanical Engineers Seminar Publication, *Future Engine and System Technologies*, Professional Engineering Publications, London, 1998.

Heywood, J.B. (1988), *Internal Combustion Engine Fundamentals,* McGraw-Hill, 1988.

Hilbig, J., Neyer, D., and Ermisch, N. (1999), "The New 1,2-L Three-Cylinder Diesel Engine from Volkswagen," VDI Berichte No. 1505, 1999, pp. 461–483.

Hochgreb, S. (1998), "Combustion-Related Emissions in SI Engines," in *Handbook of Air Pollution from Internal Combustion Engines*, Sher, E., Ed., Academic Press, Boston, 1998.

Holmes, M., Willcocks, D.A.R., and Bridgers, B.J. (1988), "Adaptive Ignition and Knock Control," SAE Paper No. 885065, Society of Automotive Engineers, Warrendale, PA, 1988.

Horie, K. and Nishizawa, K. (1992), "Development of a High Fuel Economy and High Performance Four-Valve Lean Burn Engine," Paper No. C448/014, *Combustion in Engines*, Institution of Mechanical Engineers Conf. Proc., MEP, London, 1992, pp. 137–143.

Howatson, H.M., Lund, P.G., and Todd, J.D. (1991), *Engineering Tables and Data*, Chapman and Hall, London, 1991.

Hucho, W.-H. (1978), "The Aerodynamic Drag of Cars," in *Aerodynamic Drag Mechanisms*, Sovran, G., Morel, T., and Mason, W.T., Eds., Plenum Press, New York, 1978.

Hucho, W.-H. (1998a), *Aerodynamics of Road Vehicles*, Society of Automotive Engineers, Warrendale, PA, 1998.

Hucho, W.-H. (1998b), "Introduction to Automobile Aerodynamics," in *Aerodynamics of Road Vehicles*, Fourth Ed., Hucho, W.-H., Ed., Society of Automotive Engineers, Warrendale, PA, 1998.

Hucho, W.H., Janssen, L.J., and Emmelmann, H.J. (1976), "The Optimisation of Body Details—A Method for Reducing the Aerodynamic Drag of Road Vehicles," SAE Paper No. 760185 (also in SAE PT-18), Society of Automotive Engineers, Warrendale, PA, 1976.

Ingrassia, P. and White, J.B. (1994), *Comeback: The Fall and Rise of the American Automobile Industry*, Simon and Schuster, New York, 1994.

International Energy Agency, "Energy Prices and Taxes, Quarterly Statistics," at www.iea.org/stats/files/prices.htm, 2000.

Ironside, J.M. and Stubbs, P.W.R. (1981), "Microcomputer Control of an Automotive Perbury Transmission," Third Intl. Automotive Electronics Conf., Institution of Mechanical Engineers, London, 1981, pp. 283–292.

Jackson, N.S., Stokes, J., Whitaker, P.A., and Lake, T.H. (1996), "A Direct Injection Stratified Charge Gasoline Combustion System for Future European Passenger Cars," presented at Institution of Mechanical Engineers Seminar, "Lean Burn Combustion Engines," December 3–4, 1996, London.

Jochheim, J., Hesse, D., Duesterdick, T., Engeler, W., Neyer, D., Warren, J.P., Wilkins, A.J.J., and Twigg, M.V. (1996), "A Study of the Catalytic Reduction of NOx in Diesel Exhaust," SAE Paper No. 962042, Society of Automotive Engineers, Warrendale, PA, 1996.

Johnston, R.H. (1996), "A History of Automobile Electrical Systems," in *Automotive Engineering*, September 1996, also reprinted in *The Automobile: A Century of Progress*, Society of Automotive Engineers, Warrendale, PA, 1997, pp. 151–184.

Joyce, M. (1994), "Jaguar's Supercharged 6-Cylinder Engine," in *Turbochargers and Turbocharging*, Institution of Mechanical Engineers Conf. Proc. 1994–1996.

Judge, A.W. (1924), *Modern Motor Cars*, Caxton, London, 1924.

Kawagoe, K., Suma, K., and Watanabe, M. (1997), "Evaluation and Improvement of Vehicle Roll Behavior," SAE Paper No. 970093, Society of Automotive Engineers, Warrendale, PA, 1997.

Krutz, G.W., Schueller, J.K., and Claar, P.W. II (1994), *Machine Design for Mobile and Industrial Applications*, Society of Automotive Engineers, Warrendale, PA, 1994.

Ladommatos, N., Abdelhalim, S.M., Zhao, H., and Hu, Z. (1998), "The Effects of Carbon Dioxide in Exhaust Gas Recirculation on Diesel Engine Emissions," *Proc. Institution of Mechanical Engineers*, Vol. 212, Part D, 1998, pp. 25–42.

Lancefield, T., Cooper, L., and French, B. (1996), "Design, Control and Simulation of EGR," in *Automotive Engineering*, Vol. 21, No. 1, 1996, pp. 28–31.

Larminie, J. and Dicks, A. (2000), *Fuel Cell Systems Explained*, Wiley, Chichester, 2000.

Lavoie, G.A., Heywood, J.B., and Keck, J.C. (1970), "Experimental and Theoretical Study of Nitric Oxide Formation in Internal Combustion Engines," in *Comb. Sci. and Technology*, Vol. 1, 1970, pp. 313–326.

Lee, A.Y. (1995), "Performance of Four-Wheel-Steering Vehicles in Lane Change Maneuvers," SAE Paper No. 950316, Society of Automotive Engineers, Warrendale, PA, 1995.

Lembke, M. (1988), "Motronic," in *Automotive Electric/Electronic Systems*, Adler, U. and Bauer, H., Eds., Robert Bosch, Stuttgart, 1988.

Lilly, L.R.C. (1984), *Diesel Engine Reference Book*, Butterworths, London, 1984.

Limpert, R. (1999), *Brake Design and Safety*, Society of Automotive Engineers, Warrendale, PA, 1999.

Lindeburg, M.R. (1998), *Mechanical Engineering Reference Manual for the PE Exam*, 10th Ed., Professional Publications, Inc., Belmont, CA, 1998.

Löhle, M., Gneting, R., and Mönkediek, T. (1999), "Advanced Use of CAE Tools in the Development of HVAC Systems," Paper No. C543/024/99, *Vehicle Thermal Management Systems*, Institution of Mechanical Engineers, London, 1999.

Longee, H. (1998), "The Capstone MicroTurbne™ as a Hybrid Vehicle Energy Source," SAE Paper No. 981187, Society of Automotive Engineers, Warrendale, PA, 1998.

Lord Montagu of Beaulieu (1978), "Road Vevhicles," in *A History of Technology*, Vol. VII, Williams, T.I., Ed., OUP, 1978.

Lorenz, K. and Peterreins, K. (1984), "Fuel Economy and Performance—Effects on Power Transmission," Paper No. C13/84, *Driveline '84*, Institution of Mechanical Engineers Conf. Proc. 1984-1, MEP, London, 1984.

Maly, R.R. (1984), "Spark Ignition: Its Physics and Effect on the Internal Combustion Engine," in *Fuel Economy of Road Vehicles Powered by Spark Ignition Engines*, Hilliard, J.C. and Springer, C.S., Eds., Plenum Press, New York, 1984.

Maly, R.R. and Vogel, M. (1978), "Initiation and Propagation of Flame Fronts in Lean CH4—Air Mixtures by the Three Modes of the Ignition Spark," in *17th Int. Conf. on Combustion*, The Combustion Institute, 1978, pp. 821–831.

Manzie, C., Palaninswami, M., and Watson, H.C. (1998), "Model Predictive Control of a Fuel Injection System Using a CMAC Neural Network," in *Proc. Int. Conf. on Automation, Robotics, Control, and Vision '98*, Singapore, 1998, pp. 940–944.

Martin, B. and Redinger, C.J. (1993), "42LE Electronic Four-Speed Automatic Transaxle," SAE Paper No. 930671, Society of Automotive Engineers, Warrendale, PA, 1993.

Mason, W.T. and Beebe, P.S. (1978), "The Drag Related Flow Field Characteristics of Trucks and Buses," in *Aerodynamic Drag Mechanisms,* Sovran, G., Morel, T., and Mason, W.T., Eds., Plenum Press, New York, 1978.

Massey, B.S. (1983), *Mechanics of Fluids*, Fifth Ed., Van Nostrand Reinhold, New York, 1983.

Mattavi, J.N. and Amann, C.A. (1980), *Combustion Modelling in Reciprocating Engines*, Plenum Press, New York, 1980.

May, G.S. (1975), *A Most Unique Machine*, William B. Eerdmans Publishing Co., Grand Rapids, MI, 1975.

Mercer, M., (1998), "Hybrid Car to Arrive in Europe and U.S. in 2000," in *Diesel Progress,* September/October 1998, pp. 60–61.

Merchant, R., Bradbury, I.P., Ashton, S., and Vincent, M.W. (1997), "Effect on Vehicle Performance of Changes in Automotive Diesel Fuel Composition," Institution of Mechanical Engineers Conf. "Automotive Fuels for the 21st Century," MEP, London, 1997.

Merzkirch, W. (1974), *Flow Visualisation,* Academic Press, New York, 1974.

Meyer, E.W., Green, R., and Cops, M.H. (1984), "Austin-Rover Montego Programmed Ignition System," Paper No. C446/84, *VECON '84 Fuel Efficient Power Trains and Vehicles,* Institution of Mechanical Engineers Conf. Proc. 1984-14, MEP, London, 1984.

Meyer, F. and Gerhard, A. (1988), "Alternators and Generators," in *Automotive Electric/Electronic Systems*, Adler, U. and Bauer, H., Eds., Robert Bosch, Stuttgart, 1988.

Milliken, W.F. and Milliken, D.L. (1995), *Race Car Vehicle Dynamics*, Society of Automotive Engineers, Warrendale, PA., 1995.

Milton, B. (1998), "Control Technologies in Spark Ignition Engines" in *Handbook of Air Pollution from Internal Combustion Engines*, Sher, E., Ed., Academic Press, Boston, 1998.

Moeller, E. (1951), "Luftwiderstandsinessungen am VW-Lieferwagen," in *Automobil-Technische Zeitschrift,* Vol. 53, No. 6, 1951, pp. 153–156.

Motor Trend Magazine (1996), "100 Years of the Automobile in America," Peterson Publishing, Los Angeles, CA, 1996.

Motor Trend Magazine (1999), "Best, Fastest, and Most Outrageous—The 50 Most Memorable Cars Ever Tested by Motor Trend," Peterson Publishing, Los Angeles, CA, September 1999.

Nakajima, Y., Sugihara, K., and Takagi, Y. (1979), "Lean Mixture or EGR—Which Is Better for Fuel Economy and NOx Reduction?" *Proc. Conf. on Fuel Economy and Emissions of Lean Burn Engines*, Institution of Mechanical Engineers Conf. Proc., MEP, London, 1979.

Nandi, M.K. and Jacobs, D.C. (1995), "Cetane Response of Di-Tertiary-Butyl Peroxide in Different Diesel Fuels," SAE Paper No. 952368, Society of Automotive Engineers, Warrendale, PA, 1995.

NEL (1999), *Recent Developments in Hybrid and Electric Vehicles in North America*, Department of Trade and Industry Technology Mission, March 1999.

Norbeck, J.M., Heffel, J.W., Durbin, T.D., Tabbara, B., Bowden, J.M., and Montano, M.C. (1996), *Hydrogen Fuel for Surface Transportation*, Society of Automotive Engineers, Warrendale, PA, 1996.

Olley, M. (1946), *Road Manners of the Modern Car*, Institution of Mechanical Engineers, 1946, also reprinted in Proc. of the 2000 SAE Automotive Dynamics and Stability Conf., May 15–17, 2000, Troy, MI, pp. v–xi.

Owen, K. and Coley, C. (1995), *Automotive Fuels Reference Book*, Second Ed., Society of Automotive Engineers, Warrendale, PA, 1995.

Pankhurst, R.C. and Holder, D.W. (1952), *Wind Tunnel Technique*, Pitman, London, 1952.

Parker, P. (2000a), "The Variable Valve Timing Mechanism for the Rover K16 Engine, Part I: Selection of the Mechanism and the Basis of the Design," Proc. Institution of Mechanical Engineers, Part D, Vol. 214, 2000, pp. 206–215.

Parker, P. (2000b), "The Variable Valve Timing Mechanism for the Rover K16 Engine, Part II: Application to the Engine and the Performance Obtained," Proc. Institution of Mechanical Engineers, Part D, Vol. 214, 2000, pp. 206–215.

Peterson, L.D. and Holka, T.C. (1983), "Engineering Development of the Probe IV Advanced Concept Vehicle," SAE Paper No. 831002 (also in SAE SP-545), Society of Automotive Engineers, Warrendale, PA, 1983.

Pfluger, F. (1997), "Two-Stage Turbocharging for Commercial Diesel Engines," in *Diesel Progress*, January–February 1997, pp. 18–20.

Piccone, A. and Rinolfi, R. (1998), "Fiat Third-Generation DI Diesel Engines," Institution of Mechanical Engineers Seminar Publication, *Future Engine and System Technologies*, Professional Engineering Publications, London, 1998.

Pischinger and Cartellieri, W. (1972), "Combustion System Parameters and Their Effect upon Diesel Engine Exhaust Emissions," SAE Paper No. 720756, in *SAE Transactions*. Vol. 81, Society of Automotive Engineers, Warrendale, PA, 1972.

Pischinger, F.F. (1998), "Compression-Ignition Engines—Introduction, in *Handbook of Air Pollution from Internal Combustion Engines*, Sher, E., Ed., Academic Press, Boston, 1998.

Polach, W. and Leonard, R. (1994), "Exhaust Gas Treatment," in *Diesel Fuel Injection*, Adler, U., Bauer, H., and Beer, A., Eds., Robert Bosch, Stuttgart, 1994.

Porter, F.C. (1979), "Design for Fuel Economy—The New GM Front Wheel Drive Cars," SAE Paper No. 790721, Society of Automotive Engineers, Warrendale, PA, 1979.

Pouille, J.-P., Lauga, V., Schonfeld, S., Strobel, M., and Stommel, P. (1998), "Application Strategies of Lean NOx Catalyst Systems to Diesel Passenger Cars," Institution of Mechanical Engineers Seminar Publication, *Future Engine and System Technologies*, Professional Engineering Publications, London, 1998.

Rae, J.B. (1965), *The American Automobile, A Short History*, The University of Chicago Press, Chicago, 1965.

Raimondi, A.A. and Boyd, J. (1958), "A Solution for the Finite Journal Bearing and Its Application to Analysis and Design, Parts I, II, and III," *Trans. ASLE*, Vol. 1, No. 1, in *Lubrication Science and Technology*, Pergamon, New York, 1958, pp. 159–209.

Rand, R.A.J., Woods, R., and Dell, R.M. (1998), *Batteries for Electric Vehicles*, Research Studies Press, 1988.

Redfield, R.C. (1987), *Optimal Adaptive Vehicle Suspension Design and Simulation*, Ph.D. thesis, University of California, Davis, 1987.

Ricardo, H.R. (1924), *The High-Speed Internal Combustion Engine*, Blackie, Edinburgh, 1924.

Ricardo, H. (1992), *The Ricardo Story*, Second Ed., Society of Automotive Engineers, Warrendale, PA, 1992.

Rieck, J.S., Collins, N.R., and Moore, J.S. (1998), "OBD-II Performance of Three-Way Catalysts," in *Automotive Engineering*, Vol. 106, No. 7, 1998, pp. 33–35.

Rinek, L.M. and Cowan, C.W. (1997), "U.S. Passenger Car Brake History," in *Automotive Engineering*, July 1995, also reprinted in *The Automobile: A Century of Progress*, Society of Automotive Engineers, Warrendale, PA, 1997, pp. 33–46.

Rinschler, G.L. and Asmus, T. (1997), "Powerplant Perspectives: Part I," in *Automotive Engineering*, April 1995, also reprinted in *The Automobile: A Century of Progress*, Society of Automotive Engineers, Warrendale, PA, 1997, pp. 1–12.

Rogers, G.F.C. and Mayhew, Y.R. (1967), *Engineering Thermodynamics Work and Heat Transfer*, Longman, 1967.

Sadler, M., Stokes, J., Edwards, S.P., Zhao, H., and Ladommatos, N. (1998), "Optimization of the Combustion System for a Direct Injection Gasoline Engine, Using a High-Speed In-Cylinder Sampling Valve," Institution of Mechanical Engineers Seminar Publication, *Future Engine and System Technologies,* Professional Engineering Publications, London, 1998.

SAE J201, "In-Service Brake Performance Test Procedure Passenger Car and Light-Duty Truck," March 1997, Society of Automotive Engineers, Warrendale, PA.

SAE J300, "Engine Oil Viscosity Classification," December 1999, Society of Automotive Engineers, Warrendale, PA.

SAE J304, "Engine Oil Tests," June 1999, Society of Automotive Engineers, Warrendale, PA.

SAE J357, "Physical and Chemical Properties of Engine Oils," October 1999, Society of Automotive Engineers, Warrendale, PA.

SAE J843, "Brake System Road Test Code—Passenger Car and Light-Duty Truck (A), March 1997, Society of Automotive Engineers, Warrendale, PA.

SAE J1652, "Dynamometer Effectiveness Characterization Test for Passenger Car and Light Truck Caliper Disc Brake Friction Materials," April 1995, Society of Automotive Engineers, Warrendale, PA.

SAE J2246, "Antilock Brake System Review," June 1992, Society of Automotive Engineers, Warrendale, PA.

SAE J2430, "Dynamometer Effectiveness Characterization Test for Passenger Car and Light-Truck Caliper Disc Brake Friction Products," August 1999, Society of Automotive Engineers, Warrendale, PA.

Santer, R.M. and Gleason, M.E. (1983), "The Aerodynamic Development of the Ford Probe IV Advanced Concept Vehicle," SAE Paper No. 831000, Society of Automotive Engineers, Warrendale, PA, 1983.

Schlichting, H. (1960), *Boundary Layer Theory*, McGraw-Hill, New York, 1960.

Shigley, J.E. and Mischke, C.R. (2001), *Mechanical Engineering Design*, Sixth Ed., McGraw-Hill, New York, 2001.

Smith, M. (2001), "It's Volts That Counts," in *Automotive Engineering*, Vol. 26, No. 9, 2001, pp. 52–55.

Sovran, G. (1978), in *Aerodynamic Drag Mechanisms*, Sovran, G., Morel, T., and Mason, W.T., Eds., Plenum Press, 1978.

Stone, C.R. (1988), "The Efficiency of Roots Compressors, and Compressors with Fixed Internal Compression," *Proc. I. Mech. E.*, Vol. 202, No. A3, 1988.

Stone, C.R. (1989), *Motor Vehicle Fuel Economy*, Macmillan Education, Ltd., London, 1989.

Stone, C.R. (1998), "The Efficiency of Roots Compressors, and Compressors with Fixed Internal Compression," *Proc. Institution of Mechanical Engineers,* Vol. 202, No. A3, 1998.

Stone, R. (1999), *Introduction to Internal Combustion Engines*, Third Ed., Macmillan/SAE, 1999.

Suzuki, T. (1997), "Development and Perspective of the Diesel Combustion System for Commercial Vehicles," Combustion Engine Group Prestige Lecture, Institution of Mechanical Engineers, May 22, 1997, London.

Takata, M., Ogawa, T., Kobayashi, F., Ikeda, S., and Matumoto, H. (1987), "Development of Optical Combustion Sensor for a Diesel Engine," ISATA Conf., Florence, Vol. 1, 1987, pp. 435–454.

Taylor, C.F. (1985a), *The Internal Combustion Engine in Theory and Practice, Vol. I*, MIT Press, Boston, 1985.

Taylor, C.F. (1985b), *The Internal Combustion Engine in Theory and Practice, Vol. II*, MIT Press, Boston, 1985.

Technical Manual TM 9-8000 (1985), *Principles of Automotive Vehicles*, U.S. Department of the Army, 1985.

Thompson, A.A., Lambert, S.W., and Mulqueen, S. (1997), "Prediction and Precision of Cetane Number Improver Response Equations," SAE Paper No. 972901, Society of Automotive Engineers, Warrendale, PA, 1997.

Thomson, W.T. (1988), *Theory of Vibration with Applications*, Prentice Hall, New Jersey, 1988.

Thring, R.H. (1981), "Engine Transmission Matching," SAE Paper No. 810446, Society of Automotive Engineers, Warrendale, PA, 1981.

Tschöke, H. (1994), "Distributor Injection Pumps," in *Diesel Fuel Injection*, Adler, U., Ed., Robert Bosch, Stuttgart, 1994.

Turner, J.D. and Austin, L. (2000), "A Review of Current Sensor Technologies and Applications Within Automotive and Traffic Control Systems," Proc. Institution of Mechanical Engineers, Part D, Vol. 214, 2000, pp. 589–614.

Vauxhall (1998), *Electric Vehicles FactFile*, Public Affairs Department, Vauxhall Motors, 1998.

Vauxhall (1999), *The Vauxhall Story*, Vauxhall Public Affairs Department, Vauxhall Motors, 1999.

Vincent, M.W. (1992), "Fuel Problems: Use of Modern Petrol in Older Engines," report available from Federation of Historic British Vehicle Clubs, 1992.

Wahl, A.M. (1963), *Mechanical Springs*, Second Ed., McGraw-Hill, New York, 1963.

Wakeman, A.C., Ironside, J.M., Holmes, M., Edwards, S.I., and Nutton, D. (1987), "Adaptive Engine Controls for Fuel Consumption and Emissions Reduction," SAE Paper No. 870083, Society of Automotive Engineers, Warrendale, PA, 1987.

Walker, J. (1997), "Caterpillar Presents New Engines, Plans for Growth," in *Diesel Progress*, September–October 1997, pp. 66–69.

Walker, J. (1998), Low Sulfur Fuel Stimulates Particulate Trap Sales," in *Diesel Progress*, May–June 1998, pp. 78–79.

Ward, D. (1985), "Steady at Last!" in *Motor*, January 12, 1985, pp. 14–16.

Warga, J. (1994), "Nozzles and Nozzle Holders," in *Diesel Fuel Injection*, Adler, U., Ed., Robert Bosch, Stuttgart, 1994.

Watson, N. and Janota, M.S. (1982), *Turbocharging the Internal Combustion Engine*, Macmillan, London, 1982.

Whiteley, F. (1995), "VVC for Rover K Series," in *Automotive Engineering*, Vol. 20, No. 2, 1995, pp. 52–53.

Wiedenmann, H.M., Raff, L., and Noack, R. (1984), "Heated Zirconia Oxygen Sensor for Stoichiometric and Lean Air-Fuel Ratios," SAE Paper No. 840141, Society of Automotive Engineers, Warrendale, PA, 1984.

Wielenga, T.J. and Chace, M.A. (2000), "A Study in Rollover Prevention Using Anti-Rollover Braking," SAE Paper No. 2000-01-1642, Society of Automotive Engineers, Warrendale, PA, 2000.

Winterbone, D.E. and Pearson, R.J. (1999), *Design Techniques for Engine Manifolds—Wave Action Methods for IC Engines*, PEP Ltd., 1999.

Winterbone, D.E. and Pearson, R.J. (2000), *Theory of Engine Manifold Design—Wave Action Methods for IC Engines*, PEP Ltd., 2000.

Woehrle, W.J. (1995a), "A History of the Passenger Car Tire: Part I," in *Automotive Engineering*, September 1995, also reprinted in *The Automobile: A Century of Progress*, Society of Automotive Engineers, Warrendale, PA, 1997, pp. 47–60.

Woehrle, W.J. (1995b), "A History of the Passenger Car Tire: Part II," in *Automotive Engineering*, September 1995, also reprinted in *The Automobile: A Century of Progress*, Society of Automotive Engineers, Warrendale, PA, 1997, pp. 61–72.

Womack, J.P., Jones, D.T., and Roos, D. (1991), *The Machine That Changed the World*, HarperCollins, New York, 1991.

Won, M., Chopi, S.-B., and Hedrick, J. (1998), "Air-to Fuel Ratio Control of Spark Ignition Engines Using Gaussian Network Sliding Control," in *IEEE Control Systems Technology*, 1998, pp. 678–687.

Yamaguchi, J. (1997), "Toyota Readies Gasoline/Electric Hybrid System," in *Automotive Engineering*, Vol. 105, No. 7, 1997, pp. 55–58.

Yamaguchi, J. (2000), "Shrinking Electric Cars," in *Automotive Engineering*, Vol. 108, No. 6, 2000, pp. 10–14.

Yamaguchi, K., Moroto, S., Kobayashi, K., Kawamoto, M., and Miyaishi, Y. (1996), "Development of a New Hybrid System—Dual System," SAE Paper No. 960231, Society of Automotive Engineers, Warrendale, PA, 1996.

Yamamoto, H., Kato, F., Kitagawa, J., and Machida, M. (1991), "Warm-Up Characteristics of Thin Wall Honeycomb Catalysts," SAE Paper No. 910611, Society of Automotive Engineers, Warrendale, PA, 1991.

Yanik, A.J. (1997), "The First 100 Years of Transportation History: Part I," in *Automotive Engineering*, February 1996, also reprinted in *The Automobile: A Century of Progress*, Society of Automotive Engineers, Warrendale, PA, 1997, pp. 121–132.

Yonehara, T. (2000), "Insight—The World's Most Fuel Efficient and Environmentally Friendly Car," CADETT Energy Efficiency Newsletter, No. 2, 2000, pp. 18–20.

Young, K.S. (1992), "Advanced Composites Storage Containment for Hydrogen," in *Int. J. Hydrogen Energy*, Vol. 17, 1992, p. 505.

Zeldovich, Ya.B. (1946), "The Oxidation of Nitrogen in Combustion and Explosions," in *Acta Physiochim U.R.S.S.*, Vol. 21, 1946, pp. 577–628.

Zhao, H. and Ladommatos, N. (2001), *Engine Combustion Instrumentation and Diagnostics*, Society of Automotive Engineers, Warrendale, PA, 2001.

Index

10W30 oil, 195, 197
    viscosity changes, 197*f*
10W40 oil, 195
ABS. *See* Antilock braking system
Absolute
    pressure, 58, 134, 137, 159, 173
    temperatures, 137, 159, 173, 184, 219
    velocities, 162, 263–264, 437
Absorption refrigeration system, 214, 214*f*
Acceleration
    magnitudes, 392
    of the center of gravity, 312, 385
    velocity lift, 61
Accelerator
    pedal, 136, 281
    pedal position, 173, 277, 281
Accident avoidance, 10
Ackerman
    angle, 311, 315–317, 319
    layout, 304
    steering, 304–305, 304*f*, 311, 331
    system, 304, 311
Ackerman, Rudolph, 304
A-Class, 342, 502
Acoustic quality, 346
Activation loss equation, 85–87
Active
    actuation, 395
    charcoal, 93, 144–145
    damper, 395
    damping, 391, 394–395
    DENOx Catalysts, 127
    four-wheel steering system, 334
    membrane surface, 87
    rear axle kinematics, 334
    suspensions, 391–395
    suspension system, 334, 391, 394
Active systems, 125–126, 334, 391, 394–395
Actuation pressure, 144
Actuator diaphragm, 144
Actuators, electric, 230
Additives, **45–50**, 160, 195, 201, 207, 512
    diesel, 48
    gasoline, 48
Adiabatic process, 370

Advanced Vehicle Simulator (ADVISOR)
    simulation setup screen, 492*f*
    test results screen, 493*f*
    vehicle configuration screen, 491*f*
ADVISOR. *See* Advanced Vehicle Simulator
Aerodynamics, 16, 236, **435–471**, 521
    factors influencing, 452
    significance of, 450
    drag, 236, 435–436, 450–452, 456, 469–470, 521
        effect on performance, 450*f*
        reduction effects, 451*t*
    experimental techniques, 445
    forces, 437*f*
    numerical prediction of, 463
    open vehicles aerodynamics, 461–462
    performance, numerical prediction of, 463
    resistance, 236, 450–451, 456–457
        trucks, 457*f*
    sources of drag, 436
    truck and bus aerodynamics, 456–461
        factors influencing, 456
        significance of, 456
Aero wings, 433
AFR. *See* Air-fuel ratio
AGMA. *See* American Gear Manufacturers Association
Air compressor, 94, 142, 166–167, 184–185, 368
Air conditioning
    compressor, 142
    pump, 145
    systems, 16, 189, **213–223**
        coefficient of performance (CoP), 216
        overview, 213
        performance, 222
        thermodynamics, 215
Air/coolant/air intercooler, 131
Air cylinder pressure, 109
Air density, 51, 112, 117, 131, 166–167, 173, 179, 184, 219, 438, 446, 450, 470
Air-flow
    open-top car, 462*f*
Air flow rate sensor, 136
Air-fuel
    mixtures, 19, 23, 33–34, 41, 45, 99, 115, 117, 123, 185, 479
    ratios, 30, 35, 36, 39–41, 98–99, 101, 107–117, 123–124, 133–149, 177–185, 205, 516
        control, 143
        effects of, 35*f*
        excursions, 41
        sensor, 137
Air gaps, 226
Air pressure, 20, 98, 109, 112, 129, 131, 133–134, 137, 167, 173, 181, 184, 436
Air pump, 94, 123, 145, 173

Air shroud, 112
Air standard cycle efficiency, 26, 28, 30, 98
Air temperature, 116, 131, 134, 136–137, 144, 148, 173, 184, 223, 447, 471
    sensor, 137
Air throttle, 110, 112, 114, 133–134, 136–137, 143, 172–173
Air velocity, 53–54, 161, 437–438, 446
Algorithms, 343
    closed-loop, 335
    open-loop, 335
Alignment, improper, 325t
Alkaline substances, 207
Alkene-ester copolymers, 49
Alloy cylinder heads, 58, 191
Alloy sump, 509
All-steel enclosure, 9f
All-wheel drive (AWD), **293–296**
    full-time, 295
    on-demand, 295
    part-time, 294
Alternating voltage, 227–228
Alternative vehicles, 495–524
Alternator capacity, 213
Alternators, 9, 190, **223–233**, 227, 228f
    Bosch claw-pole, 228f
    three-phase, 229f
Alternator self-energizing, 230
Alternator speed, 227, 229
Alumina matrix, 160
Aluminum
    bimetal bearings, 194
    cylinder head, 131
    matrix, 194
    oxide, 123
Aluminum-tin overlays, 195
Ambient
    pressure, 92, 129, 137, 167, 448, 470
    temperatures, 166–167, 173, 175, 184, 214, 222, 222t, 470–471
    test conditions, 97–98
American Gear Manufacturers Association, 243
American Society of Mechanical Engineers, 290
America's Sports Car, 9
Ammonia, 123, 126
Amplitude ratios, 357, 372
Anaerobic bacteria, 50
Analog
    computer, 272
    signal, 133, 135
Anchored shoes, 408
Anemometer, 136–137
Angle of incidence, 437, 439–440

Angular
    acceleration, 385–386
    displacement, 383
    velocities, 55, 62, 70, 72–73, 243, 248, 263–264, 284–286, 289, 292, 384, 419, 429
        analysis of, 248
        of wheels, 429
Annular nozzle, 112
Antifoam additive, 207
Antifoaming, 258
Antifreeze mixture, 206
Anti-friction bearings, 189–190, 190*f*
Antilock brake systems (ABS), 15–16, 146, 334, 398–399, **421–424**, 422*f*, 430
Anti-roll linkages, 376
Anti-rollover braking, 343, 398, 423
A-pillars, 453–455
    design, 455*f*
Aqueous ethylene glycol, 207
ARB. *See* Anti-rollover braking
Arc phase, 106
Arc voltage, 106
Armature windings, 225, 227
Arm suspensions, 375–382
ASME. *See* American Society of Mechanical Engineers
Aspect ratios, 424–427, 430, 432, 434
Asperities, geometry of, 201
Aspirated
    DI, 151, 157, 170
    diesel engine, 34, 97, 149
    direct injection, 97, 157, 170
    engine, 34, 45, 97, 130–132, 147, 149, 171
    spark ignition engine, 45
Atomic ratio, 33
Autocar Magazine, 1, 329
Auto-locking doors, 10
Automatic gearbox ratios, 483*t*
Automatic locking hubs, 295
Automatic transaxle, 236, 296
Automatic transmissions, 236, **255–275**, 474
    advent of, 256
    fluid couplings, 256
    hydraulics, 272
    planetary gears, 261
    Simpson drive, 267
    torque converters, 256, 265
Automobile aerodynamics, **450–455**
Autothermal reforming (AR), 90
Autovac fuel pump system, 513–514, 514*f*
Auxiliary power units, 79
Aviation Fuels, 46

Axial
    clutch, 239
    loading, 275
    space, 364
Axle weight rating, 428

Babbit metal, 193
Ballard Power Systems, 81
Ball bearings, 189–190, 238, 265, 305, 307
    deep groove, 190
Ball cages, 286
Ball joints, 286, 320, 325–327
    connections, 378
Balloon tire, 424
Ball steering gear, 307
Ball tracks, 286
Band/clutch control, 274
Bantam-weight cars, 13
Barium
    carbonate, 127
    nitrate, 127
    oxide, 124, 127
    sulfate, 127
Bartlett, William, 424
Batteries, 492, 496, 498–501, 503, 506
    performance of, 497t
    types, 490, 496
Bearings, 66, 70, **189–202**, 305, 511, 518
    anti-friction, 189–190
    copper-based, 194
    guide, 190
    housing, 161
    journal, 192
    material, 193–194
    pressure, 199–202, 237
    separation, 201
    soft-phase properties, 194
    surfaces, 66, 189, 191, 194, 200–201
    systems, 189, 233
    temperature, 202
    thin-wall, 194
    three-layer, 194
    thrust, 191
    trimetal, 194
    types, 189, 194
Belts
    construction, 276
    design, 276

Belts *(continued)*
    drives, 16, 60, 145, 189, 208, 210–211, 230, 276, 308–309
    elements, 276
    system, 16, 275
    tension, 210, 275
Belt-type continuously variable transmission (CVT), 276
Bendix system, 231
Benton Harbor Palladium, 2
Bernoulli's equation, 112
Bias-ply tires, 317, 329, 343, 424–426, 425*f*, 430, 432
Bicycle model, 313*f*
    with tractive forces, 318*f*
Bimetal
    bearings, 194–195
    construction, 194
Binomial theorem, 69
Bipolar plates, 80, 91
Black Japan enamel, 5
Blade of Crel, 162
Blocking ring, 255
BMEP. *See* Brake mean effective pressure
BMW, 334–335, 376
    rear-wheel steering system, 335*f*
Body roll angle with CG height, 389*f*
Boiler explosions, 3
Bond graph, 391–392
Bosch claw-pole alternator, 228*f*
Bounce response, 360-361
Boundary
    layer control methods, 444*f*
    lubrication, 201
        surface contact, 201*f*
    regime, 193
Bowe, Martin, 10
Bowl-in-piston design, 118
Brake mean effective pressure (bmep), 21–22, 33–36, 55, 97–98, 108–133, 147–151, 157, 170–181, 202–205, 475, 478, 513–521
Brakes, **397–434**
    antilock brake systems (ABS), 421–424
    band, 265–266, 269, 397
    calipers, 415–417, 421
    components, **403–406**
        combination valve, 405
        master cylinder, 403
        metering valve, 406
        power assistance, 404
        pressure differential switch, 406
        proportioning valve, 406
    disc brakes, 414–421
        analysis, 417

Brakes *(continued)*
    disc brakes *(continued)*
        calipers, 416
        discs, 414
        heat dissipation, 419
        pads, 416
    drum brakes, 397–398, 406–410, 407f, 412, 414, 433
        analysis of, 409
        distortion, 398f
    duo-servo, 409
    efficiency of, 22, 30, 34, 36, 45, 51, 148, 151, 168, 177–179, 202, 474, 523–524
    hydraulic principles, 402
    mechanisms, 433
    pads, 415–418, 415f, 420-421
    pedal, 403–404
    power, 178, 202, 405, 474–475
        hyperbolas, 203
        output, 180–181, 184, 203
    system design, 397
Braking
    distance, 421
    dynamics, **399–402**
    free-body diagram, 400f
    percentage, 420
    power, 343, 373, 400
    pressure, 406, 421
    systems, 397, 404, 423
    torque, 406, 415, 418, 506
Breaker-operated ignition system, 103f
Breaker-point ignition system, 4f
British auto show, 1
Bronze matrix, 194
BS4040, 512
Bus aerodynamics, **456–461**
Bypass valve, 131
    arrangement, 131f

Cabin air, 219, 462
CAD. *See* Computer aided design
Cadillac, 3, 6, 9
    V-8 of 1915, 7f
Calculus, 399
Calipers, **416–417**
    design, 417
    pistons, 421
Calorific Values, 22, 47, 47f, 51, 88, 92–93, 97–99, 147, 177, 474, 485

Cam
- geometry of, 60
- lobes, 128, 510
- period, 135
- phasing of the, 153
- profiles, 53, 62, 135, 155

Camber, 320
- angles, 320, 345, 391
- positive/negative, 320*f*
- qualities, 375

Cam-controlled acceleration, 63
Cam-driven contacts, 4
Cam-operated valve, 61*f*
Cam-over-rocker system, 60, 60*f*

Camshaft
- arrangement, 59
- cast-iron, 59
- design, 62, 512
- drive, 58, 60, 135, 509
- frequency, 64
- lobes, 191
- pickup, 134–135
- position sensor, 134
- rotation, 61
- supports, 191–192
- thrust bearing, 192*f*

Cam/tappet interface, 63–64, 64*f*
Capacitive discharge ignition (CDI) systems, 105

Carbon
- agglomerates, 45
- canister, 144–145
- deposition, 90
- particles, 45, 81

Carbon dioxide, 37, 94, 119, 158–159
- carbon dioxide/argon mixture, 158
- emissions, 123
- presence of, 158

Carbon fiber brakes, 415

Carbon monoxide, 37–39, 91–92, 94, 115, 121–122, 126–127, 147, 159–160
- emissions, 39–40, 120, 122–123, 158
- kinetics, 40
- levels, 92
- production of, 90

Carbureted engines, 48
Carter, Byron, 5
Caster angle, 323*f*

Cast iron
- block, 509
- camshafts, 59

Catalysts, 90, 101, 115, 120–122, 125–126, 147, 158–159, 230
    light-off, 115, 120, 125–126, 204
    materials, 124
    particles, 81
    performance, 124–125
    systems, 121–123, 125, 159
        conversion efficiencies, 122f
    temperature, 92, 115, 125
Catalytic
    burner, 94
    effect, 136
Caterpillar Incorporated, 488
Cathode off-gases, 94
CDI. *See* Capacitive discharge ignition
Cell voltage, 82, 88, 499
Center
    ball, 285, 325–327
    gear, 247–248, 261
    heights, 329, 384–385, 388, 390–391, 401
    link, 305, 382
    motion, 305, 384
    of gravity, 67, 313, 343–344, 354–355, 357, 359, 384, 401, 436–437
    of mass, 74, 286–287, 354
Central axis, 262
Centrifugal clutch, 243, 299
Centripetal
    acceleration, 160
    tension, 211
Ceramic
    insulator, 102
    matrices, 125
    rotors, 170
    substrates, 123
Cerium oxide, 124
Cetane-octane number ratings, 46f
Cetane rating, 46–47, 46f, 49, 175
CG. *See* Center of gravity
Chain
    slap noise, 299
    transmission, 3
Charge cooling effects, 167f
Chevrolet
    Bel Aire, 1957 performance, 15t
    Blazer, 295
    Corvette, 1956 performance, 15t
    Motor Car Company, 6
    S10, 501
Chevrolet, Louis, 6
Chicago Motor Vehicle Company, 2

Chrysler
- 300, 1955 performance, 15*t*
- 421E automatic transaxle, 297*f*, 296
    - chain transfer drive, 299
    - configuration, 296
    - control system, 299
    - planetary gear set, 296, 298*f*
    - schematic, 297*f*, 298*f*
- H cars, 296

CI engines. *See* Compression ignition engines

Closed-loop algorithm, 335

Closing velocity, 62

Cluster gears, 250–251

Clutches, 236, 254, 259, 267, 270, 274, 277–278, 281, 293, 296, 299
- activation, 238*f*
- assembly, 243, 256
- centrifugal, 243, 299
- disc, 239–240, 240*f*, 265–266
- electromagnetic, 145
- face, 240–243
- fork, 236, 239
- friction, 236–243
    - axial clutch, 239
- gear, 243, 250–253, 256, 267–271, 277–278, 293, 296–300
- one-way, 232, 265
- packs, 265–267, 272, 274, 292–293, 296–298
- pedal, 236–237, 243
- rings, 299
- surface, 241, 254
    - friction material, 241*f*
- with centrifugal weights, 243*f*

CO. *See* Carbon monoxide

Coefficient of friction, 199–202, 208, 213, 239, 241, 278, 299, 389, 402, 407, 414, 421, 423, 429–431, 433

Coil springs, 365
- clutch, 237*f*
- dimensions, 366*f*
- suspension, 365*f*

Cold-starting, 149

Cold weld, 201

Colenoid actuator, 155

Collapsible steering, 10

Combination valve, 405

Combustion
- abnormal, 48
- chambers, **150–151**, 150*f*
- equilibrium, 37
- gases, 30, 42, 206
- noise, 152, 159, 180, 185
- of the fuel, 25, 37, 178

Combustion *(continued)*
    process, 32, 39, 43, 99, 107
    products, 37–38, 38*t*, 56, 99, 127, 139, 158, 201
    reactions, 39, 42–43
    spark ignition engine, 44*f*
    speed, 31–32, 150
    stability, 36, 119, 143
    system design, 101, **113–120**
    systems, 107, 113, 123, 144, 149–150, 155, 158, 175, 512
        Honda VTEC, 115*f*
    temperatures, 42, 44, 113, 115, 157–158, 185
Common rail (CR) fuel injection systems, 156
    maximum pressure, 152*f*
Composite stiffness, 339–340
Compound gear train, 249*f*
Compressed
    air, 94, 129, 402
    charge, 19
    gas, 93, 108, 166, 206, 215
    refrigerant gas, 215
Compression ignition engines, 32–33
Compression ratios, 8, 15, 24–34, 56, 58, 97–99, 114–120, 130–131, 141, 147–151, 157, 166, 168, 170, 508–512, 517, 521–522
Compressor
    delivery
        conditions, 181
        pressure, 180, 185
        temperature, 181
    efficiency
        air density, 166*f*
        effect of, 166
        isentropic, 164–166, 169, 177, 179, 181–182, 184–186
    entry, 177, 180, 186–187, 218
    housing, 161
    map, 168, 168*f*, 180–181, 184, 187, 187*f*
    performance, 129, 168
    pressure ratio, 131, 148, 165–170, 179, 181, 184–186
    specific work, 163, 185
    wheel, 161
Computer
    modeling, 16, 436, 473, 486, 492
    simulation, 343
Computer aided design, 15, 326
Cone clutch, 254, 293
    limited slip differential, 293*f*
    schematic, 254*f*
Cone-type synchronizer, 255*f*
Connecting rods, 18, 66–68, 67*f*, 72, 80, 192–193, 508, 511
    rod length, 70
    rod length/crank-throw ratio, 74

Constant-mesh transmissions, 249
Constant velocity joint, 281, 286f, 327
Constant-volume
    combustion, 25
    fuel-air cycle, 31, 31f
Continuously variable transaxle (CTX), 276
Continuously variable transmissions (CVT), **275–281**
    belt, 276f
    Ford, 277f
    fuel economy, 485t
    operation, 478f
    performance, 481t
    Torotrak, 277
    Van Doorne, 275
Contra-rotating countershaft, 71
Control
    arms, 324, 365, 374–376, 378
    exhaust, 111, 121, 144
    rods, 333
    signals, 136, 281
    unit, 177, 183, 272, 310, 335
    valve, 59, 63, 128, 135–136, 153–154, 156, 172, 177, 218, 273–274, 308–310
        configurations, 310, 310f
Conversion efficiencies, catalyst systems, 122f
Converter
    clutch, 296, 299
    efficiency, 260–261
    housing, 260
    torque ratio, 261
Coolant, 173, 202–206, 230
    properties, 207t
    pump, 189, 204, 206, 510–511
    system, 203
    temperatures, 134, 137, 142, 167, 173, 175, 205–207
    temperature sensor, 137
Cooling systems, **202–207**
Cool running spark plugs, 102f
Copolymers of ethylene, 49
Copper-antimony particles, 193
Copper-based
    bearings, 194
    materials, 194
Copper-exchanged zeolite catalysts, 126
Copper-tin matrix, 193
Cord materials, 425
Cornering, 294, 303, 311–312, 319, 323–324, 329, 334, 337, 343, 375, 384–386, 441
    force versus slip angle, 312f
    performance, 311, 379
    stiffness, 312, 316–318, 327, 329, 343, 402, 430, 432–433
        variables, 432f

Corresponding body, 333
Corresponding reduction, 179, 483
Corrosion, 195, 206–207, 417
    corrosive environment, 81
    inhibiting properties, 206
    inhibitors, 49, 207
    resistance, 194
Corvette trailing arm suspension, 377f
Counterbalance shafts, 72f
Counter-rotating balance masses, 73f
Counter-rotating shaft, 78
Coupling efficiency, 260–261
Crankcase, 18–20, 508–509
    pressure, 19
    ventilation, 195
Crankshaft
    applications, 194
    axis, 76–77
    datum position, 135
    isometric sketch, 75f
    position, 74, 134–135
    secondary forces, 71f
    speed fluctuations, 136
Crank-slider mechanism geometry, 68–69, 69f
Crank throw, 72, 74, 78, 80
Crank webs, 68
Cross-ply tires, 425f
Cryogenic liquid, energy density, 93t
CTX. *See* Continuously variable transaxle
Current/voltage plot, 87f
Cutoff ratio, 25, 27
CV. *See* Calorific value
CV joints. *See* Constant velocity joints
CVT. *See* Continuously variable transmission
Cycle engine, four-stroke, 18f
Cylinders, 40, 64–66, 70–71, 74, 79, 104, 154, 175, 334, 409, 510
    axes, 70, 78, 99
    block, 58, 206, 509
    head, 58–59, 71, 105, 113, 150, 152, 156, 176, 191, 204, 206, 508–509, 512
    head clearance, 56
    disabled, 143
    pressures, 19–21, 41, 48, 57, 102, 107–109, 111, 113, 117, 140, 171, 310, 402, 409
    unit, 310

DaimlerChrysler, 502
Damped
    frequency of oscillation, 350
    natural frequency, 340, 350

Damped *(continued)*
    oscillatory system, 340
    vibratory system, 352
Dampers (shock absorbers), 371
Damping, 287–288, 300, 335, 339–341, 347–351, 354–355, 360, 362, 364–365, 371–372, 391–394, 424, 428
    capability, 363
    condition, 349–350
    effect of, 341, 351, 362–363, 372
    matrix, 358
    of tires, 354
    ratio, effect of, 341, 363, 372
    ratio on pitch response, 363*f*
    ratio on rollover threshold, 341*f*
Datum positions, 74, 135, 326, 348, 371
Deceleration, 61–63, 329, 399, 401, 419–420, 433, 449
de Dion suspension, 374–375, 375*f*
Delphi
    electronic unit injector, 155*f*
Deming, W. Edwards, 11
Density of the air, 51, 112, 173, 179
Department of Transportation, standards, 428
Design parameters, 200
Design team, 296
Desirable consequences, 292
Dewar Trophy, 3
Dewpoint temperature, 50
DI. *See* Direct injection
Diagonal matrix, 358
Diaphragm
    actuator, 144
    clutch, 239*f*
    spring, 238*f*
Diesel engine, 29–30, 34–35, 45, 56, 97, 121, **149–187**, 404, 502
    additives, 48
        legislation, 48
    combustion chambers, 150–151
    cycle, 25–27, 26*f*, 29–30, 97–98
        air standard, 25–30
        analysis, 23, 149
        efficiency, 27–30
    electric hybrid vehicle, 97
    emissions, 157
    emissions control, 158
        exhaust gas recirculation (EGR), 158
        particulate traps, 159
    engine combustion, 30, 34, 45, 149, 158
    engine efficiency, 23, 34, 36, 148
    fuel injection equipment, 152–157
        common rail (CR), 156

Diesel engine *(continued)*
    fuel injection equipment *(continued)*
        electronic unit injectors (EUI), 155
        pump–line–injector (PLI) systems, 153
    fuels, 45–47, 49–50, 153, 160, 175
        low-sulfur, 50
    heavy-duty, 194
    management systems, 159, 172–175, 173*f*
    operation, 126
    spark ignition, 32, 97, 126–127, 203
    technology, 97
    turbocharging, 161–172
Differentials, **290–293**
    clutch-pack-style, 292*f*
    cone-clutch type, 293*f*
    input shaft, 294
    pressures, 154
    schematic, 291*f*
    thermal expansion, 61, 124, 191
    wheel, 292–294
Diffusion-controlled combustion stage, 45
Direct current (DC)
    generator, 223–224, 224*f*, 227
    motors, 223–227, 496
        characteristics of, 225*f*, 226*f*
    voltage, 224, 227, 229
Direct injection, 120, 147, 149–150, 175
    combustion chamber, 150*f*
    diesel engines, 97, 149–150, 157, 170, 177, 180, 203
    engines, 20, 97, 101, 112, 116–117, 119–120, 149–151, 170, 177, 180, 203
    spark ignition (DISI) combustion systems, 112–113, 116, 119–120
        fuel spray calculations, 119*f*
Direct injection spark ignition (DISI)
    engines, 112–113, 116, 118–120
    performance, 118
Directional spool valve, 275*f*
Dirt
    roads, 294
    shoulder, 337
    surface, 339
Disc brakes, 16, 398–399, **414–421**, 415*f*, 433
    analysis of, 417–418
    pads, 416, 418, 420
Discharge coefficient data, 53*f*
DISI. *See* Direct injection spark ignition
Displacement compressor, 129, 214, 218–219
Distributorless ignition system, 104, 104*f*
Distributor-type fuel injection pump, 153, 175
Ditertiary butyl peroxide, 49
Diurnal temperature variations, 144

Dodge
    Intrepid, 329
    Ram all, 295
    Viper, 301, 399, 419
Doorstop, 407*f*
DOT standards. *See* Department of Transportation
Double-anchor, single-cylinder shoe arrangement, 408*f*
Double-cardan joint, 285*f*
Downforce, 433, 439, 441, 464
Drag, 10, 235, 257, 305, 318, 433–469, 471, 518, 520
    coefficient, 435–464, 471, 486, 488, 504, 518
    drag and lift, 440*f*
    effect on fuel economy, 452*f*
    factors effecting, 444*t*
    pressure and velocity distributions, 442*f*
    tractor trailer combinations, 460*f*
Drive
    axles, 236, 291
    belts, **208–213**
        connection, 230
        flat belt drives, 208
        V-belts, 212
    pulley, 128, 208, 228, 309
    spindles, 295
    system, 128, 295, 308
    torque, 232, 295, 328
    wheels, 292, 295, 404, 506
Driveability, 118, 122, 144, 476
Driveline
    torque, 16, 319, 327
    wind-up, 294
Driven
    gears, 153, 249–250, 255, 271
    pulley, 275–277
Driver's
    compartment, 403
    reaction time, 433
Driveshafts, 247, **281–290**, 300, 309, 373, 523
    angle of, 328
    Hooke's joints, 281
    shaft whirl, 286
Drum brakes, 397–398, **406–414**, 433
    radius, 412
    shoes, 408, 410
        schematic, 410*f*
Dry friction clutches, 277
Dual hybrid
    configuration, 505*f*
    fuel consumption, 524*f*
    powertrain, 524*f*

Dual master cylinder, 404*f*
Dunlop, John, 424
Duo-servo brake, 409
DuPont family, 6
Durability, 101, 120, 124, 126, 175, 502, 521
Durant, William Crapo, 6
Duryea, Charles, 2
Dynamics
    amplification, 141
    balancing, 68
    characteristics of throttle, 146
    effects, 57, 64, 421
    free-body diagram, 400
    imbalance, 287
    index, 354, 362
    index of unity, 359
    of the vehicle, 303, 345
    of traction production, 399
    response, 172, 345
    viscosity, 198, 446
    weight transfer, 401

Earl, Harley, 8
Economies of scale, 11
Efficiency data, 95, 208, 490, 498
    of engines, 30, 97, 126, 149
    of vehicle alternators, 213
EGR. *See* Exhaust gas recirculation
EHC systems. *See* Electrically heated catalysts
Eight-cylinder engines, 65, 71
    torque, 65*f*
Electric
    actuators, 230
    motors, 94, 223–224, 226, 479, 490, 496, 500, 503–504, 523
    starter, 5
Electric vehicles, 3, 97, **495–506**
    battery types, 496
        lead-acid batteries, 498
        lithium ion (Li-Ion)/lithium polymer, 499
        nickel-cadmium (NiCd), 498
        nickel-metal hydride (NiMH), 499
    dual hybrid systems, 505
    hybrid systems, 502–507
    Japanese, 502*t*
    Toyota Prius, 521
    types of, 500
    Vauxhall 14–40, 507

Electrical
    breakdown, 102
    conductivity, 50
    connections, 138, 498, 501
    energy, 213
    heating element, 138
    ignition, 160
    losses, 95, 504
    power input, 125
Electrically heated catalysts, 125*t*, 126
Electrochemical reactions, 80–81, 83
Electrode erosion, 106–107
Electro-hydraulic
    antilock brake system, 398
    device, 156
Electrolyte bypasses, 85
Electromagnetic clutch, 145
Electromotive force (EMF), 224–227
Electronic
    control, 156, 256, 277
        system, 147, 299
        unit, 136–137, 173
    fuel injection, 123
    gear selector, 299
    switching circuit, 103
Electronically modulated converter clutch (EMCC), 296
Electronic unit injector (EUI), 152, 155
    Delphi, 155*f*
    hydraulic, 156*f*
    maximum pressure, 152*f*
Electrons
    Accelerated, 105
    leakage, 86–87
Ellis, Evelyn Henry, 1
EMCC. *See* Electronically modulated converter clutch
EMF. *See* Electromotive force
Emissions
    control of, **120–127**
        catalyst light-off, 125
        durability, 124
        in diesel engines, **157-161**, 157*f*
        lean-burn NOx-reducing, "DENOx", 126
        three-way catalyst, 121
    legislation, 121*t*
        effect of, 15, 110
    sources of, **37–45**
        simple combustion equilibrium, 37
        unburned hydrocarbons (HC), 41, 45
        unburned nitrogen oxides (NOx), 41, 45
    spark ignition engine, 40*f*, 41*f*

EMS. *See* Engine management system
Enamel, black Japan, 5
Energy
    capacity, 230, 497
    density, 92–93, 498, 502
    dissipation, 400
    equation, 163, 216
    loss, 239
    storage
        density, 92–93
        hydrogen systems, 93*t*
        systems, 496*f*
    utilization, 89
Engines
    balance, 66, 71, 78–80, 202
    braking, 172, 175, 269, 275, 373, 487
        torque, 269
    combustion, 2, 20, 23, 30, 32, 42, 44, 83, 99, 116, 123, 140, 143, 150, 155, 157, 175, 185, 524
        pressures, 175
    compartment, 215, 453, 509, 512–513
    configuration, **64–70**
        balancing forces and moments, 68
        choosing number of cylinders, 64
    coolant, 134, 185, 189, 207
        temperature, 173
    cool-running, 102
    crankshaft, 66, 127, 135, 153, 193, 256–257
    data, 95, 490–491, 513
    design, 186, 508
    displacement, 61, 129–130, 488–489, 488*f*, 517
    emissions, 121, 146, 157–158, 177, 486–487, 490
    failure, 48, 405
    fires, 232, 509
    flywheel, 236
    four-stroke, 17
    frictionless, 21
    fuel cell efficiencies, 96*f*
    geometry of, 69
    indicator diagram, 23, 33, 148
    layouts, 66, 66f, 296
    lean-burn, 114, 120, 147
    lubricants, 195
    lubrication regimes, 200*f*
    management systems, **132–144**, 133*f*, 156, 173*f*
        air-fuel ratio control, 143
        exhaust gas recirculation (EGR) control, 144
        functions, 142–148
        ignition timing map, 142*f*
        strategy, 126

Engines *(continued)*
    oil, 156, 195, 204, 517
        SAE viscosity grades, 196*t*
    overspeed, 274
    performance parameters, 20
    protectors, 195
    red-line, 236
    simulations, 42
    size, 19, 21, 58, 64, 451, 474, 487, 489, 491
    structural vibrations, 140
    structure, 71, 134, 140
    swept volume, 22, 98, 119, 147–148, 177–178, 181
    test standards, 9
    torque output control, 146
    two-stroke, 17
    vacuum, 404–405
    vacuum actuator, 275
    warm-up, 125, 175, 204–206, 487, 512
    wear, 159, 172, 479, 511
Enthalpy
    of vaporization, 206
    superheated vapor region, 221*f*
Environmental Protection Agency, 487–489
Epicyclic
    annulus, 279
    gearbox, 232, 278–279, 521
    gearing, 280
    gear train, 278–280
    reduction gearbox, 232
Equations of motion, 288, 339, 341, 347, 349, 352, 355, 358, 451, 468
Equilibrium analysis, 37
Equos Research, 505
Ethylene glycol, 207
EUI systems. *See* Electronic unit injector systems
European
    auto industries, 13
    engine, 145
    legislation, 125
Evaporator
    performance, 223, 223*f*
    pressure, 218
Exciter diodes, 229-230
Exhaust
    back-pressure, 20, 58, 160
    blow-down process, 32
    gas, 18–19, 31, 37, 119, 123, 125, 127–128, 130, 137–139, 143, 160–161, 180, 185–186, 206, 213
    manifold, 57, 144, 159
    manifold pressure, 36, 57
    opening values, 39

Exhaust *(continued)*
    oxygen, 36, 123, 138–139
    oxygen sensor, 138*f*
    port, 19, 42, 204
    pressure, 20, 36, 57–58, 165
    products, 18, 107, 121
    residuals level, 36, 58
    system, 23, 111, 123, 127–128, 145, 172
    temperatures, 45, 115, 117, 126, 165, 169, 186, 204, 490
    valves, 51, 58, 117, 128, 509
    valve closure, 56–57
        closure time, 56
Exhaust gas recirculation, 58, 101, 114, 119, 121, 122-123, **144–159**, 145*f*, 173–175
    calibration, 172
    disadvantage of, 159
    level, 101, 119, 144, 158–159, 172, 175
    maps, 174*f*
    schedule, 175
    system, 123, 144–145, 159
    vacuum, 173
    valve, 133, 142, 144–145, 159, 172–173, 177
Exit gases, 92
Expansion valve, 17, 191, 214–215, 218
    thermostatically controlled, 218*f*

Face cam, 153
Fatigue life, 194
Fatigue strength, 193–194, 240
FEA. *See* Finite element analysis
Feedback system, 123, 135, 137–138, 159
    signal, 172, 175
Fifth-wheel system, 303
    steering, 303, 303*f*
Film pressures, 110, 201–202
Finger-follower system, 63
Finite element analysis, 15
Firing intervals, 66, 74, 79
First gear power flow, 251*f*
    Simpson drive, 269*f*
First law of thermodynamics, 82
Fisher Body, 8
Five-cylinder in-line engine, 79
Five-link system, 378
Fixed brake caliper, 416*f*
Flag Law, 1
Flat belt drives, 208
Flat belt pulley, 208*f*
    initial tension, 210*f*

Flexible skirt, 441*f*
Flexural stiffness, 287
Float chamber, 514
Floating brake caliper, 417*f*
Flow
    area, sport cam profile, 53*f*
    field, 440*f*
    poppet valve, 52*f*
    separation, 442–443, 443*f*, 446–448, 453–455, 457–458, 460–461, 463–464
Fluid coupling, 257*f*
    fluid flow, 258*f*
    mechanics, 97, 201, 441
    velocity, 197–198, 257–258, 446
Flywheels, 496
Foil thickness, 125
Force versus speed, 235*f*
Ford, Henry, 5–6
Ford Motor Company, 1, 5–6, 59, 110, 133, 173–174, 275, 277, 375, 488, 501
    2.5 HSDI diesel engine, 174*f*
    Bronco, 295
    continuously variable transmission, 277*f*
    CTX, 276–277
    Expedition, 300
    HSDI engine, 175
    Probe IV, 454–455, 455*f*
    Ranger EV, 501
    Ranger Supercab, 343
    T-Bird, 1958 performance, 15*t*
    V-6 Essex engine, 59*f*
Fore-and-aft location, 363
Four-bar linkages must, 305
Four-channel systems, 423
    engine, 70, 98, 103–104
        fuel consumption maps, 151*f*
    in-line engine, 66, 70, 508
    V engines, 71
Four-link suspensions, 374, 374*f*
Four-pole starter motor, 231*f*
Four-speed
    automatic transaxle, 274
    transaxle, 274, 296
    transmission, 250, 250*f*, 296
Four-steer maneuver, 343
Four-stroke engine, 17, 18, 18*f*, 180
    five-cylinder engine, 99
    gasoline engine, 98
    spark ignition engine, 36, 36*f*, 98, 147, 474
Fourth gear power flow, 253*f*
Four-valve pent-roof combustion chamber, 113

Four-wheel drive (FWD), 293–296
    active steering, 334–335
    drum brake systems, 397
    steering, 303, **330–336**
        algorithms, 335
        high-speed turns, 332
        implementation of four-wheel steering, 333
        low-speed turns, 331
    transfer case, part-time, 294*f*
Free-body spring diagram, 367*f*
Frequency ratio
    effect of, 342
    on rollover threshold, 342*f*
Friction
    belts, 208
    clutches, **236–243**
    discs, 292
    facings, 236
    loss, 248
    material, 239–241, 299, 397, 410, 416, 418
    modifiers, 195
    surface, 199, 239–242, 254, 293, 299, 407, 410, 414, 429
Frictionless engine, 21
Front-wheel drive, 285, 319, 327, 329–330, 373, 375–376
    forces, 328*f*
    influences, **327–330**
        cornering stiffness, 329
        driveline torque, 327
        tractive forces, 329
    steering, 303, 331, 335
    steering systems, 303
Fuel, 33, 46, 49–50, 87, 89–90, 111, 143, 485, 490, 502, 509, 511–512
    calorific values, 485*t*
    concentration of, 87
    consumption
        data, 175, 486, 515
        maps, four-cylinder engines 151*f*
        penalty, 123
    crossover, 83, 85–86, 91
    density, 485*t*
    droplets, 33, 41, 109–110, 112, 143–144
    economy
        of engines, 122
        penalty, 130, 236
    evaporation, 112, 150
    film, 109–110, 143, 205
    handling systems, 116
    injection equipment, **152–157**
    inlet, 18, 49, 111, 153, 156, 513–514
    jet, 112, 152

Fuel *(continued)*
 leakage, 86–87
 line, 152–153, 480
 mass, 22, 47, 51, 178, 181
 mixtures, calorific values, 47*f*
 motion, 118, 151
 outlet, 514
 pipes, 50, 154
 pressure, 33, 83, 88, 110–112, 130, 143–144, 152–154, 156
 quality, 130–131, 141, 143, 146, 175, 511
 requirements, **45–50**
 reservoir, 514
 savings, 146, 451
 spray, 45, 111, 117, 119, 151
  transport calculations, 119*f*
 tank, 110, 144, 173, 512–513
 transport, 119, 146
 use, 110, 127, 477, 503, 520, 524
 vapor, 110, 133, 142–144
Fuel-air cycle, 97
 efficiency, 30
 equivalence ratios, 40
Fuel-bound nitrogen, 42–43, 49
Fuel cells, **79–99**, 489–490, 502, 524
 and engine efficiencies, 96*f*
 consumption comparisons
  four-cylinder engines, 170*f*
 electric vehicles, 96*t*
 hydrogen systems, 93
 losses, coefficients, 88*t*
 solid polymer (SPFC), 79–92
  autothermal reforming (AR), 90
  carbon monoxide clean-up, 91
  crossover and internal currents, 85
  efficiency, 81
  efficiency activation losses, 83
  hydrogen storage, 92
  mass transfer losses, 87
  ohmic losses, 87
  oxidation (POX) reforming, 90
  response, 88
  sources of hydrogen, 88
  steam reforming (SR), 89
 system, 94*f*
 turbocharged engines, 170*f*
 voltage, 80, 83, 85, 87
 voltage model, 85*f*
Fuel injection, 25, 32, 109–112, 117, 127, 154–156
 electronic, 123
 engines, 41

Fuel injection *(continued)*
    injectors, 18, 49, 109–112, 117, 152, 154
        injector nozzle, 152, 154
    pressures, 112
    pump, 173, 183
        distributor-type, 153*f*
        single point, 110*f*
    solenoid-operated, 111*f*
    systems, 15, 49, 101, 109–112, 123, 149, 152, 156
Full-state control, 391, 394–395
Full-time four-wheel drive (4WD), 295

Gasoline
    additive requirements, 45
    additives, 48
    analysis of, 368
    clean-up unit, 94
    compression, 44, 93, 116, 135
    diffusion layer, 81
    direct injection engine, 117*f*
    exchange, 31, 135, 510
    exchange processes, 17, 19, **50–64**
        dynamic behavior of valve gear, 60
        valve flow, 50
        valve operating systems, 58
        valve timing, 55
        volumetric efficiency, 50
    ignition, 19, 48, 121, 125
    mixture, 87, 91, 123, 139, 145
    oxygen, 44, 138–139, 158
    passages, 80
    pressure, 44, 91, 118, 179, 217, 219
    prices, 13–14
        September 2000, 14*f*
    pump, 13
    spring, 369*f*
    turbine, 127–128, 180, 185, 504
    volatility changes, 513*f*
    volumes, 371
Gearbox
    output, 279
    span, 474, 482–483, 487
Gears
    analysis of, 243
    blocking ring, 254–255
    box ratios, 483*t*
    law, 248, 251, 290
    mesh schematic, 247*f*

Gears *(continued)*
    ratios, 236, 250, 262, 264–265, 268, 272, 279, 295, 299, 301, 456, 474, 479, 483, 485, 487, 492
    selector, 274
    size, 264
    theory, **243–249**
        helical spur gears, 244
        hypoid gears, 246
        spiral bevel gears, 246
        straight-tooth bevel gears, 245
        straight-tooth spur gears, 244
    train, 248–249, 248f, 251, 253, 262, 264
Gearshift lever, 255
General Motors, 6, 8, 10, 171, 487–489, 500–501, 518
    Aero Astro, 461
    EV1, 500–501, 500t
        powertrain configuration, 501f
General Motors Research Laboratories, 511
Generators, **223–233**
Gibbs energy, 82–83
GM. *See* General Motors
Governor valve, 272–274
Gravimetric
    air-fuel ratio, 33, 98, 147, 180, 185
    calorific value, 47, 98
    stoichiometric AFR, 33
Gray, Christian Hamilton, 424
Guide bearings, 190

H balance, 37
H configuration, 65
H_2SO_4, 498
Half-metric designation system, 426
Hammering effect, 59
Handling
    performance, 425
    qualities, 324, 373, 380
Hard sparking plug, 102
Hardy, Edward, 518
Harmonic, 62, 64
    forcing, 350
    loading, 341
Health grounds, 125
Heat
    buildup, 191, 260, 397, 428
    capacity, 99, 158–159, 164, 180, 206, 219
    dissipation, 260, 400, 414–415, 419
    exchanger, 83, 205, 214, 447, 512

Heat *(continued)*
    transfers, 16, 24–31, 44, 99, 115, 126, 142, 149–151, 163–168, 179, 185–187, 204, 206, 215–216, 370, 421
Heater
    control of, 175
    flow, 204
    plugs, 151, 175
Helical
    compression, 366
        spring, 366*f*
    gears, 244–246, 245
    grooves, 112
    spline, 231–232
    spur gears, 244
    teeth, 244, 246
HICAS system. *See* High capacity active control suspension
High-altitude operation, 175
High and low-regime clutches, 281
High capacity active control suspension (HICAS), 334
High-performance
    brakes, 415
    cars, 296, 376, 417, 424, 508
    engines, 55, 59, 102, 189, 194
    spark ignition engines, 55, 59
High-power solenoid, 144
High-pressure
    air, 117, 129, 215
    fuel, 152
    fuel pump, 117, 152, 156
    fuel rail, 111, 156
    gas, 92, 215
    port, 117, 129
    pump, 117, 152, 156
    side, 129
    turbine, 176
High-regime clutch, 278–279
High-speed
    combustion photography, 155
    compression ignition engines, 24
    direct injection engines, 151
    tires, 428
    turn diagram, four-wheel steering, 332*f*
Highway deaths, 1947–1983, 10–11, 11*f*
Honda
    EV Plus, 501
    integrated motor assist, 504
    Odyssey, 1999 performance, 15*t*
    VTEC
        combustion system, 115, 115*f*
        engine, 114–117
        operating regimes, 116*f*

Hooke's joints, 281, 289
Horseless carriage, 2
Horsepower ratings, 9
Hot running spark plugs, 102f
Hotchkiss
    drive, 373–374, 373f
    suspension, 373–374, 382–383
        roll center calculation, 382
Hub Drum Distortion, 398
Hub free wheels, 265
Hybrid electric vehicles, 502–507, 503f
Hydragas suspension, 369
Hydraulic
    actuated brake band, 266f
    actuation, 277, 334
    advantage, 403f
    brakes, 397
    circuits, 404, 406
    control system, 256, 272, 277
    electronic unit injector, 156, 156f
    lash adjusters, 59–60, 60f, 62
    lifter, 59
    lines, 272
    piston, 115, 260, 404
    pressure, 156, 272, 274–275, 278, 281, 308, 343, 406
        intensifier system, 156
    rams, 334
    solenoids, 274
    system, 60, 156, 256, 272, 277, 334, 397–398, 403, 423
    torque converter, 296
Hydrocarbons, 42, 45, 49, 122, 126, 139, 145, 156, 512
    emissions, 39, 41, 45, 122–123, 144, 149, 155, 157
    fuels, 32–33, 79
    oxidation kinetics, 43
Hydrodynamic lubrication, 200–201
Hydrogen, 33, 37–38, 79, 82, 87–88, 90–95, 126–127, 496, 499
    atom, 33
    compression, 94
    electrode reaction, 84
    fuel, 81, 85, 87, 89, 91–92, 94, 99
        cell systems, 93
        mixture, 87
    gas, 90–91
    ions, 81
    liquid, 92–93
    molecule, 85
    storage systems, 92–94, 499
        energy densities, 93t
    sulfide, 125

Hydroperoxides, 50
Hypoid gears, 246–247, 246f, 518
Hysteresis rubber, 428–429

Idaho National Engineering, 488
IDI engines. *See* Indirect injection engines
Idle
    speed, 136, 145, 152, 511, 517
        adjustment, 133, 142
    stability, 131
Idler gear, 248, 253, 279
Ignition, 4, 17–18, 25, 29, **32–36**, 45–48, 59, 96–99, 104–108, 122, 140–141, 148–150, 159, 168, 175, 180, 185, 230, 510
    breaker-point, 4f
    coil-on, 135
    delay period, 32, 45, 152, 155, 157–159, 180, 185
    distributorless, 104
    energy requirements, 143
    Kettering, 4f
    lamp, 229–230
    map, 142, 146, 475
    quality, 46, 146
    switch, 229, 299
    systems, 4–5, 101–105, 107, 113, 116, 123, 126, 142, 146–147
        catalyst light-off, 125
        direct injection spark ignition (DISI), 116
        durability, 124
        emissions control, 120–127
        induction tuning, 127
        lean-burn NOx-reducing catalysts, "DENOx", 126
        mixture preparation, 109–112
        overview, 101
        port injection, 113
        power boosting, 127–132
        process, 105
        selection and control, 107
        supercharging, 128
        system design, 113–120
        three-way catalyst, 121
        timing, 101–109
    temperature, 150, 160
    timing, 32–36, **101–109**, 119, 123, 126, 133–135, 140, 142, 144, 146–148, 510
        control of, 101
        effect of, 101, 108–109
        efficiency, 108f
        map, 142f
        pressure-time, 109f
        pressure-volume, 109f

Ignition *(continued)*
    transformer, 105
    tuning, 107
Impact
    stresses, 61–62, 244
    velocity, 61–62
In-cylinder
    fuel injection, 58
        equipment, 18
    injection, 18, 58, 101, 112, 119, 127
    motion, 101, 115
    pressures, 112
    sampling, 119
    temperatures, 112, 157, 180
Independent front suspension, 325, 372–375, 379–381, 388, 391
    steering geometry, 325*f*
Indirect injection engines, 149–150, 150–151, 157, 175
Induction
    geometry of, 120, 127
    system, 23, 109, 120, 128–129, 133, 137, 142, 514
    tuning, 23, 127
Inductive pickup, 134–135, 135*f*
Injection, 111, 116, 118–119, 134–135, 143, 153, 155–157, 173–175, 205, 443–444
    actuator transverse vent, 153
    combustion chambers, **150–151**
    diesel engine, 97, 170, 177, 180, 203
    engines, 25, 101, 112, 127, 149, 151, 157
    gasoline engines, 120
    of fuel, 58, 109–110, 112, 127, 153–156
    periods, 111, 143, 155
    pressures, 111–112, 118, 151–152, 156, 158, 176
    process, 156
    pump pumping plunger, 155
    spark ignition engines, 112, 116, 120
        injectors, 112
        spark ignition, 113
    systems, 109, 111–113, 116, 149, 152, 156
    timing
        cycle-by-cycle variations, 118*f*
        effect of, 118
Injector
    needle, 156
    nozzle assembly, 154*f*
    nozzles, 45, 50, 112, 152, 154
    piston bowl injector, 150
Inlet
    manifolds, 36, 49, 57–58, 107, 110–111, 134, 137, 142, 144, 159, 164, 166, 172–175, 177, 180–185, 204–205, 509, 512–514
        absolute pressure sensor, 137
        conditions, 177, 180
        temperature, 159, 183, 185, 205

Inlet *(continued)*
    ports, 111, 116, 119, 158, 509
    system, 54, 131, 205–206
    temperatures, 126–127, 159, 180, 183, 185, 205, 223, 513
    throttles, 159
    tracts, 113
    valves, 17–18, 51–54, 56–57, 59, 111, 113–115, 120, 128, 150, 513–514
        closure, 57–58, 128, 180
        closure angles, 57*f*
        disablement, 114
        opening leads, 57
        open period, 112
In-line
    crankshaft, balancing, 67*f*
    engine layout, 66*f*
    format, 65
    four-cylinder engine, 66, 70, 508
    pumps, 153
    three-cylinder engine, 71, 75–78
        primary forces, 75*f*
        primary moments, 76*f*, 77*f*
        secondary forces, 76*f*
        secondary moments, 78*f*
In-manifold mixture preparation, 57
In-service reduction, 131
Institution of Mechanical Engineers, 116, 140, 441
Intake manifold pressure, 116
Intercooling, disadvantages of, 185
Internal combustion engines, 96*t*
International Energy Agency, 14
International Standards Organization, 346
Interstate highway system, 13
Iron, 102, 134, 191, 397, 407, 500, 507–508
    alloys of, 226
ISO. *See* International Standards Organization
Isothermal compression, 93, 166
Isuzu Rodeo, 295

Jaguar
    3.4, 1957 performance, 15*t*
    AJ6 engines, comparisons, 132*t*
    four-liter engine, 130
    fuel consumptions, 132*t*
    XJ-40 multi-link suspension, 378*f*
Jaguar-Daimler Heritage Trust, 397
Japanese vehicles, electric, 502*t*
Jenatzy, Camille, 495
Joint motion, 299

Journal bearings, 192
    film pressure schematic, 202*f*
    installation, 193*f*
    lubrication of, 197

Kettering, Charles, 4
Kettering ignition system, 4*f*
Kettering's solution, 5
Kickdown
    linkage, 274
    valve, 272–274
King, C. E., 508
Kingpin axis, 305, 324
Knock
    detector, 134, 140–142
        sensor, 140
    fuel consumption, 141*f*
    piezo-electric, 140*f*
Knock-free operation, 131, 512
Knocking cylinder, 140

Laden rolling resistance, 457
Lambda, 91–92, 111, 115, 123–124, 134, 137–138, 143, 148
Lateral force, 312, 330, 381–382, 386, 421, 429–431, 436–437
    coefficients, 436–437
Lead-acid batteries, 495, 498–499, 501
    design, 230
    performance, 497*t*
Lead-bronze alloy, 194
Leaded fuel, 48, 124
Lead-free gasoline, 121
Lead oxide, 498
Leaf springs, 363
Lean-burn
    combustion systems, 123
    disadvantages of, 147
    engines, 114, 120, 147
    mode, 115, 127
    NOx, 158
    spark ignition engines, 120
Leland, Henry, 3, 5
Linear
    differential equation, 348, 463
    interpolation, 219, 221
    relationship, 222
    velocities, 242, 263–264

Lithium ion, 499
    battery performance, 497t
Lithium polymer batteries, 499–500
Load, effect of, 35
Load-carrying capacity, 195
Locking hubs, automatic, 295
Lockup torque converter, 260f
Longitudinal
    driveshaft, 373
    engine configuration, 296
    loads, 191
    radius rod, 379
    stability, 345
Low-speed
    turn diagram, 290f
        four-wheel steering, 331f
Low-sulfur fuels, 49–50, 159
Lubrication, **189-202**
    hydrodynamic, 200–201
    systems, 16, 189–190, 233, 511

Mach index, 53–54, 54f, 514
MacPherson, Earle S., 375
MacPherson strut, 375, 377f
    roll center calculation, 382
Magnetic torque, 225
Manifold walls, 110
Manitou springs, 420
Manual gearbox ratios, 483t
Manual transmissions, **249-255**, 277
    operation, 236, 243
    systems, 236
Marine engines, 25
Maserati 2000-GT, 1957 performance, 15t
Mass, effect of, 88
Master
    cylinders, 403–406, 422
    pistons, 404
Mathematical models, 146
Matthey, Johnson, 40, 122, 126
Maximum residual torque, 475, 477
Mechanical strength, 195
Membrane
    electrode assembly, 81
    materials, 87
Mercedes-Benz A-Class, 341
Metal hydrides, 93, 499
Metal-to-metal contact, 201

Metering valve, 406
Methanol-fueled fuel cell, 97
Michelin X tire, 424
Misfire, 41, 136
Mitsubishi
    engine, 117, 119–120
        GDi engine, 117, 117f, 120
            efficiency of, 120f
    MEEV-II, 502t
    system, 118
Modal expansion, 355, 358
Model T, 5–6, 338
Mode shapes, 358f
Modulus of elasticity, 193, 288
Molar
    consumption, 91
    Gibbs energy, 82
    mass, 49, 99
        alkanes, 49
        hydrocarbons, 45
Moment arms, 328–329, 390, 410
Moment of inertia, 339, 419, 486
Mono tube
    damper, 371
    types, 371
Moose test, 344
Morrisson, William, 495
Motors, **223–233**
    electric, 94, 223–224, 479, 490, 496, 503–504, 523
Multi-cylinder
    crankshaft, 68, 68f
    engines, 39, 65, 70–71, 74
Multi-
    direction input, 346
    disc clutch, 299
    grade oils, 195
    leaf spring, 364f
    link suspension, 378, 378f

National Renewable Energy Laboratory, 489
Needle-closing pressure, 154
Nernst equation, 83, 87
Neural networks, 146
Newton's law, 197, 347, 401, 465
NiCd. *See* Nickel-cadmium
Nickel, 89, 499, 501–502
Nickel-cadmium, 498–500
Nickel-cadmium battery performance, 497t

Nickel hyrdoxyoxide, 499
Nickel-metal
    battery performance, 497t
    hydride, 498–502, 504
Nickel oxide, 125
Nickel silver, 205
NiMH. *See* Nickel-metal hydride
NiOOH. *See* Nickel hyrdoxyoxide
Nissan
    Altra EV, 501
    HICAS steering system, 334
    Hypermini, 502t
    Prairie Joy, 501
Nitrile rubbers, 50
Nitrogen, 37, 42, 49, 91, 94, 158, 160, 370, 470
Nitrogen/argon mixture, 158
Nitrogen/helium mixture, 159
Nitrogen oxides (NOx), 39, 42–44, 45, 121, 126–127, 144
    absorbing material, 127
    catalysts, 123, 127, 156, 158
    control of, 114, 121, 126
    emissions, 42, 58, 108, 116–123, 146, 156–159, 168, 175, 180, 185
    emissions legislation, 144
    formation of, 45, 180, 185
    particulates, 45, 158
    ppm, 157
    reductions, 122, 126–127, 144, 158–159
    trap catalysts, 127
Noisier combustion, 158
Nose geometry, 454f
NOx. *See* Nitrogen oxides
Nylon-reinforced tires, 425

Octane rating, 8, 46–48, 46f, 141, 511–512
Off-road vehicles, 295, 320, 343, 362, 391, 435
O'Groat, John, 13
OHC. *See* Overhead camshaft
Ohmic losses, 83, 87
Ohms law, 87
Oil
    10W30, 195, 197
    10W40, 195
    cooler, 204
    film, 41, 201, 278
    film pressure, 201
    modifiers, 195
    pressure, 59, 115, 156, 201
    reservoir, 371

Olds Motor Works, 6
Olley, Maurice, 360
Onboard
    fuel reformation, 91–92, 94
    methanol reformer, 93*t*
On-demand four-wheel drive (4WD), 295
One-Intake-Valve Operation, 115–116
On-to-off ratios, 144
Open circuit voltage, 80, 82–83, 86–87
Open-loop algorithm, 335
Open vehicle aerodynamics, **461–462**
Optimization comparisons, 395*f*
Organo-metallic compounds, 48
Otto cycle, 24*f*, 26*f*
    efficiencies, 24, 26, 28, 30, 98, 148
    standard, 24–25
Overdrive performance, 479*t*
Overhead camshaft, 58
Overhead valve arrangements, 59*f*
Overrunning clutches, 267*f*
Oxidation, 39, 41–42, 89–90, 92, 94, 121, 123, 125–126, 136, 158, 160, 258
    catalyst, 42, 92, 115–116, 122–123, 136, 159–160
        diesel, 160*f*
        platinum-based, 159
Oxygen, 37–39, 44, 85–88, 90–94, 119, 122–127, 137–139, 144, 158–160, 470
    utilization, 33, 36, 158

Parallelogram steering linkages, 305*f*
Partial oxidation (POX) reforming, 90
Particulates
    in compression ignition engines, 45
    traps, 159
Part-time four-wheel drive (4WD), 294
Pascal's law, 402, 402*f*
Passive systems, 126
    rear-wheel steering, 333, 333*f*
    suspension system, 391
PbO_2. *See* Lead oxide
PCV. *See* Positive crankcase ventilation
Peltier effect, 213
Pelton wheel, 171
PEM systems. *See* Proton exchange membrane systems
Percha, Gutta, 424
Petroff's equation, 200
Petroleum-derived fuels, 46
Phase
    relationship, 70
    separation, 71, 74

Phasor
    approach, 74
    diagrams, 77
Photochemical smog, 42
Piezo-electric
    crystal, 140
    knock detector, 140f
    transducer, 137
Pike's Peak, 420–421
Pinion
    inertia of the, 231
    rack, 304, 308
Piston
    areas, 113, 371
    assembly, 66
    bowl, 112, 119
    cavity, 117
    cylinder arrangement, 59
    displacements, 118
    motion, 19, 44
    phasing of the, 79
    position, 74, 260
    speed, 22, 54, 57, 115, 150, 514, 517
    trajectories, 119
    velocity, 55, 57, 70, 372
Pitch and bounce response, 360f, 361f
Pitman
    arm, 305
    shaft, 305
Planar crankshaft, 66
Planetary
    analysis, 236, 262, 264, 271, 291
    carrier, 261, 267–268, 270–271, 291, 296, 298
    element, 265, 299
    gearbox, 521
    gears, 261, 291, 505
        analysis of, 236, 262
        operation, 265t
        schematic 263f
        set, 262f
    gear-set torque converter, 265
    motion, 290
    pinions, 261, 268
Planet-ring interface, 263
Platinum
    catalysts, 81, 94, 123–124, 126, 160
    electrodes, 137–138
Platinum/rhodium
    catalysts, 123
    system, 123

Plenum Publishing Corporation, 439–440, 459, 461
P-metric designation system, 426–427, 427f
Pneumatic (air) springs, 368
    force-deflection curve, 369f
Pneumatic tire, 424
Polygon of moments, 76, 77, 79
Polynomial functions, 62
Pomeroy, Laurence, 507
Pope, Colonel Albert A., 3
Pope Manufacturing Company, 3
Poppet valves, 51–52, 58, 144
    flow characteristics, 52f
Porsche Speedster, 1958 performance, 15t
Porsche system, 333–334
Port, geometry of, 204
Port injection, 101, 111, 113, 120
    combustion systems, 113
    efficiency of, 120f
    injectors, 112
Port Reed Valve, 19
Positive crankcase ventilation, 195
Potential energy, 106, 163, 215–216, 321, 420
Power
    boosting, **127–132**
    output, 58, 91–101, 122–123, 132, 147, 151, 177, 180, 184, 187, 225–227, 329, 480, 498, 524
    steering, 230, 308, 310
        control valve, 310f
        pump, 308–309
        pump drive, 309f
        system, 146, 308
    transmission, 235, 251, 276, 473, 504–505, 521
        capabilities, 208
Pre-engaged starter, 232f
Pressure
    chamber ball valve, 60
    dependence of entropy, 83
    differential switch, 405–406
    enthalpy
        planes, 216
        plot, 217
    indicators, 402
    loss, 93, 114, 406
    oscillations, 48, 140
    plate, 33, 112, 236–237, 239, 243, 461
    propagation delay, 154
    pulsations, 54, 127, 154
        effects of, 154
    regulator, 110–111, 273
    relief valve, 154, 206, 511
    system, 54, 93, 118, 129, 146, 156, 172, 190, 206, 274, 308, 368, 402, 406, 514
    transducer, 141, 448

Pressure/enthalpy diagrams, 217
Pressure-regulating spool valve, 274*f*
Pressure-volume, Rover M16, 21*f*
Pressurized
    gas, 93*t*
    oil, 201
Proportioning valve, 406
Proton exchange membrane (PEM) systems, 81, 85, 87, 90
Pump–line–injector (PLI) systems, 153
    maximum pressure, 152*f*
Push rod, 274, 405
    valve system, 63

Quadratic equation, 38
Quality ratings, 428
Quasi-static rollover model, 336*f*, 338*f*
Quench layer, 41

Rack and pinion steering, 308, 308*f*
Radial
    compressor, 129, 161–162, 185
    deflection, 300
    direction, 208
    engines, 79, 161
    load, 199
    shaft load, 199
    tires, 344, 424, 426, 426*f*, 430, 432
        construction of, 426
    turbine, 161, 185
    velocity, 162, 169
Radiators, 205, 260, 445
    cool outflow, 204
    flow, 204, 445
Radical flow compressor, velocity, 162*f*
Ransom Olds, 3
Rear suspension, 281, 334, 374, 379, 383, 388
Rear-wheel-drive, 281, 319, 323, 373, 474
Rear-window inclination
    influence on drag, 453*f*
Recirculating ball, 307
    gearbox, 310
    steering gear, 307, 307*f*
Reduction catalysts, 123
Reed valve, 18
Refrigerant, thermodynamic properties, 220*t*
Renault, Louis, 397

Reverse gear
    power flow, 253*f*
    Simpson drive, 271*f*
Reynolds number, 51, 438, 438*f*, 442, 446, 463–464, 470
Rhodium, 124, 127
Ricardo
    combustion system, 118, 512
    head, 6–8, 512
    turbulent head, 7*f*
Rich mixtures, 38–39, 107, 122, 124, 139
Ring gear, 232, 254–255, 261–262, 267–270, 277, 290–291, 298
Rivet heads, 416
Road load curves/fuel consumption map, 475*f*
Road noise, 379
Robert Bosch Ltd., 138
Rocker
    arms, 58-59, 63, 114–115
    axis, 63
    joint design, 299
Roll center
    analysis, 379–391
        Hotchkiss, 382
        MacPherson strut, 382
        vehicle motion, 382
        wishbone suspension, 381
    diagram
        Hotchkiss, 383*f*
        MacPherson strut, 383*f*
        negative swing arm, 381*f*
        parallel link, 382*f*
        positive swing arm, 380*f*
Roller
    bearings, 189–190, 250, 305, 518
    chain, 299
    followers, 60, 114–115
        ring, 153
    steering gear, 306
    system, 305
Rolling resistance, 456*t*
    trucks, 457*f*
Rollover, 8, 303, 336–337, 341–343, 389–390
    accident, 343
    prevention, 343
    test, 9, 9*f*, 342
    thresholds, 337–338, 340–342
        calculation of, 342*t*
Rolls-Royce, 1, 507, 511
Roots blower, pressure volume, 130*f*
Rotary fuel pump, 153, 172
Rotor arm, 104, 308

Rover
    800
        cooling system, 204*f*
        performance, 205*f*
    K16
        induction system, 129*f*
        standard & VVC, 128*t*
    M16
        pressure-volume, 21*f*
Royal Automobile Club, 3, 507
Rubber
    hardness of, 424
    natural color of, 424

SAE. *See* Society of Automotive Engineers
SAI. *See* Steering, axis inclination
Scrub radius, 321, 323
Seat belts, 10
Seating area, 155
Self-aligning torque, 321, 324
Self-energization, 406, 409, 414
    advantage of, 397
    shoe, 411*f*
Semiconductor devices, 137
Semi-trailing arm suspension, 378*f*
Sensors, 134–140
    air flow rate, 136
    air-fuel ratio, 137
    air temperature, 137
    camshaft position, 134
    coolant temperature, 137
    crankshaft speed/position, 134
    inlet manifold absolute pressure, 137
    knock detector, 140
    throttle position, 136
Shaft speeds
    input/output ratios, 284*f*
    whirl, 286
Shims, 58, 155, 237, 324
Shock absorbers, 348, 364, 371–372, 372*f*
    construction, 372
Short-long arm suspensions (SLA), 375, 376*f*
Shunt
    motor, 226
    transmission system, 278, 280
    winding, 226
Side-view model, 314*f*
Sidewall strength, 425

SI engines. *See* Spark ignition engines
Signal-to-noise ratio, 136, 141
Simple spring-mass system, 348*f*, 349*f*
Simpson drive, 267, 269–271
    drive schematic, 267*f*
    element actuation, 268*t*
Single-degree-of-freedom model (quarter car model), 347
Single-point fuel injection system, 110*f*
Six-cylinder
    engine, 65, 79
    four-stroke engine, 65
    torque, 65*f*
SLA suspension. *See* Short-long arm suspension
Slip joint, 282*f*
Slippage, 12, 243, 260, 418
Sloan, Alfred, 6
Sloper, Thomas, 424
Small-block
    Chevrolet, 9, 1955, 10*f*
Society of Automotive Engineers, 9, 195, 333, 346, 422–423, 430–431
    Handbook, 399
    viscosity grades, 195, 196*t*
SOFC. *See* Solid oxide fuel cells
Soft materials, 193
Soft-phase material, 194
Solenoid, 111, 155, 231–232
    shut-off valve, 153–154
    valve, 143–144, 153–154, 156
    valve switches, 144, 172
Solenoid-operated fuel injector, 111, 111*f*
Solid axle suspensions, 373
Solid fuels, 33
Solid oxide fuel cells, 79
Solid polymer fuel cells (SPFC), **79–92**, 81*f*, 86, 88–89, 89*f*, 91
    efficiency, 81
    electrodes of the, 80
    fuel crossover, 86*f*
    internal current losses, 86*f*
    polarization curve, 89*f*
Sommerfeld number, 200, 202
Soot
    filter assembly, diesel, 160*f*
    formation of, 33
Spark
    discharge, current and voltage, 106*f*
    energies, 143, 202
    ignition, 4, 23–24, 35, 95–96, **101–109**
        efficiency, 35*f*, 95*f*
Spark ignition engines, **101–148**
    abnormal combustion in, 48

Spark ignition engines *(continued)*
    brake power 203*f*
    combustion, 44*f*
    efficiency, 34, 36, 108, 148
    emissions, 40–41, 40*f*, 41*f*
    pre-ignition, 48
    simulation, 44
    timing, 108*f*
    turbulence, 114*f*
Spark plugs, 18, 48, 101–105, 102*f*, 107, 112–113, 116–119, 143
Spark systems, 105, 113, 116, 126, 147
Specific heat, 420
    capacities, 25, 27, 98–99, 164, 180, 187, 206, 219
    capacity ratio, 98
    of air, 98
Speed-dependent ratio, 335
Speedometer errors, 434
Speed ratio, 108, 131, 136, 145, 169, 249, 260–261, 269, 277, 284, 288, 474, 480, 482
Speed reduction, 253, 280, 476, 478, 480, 482–483
SPFC. *See* Solid polymer fuel cells
Spicer, Clarence, 518
Spider gears, 291
Spill valve, 155–156
Spindle idler arm, 305
Spiral bevel gears, 246, 246*f*, 518
Split coefficient braking, 423
Spontaneous ignition, 48
Spool valves, 156, 272, 274–275, 308, 310
Sport cam profile, 53
Spray divergence, 118
Springs, 363
    analysis, 366, 368
    deflection, 359–360, 359*f*, 362, 362*f*, 368, 387
    dimensions, formulas, 367*t*
    loading, 63, 154, 368, 370
    mass system, 347*f*
    pressure, 154–155, 237, 274, 368
    retainer, 58, 60
    steel, 364, 416
    stiffness, 63, 348–349, 363, 371, 387, 392–393
Sprung mass, free-body diagram, 386*f*
Spur gears, 244–245, 244*f*
Stability, 13, 15, 50, 295, 323, 330, 373, 398, 406, 423–424, 435, 437, 462, 464
Stanley steamers, 3
Starters
    alternator, 189, 230
    circuit, 233
    electric, 5
    motors, 226–227, 231–232
    motor shaft, 231

Starters *(continued)*
    relay, 299
    ring, 231–232
    switch, 232, 299
Static deflection, 368
Stator free-wheels, 260
Stator windings, 228, 231–232, 231*f*, 261
Steady flow energy equation 216
Steady-state
    braking, 401
    lateral acceleration, 340
    oscillation, 351
    output, 204
Steam engines, 3, 507
Steam reforming (SR), 89
Steam-powered cars, 3
Steer angle, change with speed, 317*f*
Steering
    arms, 305, 324, 329
    axes, 305, 327, 329
        inclination (SAI) 320–321, 328
    boxes, 305
    column, 5, 10, 304–308, 310
    dynamics, 16, 236, 303, **311–319**, 430
        effects of tractive forces, 318
        high-speed turning, 312
        low-speed turning, 311
    gears, 305–306, 308
        rack and pinion, 308*f*
        recirculating ball, 307*f*
        worm and roller, 306*f*
        worm and sector, 306*f*
    geometry errors, **324–327**
    linkages, 305, 324, 327, 333–334
    mechanisms, **303–310**
        power steering, 308
        rack and pinion steering, 308
        recirculating ball, 307
        worm and roller, 305
        worm and sector, 305
    moments, 323, 328–329
    reduction ratio, 307
    response, 303, 343
    systems, 16, **303–344**
Stiffness matrix, 358
Stoichiometric air-fuel ratio, 33
    air-fuel ratios, 30, 45, 123–124, 137–138
    mixtures, 33–34, 36, 38, 40, 44, 123, 125–126, 143
    near full load, 116
    reactions, 37

Stopping distance, 399, 421, 433
Straight-tooth
    bevel gears, 245, 245f
    spur gears, 244
Streamlines, 448f
Stribeck diagram, engine lubrication regimes, 200f
Substrates, 124–125
Sulfur, 50, 125, 127, 159
    content, 45, 125
    dioxide, 125
    trioxide, 125, 159
Supercharged
    engine, 20, 130–132, 148
    four-liter engine, 130
    spark ignition engine, 130, 148
Supercharger
    arrangement, 131, 131f
Surface-area-to-volume ratio, 149
Suspensions, **345–396**
    analysis of, 16, 345, 347, 354, 362, 379, 396
    design, 345–346, 358, 396
    effect of, 345
    engineers, 346, 362, 375, 396
    functions of, 353
    geometry, 319, 325, 381
    kinematics, 379
    linkages, 380, 384, 387
    motion, 281, 365
    parameters, 392
    performance, 391
    roll centers, 380f
    setup, 360
    system components, 363–372
        coil springs, 365
        dampers (shock absorbers), 371
        leaf springs, 363
        pneumatic (air) springs, 368
        springs, 363
        torsion bars, 364
    types, 363, 372–379, 380
        de Dion, 374
        four-link, 374
        Hotchkiss, 373
        independent, 375
        MacPherson struts, 375
        multi-link, 378
        short-long arm (SLA), 375
        solid axle, 373
        swing arm, 379
        trailing arm, 376

Suspensions *(continued)*
    vibrational analysis, 345, 347–363
        single-degree-of-freedom model, 347
        two-degrees-of-freedom model, 351, 354
Swash-plate design, 214
Swedish moose test, 341
Swing arm suspensions, 379-381, 379*f*

Tafel equation, 84, 84*f*, 85, 85*f*
Temperature/entropy diagram, 163
Termolecular reaction, 43
Texas, 13
Thermal
    conductivity, 206, 239
    conversion efficiency, 107
    decomposition, 45
    effects, 158–159
    efficiency, 23, 91, 98, 107, 522
    energy, 106, 213
    fatigue, 207, 240
    loading, 168, 183, 185
    management, 126
    mechanism, 42–43
    NO differential equation, 43
    NO mechanism, 42–43
    reactor, 121
    stresses, 397
Thermionic emission, 106–107
Thermodynamics, 16–17, 97
    cycle, 23, 219
    data, 99, 219
    efficiency, 23, 30, 82, 91
    equilibrium, 43
    of prime movers, 17–99
    power, 222
    principles, 16
    system, 82, 215
Thermosiphon effect, 204
Thermostatically controlled expansion valve, 218*f*
Thin-wall bearings, 194
Thomson, Robert, 424
Three-channel antilock brake system, 422*f*
Three-phase alternator, 229*f*
Three-way catalyst, 124*f*
Throttle
    actuator, 136
    application, 329
    linkage, 274

Throttle *(continued)*
    position, 133–134, 136–137, 142, 299
        sensor, 136
    response, 107, 137
    setting, 36, 99, 107, 133, 205
    valve, 58, 110, 113, 129, 131, 145, 172, 175, 272–274
Throttling loss, 114, 130–131, 143, 144
Thrust
    bearings, 191
    load on camshaft, 192*f*
Tie rods, 304–305, 308
    adjustable, 324
    ends, 304, 325–327
Timing belt, 153
Tin-antimony-copper alloy, 193
Tire-road
    contact, 382
    interface, 423, 429
Tires, **424–433**
    angles, 391, 430
    assembly, 322
    belted bias, 426
    better-performing, 424
    bias-ply, 317, 329, 343, 425–426, 430, 432
    breakaway, 390
    construction, 425, 427, 430, 433
    contact patches, 323–324, 381, 384–385, 429
    cornering stiffness, 319, 329, 430, 433
    deformation of, 419
    designations, 419, 426
    development, 424–425, 433
    force cornering
        friction versus slip, 430*f*
        lateral force versus slip angle, 431*f*
        stiffness versus inflation pressure, 432*f*
    force generation, 429
    pressure, 428–430, 433
    radius, 328, 419
    resonance, 392–394
    results, 425, 429
    scrub, 381, 391
    section, 419, 421, 427, 474
    size, 312, 430, 474
    speed ratings, 427*t*
    stiffness, 391
    technology, 15
    under-inflated, 429
    wear, 146, 311, 321, 324, 425
Titanium iron hydride, 93
Toe angles, 378

Torotrak continuously variable transmission (CVT), 275–281, 279*f*
    operating regime, 280*f*
    variator assembly, 278*f*
Torque
    capacity, 241–243, 254, 277, 280, 292, 299, 417–418, 433
    characteristics, 65*f*
    converters, 256–261, 259*f*, 265, 272, 277, 297, 299, 482, 485
        fluid flow, 259*f*
        hydrodynamic, 236, 256–257
        performance curve, 261*f*
    delivery, 239
    engine speed, 129, 176, 474, 506–507, 518
    loads, 244
    multiplication, 236, 280, 290
    of the wheels, 374
    optimum economy, 281
    output, 65, 98, 132, 143, 250, 280, 296, 486, 497
    ratio, 248–249, 251, 261, 301
    resistance, 292, 364
    transfer, 243, 262, 265, 285, 289, 293, 296
Torsion bar suspension, 364, 364*f*
Torsional vibrations, 60, 65–66
Toyota
    Corolla, 521
    e-Com, 502*t*
    Prius, 16, 495, 503, 505–507, 521, 523
        engine, 523*f*
        gearbox configuration, 521*f*
        operational modes, 506*f*
        optimum engine operating point, 523*f*
        specification, 522*t*
    RAV4, 501
Tractive
    force effects, 330*f*, 484*f*
    resistance, 450–451, 457, 468, 476
Trailing arm suspensions, 376
Transaxles, 236, 274, 296
Transient
    response, 119, 122, 127, 185
    roll model, 339*f*
Transmissibility, 352–353
Transmissions, 3, 16, 71, 146, 190, **235–301**, 329, 373, 397, 435, 444, 451–452, 504–505, 522
    automatic, 255–256, 258, 261, 474
    computer modeling, 486
        ADvanced VehIcle SimulatOR, 488
    constant-mesh, 249
    continuously variable transmissions (CVT), 479
    controller, 299
    engine size, 474
    functions of a, 281

Transmissions *(continued)*
    gearbox span, 482
    input shaft, 236, 238, 250, 257, 260, 268, 294
    manual, 249–255
        synchronizer operation, 254
        power flows, 251
    matching, **473–493**
    maximum speed, 474
    output shaft, 272, 281, 294
    overdrive ratios, 477
    ratio
        spread, 489*f*
        selection, 133
    selecting drive ratio for, 474
    shaft, 236–238, 250, 257, 260, 268, 272, 281, 294
    system, 97, 145, 260, 273, 397, 473, 477, 479, 482, 505, 522
    technology, 15
Transverse-mounted engine, 204
Treadwear
    designation, 428
    rating, 428
Trembler coil, 4*f*
Trimetal bearings, 194
Triumph, 1, 486
    Dolomite Sprint, 59
        overhead camshaft, 59*f*
Truck aerodynamics, **456–461**
Turbine
    casing, 172
    entry, 182
        conditions, 186
        temperature, 180, 182–183
    housing, 161, 256, 260
    isentropic efficiencies, 164, 180, 183, 186
    output, 180, 183–184, 186–187, 299
    overspeeding, 172
    pressure ratio, 170, 182, 184, 186
    specific work output, 180
    wheel, 161
Turbochargers, 17, 127, 130, 142, **161–172**, 161*f*, 173, 175–177, 180, 185
    analysis of, 162
    diesel engines, 34–35, 55, 97, 170, 174, 177, 180, 186–187
    disadvantages of, 187
    efficiency, 165
    engine performance, 169
    engines, 129, 145, 159, 174, 176–177, 486
    inertia of the, 170
    performance, 164
    scavenging pressure, 165*f*
    spark ignition engines, 97, 129, 186
    temperature/entropy, 163*f*

Turbulence
    effect of, 114, 464
        on spark ignition engines, 114f
Turbulent head, Ricardo, 7f
Turning, free-body diagram, 385f
Turret Tops, 8f
Twelve-cylinder engines, 65, 71
Two brake drum configurations, 409f
Two-channel antilock brake system, 422
Two-degrees-of freedom model, 392f
    half car, 354
    quarter car, 351
    rigid body vehicle, 354f
Two-stroke engines, 17, 19–20, 19f, 22–23, 51, 65, 74, 98, 147

UEGO. *See* Universal exhaust gas oxygen sensor
U-joints, 281–283, 285–286, 373, 518
    kinematics of a, 281
Unburned
    fuel, 31, 34, 41–42, 45
    gas, 43–44, 48, 108
    hydrocarbons, 39, 41–42, 45, 110, 115, 119–121, 126–127, 136, 147, 158–160
Underdamped oscillation, 350f
Under-piston scavenging, 19f
Unibody construction, 376
Unitary construction, 517f
United Kingdom, 13–14, 495, 512, 517–518
United States, 2
    Air Force Academy, 333
    Department of Energy, 489
    Department of Transportation, 428
    emissions legislation, 121, 147
    Environmental Protection Agency, 487
Universal exhaust gas oxygen sensor (UEGO), 138–139, 138f, 144
Universal joint, 282f
    angular relationships, 283f
Unsprung mass motion, 352f

V-6 engines, 58, 59, 65, 71, 79, 98, 128, 140, 519
    layout, 66f
Vacuum
    actuators, 172
    booster, 397
    check valve, 405
    port, 405
    pump, 172, 404
    signal level, 172

Vacuum-boosted hydraulic brakes, 397
Vacuum-operated
    diaphragm, 107
    power brake booster, 405, 405*f*
Valves
    acceleration, 53, 61–63
    arrangements, 59–60, 512
    axis, 59, 63
    clearances, 58
    closure, 128
    curtain area, 52
    diameter, 53, 54, 191
    duration, 53, 128
    engine, 55, 57–58, 61, 63, 114, 120, 129, 131, 156, 173, 405
    flow, 50
    gear, 18, 53, 60, 62, 64, 115, 274, 299, 310, 517
        dynamic behavior of, 60
    guides, 191*f*
    head, 59, 63
    lift data, 53*f*
    mechanism, 49, 61, 63, 128
    open duration, 53
    operating rod, 405
    operating systems, 58
    plunger, 405
    seats, 52–53, 58–62, 206
    spindle, 58
    timing, 54–56, 59, 101, 127–128, 133, 135, 142, 147, 516, 522
        diagrams, 56*f*
        diesel engine, 56*f*
        spark ignition engine, 56*f*
        tuned spark ignition engine, 56*f*
    velocity, 54, 61, 113
    volumetric efficiency, 50
Valvetrain acceleration, 53
Van Doorne continuously variable transmission (CVT), 275–276, 276*f*, 278
Vane
    compressor, 129–130, 130*f*
    pump, 153–154, 308, 309, 309*f*
Vaneless diffuser, 162, 169
Vapor
    behavior, 217
    compression, 213, 215, 215*f*
    entropy, 217, 221
    level of, 218
    region, 216–217, 221
Vaporized fuel, 110
Vaporized hydrocarbon, 89
Vauxhall 14–40
    engine performance, 515*f*

Vauxhall *(continued)*
    fuel consumption map, 516*f*
    fuel consumption variations, 520*f*
    longitudinal cross section, 510*f*
    tractive force curves, 519*f*
    transverse cross section, 509*f*
    vehicle data, 518*t*
    velocity/time history, 520*f*
V-Belts, 212
    types of, 308
    wedge action, 212*f*
Vee engine layout, 66*f*
Vehicle, electric, 3, 97, 499–500, 502–503, 506
Vehicle rollover, **337–343**
    quasi-static model, 337
    quasi-static rollover with suspension, 337
    roll model, 339
Vehicle specification example, 473*t*
Velocity
    gradient times, 199
    of the displacements, 348
    of the piston, 372
    of the planet, 263
    triangles, 162
    vector, 224
Vented rotor, 415*f*
Venturi section, 136
Vertical
    plate, use of, 460*f*, 461*f*
    stiffness, 347
Vibrational
    analysis, 288, **345–363**
    signal, 140
Vibration amplitude, 350, 372
Vinyl acetate, 49
Viscosity
    measurements, 195
    of engine oil, 197
Viscous
    couplings, 295
    dampers, 348
        coefficient of, 349
Volkswagen
    Beetle, 379, 462
    Lupo engine, 34, 176, 177*f*
        brake specific fuel consumption, 176*f*
    Microbus, 457
Voltage signal, 137, 141
Volume-to-surface-area ratio, 149

Volumetric efficiency, 23, 50–58, 94–98, 110, 113, 116, 127–129, 147–148, 151, 172, 177–183, 513–514
    inlet valve closure, 57*f*
    Mach index, 54*f*

Wahl factor, 368
Wankel engines, 17
Waste-gate control, 172*f*
Weight transfer with center of gravity height, 388*f*
Weissach axle, 333
Well-to-wheel efficiency, 93, 95, 96*f*, 96*t*
Wheel and tire assembly, 322*f*
Wheels
    alignment, 303, **320–324**
        camber, 320
        caster, 323
        steering axis inclination (SAI), 320
        toe, 321
    arc, 321
    axis, 323, 449
    back, 321
    cylinder, 408–409
    self-align, 323
    speed variation, 294
Whirl speed, 288
White metals, 193–194
    bearings, 193
Wilson, Alexander, 507
Wilson, Woodrow, 1
Windscreen, 455
Windshield wipers, 226, 453
Wind velocity, 436–437, 449
Winton gasoline-powered vehicles, 3
Wishbone-type suspensions, 380
    suspension, 375–376, 376*f*, 381
        roll center calculation, 381
World War II, 9, 375
Worm and roller steering gear, 305, 306*f*
Worm and sector, 305
Worm gear, 305–307
Worm systems, 305
Worm threads, 307
Worm-type
    mechanisms, 304
    steering systems, 308
    systems, 305, 307–308
Worm/wheel configuration, 246
Worth, William O., 2

Yaw
   angle, 437, 439, 439f, 449, 456, 458, 460
   control system diagram, 423f
   rate versus time, 336f
Yttrium oxide, 138

ZEBRA battery, 502
Zeldovich mechanism, 42
Zener diode, 230
Zeolite chemistry, 127

About the Author

Richard Stone is a Reader in Engineering Science in the Department of Engineering Science at Oxford. He was appointed to a lectureship at Oxford in 1993, and for 11 years prior to that appointment, he was a Lecturer/Senior Lecturer at Brunel University.

Dr. Stone's main interest is combustion in spark ignition engines, but he also has interests in fuel cells and the measurement of laminar burning velocities in zero gravity. He is undertaking a longitudinal study of vehicle technology that commenced with 1970s technology 20 years ago but now has been extended back to the 1920s, with completion now projected in 2020.

Dr. Stone has written approximately 90 papers, most in the areas of engine combustion and instrumentation. He is well known for his book, *Introduction to Internal Combustion Engines* (third edition, 1999).

Jeff Ball is currently a Senior Engineer for Knott Laboratory in Centennial, Colorado. Previously, he was an Associate Professor, Instructor Pilot, and Deputy Department Head in the Department of Engineering Mechanics at the U.S. Air Force Academy. He holds a Bachelor of Science degree in Engineering Mechanics from the U.S. Air Force Academy, a Masters of Engineering degree from Purdue University, and a Doctor of Philosophy in Engineering Science degree from the University of Oxford.

Dr. Ball began teaching in 1993. For 10 years prior to teaching, he was a fighter pilot, instructor pilot, and flight examiner in the F-4D, F-4E, and AT-38B aircraft. His current research involves alternative powerplants for unmanned aerial vehicles.

Dr. Ball has taught courses in statics, dynamics, strength of materials, thermodynamics, machine design, industrial design, and automotive systems analysis.